Transition Metals
in the Synthesis
of Complex
Organic Molecules

SECOND EDITION

Transition Metals in the Synthesis of Complex Organic Molecules

Louis S. Hegedus
COLORADO STATE UNIVERSITY

University Science Books
Sausalito, California

University Science Books
55D Gate Five Road
Sausalito, CA 94965

Fax: (415) 332-5393
www.uscibooks.com

Production manager: *Susanna Tadlock*
Design consultant: *Robert Ishi*
Index editor: *Judith A. Douville*
Printer and binder: *Maple-Vail Book Manufacturing Group*

This book is printed on acid-free paper.

Library of Congress Cataloging-in-Publication Data

Hegedus, Louis S.
    Transition metals in the synthesis of complex organic molecules /
Louis S. Hegedus. — 2nd ed.
        p.   cm.
    Includes bibliographical references and index.
    ISBN 1-891389-04-1 (acid-free paper)
    1. Organic compounds—Synthesis. 2. Organometallic compounds.
3. Transition metal compounds.   I. Title.
QD262.H35 1999
547'.2—dc21   99-10276
                                                                      CIP

Printed in the United States of America
10 9 8 7 6 5 4 3 2

Contents

Preface

Much has happened in the six years since the first edition of this book appeared. The use of transition metals in organic synthesis has finally achieved general acceptance within the synthetic organic community, and it is rare to find complex total syntheses that don't involve at least a few transition-metal mediated key steps. In addition, industry has lost much of its aversion to using homogeneous catalysis. Finally, the use of transition metals in the synthesis of polymers and materials, and in solid-phase combinatorial chemistry has also increased dramatically. As a consequence, this second edition, which covers the literature through July of 1998, is substantially different from the first edition. It contains over 500 new references, with particular emphasis on recent reviews, and a very large number of new, more complex examples of processes introduced in the first edition. In addition, several new topics are covered (see below).

Chapters 1 and 2, dealing with formalisms and mechanisms ("the rules") remain unchanged. Chapter 3 details the advances and current status of homogeneous hydrogenation with emphasis on asymmetric hydrogenation. Although a myriad of new catalysts have been developed, the range of substrates reducible with high asymmetric induction has only been expanded marginally, indicating how difficult a problem asymmetric hydrogenation is. Chapter 4 treats the very extensive chemistry of σ-alkylmetal complexes, and the bulk of this chapter is new because of the warmth with which synthetic organic chemists have embraced Pd catalyzed reactions – Heck, Stille, Hiyama, and Suzuki couplings in particular. Chapter 5 treats metal carbonyl chemistry, and as was the case with the first edition, little new has transpired, and this chapter was only marginally modified. Chapter 6 deals with metal carbene chemistry, an area which has seen a great deal of recent activity. In addition to the many advances in Group 6 carbene chemistry and Rh(II) catalyzed decomposition of diazocompounds, ring closing metathesis has been discovered by synthetic chemists in a major way, and a new section on this reaction chemistry has been added. Chapter 7 updates the reaction chemistry of metal alkene, diene and dienyl complexes. In addition, a new section on metal-catalyzed cycloadditions has been added. Chapter 8 treats metal alkyne chemistry. Although new catalysts and conditions for both the Pauson-Khand reaction and alkyne cyclotrimerization have been developed, no major advances have occurred. Chapter 9 covers η^3-allyl chemistry, primarily that of palladium. Asymmetric induction has been studied intensively in this context, and a more systematic treatment is presented in this revised edition. A section on new η^3-allyl iron complex chemistry has also been added. The final chapter considers arene complex chemistry. Major advances in the area of asymmetric induction as well as the chemistry of Ru, Fe, and Mn arenes have been made and are described in this chapter. Newly developed chemistry of η^2-areneosmium complexes is also presented.

The manuscript was typed in Microsoft Word 6.0, and the figures and equations executed in CSC ChemDraw 3.1 by Michelle Swanson.

Louis S. Hegedus
COLORADO STATE UNIVERSITY

Transition Metals in the Synthesis of Complex Organic Molecules

CHAPTER 1

Formalisms, Electron Counting, Bonding (How Things Work)

1.1　Introduction

Why should an organic chemist even consider using transition metals in complex syntheses? There are many reasons. Virtually every organic functional group will coordinate to some transition metal, and upon coordination, the reactivity of that functional group is often dramatically altered. Electrophilic species can become nucleophilic and vice versa, stable compounds can become reactive and highly reactive compounds can become stabilized. Normal reactivity patterns of functional groups can be inverted, and unconventional (*impossible*, under "normal" conditions) transformations can be achieved with facility. Highly reactive, normally unavailable reaction intermediates can be generated, stabilized, and used as efficient reagents in organic synthesis. Most organometallic reactions are highly specific, able to discriminate between structurally similar sites, thus reducing the need for bothersome "protection-deprotection" sequences that plague conventional organic synthesis. Finally, by careful selection of substrate and metal, multistep cascade sequences can be generated to form several bonds in a single process in which the metal "stitches together" the substrate.

However wonderful this sounds, there are a number of disadvantages as well. The biggest one is that the use of transition metals in organic synthesis requires a new way of thinking, as well as a rudimentary knowledge of how transition metals behave. In contrast to carbon, metals have a number of stable, accessible oxidation states, geometries, and coordination numbers, and their reactivity towards organic substrates is directly related to these features. The structural complexity of the transition metal species involved is, at first, disconcerting. Luckily, a few, easily mastered formalisms provide a logical framework upon which to organize and systematize this large amount of information, and this chapter is designed to provide this mastery.

The high specificity, cited as an advantage above, is also a disadvantage, in that specific reactions are not very general, so that small changes in the substrate can turn an efficient reaction into one that does not proceed at all. However, most transition metal systems carry "spectator" ligands — ligands which are coordinated to the metal but are not directly involved in the reaction — in addition to the organic substrate of interest, and these allow for the fine tuning of reactivity, and an increase in the scope of the reaction.

Finally, organometallic mechanisms are often complex and poorly understood, making both the prediction and the rationalization of the outcome of a reaction difficult. Frequently, there is a manifold of different reaction pathways of similar energy, and seemingly minor changes in the features of a reaction cause it to take an entirely unexpected course. By viewing this as an opportunity to develop new reaction chemistry rather than an obstacle to performing the desired reaction, progress can be made.

1.2   Formalisms

Every reaction presented in this book proceeds in the coordination sphere of a transition metal, and it is the precise electronic nature of the metal that determines the course and the outcome of the reaction. Thus, a very clear view of the nature of the metal is critical to an understanding of its reactivity. The main features of interest are (1) the oxidation state of the metal, (2) the number of d electrons on the metal in the oxidation state under consideration, (3) the coordination number of the metal, and (4) the availability (or lack thereof) of vacant coordination sites on the metal. The simple formalisms presented below permit the easy determination of each of these characteristics. A caveat, however, is in order. These formalisms are just that — formalisms — not reality, not the "truth", and in some cases, not even chemically reasonable. However, by placing the entire organic chemistry of transition metals within a single formalistic framework, an enormous amount of disparate chemistry can be systematized and organized, and a more nearly coherent view of the field is thus available. As long as it remains consistently clear that we are dealing with formalisms, exceptions will not cause problems.

a. Oxidation State

The oxidation state of a metal is defined as the charge left on the metal atom after *all* ligands have been removed *in their normal, closed-shell, configuration* — that is, *with their electron pairs*. The oxidation state is *not* a physical property of the metal, and it cannot be measured. It is a formalism which helps us in counting electrons, but no more. Typical examples are shown in Figure 1.1. (An alternative formalism, common in the older literature and still prevalent among Europeans, removes each covalent, anionic ligand as a *neutral* species with a single electron (homolytically), leaving the other electron on the metal. Although this results in a different *formal* oxidation state for the metal, it leads to the same conclusions as to the total number of electrons in the bonding shell, and the degree of coordination saturation.)

It must be emphasized that the *chemical* properties of the ligands are not always consonant with the oxidation state formalism. In metal hydrides, the hydride ligand is *always* formally considered to be H⁻, even though some transition metal "hydrides" are strong acids! Despite this, the formalism is still useful.

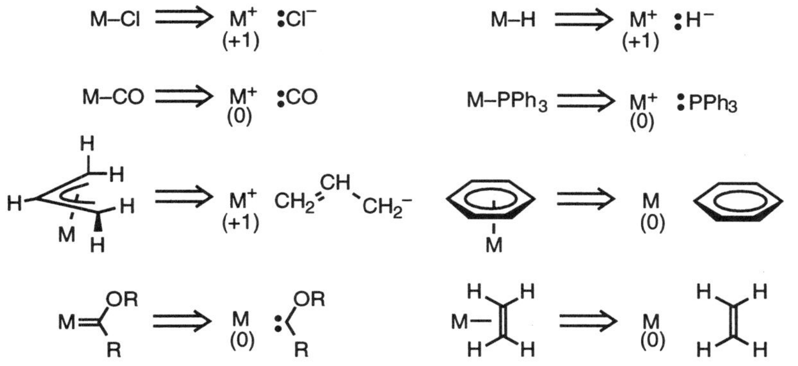

Figure 1.1   Examples of Oxidation State Determination

b. *d*-Electron Configuration, Coordination Saturation and the 18-Electron Rule

Having assigned the oxidation state of the metal in a complex, the number of *d* electrons on the metal can easily be assessed by referring to the periodic table. Figure 1.2 presents the transition elements along with their *d*-electron count. The transition series is formed by the systematic filling of the *d* orbitals. Note that these

Group number		4	5	6	7	8	9	10	11
First row	3*d*	Ti	V	Cr	Mn	Fe	Co	Ni	Cu
Second row	4*d*	Zr	Nb	Mo	Tc	Ru	Rh	Pd	Ag
Third row	5*d*	Hf	Ta	W	Re	Os	Ir	Pt	Au
	0	4	5	6	7	8	9	10	---
	I	3	4	5	6	7	8	9	10
Oxidation state	II	2	3	4	5	6	7	8	9
	III	1	2	3	4	5	6	7	8
	IV	0	1	2	3	4	5	6	7

d^n

Figure 1.2   *d*-Electron Configuration for the Transition Metals as a Function of Formal Oxidation State

electron configurations differ from those presented in most elementary texts in which the 4*s* level is presumed to be lower in energy than the 3*d*, and is filled first. Although this is the case for the *free atom* in the elemental state, these two levels are quite close in energy, and, for the *complexes* — which are not free metal atoms, but rather metals surrounded by ligands — discussed in this text, the assumption that the outer electrons are *d* electrons is a good approximation. By referring to the periodic table (or preferably, by remembering the positions of the transition metals) the *d*-electron count for any transition metal in any oxidation state is easily found.

The *d*-electron count is critical to an understanding of transition metal organometallic chemistry because of the **18-electron rule**, which states "in mononuclear, diamagnetic complexes, the total number of electrons in the bonding shell (the sum of the metal *d* electrons plus those contributed by the ligands) never exceeds 18" (at least not for very long — see below). This 18-electron rule determines the *maximum* allowable number of ligands for any transition metal in any oxidation state. Compounds having the maximum allowable number of ligands — having 18 electrons in the bonding shell — are said to be coordinatively saturated — that is, there are *no* remaining coordination sites on the metal. Complexes *not* having the maximum number of ligands allowed by the 18-electron rule are said to be coordinatively unsaturated — that is, they have vacant coordination sites. Since vacant sites are usually required for catalytic processes (the substrate must coordinate before it can react), the degree of coordination is central to many of the reactions presented below.

c. Classes of Ligands

The very large number of ligands that are involved in organotransition metal chemistry can be classified into three families: (1) *formal* anions, (2) *formal* neutrals, (3) *formal* cations. These families result from the oxidation state formalism requiring the removal of ligands in their closed-shell (with their pair(s) of electrons) state. Depending on the ligand, it can be either electron-donating or electron-withdrawing, and its specific nature has a profound effect on the reactivity of the metal center. Ligands with additional unsaturation may coordinate to more than one site, and thus contribute more than two electrons to the total electron count. Examples of each of these will be presented below, listed in approximate order of decreasing donor ability.

Formal anionic ligands which act as two electron donors are:

$$R^- > Ar^- > H^- > R\overset{\displaystyle O}{\overset{\|}{C}}(\text{-}) > halide^- \approx CN^-$$

These ligands fill one coordination site through one-point of attachment and are called "monohapto" ligands, designated as η^1. The allyl group, $C_3H_5^-$, can act as a monohapto (η^1) two electron donor, or a trihapto (η^3) (π-allyl) four electron donor (Figure 1.3). In the latter case, *two* coordination sites are filled, and all three carbons are bonded to the metal, but the ligand as a whole is still a formal mono anion. The cyclopentadienyl ligand, $C_5H_5^-$, most commonly bonds in an η^5-fashion, filling three coordination sites and acting as a six electron donor, although η^3 (four electrons, two site, equivalent to η^3-allyl) and η^1 coordinations are known. The cyclohexadienyl ligand, produced from nucleophilic attack on η^6-arene metal complexes (Chapter 10), is almost invariably a six electron, mono-negative, η^5-ligand which fills three coordination sites. These ligands are illustrated in Figure 1.3.

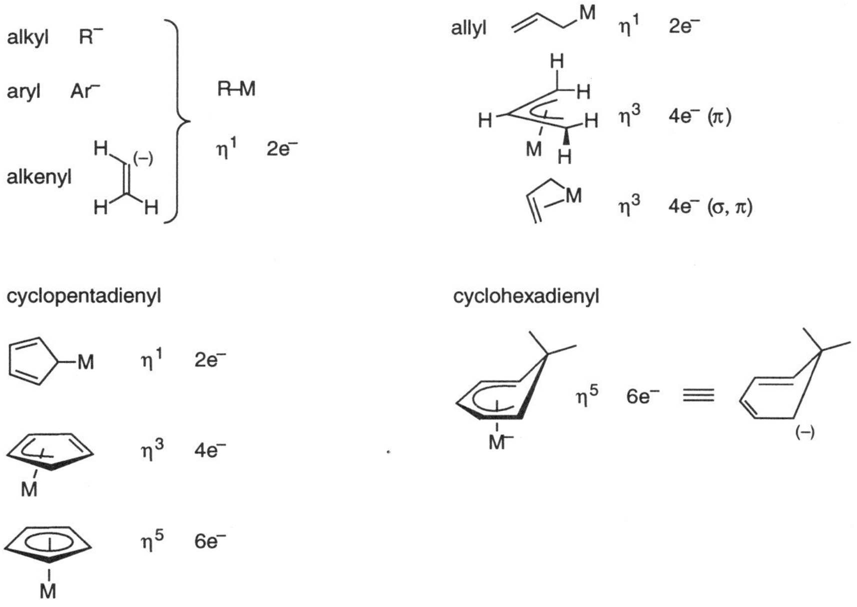

Figure 1.3 Bonding Modes for Mononegative Anionic Ligands

Formal neutral ligands abound, and they encompass not only important classes of "spectator" ligands — ligands such as phosphines and amines introduced to moderate the reactivity of the metal, but which are not *directly* involved in the reaction under consideration — but also ligands such as carbon monoxide, alkenes, alkynes, and arenes, which are often *substrates* (Figure 1.4). Ligands such as phosphines and amines are good σ-donors in organometallic reactions, and increase the electron density at the metal, while ligands such as carbon monoxide, isonitriles, and olefins are π-acceptors, and decrease the electron density at the metal. The reason for this is presented in the next section.

Formally cationic ligands are much less common, since species that bear a full formal positive charge *and* a lone pair of electrons are rare. The nitrosyl group is one of these, being a cationic two electron donor. It is often used as a spectator ligand, or to replace a carbon monoxide, to convert a neutral carbonyl complex to a cationic nitrosyl complex.

With all of the above information in hand, it is now possible to consider virtually any transition metal complex, assign the oxidation state of the metal, assess the total number of electrons in the bonding shell, and decide if that complex

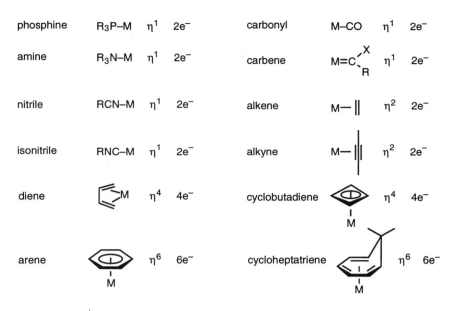

phosphine	R_3P–M	η^1	$2e^-$		carbonyl	M–CO	η^1	$2e^-$
amine	R_3N–M	η^1	$2e^-$		carbene		η^1	$2e^-$
nitrile	RCN–M	η^1	$2e^-$		alkene		η^2	$2e^-$
isonitrile	RNC–M	η^1	$2e^-$		alkyne		η^2	$2e^-$
diene		η^4	$4e^-$		cyclobutadiene		η^4	$4e^-$
arene		η^6	$6e^-$		cycloheptatriene		η^6	$6e^-$

Figure 1.4   Bonding Modes for Neutral Ligands

is coordinatively saturated or unsaturated. For example, the complex $CpFe(CO)_2(C_3H_7)$ is a stable, neutral species, containing two formally mononegative ligands, the n-propyl (C_3H_7) group, an η^1, $2e^-$ donor ligand, and the Cp ligand, an η^3 $6e^-$ donor ligand, and two neutral carbon monoxide ligands which contribute two electrons each. Since the overall complex is neutral, and has two mononegative ligands, the iron must have a formal 2+ charge, making it Fe^{II}, d^6. For an overall electron count, there are six electrons from the metal, a total of four from the two CO's, two electrons from the propyl group and six electrons from the Cp group, for a grand total of $18e^-$. Thus, this complex is an Fe^{II}, d^6, coordinatively saturated complex. Other examples, with explanatory comments, are given in Figure 1.5.

The last example in Figure 1.5 illustrates one of the difficulties in the strict application of formalisms. Strictly speaking, this treatment of the cyclohexadienyliron tricarbonyl cation is the formally correct one, since the rules state that each ligand shall be removed *in its closed shell form*, making the cyclohexadienyl ligand a six electron donor *anion*. However, this particular complex is prepared by a *hydride* abstraction from the neutral cyclohexadieneiron tricarbonyl, making the assumption that the cyclohexadienyl ligand is four electron donor cation reasonable. Further, the cyclohexadienyl group in the complex is quite reactive towards nucleophilic attack (Chapter 7). The true situation is impossible to establish. In reality, the plus charge resides neither exclusively on the metal, as the first, formally correct, treatment would assume, nor completely on the cyclohexadienyl ligand from which the hydride was abstracted, but rather is distributed throughout the entire metal-ligand array. Thus *neither* treatment represents the true situation, while *both* treatments come to the conclusion that the complex is an $18e^-$ saturated complex. The power of the formalistic treatment is that

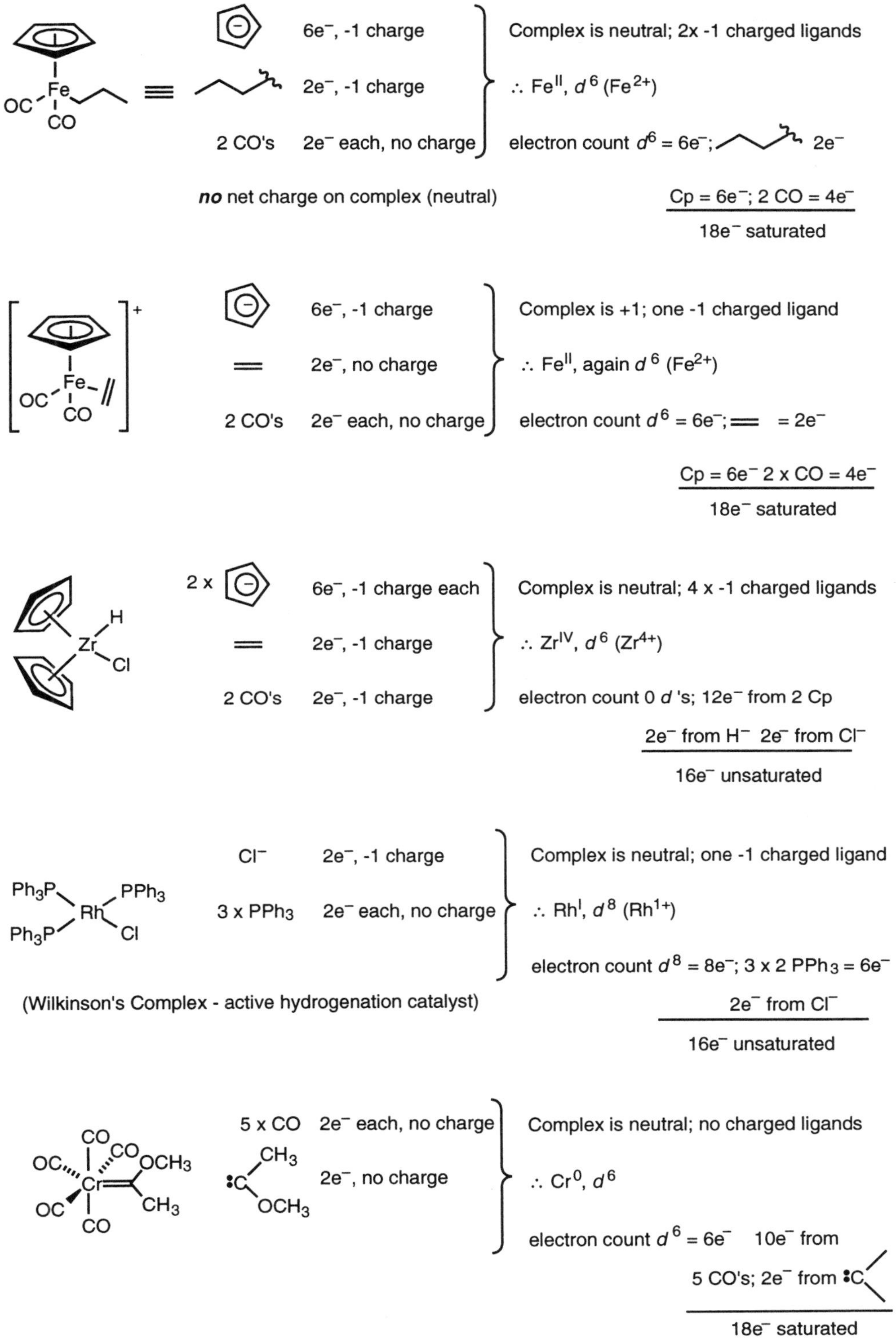
Fe
OC
CO
6e⁻, -1 charge
2e⁻, -1 charge
2 CO's
2e⁻ each, no charge
Complex is neutral; 2x -1 charged ligands
∴ Fe^{II}, d^6 (Fe^{2+})
electron count d^6 = 6e⁻; 2e⁻
Cp = 6e⁻; 2 CO = 4e⁻
18e⁻ saturated
no net charge on complex (neutral)
+
Fe
OC
CO
6e⁻, -1 charge
2e⁻, no charge
2 CO's
2e⁻ each, no charge
Complex is +1; one -1 charged ligand
∴ Fe^{II}, again d^6 (Fe^{2+})
electron count d^6 = 6e⁻; = 2e⁻
Cp = 6e⁻ 2 x CO = 4e⁻
18e⁻ saturated
H
Zr
Cl
2 x
6e⁻, -1 charge each
2e⁻, -1 charge
2 CO's
2e⁻, -1 charge
Complex is neutral; 4 x -1 charged ligands
∴ Zr^{IV}, d^6 (Zr^{4+})
electron count 0 d 's; 12e⁻ from 2 Cp
2e⁻ from H⁻ 2e⁻ from Cl⁻
16e⁻ unsaturated
Ph_3P
PPh_3
Rh
Ph_3P
Cl
Cl⁻
2e⁻, -1 charge
3 x PPh_3
2e⁻ each, no charge
Complex is neutral; one -1 charged ligand
∴ Rh^{I}, d^8 (Rh^{1+})
electron count d^8 = 8e⁻; 3 x 2 PPh_3 = 6e⁻
2e⁻ from Cl⁻
16e⁻ unsaturated
(Wilkinson's Complex - active hydrogenation catalyst)
CO
OC
CO
OCH_3
Cr
OC
CH_3
CO
5 x CO
2e⁻ each, no charge
CH_3
:C
OCH_3
2e⁻, no charge
Complex is neutral; no charged ligands
∴ Cr^0, d^6
electron count d^6 = 6e⁻ 10e⁻ from
5 CO's; 2e⁻ from :C
18e⁻ saturated

Figure 1.5 Electron Counting, Oxidation State
(continued on next page)

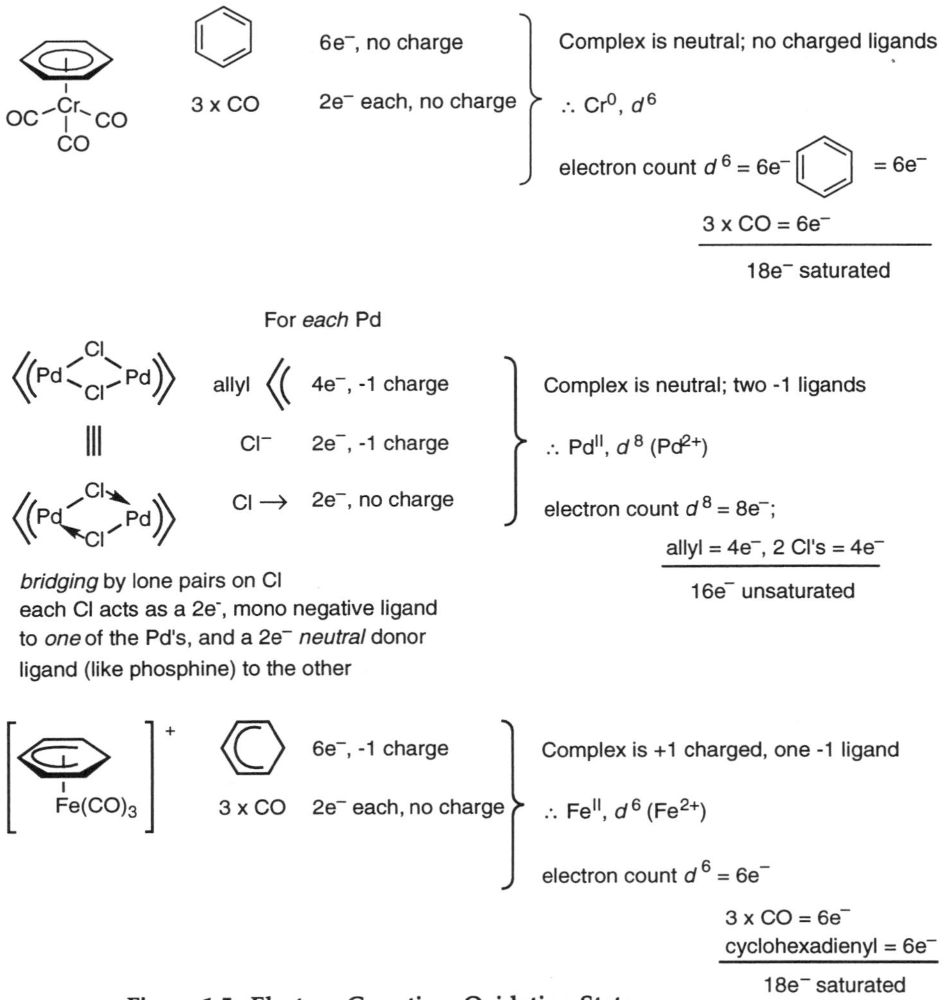

6e⁻, no charge

3 × CO   2e⁻ each, no charge

Complex is neutral; no charged ligands

∴ Cr^0, d^6

electron count d^6 = 6e⁻   = 6e⁻

3 × CO = 6e⁻
_____
18e⁻ saturated

For *each* Pd

allyl   4e⁻, -1 charge

Cl⁻   2e⁻, -1 charge

Cl →   2e⁻, no charge

bridging by lone pairs on Cl
each Cl acts as a 2e⁻, mono negative ligand
to *one* of the Pd's, and a 2e⁻ *neutral* donor
ligand (like phosphine) to the other

Complex is neutral; two -1 ligands

∴ Pd^{II}, d^8 (Pd^{2+})

electron count d^8 = 8e⁻;

allyl = 4e⁻, 2 Cl's = 4e⁻
_____
16e⁻ unsaturated

$Fe(CO)_3$

6e⁻, -1 charge

3 × CO   2e⁻ each, no charge

Complex is +1 charged, one -1 ligand

∴ Fe^{II}, d^6 (Fe^{2+})

electron count d^6 = 6e⁻

3 × CO = 6e⁻
cyclohexadienyl = 6e⁻
_____
18e⁻ saturated

Figure 1.5 Electron Counting, Oxidation State
(*continued from preceding page*)

it is consistent, and does *not* require a knowledge of how the complex is synthesized or how it reacts. As long as formalisms are not taken as a serious representation of reality, but rather treated as a convenient way of organizing and unifying a vast amount of information, they will be useful. To quote Roald Hoffmann "Formalisms are convenient fictions which contain a piece of the truth . . . and it is so sad that people spend a lot of time arguing about the deductions they draw . . . from formalisms without worrying about their underlying assumptions."[1]

1.3   Bonding Considerations[2]

The transition series is formed by the systematic filling of the *d* shell which, in complexes, is of lower energy than the next *s*, *p* level. Thus, transition metals have partially filled *d* orbitals, and vacant *s* and *p* orbitals. In contrast, most of the ligands have filled "*sp^n*" hybrid orbitals, and, for unsaturated organic ligands, vacant anti-bonding π* orbitals. The *d* orbitals on the metal have the same symmetry *and* similar energy as the antibonding π* orbitals of the unsaturated ligands (Figure 1.6). Transition metals participate in two types of bonding, often simultaneously. σ-*Donor* bonds are formed by the overlap of filled *sp^n* hybrid orbitals on the ligand

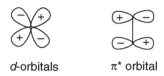

d-orbitals   π* orbital

Figure 1.6   Symmetry of *d* Orbitals and π* Orbitals

(including π-bonding orbitals of unsaturated hydrocarbons) with vacant "*dsp*" hybrid orbitals on the metal. Ligands which are primarily σ-donors, such as R_3P, R_3N, H^-, and R^-, *increase* the electron density on the metal. Unsaturated organic ligands, such as alkenes, alkynes, arenes, carbon monoxide and isonitriles, which have π*-antibonding orbitals, bond somewhat differently. They, too, form σ-donor bonds by overlap of their filled π-bonding orbitals with the vacant "*dsp*" hybrid orbitals of the metal. In addition, *filled* d *orbitals* of the metal can overlap with the *vacant* π*-antibonding orbital of the ligand, back donating electron density from the metal to the ligand (Figure 1.7). Thus, unsaturated organic ligands can act as π-acceptors or π-acids, *decreasing* electron density on the metal. Because of these two modes of bonding — σ-donation and π-accepting "back bonding" — metals can act

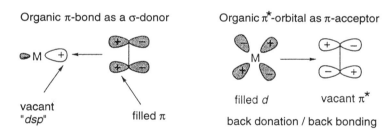

Organic π-bond as a σ-donor   Organic π*-orbital as π-acceptor

vacant   filled π   filled *d*   vacant π*
"*dsp*"   back donation / back bonding

Figure 1.7   Details of π-Bonding

as an electron sink for ligands, either supplying or accepting electron density. As a consequence, the electron density on the metal, and hence its reactivity, can be modulated by varying the ligands around the metal. This offers a major method for fine-tuning the reactivity of organometallic reagents.

There are two modes of π-back bonding and two types of π-acceptor ligands: (1) longitudinal acceptors, such as carbon monoxide, isonitriles and linear nitrosyls, and (2) perpendicular acceptors, such as alkenes and alkynes (Figure 1.8).

There is ample physical evidence for π-back bonding with good π-acceptor ligands. Carbon monoxide, which always acts as an acceptor, experiences a lengthening of the CO bond and a decrease in CO stretching frequency in the infrared

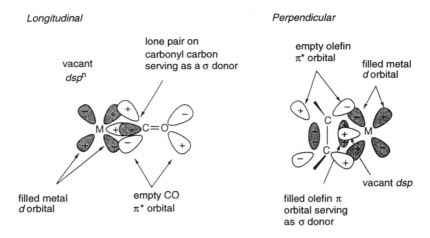

Figure 1.8 Types of π-Acceptors

spectrum upon complexation, both indications of the population of a π* orbital on CO, decreasing the CO bond order. The situation is a little more complex with olefins. With electrophilic metals, such a Pd^{2+} or Pt^{2+}, olefins are primarily σ-donor ligands, and the C=C bond length in such complexes is virtually the same as for the free olefin. With electron-rich metals, such as Pd(0), substantial back bonding results, the olefin C–C bond is lengthened, and the hybridization at the olefinic carbons changes towards sp^3. The degree of π-back bonding has a serious effect on the reactivity of π-acceptor ligands. Organometallic processes involving reactions of π-acceptor ligands are the next topic for consideration.

1.4 Structural Considerations

Although the focus of this book is the *reactivity* of metal-bound ligands, it is important to appreciate the geometries adopted by organometallic complexes. For 18-electron (saturated) complexes these geometries are determined by steric factors, such that four coordinate complexes such as $Ni(CO)_4$ and $Pd(PPh_3)_4$, are tetrahedral, five coordinate complexes such as $Fe(CO)_5$ and $(diene)Fe(CO)_3$, are trigonal bipyramidal, and six coordinate complexes such as $Cr(CO)_6$, Ir^{III} complexes, and Rh^{III} complexes, are octahedral.

In contrast, the d^8 complexes of Pd^{II}, Pt^{II}, Ir^{I}, and Rh^{I} find it energetically favorable to form square planar, four coordinate unsaturated $16e^-$ complexes. The planar geometry makes the d $x_2–y_2$ orbital so high in energy that it remains unoccupied in stable complexes, but enables such complexes to undergo facile ligand substitution by an associative mechanism. This, and other mechanistic considerations are the topic of the next chapter.

References

1. Sailland, J-V.; Hoffmann, R. *J. Am. Chem. Soc.* **1984**, *106*, 2006.
2. For a more rigorous treatment of bonding see: Collman, J.P.; Hegedus, L.S.; Finke, R.O.; Norton, J.R. *Principles and Applications of Organotransition Metal Chemistry.* University Science Books, Mill Valley, CA, 1987, pp. 21-56.

CHAPTER 2

Organometallic Reaction Mechanisms

2.1   Introduction

The focus of this entire book is the *organic* chemistry of organometallic complexes, and the effect coordination to a transition metal has on the reactivity of organic substrates. In transition metal catalyzed reactions of organic substrates, it is the chemistry of the *metal* that determines the course of the reaction, and a rudimentary understanding of the mechanisms by which transition metal organometallic complexes react is critical to the rational use of transition metals in organic synthesis.

Mechanistic organometallic chemistry is both more complex and less developed than mechanistic organic chemistry, although many parallels exist. The organic chemist's reliance on "arrow pushing" to rationalize the course of a reaction is equally valid with organometallic processes, provided the appropriate rules are followed. Some organometallic mechanisms are understood in exquisite detail, while others are not understood at all. The field is very active, and new insights abound. However, this book will treat the topic only at a level required to permit the planning and execution of organometallic processes, and to facilitate the understanding of the literature in the field. More specialized treatises should be consulted for a more thorough treatment.[1]

2.2   Ligand Substitution Processes

Ligand substitution (exchange) processes (Eq. 2.1) are central to virtually all organometallic reactions of significance for organic synthesis. For catalytic processes,

Eq. 2.1

$$M-L \; + \; L' \; \rightleftharpoons \; M-L' \; + \; L$$

it is common that a stable catalyst precursor must lose a ligand, coordinate the substrate, promote whatever reaction is being catalyzed and then release the substrate. Three of the four steps are ligand exchange processes. In other instances, it may be necessary to replace an innocent (spectator) ligand in an organometallic complex to adjust the reactivity of a coordinated substrate (Eq. 2.2) or to stabilize an unstable intermediate for isolation or mechanistic studies (Eq. 2.3). Thus, a fundamental understanding of ligand exchange processes is central to the utilization of organometallic complexes in synthesis.

Eq. 2.2

unreactive toward
nucleophiles

reactive toward
nucleophiles

Eq. 2.3

requires HCl/HNO$_3$
for oxidation

unstable
to air

requires HNO$_3$
for oxidation

Ligand exchange processes at metal centers share many characteristics with organic nucleophilic displacement reactions at carbon centers, in that the outgoing ligand departs with its pair of electrons, while the incoming ligand attacks the metal with *its* pair of electrons. An obvious difference is that metals can have coordination numbers larger than four, and stable, coordinatively unsaturated metal complexes are common. Thus, the *details* of ligand exchange are somewhat different, and more varied, than typical organic nucleophilic displacement reactions. However ligand substitution processes can be classified in much the same way as organic nucleophilic displacement reactions, and can involve either two-electron or one-electron processes, in associative (S$_N$2-like) or dissociative (S$_N$1-like) reactions (Figure 2.1).

By far the most extensively studied ligand exchange reactions are those of coordinatively unsaturated, 16 electron, d^8, square planar complexes of Ni^{II}, Pd^{II}, Pt^{II}, Rh^I and Ir^I. These typically involve two-electron, associative processes, which resemble S$_N$2 reactions, but with some differences. The platinum system shown in Eq. 2.4 is typical.[2] Because the metal is coordinatively unsaturated, apical attack at the vacant site occurs, producing a square pyramidal, saturated intermediate. This rearranges via a trigonal bipyramidal intermediate (analogous to the transition state in organic S$_N$2 reactions) to the square pyramidal intermediate with the leaving group axial; the leaving group then leaves. The rate is second order, and depends upon the

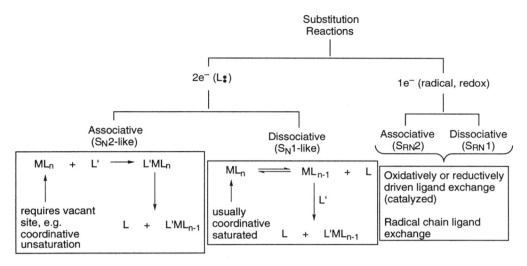

Figure 2.1   Classification of Ligand Substitution Processes

Eq. 2.4

metal, in the order Ni > Pd >> Pt, with a range of ~10^6. This is thought to reflect the relative abilities of the metals to form five-coordinate, 18-electron intermediates. As a consequence of this rate difference, if a reaction depends on ligand exchange, platinum-catalyzed processes may be too slow to be synthetically useful. The rate also depends on the incoming ligand, Y, in the order R_3P > Py > NH_3, Cl^- > H_2O > OH^-, spanning a range of ~10^5. Further the rate depends on the leaving group X, in the order NO_3^- > H_2O > Cl^- > Br^- > I^- > N_3^- > SCN^- > NO_2^- > CN^-. The rate depends on the ligand *trans* to the ligand being displaced, in the order R_3Si^- > H^- ≈ CH_3^- ≈ CN^- ≈ olefins ≈ CO > PR_3 ≈ NO_2^- ≈ I^- ≈ SCN^- > Br^- > Cl^- > RNH_2 ≈ NH_3 > OH^- > NO_3^- ≈ H_2O. A practical consequence of this "*trans* effect" is the ability to activate ligand exchange by replacing the ligand *trans* to the one to be displaced.

In contrast, 18-electron *saturated* systems undergo ligand exchange reactions much more slowly, and then by a dissociative, S_N1-like process. That is, a coordination site must be vacated before ligand exchange can occur. Typical complexes which undergo ligand exchange by this process are the saturated metal carbonyl complexes, such as $Ni(CO)_4$ ($M°$, d^{10}, saturated), $Fe(CO)_5$ ($M°$, d^8, saturated) and $Cr(CO)_6$ ($M°$, d^6, saturated). Nickel tetracarbonyl is the most labile and most reactive toward ligand exchange (Eq. 2.5). The rate of exchange is first order, and proportional to the concentration of nickel carbonyl. With ligands other

Eq. 2.5

$$Ni(CO)_4 \overset{slow}{\rightleftharpoons} Ni(CO)_3 \; + \; CO \xrightarrow[\text{fast}]{L} LNi(CO)_3$$

tetrahedral
labile

$$Rate \; \alpha \; [Ni(CO)_4] \quad - \quad 1st \; order$$

than CO, the rate of ligand exchange can be accelerated by bulky ligands, since loss of one ligand leads to release of steric strain. This is best illustrated by the nickel(0) (tetrakis)phosphine complexes seen in Table 2.1. A measure of the bulk of a phosphine is its "cone angle" (Figure 2.2).[3] As the size of the phosphine increases, the equilibrium constant for loss of a ligand changes by more than 10^{10}. A practical consequence is that, in catalytic processes involving loss of a phosphine ligand, reactivity can be "fine tuned" by alteration of the phosphine ligand.

Table 2.1   Ligand Dissociation as a Function of Phosphine Cone Angles

$$L_4Ni \underset{\text{PhH} \; 25°}{\overset{K_D}{\rightleftharpoons}} L_3Ni-L \; + \; L$$

L =	P(OEt)$_3$	P(O-p-tolyl)$_3$	P(O-iPr)$_3$	P(O-o-tolyl)$_3$	PPh$_3$
Cone angle	109°	128°	130°	141°	145°
K_D	$<10^{-10}$	6×10^{-10}	2.7×10^{-5}	4×10^{-2}	No NiL$_4$

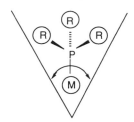

Figure 2.2   Cone Angle

In contrast to nickel tetracarbonyl, iron pentacarbonyl (trigonal bipyramidal) and chromium hexacarbonyl (octahedral) are *not* labile, and are quite inert to ligand substitution, and either heat, sonication, or irradiation (light) is required to affect ligand substitution in these systems. Carbon monoxide displacement in these substitutionally inert complexes can also be chemically promoted. Tertiary amine oxides irreversibly oxidize metal-bound carbon monoxide to carbon dioxide, which readily dissociates. Since amines are not strong ligands for metals in low oxidation states, a coordination site is opened (Eq. 2.6).[4] Amine oxides are not particularly powerful oxidants for organic compounds, and have been used to remove metal carbonyl fragments — by "chewing off" the CO's—from organic ligands reluctant to

Eq. 2.6

$$(CO)_4Fe\!-\!C\!\equiv\!O \longleftrightarrow (CO)_4Fe\!=\!C\!=\!O \;+\; R_3\overset{+}{N}\!-\!O^- \longrightarrow \left[(CO)_4Fe\overset{(-)}{\underset{}{-}}\overset{O}{\overset{\|}{C}}\!-\!O\!-\!\overset{+}{N}R_3 \right]$$

$$LFe(CO)_4 \xleftarrow[\text{fast}]{L} Fe(CO)_4 \;+\; CO_2 \;+\; R_3N$$

leave on their own. This has been used extensively in organic synthetic applications of cationic dienyliron carbonyl complexes (Chapter 7). Tributylphosphine oxide can *catalyze* CO exchange, presumably by a similar mechanism (Eq. 2.7),[5] although in this case the strength of the P–O bond prevents fragmentation and oxidation of CO.

Eq. 2.7

$$Fe(CO)_5 + R_3PO \longrightarrow (CO)_4Fe\overset{(-)}{\underset{}{-}}\overset{O}{\overset{\|}{C}}\!-\!O\!-\!\overset{+}{P}R_3$$

strong P–O bond

$$LFe(CO)_4 \xleftarrow{L} Fe(CO)_4 \;+\; CO \;+\; R_3PO$$

Ligand exchanges can also be promoted by one-electron oxidation or reduction of 18e⁻ systems, transiently generating the much more labile 17e⁻ or 19e⁻ systems. However, to date these processes have found little application in the use of transition metals in organic synthesis.

Not all 18e⁻ complexes undergo ligand substitution by a dissociative mechanism. In these cases, an exchange phenomenon called "slippage" is important. Ligands such as η^3-allyl, η^5-cyclopentadienyl, and η^6-arene normally fill multiple coordination sites. However, even in saturated compounds, they can open coordination sites on the metal by "slipping" to a lower hapticity (coordination number). This is illustrated in Figure 2.3. This "slippage" can account for a number

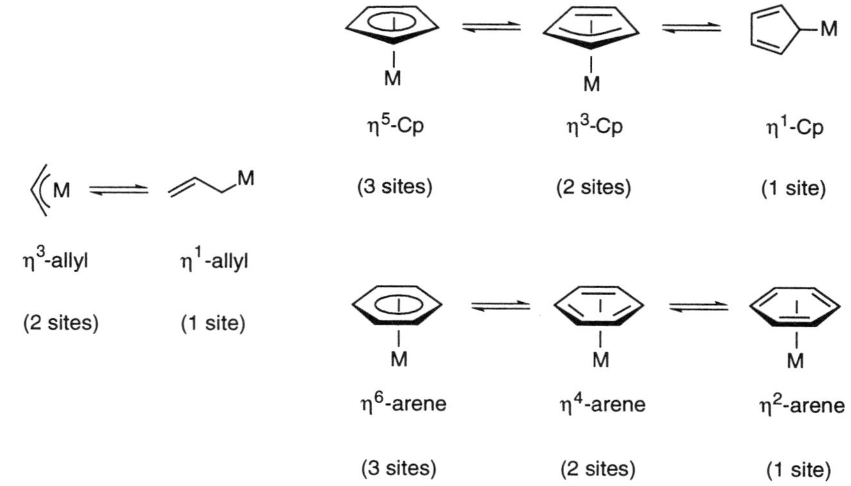

η⁵-Cp
(3 sites)

η³-Cp
(2 sites)

η¹-Cp
(1 site)

η³-allyl
(2 sites)

η¹-allyl
(1 site)

η⁶-arene
(3 sites)

η⁴-arene
(2 sites)

η²-arene
(1 site)

Figure 2.3 Slippage in Polyhapto Complexes

of otherwise difficult to explain observations, including apparent associative S_N2-like ligand exchange with coordinatively saturated complexes (for example, Eq. 2.8). With arene complexes in particular, the first step ($\eta^6 \rightarrow \eta^4$) requires the most energy, since aromaticity is disrupted. With fused arene ligands such as naphthalene or indene, this step is easier, and rate enhancements of 10^3-10^8 have been observed.

Eq. 2.8

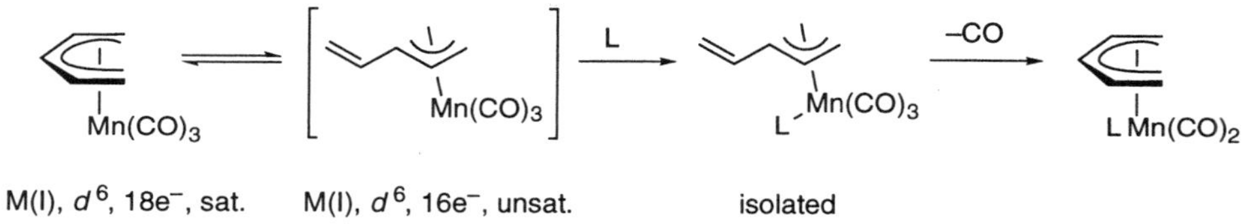

M(I), d^6, 18e⁻, sat.   M(I), d^6, 16e⁻, unsat.   isolated

2.3   Oxidative Addition/Reductive Elimination[6]

Oxidative addition/reductive elimination processes are central to a vast array of synthetically useful organometallic reactions (e.g., the "Heck" reaction, Chapter 4), and occur because of the ability of transition metals to exist in several different oxidation states, in contrast to nontransition metal compounds, which usually have closed shell configurations. In fact, it is this facile shuttling between oxidation states that makes transition metals so useful in organic synthesis. The terms oxidative addition and reductive elimination are generic, describing an overall transformation but not the specific mechanism by which that transformation occurs. There are many mechanisms for oxidative addition, some of which are presented below. The general transformation is described in Eq. 2.9.

Eq. 2.9

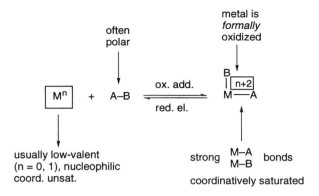

A coordinatively unsaturated transition metal, usually in a low (0, +1) oxidation state and hence relatively electron rich and nucleophilic, undergoes reaction with some substrate A–B, to form a new complex A–M–B in which the metal has formally inserted into the A–B bond. Because of the oxidation state formalism, which demands each ligand be removed with its pair of electrons (although A and B really contribute only one electron each to A–M–B), the metal is formally oxidized in the A–M–B adduct. Since the "oxidant" A–B has added to the metal, the process is termed "oxidative addition." In general, ligands such as R_3P, R^- and H^-, which are good σ-donors and increase the electron density at the metal, facilitate oxidative addition, whereas ligands such as CO, CN^- and olefins, which are good π-acceptors and decrease electron density at the metal, suppress oxidative addition. The reverse of oxidative addition is reductive elimination, and both the transformation and terminology are obvious. Oxidative addition-reductive elimination processes are central to the use of transition metals in organic synthesis because one *need not reductively eliminate the same two groups that were oxidatively added to the metal.* This is the basis of a very large number of cross-coupling reactions (Chapter 4).

The most commonly-encountered systems are Ni^0, $Pd^0 \rightarrow Ni^{II}$, Pd^{II} ($d^{10} \rightarrow d^8$) and Rh^I, $Ir^I \rightarrow Rh^{III}$, Ir^{III} ($d^8 \rightarrow d^6$). The most extensively-studied system and the one that contributed heavily to the development of many of the formalisms presented above is Vaska's compound, the lemon yellow, Ir^I, d^8, 16 electron, coordinatively unsaturated complex which undergoes oxidative addition with a wide array of substrates to form stable Ir^{III}, d^6, 18 electron saturated complexes (Eq. 2.10).

Eq. 2.10

Ph3P, Ir, CO, Cl, PPh3 + A–B ⇌ Ph3P, A, Ir, CO, Cl, PPh3, B (or *cis*)

Ir^I, d^8, 16e⁻, unsat.
Vaska's compound

Ir^{III}, d^6, 18e⁻, sat.

The range of addenda that participate in oxidative addition reactions to transition metal complexes is large, and contains many members useful in organic synthesis. There are three general classes of addend: polar electrophiles, nonpolar substrates and multiple bonds (Figure 2.4). Note that many of these result in the formation of metal-hydrogen and metal-carbon bonds, and in the "activation" of a range of organic substrates.

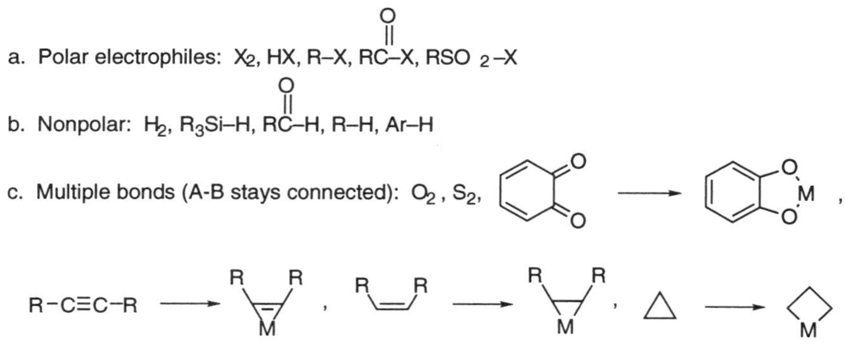

a. Polar electrophiles: X_2, HX, R–X, $RC(=O)–X$, RSO_2–X

b. Nonpolar: H_2, R_3Si–H, $RC(=O)$–H, R–H, Ar–H

c. Multiple bonds (A-B stays connected): O_2, S_2,

R–C≡C–R

Figure 2.4 Classes of Substrates for Oxidative Addition Reactions

There are several documented mechanisms for oxidative addition, and the one followed depends heavily on the nature of the reacting partners. They include (1) concerted, associative, one step "insertions" into A–B by M, (2) ionic, associative, two step S_N2 reactions, and (3) electron-transfer - radical chain mechanisms.

Concerted oxidative additions are best known for nonpolar substrates, particularly the oxidative addition of H_2 (central to catalytic hydrogenation), and oxidative addition into C–H bonds (hydrocarbon activation). The process is thought to involve prior coordination of the H–H or C–H bond to the metal in an "agostic" (formally a two electron, three center bond) fashion, followed by *cis* insertion (Eq. 2.11). This idea is supported by the isolation of molecular hydrogen

Eq. 2.11

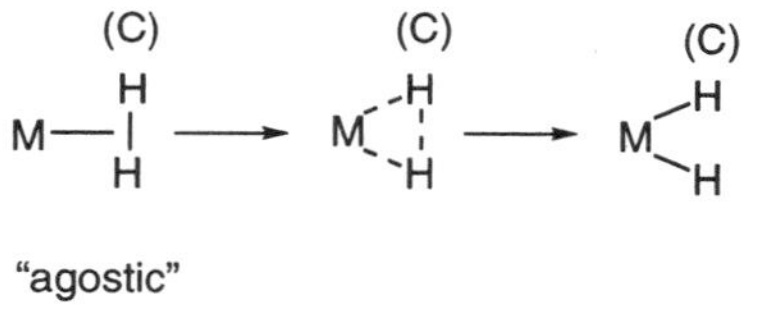

"agostic"

complexes of, for example, tungsten (Eq. 2.12)[7] in which just this sort of bonding is observed. This same process is likely but not yet demonstrated for oxidative addition of M^0 d^{10} complexes with sp^2 halides, a process that goes with retention of stereochemistry at the olefin (Eq. 2.13).

Oxidative addition via S_N2-type processes is most often observed with strongly nucleophilic, low-valent metals and classic S_N2 substrates such as primary

Eq. 2.12

$$L_2W(CO)_3 \ + \ H_2 \longrightarrow$$

Eq. 2.13

$$L_4Pd \rightleftharpoons L \ + \ L_3Pd \xrightarrow{} \left[\ \right] \longrightarrow$$

and secondary organic halides and tosylates. An early example is shown in Eq. 2.14,[8] and its features are typical of metals which react by S_N2-type processes. The rate law is second order, first order metal and first order in substrate, and polar sol-

Eq. 2.14

$$+ \ CH_3X \xrightarrow{slow} \left[\ \right]^+ X^- \xrightarrow{fast}$$

(Ir^I, d^8, 16e$^-$)   (Ir^{III}, d^6, 18e$^-$)

X = I > Br > Cl

L = Et$_3$P > Et$_2$PPh > EtPPh$_2$ > PPh$_3$ (30 fold rate difference)

vents accelerate the reaction, as expected for polar intermediates. The order of substrate reactivity is I > Br > Cl as expected for S_N2-type reactions. The rate is also dependent on the nature of the spectator phosphine ligands, and, in general, increases with increasing basicity (ability to donate electrons to the metal) of the phosphine. This is an important observation, since it confirms the claim that reactivity of transition metal complexes can be "fine tuned" by changing the nature of the spectator ligands. This feature is exceptionally valuable when using transition metals in complex total synthesis, since the inherent specificity of organometallic reactions can be applied over a wide range of substrates.

A synthetically more useful complex that reacts by an S_N2-type mechanism is Collman's reagent, $Na_2Fe(CO)_4$ (Eq. 2.15).[9] This very potent nucleophile is prepared by reducing the Fe^0 complex $Fe(CO)_5$ with sodium-benzophenone ketyl to produce the Fe^{2-} complex. This is an unusually low oxidation state for iron, but, because carbon monoxide ligands are powerful π-acceptors and strongly electron-withdrawing, they stabilize negative charge, and highly reduced metal carbonyl complexes are common. The reaction of this complex with halides has all the features common to S_N2 chemistry (Eq. 2.15), and the S_N2 mechanism is the most likely pathway for the reaction of d^8–d^{10} transition metal complexes with 1°, 2°, allylic and benzylic halides.

Eq. 2.15

$$Na_2Fe(CO)_4 \ + \ RX \longrightarrow [RFe(CO)_4]Na \ + \ NaX$$

Rate ∝ [RX][Fe(CO)$_4$$^{2-}$] for RX 1° > 2°   X = I > Br > OTs > Cl

Stereochemistry at carbon - clean inversion

However, in many cases there is a competing radical-chain process which can affect the same overall transformation, but by a completely different mechanism. For example, Vaska's complex (Eq. 2.14) is unreactive towards organic bromides and even 2° iodides when the reaction is carried out with the strict exclusion of air. However, traces of air or addition of radical initiators promotes a smooth reaction with these substrates, via a radical-chain mechanism (Eq. 2.16).[10] In this case, racemization is observed, as expected. The existence of an accessible radical-chain mechanism for oxidative addition explains the empirical observation that traces of air sometimes promote the reaction of complexes which normally require strict exclu-

Eq. 2.16

$$RX \xrightarrow[\bullet O_2R]{h\nu \ or} R\bullet \qquad\qquad \text{initiation}$$

$$R\bullet \ + \ L_2Ir^I(CO)Cl \longrightarrow L_2Ir^{II}(R)(CO)Cl$$

$$L_2Ir^{II}(R)(CO)Cl \ + \ RX \longrightarrow L_2Ir^{III}(R)(CO)(Cl)X \ + \ R\bullet$$

sion of air, and serves as a warning that the mechanism by which a reaction proceeds may be a function of the care with which it is carried out.

A more synthetically-useful example of a radical-chain oxidative addition is the nickel catalyzed coupling of aryl halides to biaryls, which was originally thought to proceed by two sequential two-electron oxidative additions. However, careful studies showed this was not the case. Rather, a complicated radical-chain process involving $Ni^I \rightarrow Ni^{III}$ oxidative additions is the true course of the reaction (Eq. 2.17).[11]

Reductive elimination is the reverse of oxidative addition and is usually the last step of many catalytic processes. It is exceptionally important for organic synthetic applications because *it is the major way in which transition metals are used to make carbon-carbon and carbon-hydrogen bonds!* In spite of its importance, reductive elimination has not been extensively studied. It is known that the groups to be eliminated must occupy *cis* positions on the metal, or must rearrange to *cis* if they are *trans* (Eq. 2.18).[12] Reductive elimination can be promoted. Anything which reduces the electron density at the metal facilitates reductive elimination. This can be as

Eq. 2.17

$$L_nNi + ArX \longrightarrow ArAr + NiX_2$$

$$Ni^0 + ArX \longrightarrow Ar-NiX^{II} \xrightarrow{ArX} Ar-Ni^{IV}-X \longrightarrow Ar-Ar$$

(with Ar above and X below the Ni^{IV} center)

original (incorrect) mechanism

actually a *radical chain*

initiation

$$ArX + L_3Ni^0 \xrightarrow[\text{ox. add.}]{2e^-} Ar-Ni^{II}-X$$

(with L above and L below the Ni^{II})

$$L_2Ni^{II}Ar + ArX \xrightarrow{e^- \text{ transfer}} [ArNi^{III}XL_2]^{\cdot+}[ArX]^{\cdot-}$$

(with X on the $L_2Ni^{II}Ar$)

$$[ArNi^{III}XL_2]^{\cdot+}[ArX]^{\cdot-} \longrightarrow \boxed{Ni^IX} + \ldots ?$$

$$\text{or } L_nNi^0 + ArX \xrightarrow{e^- \text{ transfer}} [L_nNi^{\cdot+}][ArX]^{\cdot-} \longrightarrow \boxed{Ni^IX} + Ar\bullet \ldots$$

chain

$$Ni^IX + ArX \longrightarrow ArNi^{III}X_2$$

$$ArNi^{III}X_2 + ArNi^{II}X \rightleftharpoons Ar_2Ni^{III}X + Ni^{II}X_2$$

$$Ar_2Ni^{III}X \longrightarrow ArAr + Ni^IX$$

Eq. 2.18

(Pd complex with Ph$_2$P–CH$_2$CH$_2$–PPh$_2$ chelate and two Me groups) $\xrightarrow{\text{fast}}$ CH_3-CH_3 but (binaphthyl-bridged Pd complex: $Ph_2P-Pd-PPh_2$ with two Me groups) $\xrightarrow{\Delta}$ No Reaction

simple as the dissociation of a ligand, either spontaneously or by the application of heat or light. Both chemical and electrochemical oxidation promotes reductive elimination (although oxidatively-driven reductive elimination sounds like an oxymoron). Of practical use is the observation that the addition of strong π-acceptor ligands, such as carbon monoxide, maleic anhydride, quinones or tetracyanoethylene, promotes reductive elimination.

2.4 Migratory Insertion/β-Hydride Elimination

The third organometallic mechanism of interest for organic synthesis is the process of migratory insertion (Eq. 2.19), in which an unsaturated ligand — usually CO, RNC, olefins, or alkynes — *formally* inserts into an adjacent (*cis*) metal-ligand

Eq. 2.19

$$\begin{matrix} X \\ | \\ M\!-\!Y \end{matrix} \rightleftharpoons M\!-\!Y\!-\!X \overset{L}{\rightleftharpoons} \begin{matrix} L \\ | \\ M\!-\!Y\!-\!X \end{matrix}$$

Y = CO, RNC, C=C, C≡C   X = H, R, etc.

bond. When this adjacent ligand is hydride, or σ-alkyl, the process forms a new carbon-hydrogen or carbon-carbon bond, and in the case of alkenes, a new metal-carbon bond. The term "insertion" is somewhat misleading, in that the process actually proceeds by *migration* of the adjacent ligand to the metal-bound unsaturated species, generating a vacant site. Insertions are usually reversible. If the migrating group has stereochemistry, it is usually retained in both the insertion step and the reverse.

The insertion of carbon monoxide into a metal-carbon σ-bond is one of the most important processes for organic synthesis because it permits the direct introduction of molecular carbon monoxide into organic substrates, to produce aldehydes, ketones, and carboxylic acid derivatives. The reverse process is also important, as it allows for the decarbonylation of aldehydes to alkanes and acid halides to halides.

The general process of CO insertion, shown in Eq. 2.20, involves reversible migration of an R group to a *cis* bound CO, as the rate determining step. The vacant site generated by this migration is then filled by some external ligand present.

Eq. 2.20

$$\begin{matrix} M\!-\!R \\ | \\ CO \end{matrix} \underset{cis}{\overset{}{\diagdown}} \quad \underset{k_{-1}}{\overset{k_1}{\rightleftharpoons}} \quad O\!\!=\!\!\!\overset{M}{\underset{}{\diagup}}\!\!R \quad \underset{-L}{\overset{L}{\rightleftharpoons}} \quad O\!\!=\!\!\!\overset{M-L}{\underset{}{\diagup}}\!\!R$$

Similarly, the reverse process requires a vacant coordination site *cis* to the acyl ligand. Both the migratory insertion and the reverse deinsertion occur with retention of configuration of the migrating group and of the metal, if the metal center is chiral. The migratory aptitude is: Et > Me > PhCH$_2$ > η^1 allyl > vinyl ≥ aryl, ROCH$_2$ > propargyl > HOCH$_2$. Hydride (H), CH$_3$CO, and CF$_3$ usually do not migrate, and heteroatoms such as RO$^-$ and R$_2$N$^-$ rarely migrate. Lewis acids often accelerate CO insertion reactions, as does oxidation, although in this case the metal acyl species often continues to react, and is oxidatively cleaved from the metal.

Alkenes also readily insert into metal-hydrogen bonds (a key step in catalytic hydrogenation of olefins) and metal-carbon σ-bonds, resulting in alkylation of the olefin (Eq. 2.21). The reverse of hydride insertion is β-hydrogen elimination, a very common pathway for decomposition of σ-alkylmetal complexes. Olefin insertions share many features. The olefin *must* be coordinated to the metal prior to insertion, and must be *cis* to the migrating hydride or alkyl group. The migration generates a vacant site on the metal, and for the reverse process, β-hydride elimination (β-alkyl

Eq. 2.21

elimination does not occur) a vacant *cis* site on the metal is required. Both the metal and the migrating group add to the same face of the olefin. When an alkyl group migrates, its stereochemistry is maintained. A rough order of migrating aptitude is: H >> R, vinyl, aryl > RCO >> RO, R_2N. Again, heteroatom groups migrate only with difficulty, since the extra lone pair(s) on these ligands can form multiple bonds to the metal (Eq. 2.22).

Eq. 2.22

Alkynes also insert into metal-hydrogen and metal-carbon bonds, again with *cis* stereochemistry and with retention at the migrating center (Eq. 2.23). This has been less studied, and is complicated by the fact that alkynes often insert into the product σ-vinyl complex, resulting in oligomerization.

Eq. 2.23

2.5   Nucleophilic Attack on Ligands Coordinated to Transition Metals

The reaction of nucleophiles with coordinated, unsaturated organic ligands is one of the most useful processes for organic synthesis.[13] Unsaturated organic compounds such as carbon monoxide, alkenes, alkynes and arenes are electron rich, and are usually quite unreactive towards nucleophiles. However, complexation to electron-deficient metals inverts their normal reactivity, making them generally subject to nucleophilic attack, and opening up entirely new synthetic transformations. The more electron deficient a metal center is, the more reactive towards nucleophiles its associated ligands become. Thus, cationic complexes and complexes having strong π-accepting spectator ligands such as carbon monoxide are especially reactive in these processes.

Metal-bound carbon monoxide is generally reactive towards nucleophiles, providing a major route for the incorporation of carbonyl groups into organic substrates. Neutral metal carbonyls usually require strong nucleophiles such as organolithium reagents, and the reaction (a) produces anionic metal acyl complexes (acyl "ate" complexes) (Eq. 2.24). (For the purposes of electron bookkeeping, the resonance structure denoting both the σ-donor bond and the π-acceptor bond in metal-bound CO's is used. CO is still a formal two-electron donor ligand.) Many of these acyl "ate" complexes are quite stable and can be isolated and manipulated.

Eq. 2.24

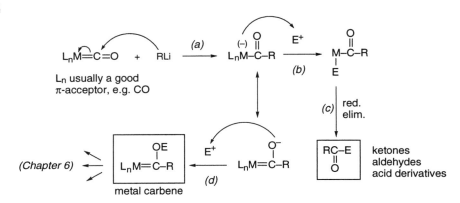

Electrophiles can react at the metal (b) (analogous to C-alkylation of an enolate) giving a neutral acyl/electrophile complex, which can undergo reductive elimination (c) to produce ketones, aldehydes, or carboxylic acid derivatives. Alternatively, reaction of electrophiles at oxygen (d) (analogous to O-alkylation of an enolate) produces heteroatom-stabilized carbene complexes, which have a very rich organic chemistry in their own right (Chapter 6). The site of electrophilic attack depends both on the metal and the electrophile. Hard electrophiles such as $R_3O^+BF_4^-$, ROTf and $ROSO_2F$ undergo reaction at oxygen,[14] while softer electrophiles undergo reaction at the metal.[15]

Alkoxides and amines attack metal carbonyls, particularly the more reactive cationic ones (Eq. 2.25), to produce alkoxycarbonyl or carbamoyl complexes,[16]

Eq. 2.25

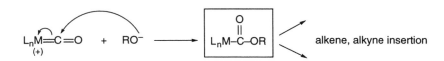

species which are involved in a variety of metal-assisted carbonylation reactions via insertion into the metal-acyl carbon bond.

Hydroxide ion attacks metal carbonyls to produce (normally) unstable metal carboxylic acids, which usually decompose to anionic metal hydrides and carbon dioxide[17] (Eq. 2.26).

Eq. 2.26

$$L_nM=C=O \quad + \quad HO^{(-)} \longrightarrow L-M\overset{O}{\underset{(-)}{\diagdown}}\quad \longrightarrow \quad LMH^{(-)} \quad + \quad CO_2$$

Electrophilic carbene complexes undergo nucleophilic attack at the carbene carbon, resulting in either heteroatom exchange (Eq. 2.27)[14] or the formation of stable adducts (Eq. 2.28).[18]

Eq. 2.27

$$(CO)_5Cr\overset{\delta-}{\underset{}{=}}\overset{\delta+ OMe}{\underset{R}{}} \quad + \quad Nuc^- \longrightarrow \left[(CO)_5Cr\overset{OMe}{\underset{Nuc}{\diagup R}} \right] \longrightarrow (CO)_5Cr=\overset{Nuc}{\underset{R}{}}$$

$$Nuc^- = RS^-, R_2NH, PhLi, \text{ etc.}$$

Eq. 2.28

$$\left[OC-Fe\overset{OMe}{\underset{H}{=}} \right]^+ \quad + \quad CH_3Li \longrightarrow OC-Fe\overset{OMe}{\underset{H}{\diagup Me}}$$

stable

Nucleophilic cleavage of metal carbon σ-bonds is another process of interest in organic synthesis, since it is frequently involved in freeing an organic substrate from the metal. Many metal-acyl complexes undergo direct cleavage by alcohols or amines, a key step in many metal-catalyzed acylation reactions (Eq. 2.29).[19] Others

Eq. 2.29

$$\overset{O}{ArC}-\overset{L}{\underset{L}{Pd}}-X \quad \overset{R\ddot{O}H}{\longrightarrow} \quad \overset{O}{ArC}-OR \quad + \quad L_2Pd \quad + \quad HX$$

are more stable and require an oxidative cleavage (Eq. 2.30).[20] σ-Alkyliron complexes undergo oxidative cleavage, and this process has been studied in some detail (Eq. 2.31).[21] Oxidation of the metal makes it a better leaving group, and promotes displacement with clean inversion.

Eq. 2.30

$$\text{(structure)} \xrightarrow[\text{MeOH}]{\text{Br}_2} \text{(structure)}$$

Eq. 2.31

$$\text{(structures)} \xrightarrow[\text{or I}_2]{\text{Br}_2} \text{(structure)} \quad \text{via} \quad \left[\text{(structure)} \right] \xrightarrow[\text{S}_\text{N}2]{\text{Br}^-}$$

Another nucleophilic reaction of some consequence to organic synthesis is the self exchange reaction. Oxidative addition of organic halides to low-valent metals is known to go with inversion of configuration at carbon. However, occasionally racemization or partial racemization is observed. This is thought to result from nucleophilic displacement of the metal(II) in the σ–alkyl complex by the nucleophilic metal(0) starting complex (Eq. 2.32).[22] Since the reaction is "thermally neutral" this need not be a difficult process, and it should be considered whenever unexpected racemization is observed.

Eq. 2.32

$$L_4Pd + \text{(structure)} \xrightarrow[\text{inv.}]{} \text{(structure)} \xleftrightarrow[\text{inv.}]{L_4Pd} \text{(structure)}$$

Nucleophilic attack on complexed π-unsaturated hydrocarbons is among the most useful of organometallic processes for organic synthesis, since it is exactly the opposite of "normal" reaction chemistry for these substrates. Nucleophilic attack on 18 electron, cationic complexes has been particularly well-studied,[23] and, based upon a large number of examples, the following order of reactivity for kinetically controlled reactions of this class of complexes was derived:

$$\| > \text{(structure)} > \text{(structure)} > \text{(structure)} > \text{(structure)} \gg \text{(structure)} > \text{(structure)} > \text{(structure)}$$

Note that the cyclopentadienyl ligand (Cp) is among the least reactive, and, as a consequence, is often used as a spectator ligand.

The typical reaction for electrophilic metal complexes of olefins, particularly those of Pd[II], Pt[II] and Fe[II], is nucleophilic attack on the olefin.[24] In most cases, the nucleophile attacks the more-substituted position of the olefin, from the face opposite the metal (*trans*), forming a carbon-nucleophile bond and a carbon-metal bond (Eq. 2.33). A few nucleophiles, such as chloride, acetate, and nonstabilized carbanions, can also attack first at the metal. This is followed by insertion of the

Eq. 2.33

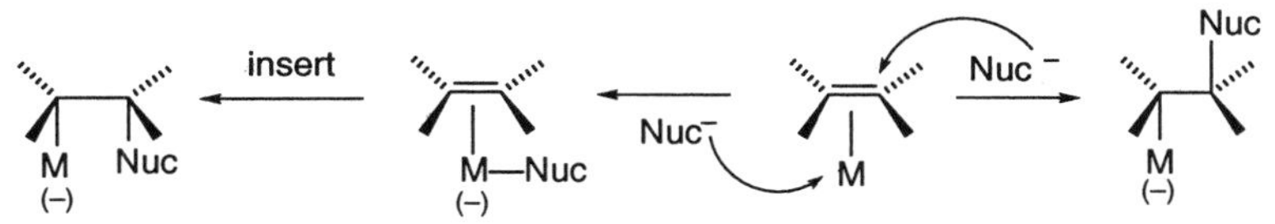

$$Nuc^- = Cl^-, AcO^-, ROH, RNH_2, RLi$$

olefin, resulting in an overall *cis* addition, with attack at the less-substituted position. This chemistry is normally restricted to mono and 1,2-disubstituted olefins, since more highly substituted olefins coordinate only weakly, and displacement of the olefin by the nucleophile competes.

Alkynes complexed to electrophilic metals also undergo nucleophilic attack, although this is a much less-studied process, since stable alkyne complexes of electrophilic metals are rare. (These metals tend to oligomerize alkynes.) Cationic iron complexes of alkynes undergo clean nucleophilic attack from the face opposite the metal, to give stable σ-alkenyl iron complexes (Eq. 2.34).[25] This process has not been used in organic synthesis to any extent.

Nucleophilic attack on η^3-allyl metal complexes, particularly those of Pd^{II}, has been extensively developed as an organic synthetic method and is discussed in

Eq. 2.34

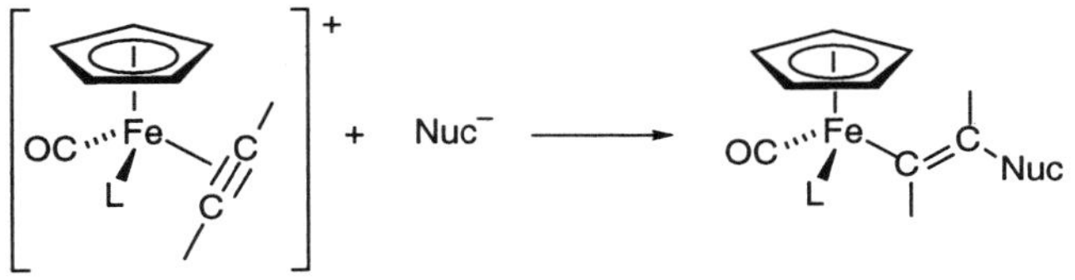

$$Nuc^- = PhS^-, CN^-, {}^-CH(CO_2Et)_2 [Ph^-, Me^-, H_2C=CH \ MeC\equiv C^- \ from \ R_2Cu(CN)Li_2]$$

$$L = PPh_3, P(OPh)_3$$

detail in Chapter 9. The general features are shown in Eq. 2.35. Nucleophilic attack occurs from the face opposite the metal. From a synthetic point of view, the most important reaction involving η^3-allyl complexes is the Pd^0-catalyzed reaction of

Eq. 2.35

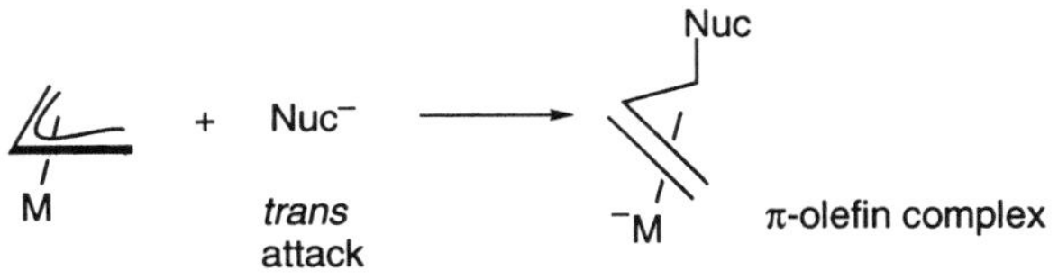

allylic substrates, a process which proceeds with overall retention (two inversions) of configuration at carbon (Eq. 2.36).[26] In certain rare cases nucleophilic attack can occur at the central carbon of an η^3-allyl complex, generating a metallacyclobutane. In the case of the cationic molybdenum complex in Eq. 2.37, the complex was stable and isolated.[27]

Eq. 2.36

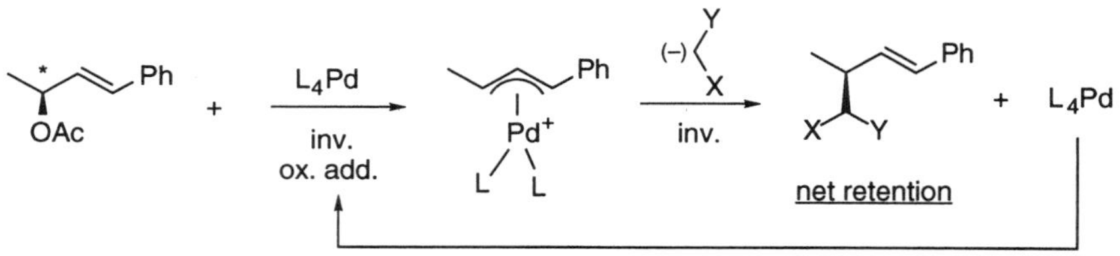

Eq. 2.37

Although extended Hückel calculations[28] suggested that attack at the central carbon of η^3-allylpalladium complexes was not feasible, under appropriate conditions (α-branched ester enolates, polar solvent) just such a process occurred smoothly, to produce cyclopropanes (Eq. 2.38).[29a,b] The proposed palladiacyclobutane intermediate, stabilized by TMEDA, has recently been isolated and structurally characterized.[29c]

Eq. 2.38

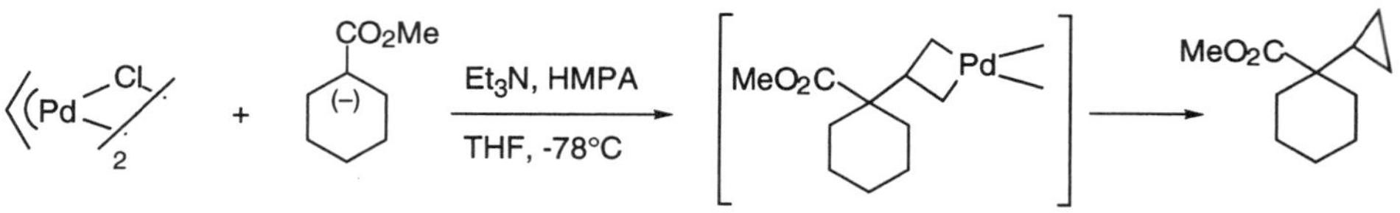

Neutral η^4-dieneiron tricarbonyl complexes undergo reaction with strong nucleophiles (e.g. LiCMe$_2$CN) at both the terminal position, to produce an η^3-allyl complex, and an internal position, to produce a σ-alkyl-η^2-olefin complex (Eq. 2.39).[30] In contrast, the much more electrophilic palladium chloride promotes exclusive terminal attack to produce stable η^3-allylpalladium complexes, which can be further functionalized (Chapter 9) (Eq. 2.40).[31] Again, nucleophilic attack occurs from the face opposite the metal.

Eq. 2.39

Eq. 2.40

In contrast to neutral η^4-diene complexes, cationic η^5-dienyl complexes are broadly reactive towards nucleophiles, and iron tricarbonyl complexes of the η^5-cyclohexadienyl ligand have found extensive use in organic synthesis. Nucleophiles ranging from electron-rich aromatic compounds through organocopper species readily add to these complexes, attacking, as usual, from the face opposite the metal, to produce stable η^4-diene complexes (Eq. 2.41).[32] With substituted cyclohexadienyl ligands, the site of nucleophilic attack is generally governed by electronic factors. By coupling this nucleophilic addition to η^5-dienyl complexes with the generation of η^5-dienyl complexes by hydride abstraction from η^4-diene complexes, regio- and stereospecific polyfunctionalization of this class of ligands has been achieved (Chapter 7).

Eq. 2.41

Arenes normally undergo electrophilic attack and, except in special cases, are quite inert to nucleophilic attack. However, by complexation to electron-deficient metal fragments, particularly metal carbonyls (recall that CO groups are strongly electron withdrawing), arenes become generally reactive toward nucleophiles. By far the most extensively studied complexes are the η^6-arenechromium tricarbonyl complexes (Eq. 2.42).[33] Again, the nucleophile attacks from the face opposite the metal to give a relatively unstable anionic η^5-cyclohexadienyl complex, which itself has a rich organic chemistry (Chapter 10). Oxidation regenerates the arene, while protolytic cleavage gives the cyclohexadiene.

Eq. 2.42

The reactions of nucleophiles with metal-coordinated ligands are summarized in Table 2.2 along with the d-electron count and formal oxidation states for reactants and products. Although there is a profound reorganization of electrons about the metal, most of these processes result in no change in the formal oxidation state of the metal, differentiating them from the oxidative addition/reductive elimination family of reactions. This is because in each case, a formally neutral ligand is converted to a formally mononegative ligand in the process — that is, nucleophilic attack reduces the ligand, not the metal. The two exceptions are the η^3-allyl and the η^5-dienyl complexes, in which formally negative ligands are converted to formally neutral ligands by the process and the metal is reduced.

Table 2.2   Summary of Nucleophilic Reactions on Transition Metals

Table 2.2 continued

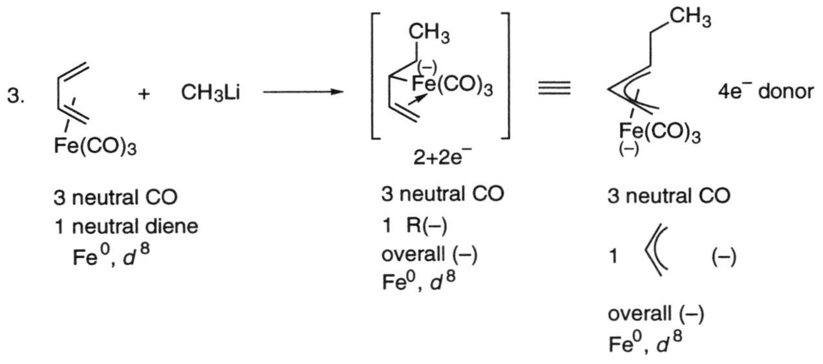
3.
+ CH3Li
CH3
(−)
Fe(CO)3
2+2e⁻
CH3
4e⁻ donor
Fe(CO)3
(−)
3 neutral CO
1 neutral diene
Fe0, d8
3 neutral CO
1 R(−)
overall (−)
Fe0, d8
3 neutral CO
1 (−)
overall (−)
Fe0, d8

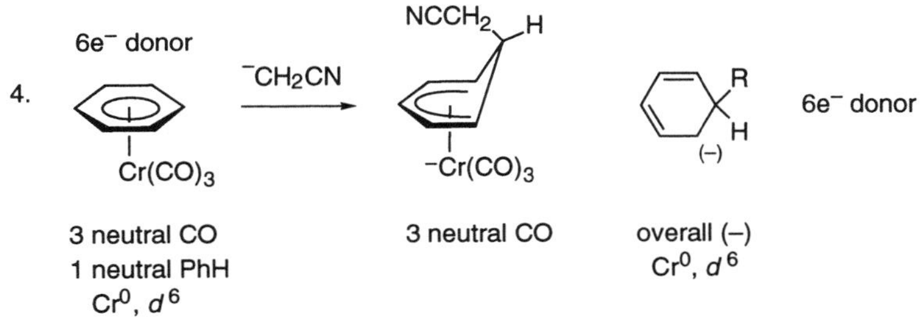
6e⁻ donor
4.
Cr(CO)3
⁻CH2CN
NCCH2 H
−Cr(CO)3
R
H
(−)
6e⁻ donor
3 neutral CO
1 neutral PhH
Cr0, d6
3 neutral CO
overall (−)
Cr0, d6

but

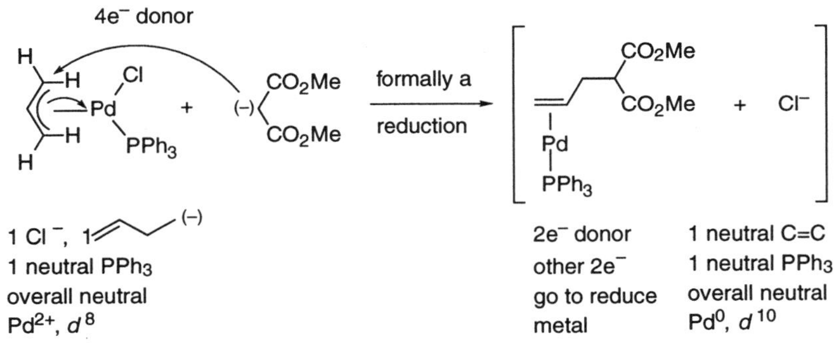
4e⁻ donor
H
H
Cl
Pd
H
PPh3
H
+
(−)
CO2Me
CO2Me
formally a
reduction
CO2Me
CO2Me
Pd
PPh3
+ Cl⁻
1 Cl⁻, 1 (−)
1 neutral PPh3
overall neutral
Pd2+, d8
2e⁻ donor
other 2e⁻
go to reduce
metal
1 neutral C=C
1 neutral PPh3
overall neutral
Pd0, d10

2.6 Transmetallation

Transmetallation is a process of increasing importance in the area of transition metals in organic synthesis, but has been little studied and is not well understood. The general phenomenon involves the transfer of an R group from a main group organometallic compound to a transition metal complex (Eq. 2.43).[34] When combined with reactions which introduce an R group into the transition metal complex such as oxidative addition or nucleophilic attack on alkenes, efficient carbon-carbon bond forming (cross coupling) reactions ensue. In cases such as these, the transmetallation step is almost always the rate limiting step and when catalytic cycles involving a transmetallation step fail, it is usually this step which needs attention.

Eq. 2.43

$$RM \ + \ M'X \ \underset{}{\overset{K}{\rightleftharpoons}} \ RM' \ + \ MX$$

M = Zn, Zr, B, Hg, Si, Sn, Ge
M' = transition metal

In order for the transmetallation step to proceed, the main group organometallic M must be more electropositive than the transition metal M'. However, since this is an equilibrium, if RM' is irreversibly consumed in a subsequent step, the process can be used even if it is not favorable (K is small). This, coupled with uncertain accuracy of numbers for electronegativity, make experiment the wisest course.

An often neglected fact is that Eq. 2.43 *is* an equilibrium, and that both partners must profit, thermodynamically, from the process. Therefore, the nature of X may well be as important as the nature of M or M', and transmetallation procedures can often be promoted by adding appropriate counterions, X (Chapter 4).

2.7 Electrophilic Attack on Transition Metal Coordinated Ligands

Reactions of coordinated organic ligands with electrophiles has found substantially less use in organic synthesis than have reactions with nucleophiles. Perhaps the most common electrophilic reaction is the electrophilic cleavage of metal-carbon sigma bonds, as a method to free an organic substrate from a metal template. If the metal complex has *d* electrons, electrophilic attack usually occurs at the metal, a formal oxidative addition. This is followed by reductive elimination to result in cleavage with retention of configuration at carbon (Eq. 2.44). If a sufficiently good nucleophile is present (e.g. Br⁻ from Br₂), cleavage with *inversion* may result. An

Eq. 2.44

$$R\!-\!M \ + \ E^+ \ \longrightarrow \ \overset{\overset{E}{|}}{R\!-\!M^+} \ \underset{el.}{\overset{red.}{\longrightarrow}} \ R\!-\!E$$

retention

$$\xrightarrow{Nuc^-} \ R\!-\!E$$

inversion

(E⁺ = H⁺, D⁺, Br⁺, I⁺)

example of this kind of cleavage has already been discussed (Eq. 2.31). Even σ-alkyl complexes of metals having no *d* electrons undergo facile electrophilic cleavage, and usually with retention of configuration. In these cases, direct electrophilic attack at the metal-carbon σ-bond (S_E2) is the likely process (Eq. 2.45).[35]

Eq. 2.45

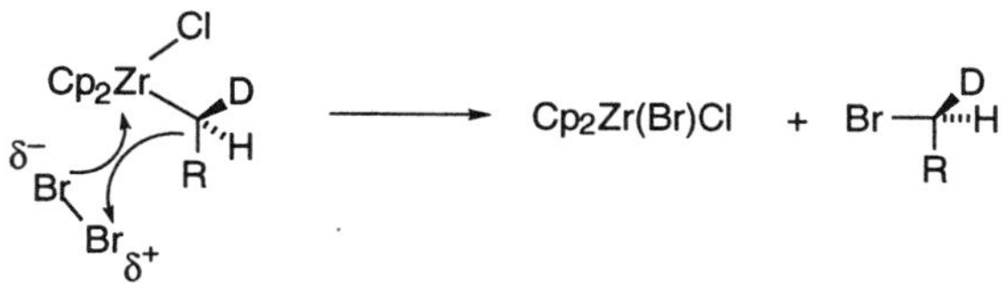

$$Zr^{IV}, d(0)$$

Electrophiles also can react at *carbon* in organometallic complexes, and a variety of useful processes involve this type of reaction. These are summarized in Table 2.3. Some, particularly reactions of electrophiles with η^1-allyl complexes (γ attack) and hydride abstraction from η^4-diene complexes, are quite useful in organic synthesis and are discussed in detail later.

Table 2.3   Electrophilic Attack on Transition Metal Coordinated Organic Ligands

1. Electrophilic cleavage of σ-alkyl metal bonds

$$R-M \ + \ E^+ \longrightarrow R-E \quad \textit{retention} \text{ at R}$$

2. Attack at α-position

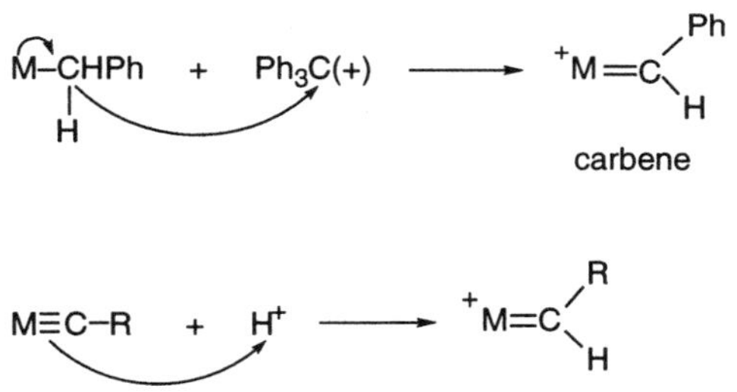

3. Attack at β-position

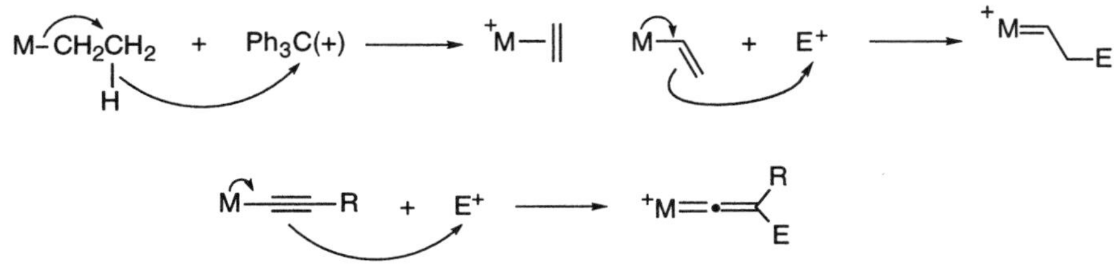

Table 2.3 Continued

4. Attack at   γ-position

M⌒▽═ + E⁺ ⟶ ⁺M—‖⌒E

M⌒▽═ + A‖B ⟶ ⁺M—‖⌒(A / −B) ⟶ M—⌐(A / B)

5. Attack on coordinated polyenes

(cyclohexadienyl complex, H,H) + Ph₃C(+) ⟶ (arene complex M) ≡ (arene complex M⁺)

References

1. (a) Collman, J.P.; Hegedus, L.S.; Norton, J.R.; Finke, R.G. *Principles and Applications of Organotransition Metal Chemistry*, 2nd ed. University Science Books, Mill Valley, CA, 1987, pp. 235-458. (b) Twigg, M.V., ed. *Mechanisms of Inorganic and Organometallic Reactions*. Plenum, New York, annually since 1983, see part 3 in each volume.
2. Odell, L.A.; Raethel, H.A. *Chem. Comm.* **1968**, 1323; **1969**, 87.
3. Tolman, C.A. *Chem. Rev.* **1977**, 77, 313.
4. Albers, M.O.; Coville, N.J. *Coord. Chem. Rev.* **1984**, 53, 227.
5. (a) Darensbourg, D.J.; Walker, N.; Darensbourg, M. *J. Am. Chem. Soc.* **1980**, 102, 1213. (b) Darensbourg, D.J.; Darensbourg, M.Y.; Walker, N. *Inorg. Chem.* **1981**, 20, 1918. (c) Darensbourg, D.J.; Ewen, J.A. *Inorg. Chem.* **1981**, 20, 4168.
6. For reviews see: (a) Collman, J.P.; Roper, W.R. *Adv. Organomet. Chem.* **1968**, 7, 53. (b) Collman, J.P. *Acc. Chem. Res.* **1968**, 1, 136. (c) Stille, J.R.; Lau, K.S.Y. *Acc. Chem. Res.* **1977**, 10, 434. (d) Stille, J.K. in *The Chemistry of the Metal-Carbon σ-Bond*, vol. 2, Hartley, F.R.; Patai, S., eds., Wiley, New York, 1985, pp. 625-787.
7. (a) Sweany, R.L. *J. Am. Chem. Soc.* **1985**, 107, 2374. (b) Upmacis, R.K.; Gadd, G.E.; Poliakoff, M.; Simpson, M.B.; Turner, J.J.; Whyman, R.; Simpson, A.F. *J. Chem. Soc. Chem. Comm.* **1985**, 27.
8. Chock, P.B.; Halpern, J. *J. Am. Chem. Soc.* **1966**, 88, 3511.
9. Collman, J.P.; Finke, R.G.; Cawse, J.N.; Brauman, J.I. *J. Am. Chem. Soc.* **1977**, 99, 2515.
10. (a) Labinger, J.R.; Osborn, J.A. *Inorg. Chem.* **1980**, 19, 3230. (b) Labinger, J.A.; Osborn, J.A.; Coville, N..J. *Inorg. Chem.* **1980**, 19, 3236. (c) Jenson, F.R.; Knickel, B. *J. Am. Chem. Soc.* **1971**, 93, 6339.
11. Kochi, J.K. *Pure Appl. Chem.* **1980**, 52, 571.
12. Gillie, A.; Stille, J.K. *J. Am. Chem. Soc.* **1980**, 102, 4933.
13. For a review see: Hegedus, L.S. "Nucleophilic Attack on Transition Metal Organometallic Compounds" in *The Chemistry of the Metal-Carbon σ-Bond, vol. 2*, Hartley, F.R.; Patai, S., eds., Wiley, New York, 1985, pp. 401-512.
14. Dötz, K.H. *Angew. Chem. Int. Ed. Engl.* **1984**, 23, 587.
15. Collman, J.P. *Acc. Chem. Res.* **1975**, 8, 342.
16. Angelici, R.L. *Acc. Chem. Res.* **1972**, 5, 335.
17. Ungermann, C.; Landis, V.; Moya, S.A.; Cohen, H.; Walker, H.; Pearson, R.G.; Rinker, R.G.; Ford, P.C. *J. Am. Chem. Soc.* **1979**, 101, 5922 and references therein.
18. Casey, C.P.; Miles, W.M. *J. Organomet. Chem.* **1983**, 254, 333.
19. Heck, R.F. *Pure Appl. Chem.* **1978**, 50, 691.

20. Hegedus, L.S.; Anderson, O.P.; Zetterberg, K.; Allen, G.; Siirala-Hansen, K.; Olsen, D.J.; Packard, A.B. *Inorg. Chem.* **1977**, *16*, 1887.
21. Whitesides, G.M.; Boschetto, D.J. *J. Am. Chem. Soc.* **1971**, *93*, 1529.
22. Lau, K.S.Y..; Fries, R.M.; Stille, J.K. *J. Am. Chem. Soc.* **1974**, *96*, 4983. Similar loss of stereochemistry by Pd self-exchange has been observed in η^3-allyl palladium chemistry: Granberg, K.L.; Bäckvall, J-E. *J. Am. Chem. Soc.* **1992**, *114*, 6858.
23. Davies, S.G.; Green, M.L.H.; Mingos, D.M.P. *Tetrahedron* **1978**, *34*, 3047.
24. (a) Lennon, P.M.; Rosan, A.M.; Rosenblum, M. *J. Am. Chem. Soc.* **1977**, *99*, 8426. (b) Hegedus, L.S.; Akermark, B.; Zetterberg, K.; Olsson, L.F. *J. Am. Chem. Soc.* **1984**, *106*, 7122. (c) Hegedus, L.S.; Williams, R.E.; McGuire, M.A.; Hayashi, T. *J. Am. Chem. Soc.* **1980**, *102*, 4973. (d) For a review see, "Nucleophilic Attack on Alkene Complexes", McDaniel, K.F. in "Comprehensive Organometallic Chemistry II, Abel, E.W.; Stone, F.G.A.; Wilkinson, G.; Eds., Pergamon, Oxford, UK, 1995, Vol. 12, 601-622.
25. Reger, D.L.; Belmore, K.A.; Mintz, E.; McElligott, P.J. *Organometallics* **1984**, *3*, 134; 1759.
26. (a) Hayashi, T.; Hagihara, T.; Konishi, M.; Kumada, M. *J. Am. Chem. Soc.* **1983**, *105*, 7767. (b) For a review see: Harrington, P.J., "Transition Metal Allyl Complexes" in "Comprehensive Organometallic Chemistry II, Abel, E.W.; Stone, F.G.A.; Wilkinson, G.; Eds., Pergamon, Oxford, UK, 1995, Vol. 12, 797-904.
27. Periana, R.A.; Bergman, R.G. *J. Am. Chem. Soc.* **1984** *106*, 7272.
28. Curtis, M.D.; Eisenstein, O. *Organometallics* **1984**, *3*, 887.
29. (a) Hegedus, L.S.; Darlington, W.H.; Russel, C.E. *J. Org. Chem.* **1980**, *45*, 5193. (b) Carfagna, C.; Mariani, L.; Musco, M.; Sallese, G.; Santi, R. *J. Org. Chem.* **1991**, *56*, 3924. (c) Hoffmann, H.M.R.; Otte, A.R.; Wilde, A.; Menzer, S.; Williams, D.J. *Angew. Chem. Int. Ed. Engl.* **1995**, *34*, 100. For a consideration of ligand effects on the site of nucleophilic attack on η^3-allylpalladium complexes see: (d) Aranyos, A.; Szabo, K.J.; Castaño, A.M.; Bäckvall, J-E. *Organomet.* **1997**, *16*, 1997 1058.
30. (a) Semmelhack, M.F.; Le, H.T.M. *J. Am. Chem. Soc.* **1984**, *106*, 2715. (b) Semmelhack, M.F.; Herndon, J.W. *Organometallics* **1983**, *2*, 363.
31. Bäckvall, Jan-E.; Nyström, Jan-E.; Nordberg, R.E. *J. Am. Chem. Soc.* **1985**, *107*, 3676.
32. (a) Pearson, A.J. *Comp. Org. Synth.* **1991**, *4*, 663. (b) Pearson, A.J., "Nucleophilic Attack on Diene and Dienyl Complexes", in Comprehensive Organometallic Chemistry II, Abel, E.W.; Stone, F.G.A.; Wilkinson, G.; Eds., Pergamon, Oxford, UK, 1995, Vol. 12, 637-684.
33. (a) Semmelhack, M.F. *Comp. Org. Synth.* **1991**, *4*, 517. (b) Semmalhack, M.F., "Transition Metal Arene Complexes" in "Comprehensive Organometallic Chemistry II, Abel, E.W.; Stone, F.G.A.; Wilkinson, G.; Eds., Pergamon, Oxford, UK, 1995, Vol. 12, 979-1038.
34. Negishi, E-I. *Organometallics in Organic Synthesis.* Wiley, New York, 1980.
35. Labinger, J.A.; Hart, D.W.; Seibert, W.E., III; Schwartz, J. *J. Am. Chem. Soc.* **1975**, *97*, 3851.

CHAPTER 3

Synthetic Applications of Transition Metal Hydrides

3.1   Introduction

Transition metal hydrides are an important class of organometallic complex, primarily because of their role in homogeneous hydrogenation, hydroformylation, and hydrometallation reactions. The reactivity of transition metal hydrides depends very much on the metal and the other ligands, and ranges all the way from hydride donors to strong protic acids![1] However, their major use in synthesis relies on neither of these characteristics, but rather their propensity to insert alkenes and alkynes into the M-H bond, generating σ-alkylmetal complexes for further transformations.

Transition metal hydrides can be synthesized in a number of different ways (Figure 3.1), but the most commonly used method is oxidative addition. Transition metals are unique in their ability to activate hydrogen under very mild conditions, a feature which accounts for their utility in catalytic hydrogenation.

3.2   Homogeneous Hydrogenation

When synthetic chemists are faced with the reduction of a carbon-carbon double bond, their first choice is, almost invariably, *heterogeneous* catalysis, using something like palladium on carbon. Heterogeneous catalysts are convenient to handle, efficient, and easy to remove by filtration after the reaction is complete. In contrast, homogeneous catalysts are sometimes difficult to handle, sensitive to impurities, such as traces of oxygen, have a tendency to cause olefin isomerization, and are difficult to recover, because they are soluble. However, their enormous advantage, selectivity, mitigates these shortcomings.

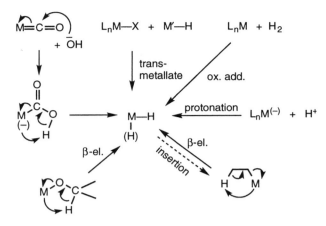

Figure 3.1   Syntheses of Transition Metal Hydrides

There are two general classes of homogeneous hydrogenation catalysts — monohydrides and dihydrides — and they react by different mechanisms and have different specificities. The best-studied monohydride catalyst is the lemon yellow, air stable, crystalline solid $Rh(H)(PPh_3)_3(CO)$, and the mechanism by which it hydrogenates olefins is shown in Figure 3.2. It is *completely* specific for terminal olefins, and will tolerate internal olefins, aldehydes, nitriles, esters, and chlorides elsewhere in the molecule. Its major limitation is that olefin isomerization competes with hydrogenation, and the isomerized olefin cannot be reduced by the catalyst.

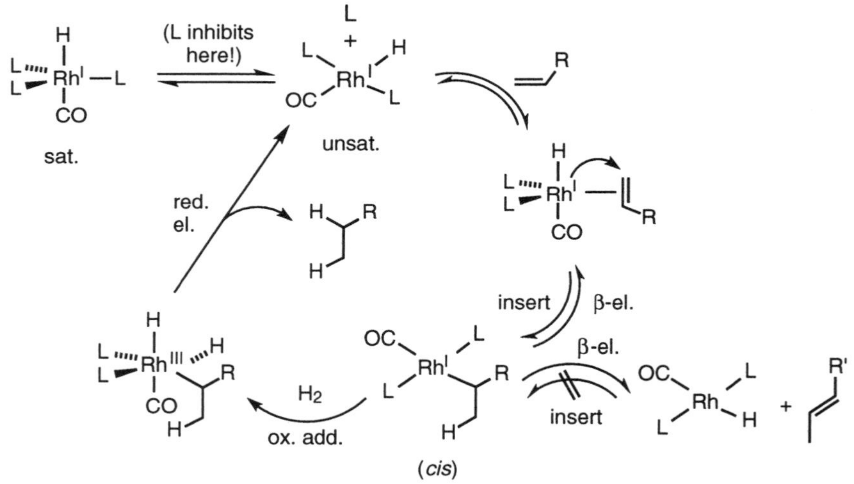

Figure 3.2   Reduction of Alkenes by Monohydride Catalysts

The starting complex is a coordinatively saturated, Rh(I) d^8 complex and the first step requires loss of a phosphine (ligand dissociation). As a consequence, added phosphine inhibits catalysis. Coordination of the substrate alkene to the unsaturated metal center is followed by migratory insertion, to produce an unsaturated σ-alkylrhodium(I) complex. The regiochemistry of insertion is not

40   •   **Transition Metals in the Synthesis of Complex Organic Molecules**

known, and the rhodium may occupy either the terminal or internal position. Oxidative addition of hydrogen to this σ-alkylrhodium(I) complex, followed by reductive elimination of R–H, results in reduction of the alkene and regeneration of the catalytically active, unsaturated rhodium(I) monohydride. Irreversible olefin isomerization occurs if a 2° σ-alkylrhodium(I) intermediate undergoes β-hydrogen elimination (the reverse of migratory insertion) *into* the alkyl chain, producing an internal olefin and the catalytically active rhodium(I) monohydride. Because this catalyst cannot reduce internal olefins, this isomerization is irreversible, and competitive. Although this particular monohydride catalyst is little-used in synthesis, the monohydride pathway may be important in the ruthenium(II) asymmetric hydrogenation catalysts discussed below.

Dihydride catalysts are much more versatile, and much better understood. The most widely used catalyst precursor is the burgundy red solid RhCl(PPh$_3$)$_3$, Wilkinson's complex. The mechanism of hydrogenation of alkenes by this catalyst has been studied in great detail, and is quite complex (Figure 3.3).[2] Although Wilkinson's compound is already coordinatively unsaturated (Rh^I, d^8, 16e$^-$) the kinetically active species is the 14 electron complex RhCl(PPh$_3$)$_2$, with, perhaps, solvent loosely associated. In this case, oxidative addition of H$_2$ is the initial step, followed by coordination of the alkene. Migratory insertion is the rate-limiting step, followed by a fast reductive elimination of the alkane and regeneration of the active catalyst. In addition to the catalytic cycle, several other equilibria are operating, complicating the system.

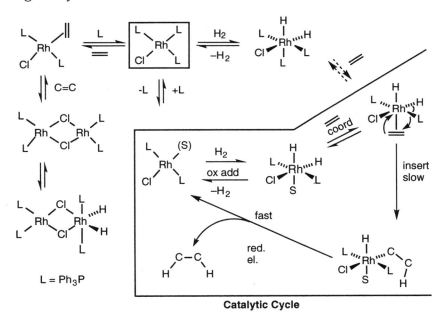

Figure 3.3   Reduction of Alkenes by Wilkinson's Catalyst

Wilkinson's catalyst is attractive for synthesis because it is selective and efficient, tolerant of functionality, and, in contrast to monohydride catalysts, does not promote olefin isomerization. The reactivity towards olefins parallels their coordination ability:

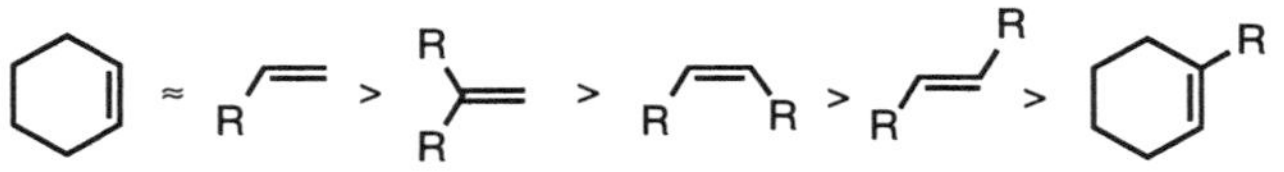

with a 50-fold difference in rate over this range of alkenes. Noncyclic tri- and tetrasubstituted olefins are not reduced, and ethylene, which coordinates too strongly, poisons the catalyst. Alkynes are rapidly reduced. Nitro olefins are reduced to nitro alkanes without competitive reduction of the nitro goup.

There are several limitations of reductions using Wilkinson's catalyst. In solution, it catalyzes the oxidation of triphenylphosphine, which in turn, leads to destruction of the complex, so air cannot be tolerated. Strong ligands such as ethylene, thiols, and very basic phosphines poison the system. Carbon monoxide, and substrates prone to decarbonylation, such as acid halides, are not tolerated. Notwithstanding these limitations, Wilkinson's catalyst is quite useful in synthesis, and is usually the first choice for routine homogeneous hydrogenations.

Of more interest to synthetic chemists is asymmetric homogeneous hydrogenation,[3] wherein prochiral olefins are reduced to enantiomerically-enriched products. This topic has been studied with excruciating detail, and when it works, it is spectacular! However, the requirements for successful asymmetric hydrogenation are so stringent that, until very recently, the process was limited to a very narrow range of substrates, and is still not universal.

The catalysts used are cationic rhodium(I) complexes of optically active chelating diphosphines, such as DIOP, DIPAMP, CHIRAPHOS, BPPM and BINAP (Figure 3.4), generated *in situ* by reduction of the corresponding diphosdiolefin complex (Eq. 3.1). These catalysts are extremely efficient in the asymmetric hydrogenation of a narrow range of prochiral olefins which can bind in a bidentate,

Eq. 3.1

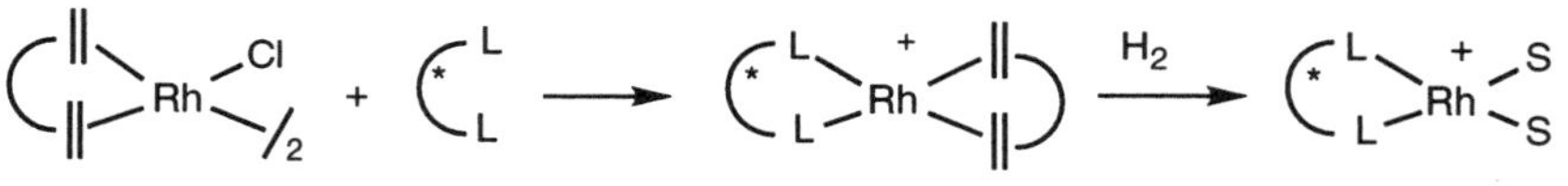

S = Solvent

chelate manner. By far the best substrates are Z-α-acetamidoacrylates, which are reduced to optically active α-amino acids. This system has been studied in minute detail, and several key intermediates in the proposed catalytic cycle have been fully characterized by ^{1}H, ^{13}C, ^{31}P NMR spectroscopy and by X-ray crystallography.[4] All indicate that the substrate olefin coordinates in a bidentate manner, with the amide oxygen acting as the second ligand, and that two diastereoisomeric olefin complexes

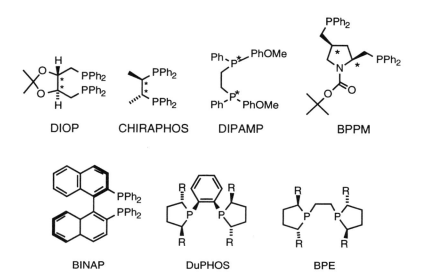

Figure 3.4 Optically Active Phosphines

can be formed. The only problem with all of these studies was that the diastereoisomer detected would lead to the opposite absolute configuration of the product than was observed! Careful kinetic studies[4b-d] resolved this problem, and led to the following (after the fact) obvious observation that, if you can detect intermediates in a catalytic process, they probably are not involved. That is, anything that accumulates sufficiently to be detected is kinetically inactive; kinetically active species just react, converting starting material to product, and are never present in any appreciable amount.

The mechanism for asymmetric hydrogenation of Z-acetamidoacrylates is shown in Figure 3.5. Its importance to synthetic chemists lies not with its details, but rather the clear illustration it provides of how closely balanced a series of complex steps must be to achieve efficient asymmetric hydrogenation. It is for this reason that the range of prochiral olefins asymmetrically hydrogenated by these catalysts is so narrow, and why extrapolation to different substrates is difficult.

There are two features of the process critical to high asymmetric induction; the two diastereoisomers must undergo reaction at substantially different rates, and they must equilibrate rapidly. The rate-limiting step is the oxidative addition of hydrogen to the rhodium(I) olefin complex, and one (the minor) diastereomer undergoes this reaction at a rate 10^3 greater than the other. Provided the two diastereoisomeric olefin complexes can equilibrate rapidly, high asymmetric induction can be observed. Interestingly, higher hydrogen pressure and lower temperatures decrease the enantioselectivity, since both interfere with this equilibration.

The above catalysts are primarily effective with Z-acetamidoacrylates. The E isomer reduces to give the opposite enantiomer of the amino acid, but E–Z isomerization during reduction often compromises the enantioselectivity of reduction of E-acetamidoacrylates. In contrast, related catalyst containing the phosphole

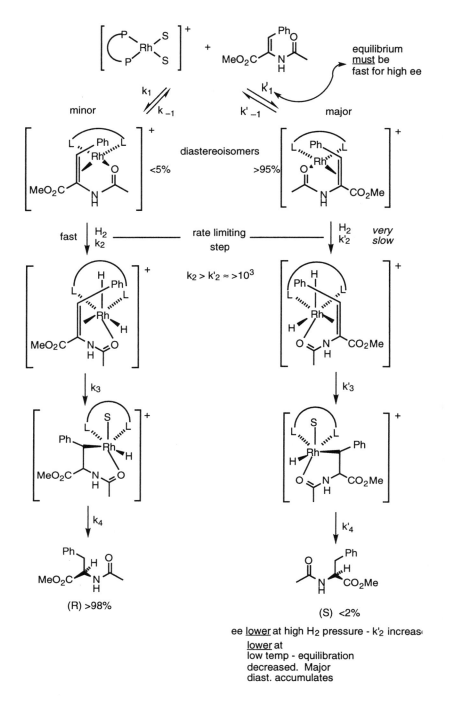

Figure 3.5 Mechanism for Asymmetric Catalytic Hydrogenation of Z-Acetamidoacrylates

ligands DUPHOS or BPE ligands reduce both E and Z acetamidoacrylates with equally high enantioselection, and, for a given catalyst configuration, give the same product absolute configuration regardless of olefin geometry. In addition, β,β-disubstituted enamides are reduced to β-branched α-amino acids with high enantioselectivity by these catalysts.[5] Examples are presented in Eq. 3.2.

Eq. 3.2

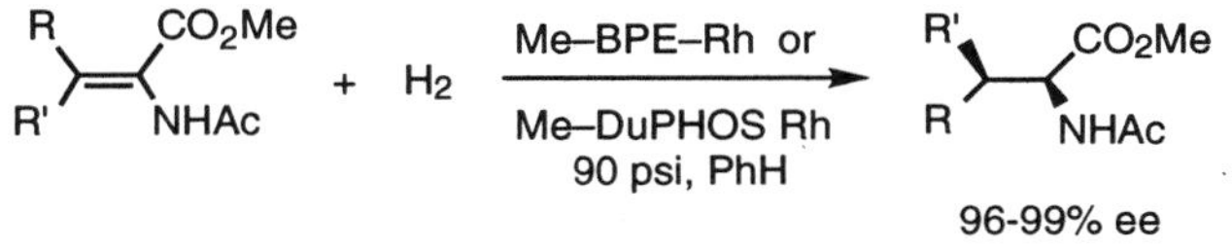

$$R = R' = Et, nPr, -(CH_2)-, -(CH_2)_5-, -(CH_2)_6-$$
$$R = Bn, R' = Me$$

A much more broadly useful class of asymmetric hydrogenation catalysts, which also relies upon bidentate chelate coordination of the olefinic substrate, is the ruthenium(II) BINAP system.[6] These complexes reduce a wide range of substrates with high asymmetric induction, relying on the rigidity of complexation confered by chelation to assure high enantioselectivity. For examples, α,β-unsaturated carboxylic acids were reduced selectively and efficiently (Eq. 3.3),[7] and a very wide range of functionality was tolerated. The antiinflammatory Naproxen was produced with 97% ee in a high-pressure reduction in methanol, as was a thienamycin precursor, albeit with lower enantioselectivity.

Eq. 3.3

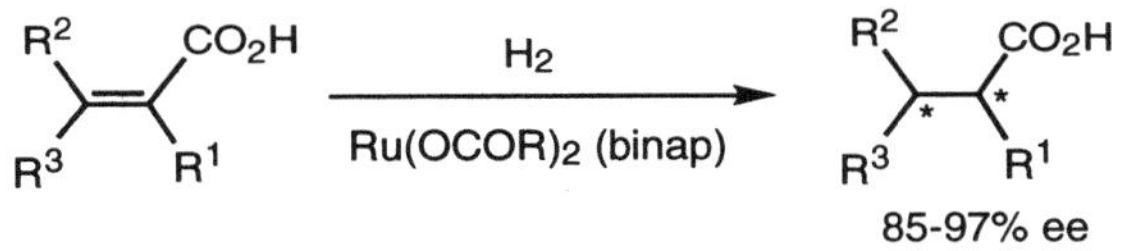

also

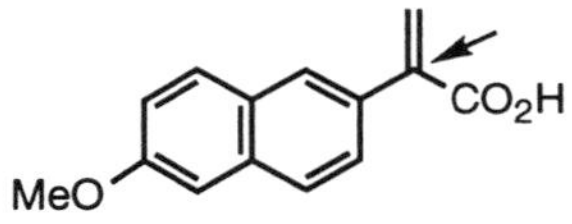

97% ee (Naproxen)

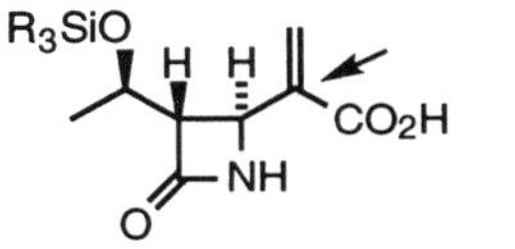

74% de (Thienamycin)

R¹	R²	R³	ee
Me	Me	H	91
H	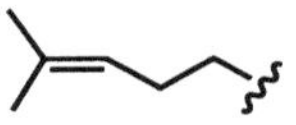 (propenyl)	Me	87
H	Me	Ph	85
Ph	H	H	92
H	HOCH$_2$	Me	93
H	CH$_3$	COOCH$_2$CMe	95

N-Acylenamines were also efficient coordinating groups and 1,2,3,4-tetrahydroisoquinoline derivatives including morphine, benzomorphans, and morphinans were produced in 95-100% ee by reduction under 1-4 atm of hydrogen in methanol (Eq. 3.4).[8]

Eq. 3.4

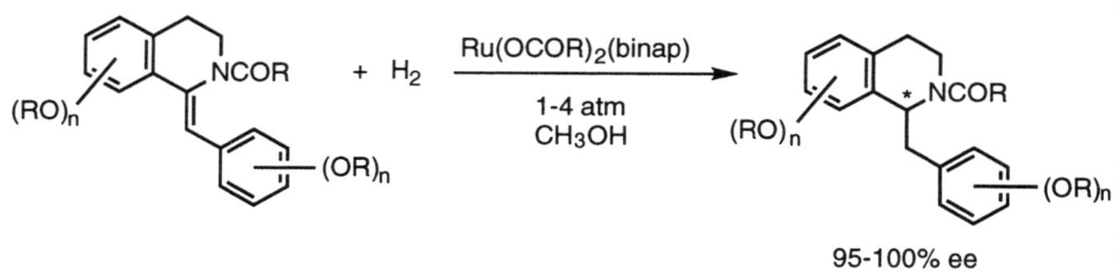

95-100% ee

Allylic and homoallylic alcohols also undergo selective reduction with high enantioselectivity (Eq. 3.5).[9] Geraniol was hydrogenated to citronellol with high ee, with no net reduction of the remote double bond (Eq. 3.6). Either geometrical isomer could be converted to either enantiomer by appropriate choice of BINAP ligand. Homogeraniol was also reduced selectively, but *bis*-homogeraniol was inert. This indicates that bidentate, chelate coordination of the olefinic substrates is required for *reactivity* as well as enantioselectivity. This feature was used to synthesize dolichols, wherein only the allylic alcohol olefin was reduced (Eq. 3.7).[10]

Eq. 3.5

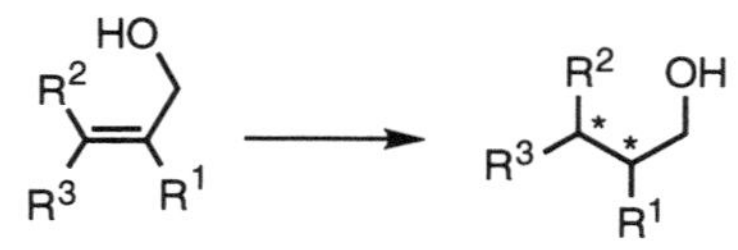

Eq. 3.6

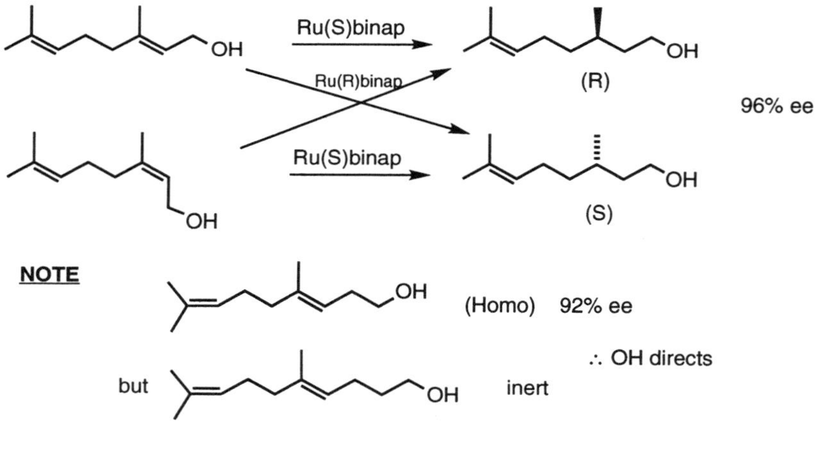

Eq. 3.7

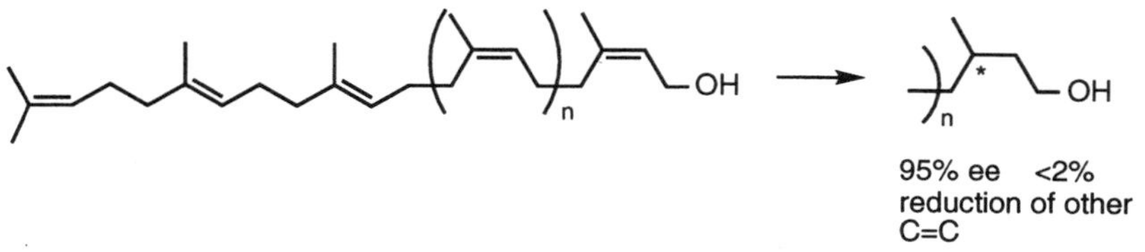

95% ee   <2% reduction of other C=C

Racemic secondary allylic alcohols could be efficiently resolved by carrying out a partial hydrogenation over a ruthenium BINAP catalyst (Eq. 3.8).[11] A wide variety of allylic alcohols undergo this reaction efficiently.

Eq. 3.8

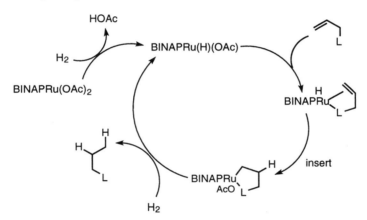

Although the overall transformation using these ruthenium BINAP catalysts is the same as that using rhodium/chiral ligand catalyst systems, the mechanism is totally different,[12] in that ruthenium remains in the same (+2) oxidation state throughout the catalytic cycle.

Figure 3.6   Hydrogenation Mechanism for RuBINAP Catalysts

Most homogeneous hydrogenation catalysts are specific for carbon-carbon double bonds, and other multiply-bonded species are inert. In marked contrast, halogen-containing ruthenium-BINAP complexes[13] catalyze the efficient asymmetric reduction of the carbonyl group, provided there is a heteroatom in the α, β, or γ position to provide the requisite second point of attachment to the catalyst (Eq. 3.9).

Eq. 3.9

n = 1-3
X = OH, OMe, CO$_2$Me, NMe$_2$, Br, COSMe, CONMe$_2$
C = sp^2, sp^3

β-Keto ester compounds were efficiently reduced to β-hydroxy esters by this system (Eq. 3.10).[14] Originally, high pressures of hydrogen (50-100 atm) were required, but the simple expediency of adding acid co-catalysts allows efficient catalysis under

Eq. 3.10

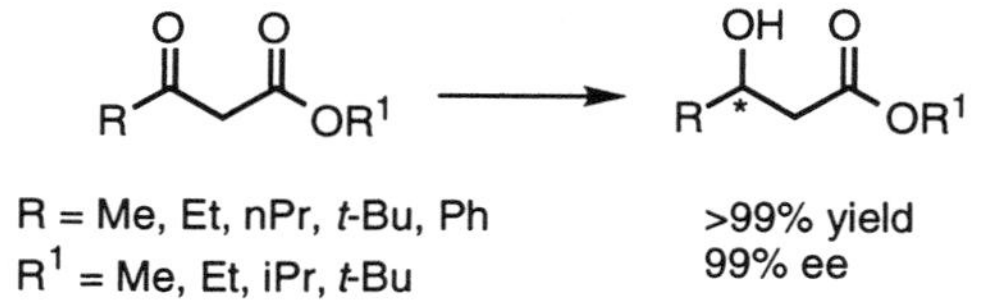

O O
R OR1
OH O
R * OR1
R = Me, Et, nPr, t-Bu, Ph
R1 = Me, Et, iPr, t-Bu
>99% yield
99% ee

much milder conditions.[15] This reduction has found extensive application in total synthesis (Figure 3.7).[16] Racemic β-keto esters were reduced with high selectivity for each diastereoisomer (Eq. 3.11). However, when the reaction was carried out under conditions which ensured rapid equilibration of the α-position, high yields of a single diastereoisomer were obtained (Eq. 3.12).[17]

Eq. 3.11

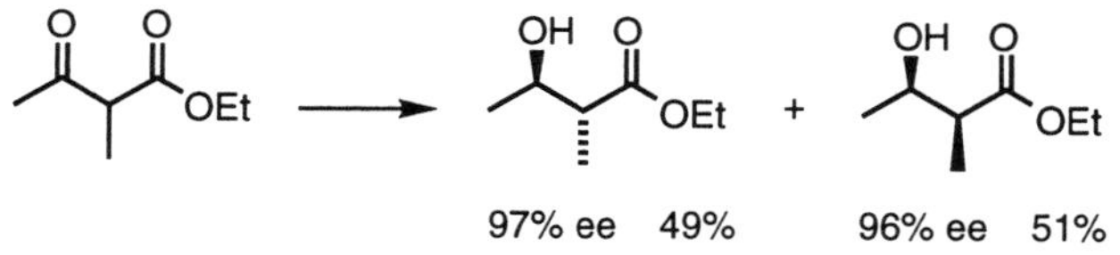

O O
OEt
OH O
OEt
+
OH O
OEt
97% ee 49%
96% ee 51%

Eq. 3.12

OH O
R1 OR3
R2
same
Slow
O O
R1 OR3
R2
O O
R1 OR3
R2
H2, cat.
Fast
OH O
R1 OR3
R2
>95:5 syn/anti
92-98% ee
R1 = Me,
O
O
R2 = NHAc, NHCO2Bn, CH2NHAc
cat. = (R) RuBINAP

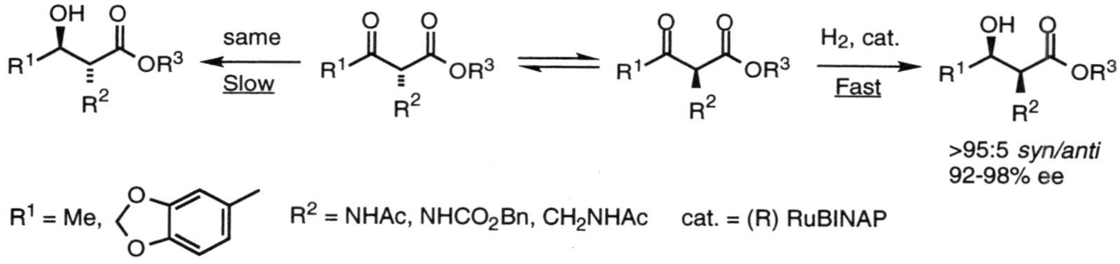

This asymmetric reduction is not restricted to β-ketoesters, but rather proceeds efficiently with a wide range of carbonyl compounds having appropriately disposed directing groups.[18]

BPE-ruthenium dibromide catalyst were also efficient in the reduction of β-ketoesters.[19] The reduction proceeded under mild conditions (35°C, 60 psi H_2, MeOH(H_2O). The normally high enantiomeric excess observed were much-reduced with halogen-containing substrates, and t-butyl or aryl substituents resulted in low conversion. These catalysts were also efficient when used for dynamic resolution of chiral racemic β-ketoesters (Eq. 3.12), for the reduction of enamides to amines,[20] and the reduction of enol esters to α-hydroxy esters.[21]

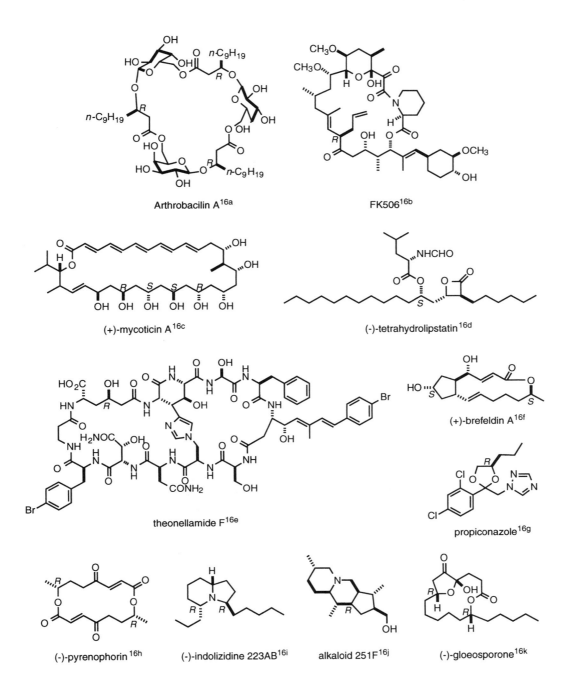

Figure 3.7 BINAP -Ru Catalyzed Asymmetric Hydrogenation in Total Synthesis. The Centers Generate By Reduction are Labeled R or S (from ref. 6b).

Another synthetically useful class of hydrogenation catalyst is the "super unsaturated" iridium or rhodium complex catalyst system, known as Crabtree's catalysts.[22] These are generated *in situ* by reduction of cationic cyclooctadiene iridium(I) complexes in the "noncoordinating" solvent, dichloromethane (Eq. 3.13).

Eq. 3.13

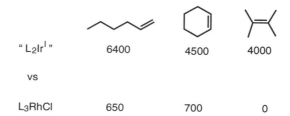

(Cy = cyclohexyl)

The solvent is apparently so loosely coordinated that it is replaced by virtually any olefinic substrate. These catalysts are exceptionally efficient for the reduction of tetrasubstituted olefins, substrates most hydrogenation catalysts will not reduce. A comparison of Crabtree's catalyst with Wilkinson's catalyst is shown in Figure 3.8. Nonpolar solvents must be used with Crabtree's catalysts. Acetone and ethanol act as competitive inhibitors, and greatly reduce catalytic activity.

" L_2Ir^I "	6400	4500	4000
vs			
L_3RhCl	650	700	0

Figure 3.8   Comparison of Wilkinson's and Crabtree's Catalysts

The observation that Crabtree's catalyst coordinates to alcohols has been used to an advantage in synthesis, by using adjacent "hard" ligands in the substrate to coordinate to the catalyst and deliver it from a single face of the alkene. Equation 3.14 compares Crabtree's catalyst to the normally used palladium on carbon. The iridium catalyst was delivered exclusively from the same face of the olefin as that

Eq. 3.14

Pd/C	20	:	30
L_2Ir^I	99.9	:	<0.1

occupied by the OH group, while Pd/C was nonselective.[23] This same effect was seen with a much more complex substrates (Eq. 3.15)[24] and (Eq. 3.16).[25] In all cases, the rate of hydrogenation was slower than that for substrate without coordinating functional groups.

Eq. 3.15

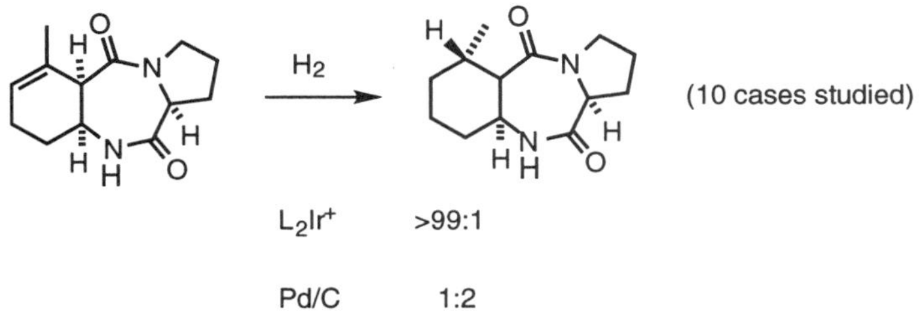

(10 cases studied)

L₂Ir⁺   >99:1

Pd/C   1:2

Eq. 3.16

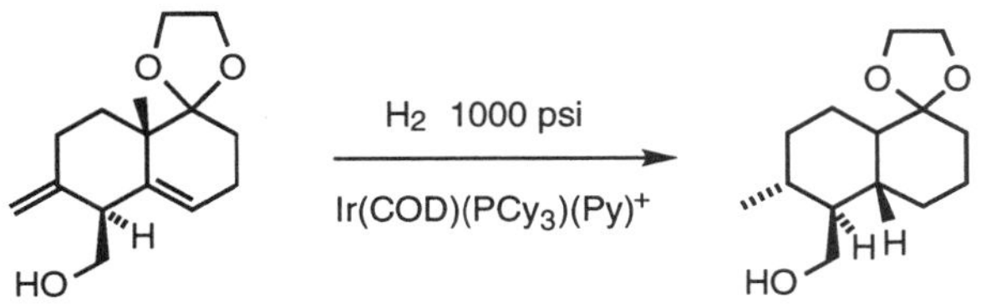

Another catalytic hydrogenation that involves a somewhat different form of "two point" attachment is the 1,4-reduction of 1,3-dienes to Z-alkenes in the presence of group 6 catalyst precursors such as $L_3Cr(CO)_3$ where L_3 is an arene,[26] $(CH_3CN)_3$ or $(CO)_3$[27] (Eq. 3.17). This reduction is restricted to dienes which can achieve cisoid conformations, implying the intermediacy of a chelate diene complex

Eq. 3.17

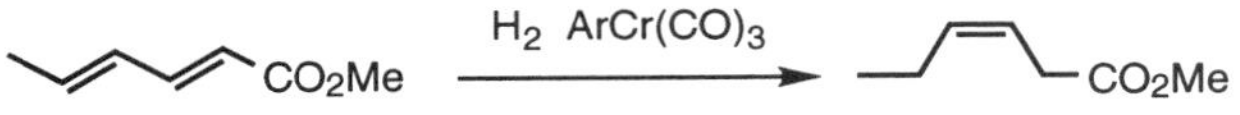

in the process. These same complexes catalyze the reduction of α,β-unsaturated carbonyl compounds, but again only those which can achieve an S-cis conformation (Eq. 3.18).[28] In the absence of hydrogen, these complexes catalyze the rearrangement of (1Z)-1-[silyloxymethyl]butadiene to silyl dienol ethers with high stereoselectivity (Eq. 3.19).[29]

Eq. 3.18

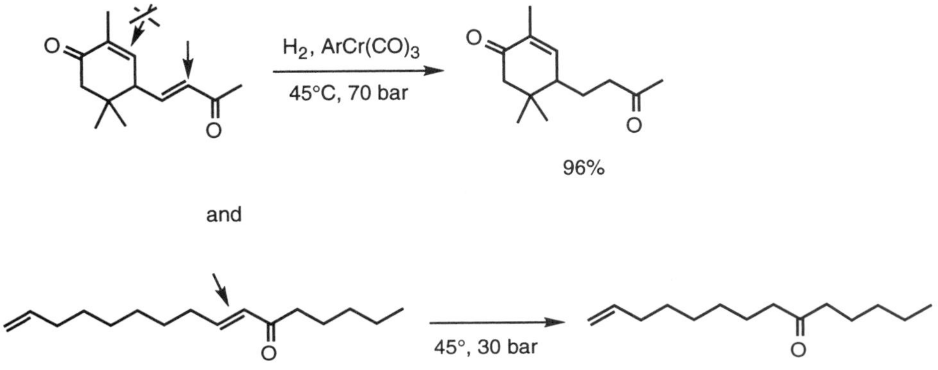

96%

and

45°, 30 bar

Eq. 3.19

97%

3.3   Other Reductions

In all of the reactions presented above, molecular hydrogen was the source of hydrogen for the reduction, and the ability of transition metals to break the hydrogen-hydrogen bond in an oxidative addition reaction was central to the process. However, metal hydrides are produced by several other pathways, and some of these have proven synthetically useful.

Treatment of iron pentacarbonyl with KOH produces an iron hydride which efficiently reduces α,β-unsaturated aldehydes, ketones, esters, lactones, and nitriles to saturated derivatives, with no reduction of the carbonyl or nitrile group (Eq. 3.20).[30] The reaction proceeds by rapid, irreversible addition of H–Fe across the

Eq. 3.20

double bond, followed by protonolysis.[31] Thus, in this reduction, both hydrogens came from water. The related binuclear complex $NaHFe_2(CO)_8$ affects similar chemistry but the reaction proceeds by a somewhat different mechanism.[29] Both systems are inefficient with β-substituted or sterically hindered enones.

The hexamer $[(Ph_3P)CuH]_6$ reduces conjugated enones to saturated ketones stoichiometrically, presumably via a conjugate addition to produce a copper enolate intermediate.[32,33] Interestingly, under modest hydrogen pressure in the presence of excess phosphine, this hexamer catalyzed this reduction.[32]

Yet another efficient system for the reduction of conjugated enones is the combination of a palladium(0) catalyst and tributyltin hydride (Eq. 3.21).[34] The mechanism of the process is not known. A possibility is shown in Eq. 3.22.

Eq. 3.21

(lactone not
reduced)

Eq. 3.22

3.4   Other Reactions of Transition Metal Hydrides

Transition metal hydrides are involved in a host of other synthetically useful reactions, most of which proceed by "hydrometallation" (insertion into a metal-hydrogen bond) of an alkene or alkyne to produce σ-alkylmetal complexes. It is the subsequent reactions of these σ-alkylmetal species that provide the synthetically useful transformations, and these will be discussed in the next chapter.

References

1. Moore, E.J.; Sullivan, J.M.; Norton, J.R. *J. Am. Chem. Soc.* **1986**, *108*, 2257.
2. Halpern, J.; Okamoto, T.; Zakhariev, A. *J. Mol. Catal.* **1976**, *2*, 65.

3. For reviews see: (a) Noyori, R. "Asymmetric Catalysis in Organic Synthesis", John Wiley and Sons, New York, 1994, pp. 16-94. (b) Ojima, I. "Transition Metal Hydrides: Hydrocarboxylation Hydroformylation and Asymmetric Hydrogenation" in "Comprehensive Organometallic Chemistry II", Abel, E.W.; Stone, F.G.A.; Wilkinson, G., Eds., Pergamon, Oxford, UK, 1995, Vol. 12, pp. 9-38.

4. (a) Brown, J.M.; Chaloner, P.A. *J. Am. Chem. Soc.* **1980**, *102*, 3040. (b) Brown, J.M.; Parker, D. *Organometallics* **1982**, *1*, 950. (c) Halpern, J. *Science* **1982**, *217*, 401. (d) Halpern, J. *Acct. Chem. Res.* **1982**, *15*, 332.

5. (a) Burke, M.J.; Gross, M.F.; Harper, T.G.P.; Kalberg, C.S.; Lee, J.R.; Martinez, J.P. *Pure Appl. Chem.* **1996**, *68*, 37. (b) Burk, M.J.; Wang, Y.M.; Lee, J.R. *J. Am. Chem. Soc.* **1996**, *118*, 5142. (c) Burk, M.J.; Gross, M.F.; Martinez, J.P. *J. Am. Chem. Soc.* **1995**, *117*, 9375.

6. For reviews see: (a) Noyori, R.; Takaya, H. *Acct. Chem. Res.* **1990**, *23*, 345. (b) Noyori, R. *Acta Chem. Scand.* **1996**, *50*, 380. (c) For a synthesis of the catalyst see: Takaya, H.; Akatagawa, S.; Noroyi, R. *Org. Synth.* **1988**, *67*, 20.

7. Ohta, T.; Takaya, H.; Kitamura, M.; Nagai, K.; Noyori, R. *J. Org. Chem.* **1987**, *52*, 3174.

8. (a) Noyori, R.; Ohta, M.; Hsiao, Y.; Kitamura, M. *J. Am. Chem. Soc.* **1986**, *108*, 7117. (b) Kitamura, M.; Hsiao, Y.; Ohta, M.; Tsukamoto, M.; Ohta, T.; Takaya, H.; Noyori, R. *J. Org. Chem.* **1994**, *59*, 297. (c) Uematsu, N.; Fujii, A.; Hishiguchi, S.; Ikoriza, T.; Noyori, R. *J. Am. Chem. Soc.* **1996**, *118*, 4916.

9. Takaya, H.; Ohta, T.; Sayo, N.; Kumobayashi, H.; Akutagawa, S.; Inoue, S-I.; Kasahara, I.; Noyori, R. *J. Am. Chem. Soc.* **1987**, *109*, 1596.

10. Imperiali, B.; Zimmerman, J.W. *Tetrahedron Lett.* **1988**, *29*, 5343.

11. Kitamura, M.; Kasahara, I.; Manabe, K.; Noyori, R.; Takaya, H. *J. Org. Chem.* **1988**, *53*, 708.

12. (a) Ohta, T.; Takaya, H.; Noroyi, R. *Tetrahedron Lett.* **1990**, *31*, 7189. (b) Ashby, M.T.; Halpern, J. *J. Am. Chem. Soc.* **1991**, *113*, 589.

13. (a) Mashima, K.; Kusano, K.; Sato, N.; Matsumura, Y.; Nozaki, K.; Kumbayashi, H.; Sayo, N.; Hori, Y.; Ishizaki, T.; Akutagawa, S.; Takaya, H. *J. Org. Chem.* **1994**, *59*, 3064. (b) Kitamura, M.; Tokamages, M.; Ohkuma, T.; Noyori, Y. *Org. Synth.* **1992**, *71*, 1.

14. Noyori, R.; Ohnkuma, T.; Kitamura, M.; Takaya, H.; Sayo, N.; Kamobayashi, H.; Akutagawa, S. *J. Am. Chem. Soc.* **1987**, *109*, 5856.

15. King, S.A.; Thompson, A.S.; King, A.O.; Verhoeven, T.R. *J. Org. Chem.* **1992**, *57*, 6689.

16. (a) Garcia, D.M.; Yamada, H.; Hatakeyama, S.; Nishizawa, M. *Tetrahedron Lett.* **1994**, *35*, 3325. (b) Nakatsuka, M.; Ragan, J.A.; Sammakia, T.; Smith, D.B.; Uehling, D.E.; Schreiber, S.L. *J. Am. Chem. Soc.* **1990**, *112*, 5583. (c) Poss, C.S.; Rychnovsky, S.D.; Schreiber, S.L. *J. Am. Chem. Soc.* **1993**, *115*, 3360. (d) Case-Green, S.C.; Davies, S.G.; Hedgecock, C.J. *Synlett* **1991**, 781. (e) Tohda, K.; Hamada, Y.; Shioiri, T. *Synlett* **1994**, 105. (f) Taber, D.F.; Silverberg, L.J.; Robinson, E.D. *IJ. Am. Chem. Soc.* **1991**, *113*, 6639. (g) Baldwin, J.E.; Adlington, R.M.; Ramcharitar, S.H. *Synlett* **1992**, 875. (h) Taber, D.F.; Deker, P.B.; Silverberg, L.J. *J. Org. Chem.* **1992**, *57*, 5990. (I) Taber, D.F.; You, K.K. *J. Am. Chem. Soc.* **1995**, *117*, 5757. (j) Schreiber, S.L.; Kelly, S.E.; Porco, J.A., Jr.; Sammakia, T.; Suh, E.M. *J. Am. Chem. Soc.* **1988**, *110*, 6210.

17. (a) Noyori, R.; Ikeda, T.; Ohkuma, T.; Widhalm, M.; Kitamura, M.; Takaya, H.; Akutagawa, S.; Sayo, N.; Saito, T.; Taketomi, T.; Kumobayashi, H. *J. Am. Chem. Soc.* **1989**, *111*, 9134. (b) Noyori, R.; Tokunaga, M.; Kitamura, M. *Bull Chem. Soc., Jpn.* **1995**, *68*, 36.

18. Kitamura, M.; Ohnkuma, T.; Inoue, S.; Sayo, N.; Kumobayashi, H.; Akutagawa, S.; Ohta, T.; Takaya, H.; Noyori, R. *J. Am. Chem. Soc.* **1988**, *110*, 629.

19. Burk, M.J.; Harper, T.G.P.; Kalberg, C.S. *J. Am. Chem. Soc.* **1995**, *117*, 4423.

20. (a) Burk, M.J.; Wang, Y.M.; Lee, J.R. *J. Am. Chem. Soc.* **1996**, *118*, 5143. (b) Burk, M.J.; Allen, J.G.; Kiesman, W.F. *J. Am. Chem. Soc.* **1998**, *120*, 657.

21. (a) Burk, M.J. *J. Am. Chem. Soc.* **1991**, *113*, 8518. (b) Burk, M.J.; Allen, J.G.; Kiesman, W.F. *J. Am. Chem. Soc.* **1998**, *120*, 4345.

22. Crabtree, R. *Acct. Chem. Res.* **1979**, *12*, 331.

23. Crabtree, R.H.; Davis, M.W. *Organometallics* **1983**, *2*, 681.

24. Schultz, A.G.; McCloskey, P.J. *J. Org. Chem.* **1985**, *50*, 5905.

25. Watson, A.T.; Park, K.; Wiemer, D.F.; Scott, W.J. *J. Org. Chem.* **1995**, *60*, 357.

26. (a) Tucker, J.R.; Riley, D.P. *J. Organomet. Chem.* **1985**, *279*, 49. (b) LeMaux, P.; Jaouen, G.; Saillard, J-Y. *J. Organomet. Chem.* **1981**, *212*, 193.

27. (a) Wrighton, M.S.; Schroeder, MA. *J. Am. Chem .Soc.* **1973**, *95*, 5764. (b) Mirbach, M.J.; Tuyet, N.P.; Saus, A. *J. Organomet. Chem.* **1982**, *236*, 309.

28. Sodeoka, M.; Shibasaki, M. *J. Org. Chem.* **185**, *50*, 1147; for a review see Sodeoka, M.; Shibasaki, M. *Synthesis* **1993**, 643.

29. Sodeoka, M.; Yamada, H.; Shibasaki, M. *J. Am. Chem. Soc.* **1990**, *112*, 4907.

30. Noyori, R.; Umeda, I.; Ishigami, T. *J. Org. Chem.* **1972**, *37*, 1542.

31. Collman, J.P.; Finke, R.G.; Matlock, P.L.; Wahren, R.; Komoto, R.G.; Brauman, J.I. *J. Am. Chem. Soc.* **1978**, *100*, 1119.
32. Mahoney, W.S.; Brestensky, D.M.; Stryker, J.M. *J. Am. Chem. Soc.* **1988**, *110*, 291.
33. Mahoney, W.S.; Stryker, J.M. *J. Am. Chem. Soc.* **1989**, *111*, 8818.
34. Keinan, E.; Gleize, P.A. *Tetrahedron Lett.* **1982**, *23*, 477.

CHAPTER 4

Synthetic Applications of Complexes Containing Metal-Carbon σ-Bonds

4.1   Introduction

Transition metal complexes containing metal-carbon σ-bonds are central to a majority of transformations in which transition metals are used to form carbon-carbon and carbon-heteroatom bonds, and thus are of supreme importance for organic synthesis. The transition metal-to-carbon σ-bond is usually covalent rather than ionic, and this feature strongly moderates the reactivity of the bound organic group, restricting its' reactions to those accessible to the transition metal (e.g., oxidative addition, insertion, reductive elimination, β-hydrogen elimination transmetallation). That is, the transition metal is much more than a sophisticated counterion for the organic group; it is the major determinant of the chemical behavior of that organic group.

σ-Carbon-metal complexes can be prepared by a number of methods, summarized in Figure 4.1. This variety makes virtually every class of organic compound a potential source of the organic group in these complexes, and thus subject to all of the carbon-carbon bond forming reactions of σ-carbon-metal complexes. The use of these processes in organic synthesis is the subject of this chapter.

4.2   σ-Carbon-Metal Complexes from the Reaction of Carbanions and Metal Halides: Organocopper Chemistry

Starting with the early observation that small amounts of copper(I) salts catalyzed the 1,4-addition of Grignard reagents to conjugated enones,[1] organocopper chemistry has been warmly accepted by synthetic chemists and vastly widened in scope by organometallic chemists. As a broad class, they are among the most

1. Preparation

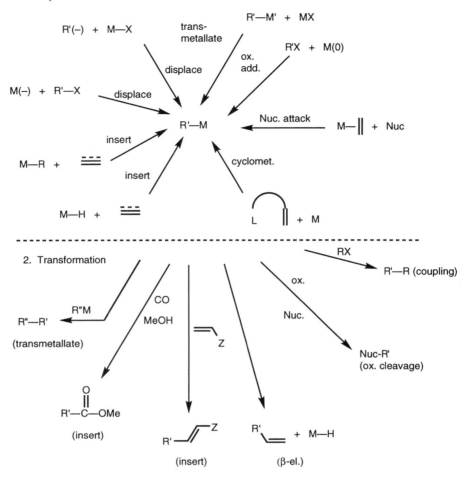

Figure 4.1 Preparation and Transformation of σ-Carbon-Metal Complexes

extensively used organometallic reagents, and the variety of species and the transformations they effect is staggering.[2,3]

The simplest and most extensively used complexes are the lithium diorganocuprates, R_2CuLi, soluble, thermally unstable complexes generated *in situ* by the reaction of copper(I) iodide with two equivalents of an organolithium reagent (Eq. 4.1). (The *mono* alkyl copper complexes, RCu, are yellow, insoluble, oligomeric complexes, little used in synthesis.) These reagents efficiently alkylate a variety of halides (Eq. 4.2). Their low basicity favors displacement over competing elimination processes.

Eq. 4.1

$$2RLi \ + \ CuI \ \longrightarrow \ R_2CuLi \ + \ LiI$$

Eq. 4.2

$$R_2CuLi \ + \ R^1X \ \longrightarrow \ R\!-\!R^1$$

The range of useful halide substrates is very broad. The order of reactivity of alkyl (sp^3) halides is primary > secondary >> tertiary, with iodides more reactive than bromides and chlorides. Alkyl tosylates also undergo substitution by diorganocuprates at a rate comparable to iodides. Alkenyl halides and triflates (enol triflates derived from ketones[4]) are also very reactive toward lithium diorganocuprates, and react with essentially complete retention of geometry of the double bond, a feature which has proved useful in organic synthesis. Aryl iodides and bromides are alkylated by lithium organocuprates, although halogen-metal exchange is sometimes a problem. (RCu reagents are sometimes more efficient in this reaction.) Benzylic, allylic, and propargylic halides also react cleanly. Propargyl halides produce primarily allenes (via S_N2'-type reactions), but allylic chlorides and bromides react without allylic transposition. In contrast, allylic acetates react mainly with allylic transposition. Acid halides are converted to ketones by lithium dialkyl-cuprates, although alkyl hetero (mixed) cuprates are more efficient in this process. Finally, epoxides undergo ring opening resulting from alkylation at the less-substituted carbon. Consideration of a great deal of experimental data leads to the following order of reactivity of lithium diorganocuprates with organic substrates: acid chlorides > aldehydes > tosylates ≈ epoxides > iodides > ketones > esters > nitriles. This is only a generalization; many exceptions are caused by unusual features in either the organocopper reagent or the substrate.

Another extensively used process in the synthesis of complex organic molecules is the conjugate addition of lithium diorganocuprates to enones (Eq. 4.3).[5] The range of R groups available for this useful reaction is similar to that for the alkylation reaction. Virtually all types of sp^3 hybridized dialkylcuprate reagents react cleanly, including straight-chain and branched primary, secondary, and tertiary alkyl complexes. Phenyl and substituted arylcuprates react in a similar fashion, as do vinylcuprates. Allylic and benzylcuprates also add 1,4 to conjugated enones, but they are no more efficient than the corresponding Grignard reagents. Acetylenic copper reagents do *not* transfer an alkyne group to conjugated enones, a feature used to advantage in the chemistry of mixed cuprates.

Eq. 4.3

$$R_2CuLi \ + \ CH_2\!=\!CHCCH_3 \ \longrightarrow \ RCH_2CH_2CCH_3$$

The range of α,β-unsaturated carbonyl substrates that undergo 1,4-addition with diorganocuprates is indeed broad. Although the reactivity of any particular system depends on many factors, including the structure of the copper complex and the substrate, sufficient data are available to allow some generalizations. Conjugated ketones are among the most reactive substrates, combining very rapidly with diorganocuprates (the reaction is complete in less than 0.1 s at 25°C) to produce the 1,4 adduct in excellent yield. Alkyl substitution at the α, α´, or β positions of the parent enone causes only slight alterations in the rate and course of this conjugate addition with acyclic enones, although this same substitution, as well as remote substitution, does have stereochemical consequences with cyclic enone systems.

Conjugated esters are somewhat less reactive than conjugated ketones. With such esters, α,β- and β,β-disubstitution decreases reactivity drastically. Conjugated carboxylic acids do not react with lithium diorganocuprates, and conjugated aldehydes suffer competitive 1,2-additions. Conjugated anhydrides and amides have not been studied to any extent.

Conjugate additions of organocuprates to enones initially produces an enolate, and further reaction with an electrophile is possible, although often times it is slow and inefficient. This problem has been overcome, in the context of prostaglandin synthesis, by use of a butylphosphine-stabilized copper reagent, and by conversion of the unreactive, initially formed copper (or lithium) enolate into a more reactive tin enolate by addition of triphenyl tin chloride (Eq. 4.4).[6]

Eq. 4.4

Although lithium dialkylcuprates are very useful reagents, they suffer several serious limitations, the major one being that only one of the two alkyl groups is transferred. Because of the instability of the reagent, a threefold to fivefold excess (a sixfold to tenfold excess of R) is often required for complete reaction. When the R group is large or complex, as it often is in prostaglandin chemistry, this lack of efficiency is intolerable. Cuprous alkoxides, mercaptides, cyanides, and acetylides are considerably more stable than cuprous alkyls, and, furthermore, are relatively unreactive. By making mixed alkyl heterocuprates with one of these nontransferable, stabilizing groups, this problem can be avoided.[7] Among the heterocuprates, lithium phenylthio(alkyl)cuprates, lithium t-butoxy(alkyl)cuprates, and lithium 2-thienylcuprates[8] are the most useful, and are stable at about -20°C to 0°C.

Mixed alkyl heterocuprates are the most efficient reagents for the conversion of acid chlorides to ketones. Although large excesses of lithium dialkylcuprates are required for this reaction, and secondary and tertiary organocuprates never work well, only a 10 percent excess of these mixed cuprates is required to provide high yields of ketones, even in the presence of remote halogen, keto, and ester groups. In addition, although aldehydes are unstable toward lithium dialkylcuprates even at -90°C, equal portions of benzaldehyde and benzoyl chloride react with one equivalent of lithium phenylthio(t-butyl)cuprate to produce pivalophenone in 90% yield with 73% recovery of benzaldehyde. The lithium phenylthio(t-alkyl)cuprate is also the reagent of choice for the alkylation of primary alkyl halides with t-alkyl groups. Secondary and tertiary halides fail to react with this reagent, however.

Perhaps most importantly, these mixed cuprate reagents, especially the acetylide and 2-thienyl complexes, are extremely efficient in the conjugate alkylation of α,β-unsaturated ketones. Since it is this reaction that is used most extensively in the synthesis of complex molecules, the development of mixed cuprates was an important synthetic advance.

Recently, organocopper chemistry has been developed far beyond these original boundaries. Thermally stable cuprates have been synthesized by using sterically hindered nontransferable ligands such as diphenylphosphide (Ph_2P) and dicyclohexylamide (Cy_2N) to prepare $RCu(L)Li$.[9] These reagents undergo typical organocuprate reactions, and are stable at 25°C for over an hour. Even more impressive is the very hindered reagent $RCu[P(tBu)_2]Li$, which is stable in THF at reflux for several hours, but still enjoys organocuprate reactivity (Eq. 4.5),[10a] and the more easily prepared β-silyl organocuprates $RCuCH_2SiR_3$.[10b]

Eq. 4.5

Treatment of copper(I) cyanide with two equivalents of an organolithium reagent produces the "higher order" cuprate "$R_2Cu(CN)Li_2$," which is more stable than R_2CuLi, and is particularly effective for the reaction with normally unreactive secondary bromides or iodides, as well as for 1,4-addition to conjugated ketones and esters. (The nature and even the existence of higher order cuprates is still a matter of controversy.[11] Whatever the reagent is, its reactivity differs substantially from that of simple diorganocuprates.) By taking advantage of the observation that mixed diaryl "higher-order" diarylcuprates do not scramble aryl groups at low (<-100°C) temperatures, an efficient synthesis of unsymmetrical biaryls was developed (Eq.

4.6).[12] This chemistry was used in efficient syntheses of axially dissymmetric biaryls (Eq. 4.7)[13] and in an approach to calphostins (Eq. 4.8).[14]

Eq. 4.6

$$ArLi + CuCN \xrightarrow{-78°C} ArCu(CN)Li \xrightarrow[-125°C]{Ar'Li} [ArCuAr']CNLi_2 \xrightarrow[-125°C]{O_2} ArAr'$$

78-90%
(>96% cross coupling)

Ar = Ph, oMeOPh, mMeOPh, pMeOPh, (2Cl, 4CF₃)Ph, 1 Naphth.

Ar' = oMeOPh, pMeOPh, oMePh, (3F,4Me)Ph

Eq. 4.7

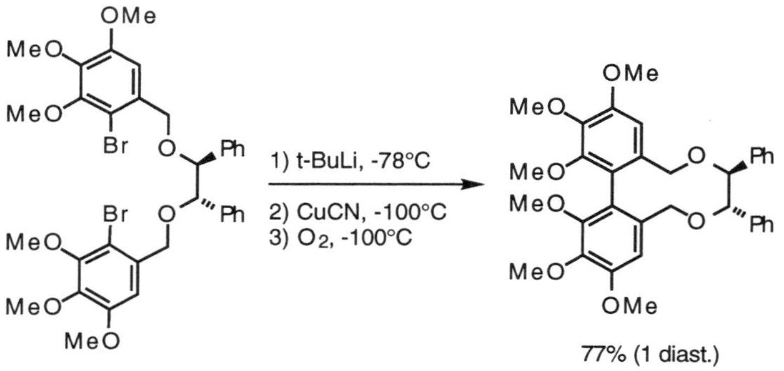

1) t-BuLi, -78°C
2) CuCN, -100°C
3) O₂, -100°C

77% (1 diast.)

Eq. 4.8

1) t-BuLi, -78°C
2) CuCN—TMEDA
3) O₂, -78°C

70%
8:1

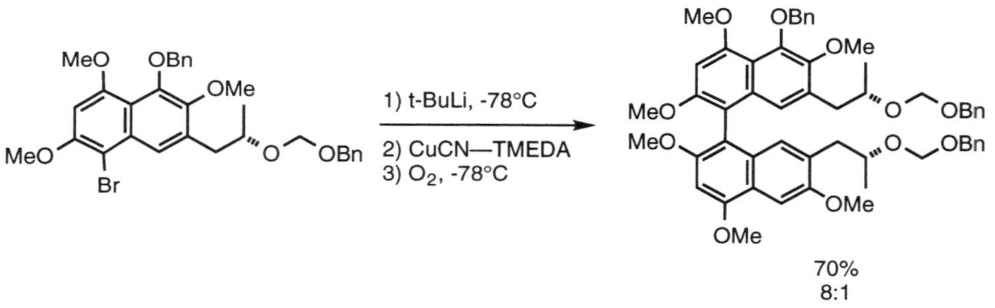

Addition of Lewis acids such as $BF_3 \bullet OEt_2$ or TMSCl to organocopper compounds dramatically increases their reactivity and broadens their synthetic utility. Addition of BF_3 to the normally unreactive RCu reagents produces species which are highly reactive for the alkylation of allylic substrates[15] as well as conjugate additions to enones, including α,β-unsaturated carboxylic acids, themselves unreactive towards R_2CuLi.[16] The reagent $RCu \bullet BF_3$ monoalkylates ketals with displacement of an alkoxy group, a very unconventional transformation (Eq. 4.9).[17] Optically active allylic acetals undergo S_N2' cleavage with very high diastereoselectivity (Eq. 4.10).[18]

Eq. 4.9

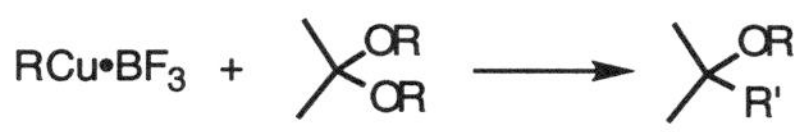

$$RCu \cdot BF_3 \; + \; \chememph{} \longrightarrow \chememph{}$$

Eq. 4.10

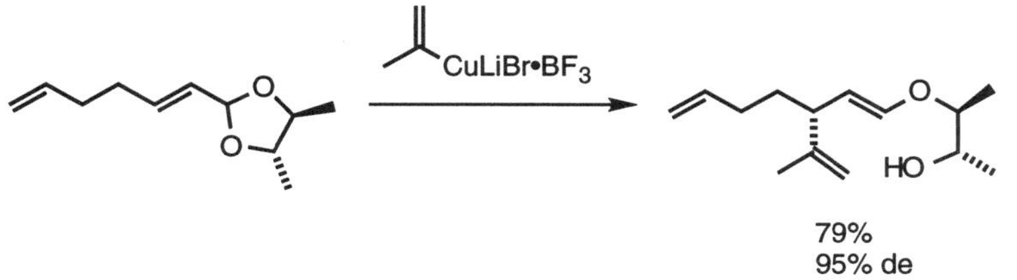

These Lewis acid modified cuprate reagents also efficiently add 1,4- to conjugated enones, and effective asymmetric induction in this process has been achieved in at least two different ways. The most widely studied system involves the use of unsaturated optically active sultam amides as substrates (Eq. 4.11).[19] In most cases, both high yields and high diastereomeric excesses are observed. Similar efficiencies are seen with naphthyl substituted camphor esters and trimethylsilyl activation (Eq. 4.12).[20]

Eq. 4.11

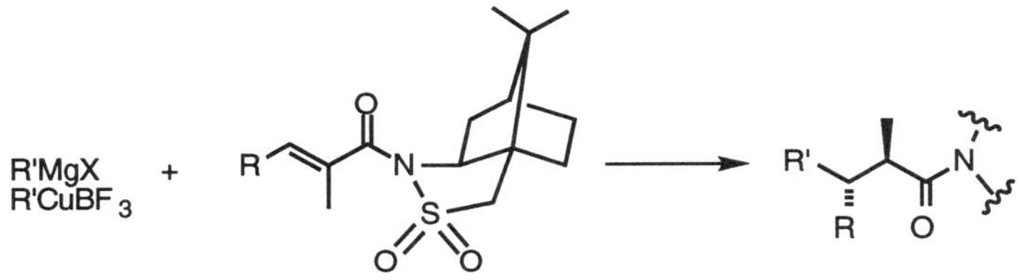

Eq. 4.12

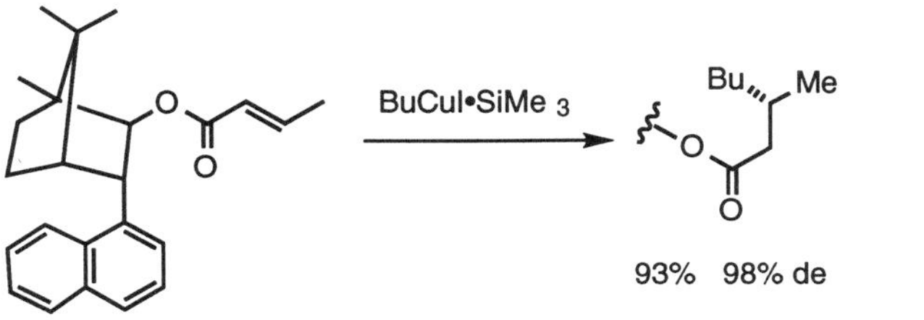

Copper-catalyzed additions of organolithium (Eq. 4.13)[21a,b] or organozinc reagents[21c] to conjugated enones in the presence of optically active ligands offers a potentially more general approach to asymmetric induction, although this methodology has not been broadly developed, and the range of effective substrates is small.

Addition of $BF_3 \cdot OEt_2$ to diorganocuprates, R_2CuLi, also dramatically enhances their reactivity, and permits the alkylation of very sterically hindered unsaturated ketones,[22] the ring opening of aziridines[23] and diastereoselective alkylation of chiral ketals as in Eq. 4.10.[24]

Eq. 4.13

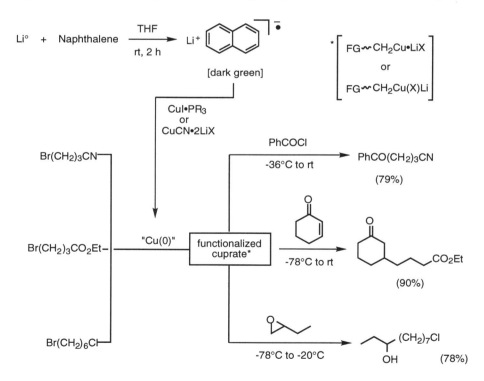

All of the organocopper chemistry discussed above relies upon the generation of the reactive organocopper species from either a Grignard reagent or an organolithium compound. This puts a serious limitation on the nature of the functional groups contained in the organocopper species, since they must be stable to organolithium reagents. Several solutions to the problem have recently been developed.

Reduction of CuCN•2LiBr or CuI•PR$_3$ with lithium naphthalenide at -100°C produces a highly active copper which can react with *functionalized* organic halides (I > Br >> Cl) to generate *functionalized* organocopper species in solution. Addition of electrophiles directly to these solutions produces alkylation products in excellent yield (Figure 4.2).[25] By this process, ester groups, nitriles and both aryl and alkyl

Figure 4.2 Preparation and Reactions of Functionalized Organocopper Species

chlorides are inert, and cuprates containing these functional groups can be generated and transferred. This class of reagent efficiently cyclized ω-haloepoxides (Eq. 4.14), alkylated imines with functionalized allylic halides (Eq. 4.15)[36] and even coupled aryl halides (prone to benzyne formation when lithiation is attempted), to reactive electrophiles (Eq. 4.16).[27]

Eq. 4.14

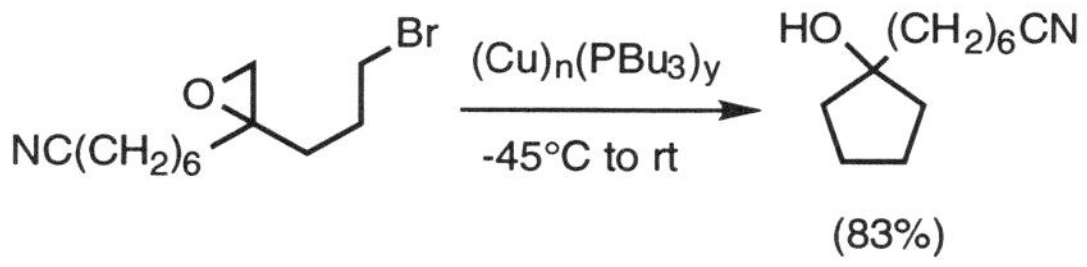

Eq. 4.15

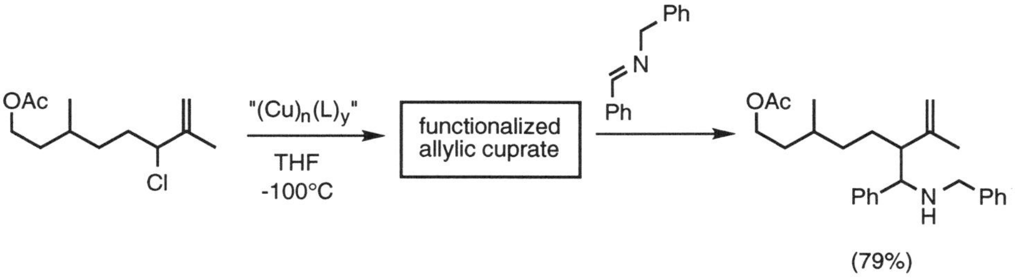

Eq. 4.16

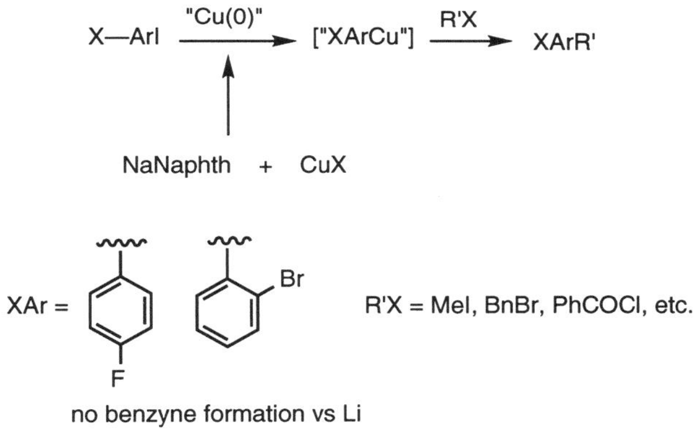

Another route to functionalized organocopper reagents involves transmetallation to copper from less reactive organometallic reagents, particularly organozinc compounds.[28] Organozinc halides can be made directly from organic halides and zinc metal or diethyl zinc, or from treatment of organolithium species with zinc chloride. Many functional groups are quite stable to these organometallic reagents. Treatment of CuCN•2LiCl with functionalized organozinc halides produces functionalized organocuprates, which then undergo normal cuprate coupling processes, introducing *functionalized* R groups into the substrate (Figure 4.3). The process has been made catalytic in copper.[29]

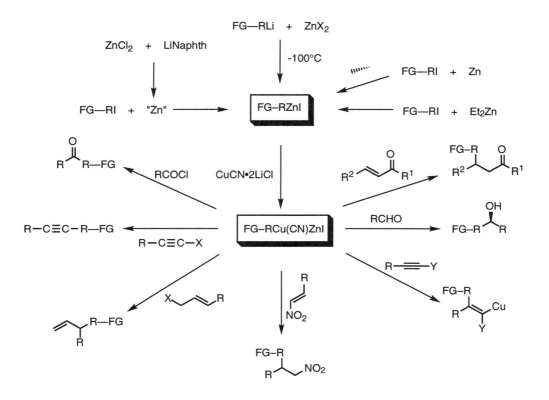

Figure 4.3 Reactions of Functionalized Organocopper Reagents

A particularly nice application of this type of zinc-copper chemistry is in the synthesis of cyclopentenones from zinc homoenolates and ynones, catalyzed by copper salts (Eqs. 4.17 and 4.18).[30] This reaction has been used in the synthesis of ginkolides (Eq. 4.19).[31]

Other functionalized organocopper complexes are also useful. α-Alkoxytin reagents are easily prepared by the reaction of trialkyl stannyllithium reagents with aldehydes. Although tin will not transmetallate to copper, the α-alkoxyalkyl group is readily cleaved from tin by butyllithium producing the α-alkoxyalkyllithium

Eq. 4.17

Eq. 4.18

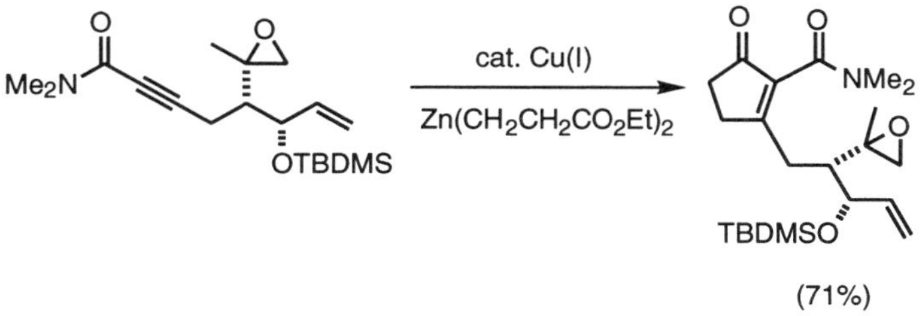

(71%)

Eq. 4.19

(52%)

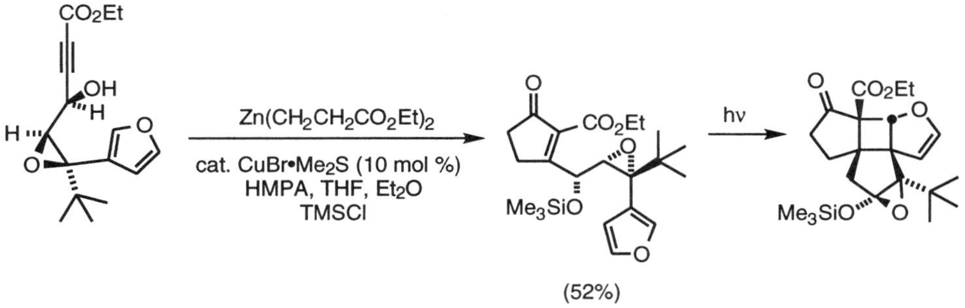

reagent. Addition of copper(I) cyanide followed by the typical array of electrophiles results in introduction of an α-alkoxyalkyl group (Eq. 4.20).[33]

Eq. 4.20

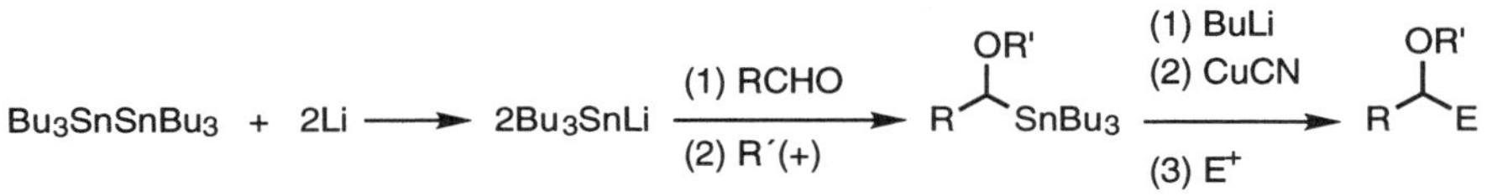

Although dialkyl or alkyl heterocuprates are by far the most extensively used organocopper reagents, the simple organocopper complexes RCuMX perform much of the same chemistry, and are clearly superior for certain types of reactions, particularly the addition of RCuMgX$_2$ to terminal alkynes. R$_2$CuLi complexes often abstract the acidic acetylenic proton from terminal alkynes, and pure RCu or RCuLiX do not add at all. However, RCuMgX$_2$ complexes generated from Grignard reagents and copper(I) iodide cleanly add *syn* to simple terminal alkynes to produce alkenylcuprates both regio- and stereospecifically ("carbocupration"). Note that the acidic acetylenic C–H bond is left intact. These alkenyl complexes enjoy the reactivity common to organocopper species and react with a variety of organic substrates (Figure 4.4).[33] Functionalized cuprates generated as in Figure 4.3 also "carbocuprate" alkynes, producing functionalized alkenylcuprates for use in synthesis (Eq. 4.21).[34] When carried out intramolecularly exocyclic alkenes are produced (Eq. 4.22).

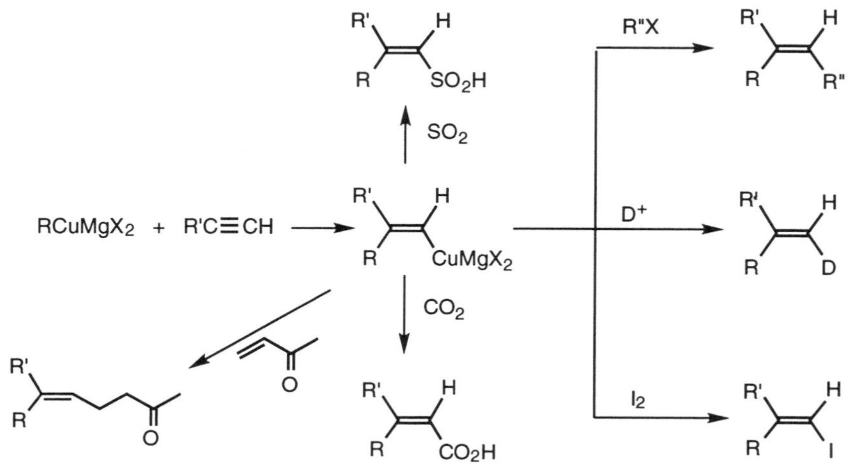
RCuMgX2 + R'C≡CH
CuMgX2
SO2H
SO2
CO2
CO2H
R"X
D+
I2

Figure 4.4 Carbocupration of Alkynes

Eq. 4.21

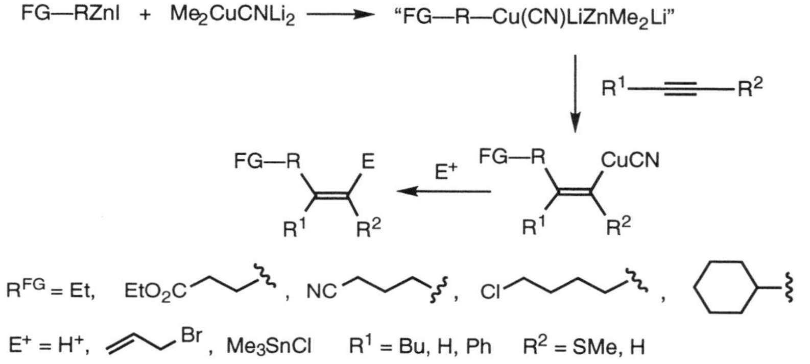
FG—RZnI + Me2CuCNLi2
"FG—R—Cu(CN)LiZnMe2Li"
R1—≡—R2
E+
CuCN
RFG = Et, EtO2C, NC, Cl
E+ = H+, Br, Me3SnCl R1 = Bu, H, Ph R2 = SMe, H

Eq. 4.22

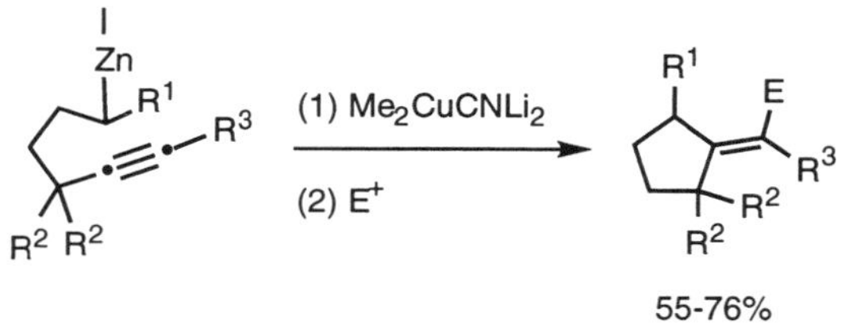
(1) Me2CuCNLi2
(2) E+
55-76%

4.3 σ-Carbon-Metal Complexes from Insertion of Alkenes and Alkynes into Metal-Hydrogen Bonds

The insertion of an alkene or alkyne into a metal-hydrogen bond to form a σ-carbon-metal complex is a key step in a number of important processes, including catalytic hydrogenation and hydroformylation. This insertion is an equilibrium, with β-hydride elimination being the reverse step. With electron-rich metals in low oxidation states [Rh(I), Pd(II)] the equilibrium lies far to the left, presumably because π-acceptor (electron withdrawing) olefin ligands stabilize these electron-rich metals more than the corresponding σ-donor alkyl ligands resulting from insertion (Eq. 4.23).

Eq. 4.23

The situation is the reverse with electron-poor metals in high oxidation states (e.g., d° Zr(IV)). In these cases σ-donor alkyl ligands are stabilizing (and π-accepting olefin ligands destabilizing) and the equilibrium lies to the right. From a synthetic point of view, the position of the equilibrium is not important, as long as the σ-carbon-metal complex formed by insertion is more reactive than the olefin-hydride complex. σ-Carbon-metal complexes of both electron-rich and electron-poor metals, formed from metal hydrides and olefins, have been used in the synthesis of complex molecules.

An example of the former is the palladium(II)-catalyzed cycloisomerization of enynes (Eq. 4.24).[35] Although the mechanism of this reaction is not yet known, it is likely to involve *in situ* generation of a palladium(II) hydride species, which "inserts" (hydrometallates) the alkyne to give a σ-alkenylpalladium(II) complex with an adjacent alkene group. Coordination of this alkene, followed by insertion into the σ-alkenylpalladium complex, forms a carbon-carbon bond (the ring) and a σ-*alkyl*-palladium(II) complex having two sets of β-hydrogens. β-Hydrogen elimination can occur in either direction, generating the cyclic diene and regenerating the palladium(II) hydride species to reenter the catalytic cycle. The direction of β-elimination can be influenced by the addition of ligands (Eq. 4.25), although the reason for this is not entirely clear.

Eq. 4.24

Eq. 4.25

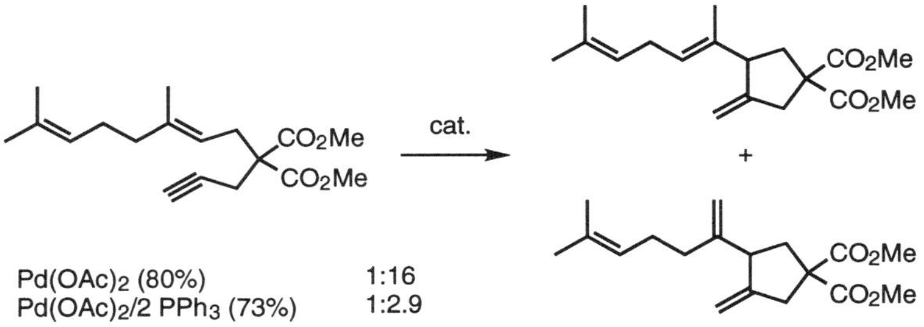

Pd(OAc)₂ (80%)   1:16
Pd(OAc)₂/2 PPh₃ (73%)   1:2.9

The reaction tolerates a range of functional groups (Eq. 4.26), as well as structural complexity in the substrate (Eq. 4.27).[36] When carried out in the presence of an appropriate reducing agent, the ultimate σ-alkylpalladium(II) complex can be reduced prior to β-elimination, giving monoenes rather than dienes (Eqs. 4.28 and 4.29).[37] In the presence of organotin reagents transmetallation/reductive elimination alkylates the final product (Eq. 4.30).[38]

Eq. 4.26

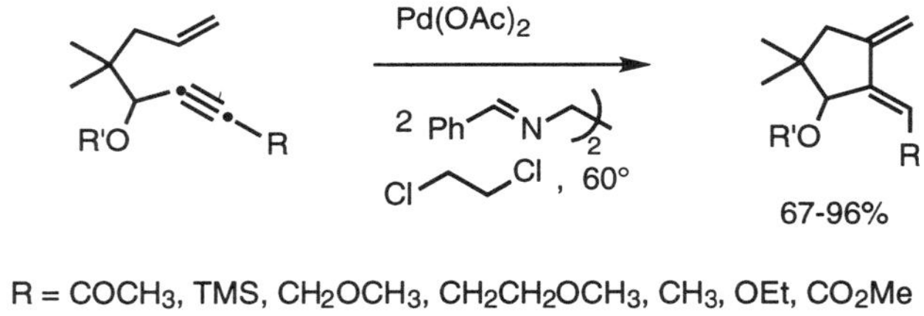

R = COCH₃, TMS, CH₂OCH₃, CH₂CH₂OCH₃, CH₃, OEt, CO₂Me

Eq. 4.27

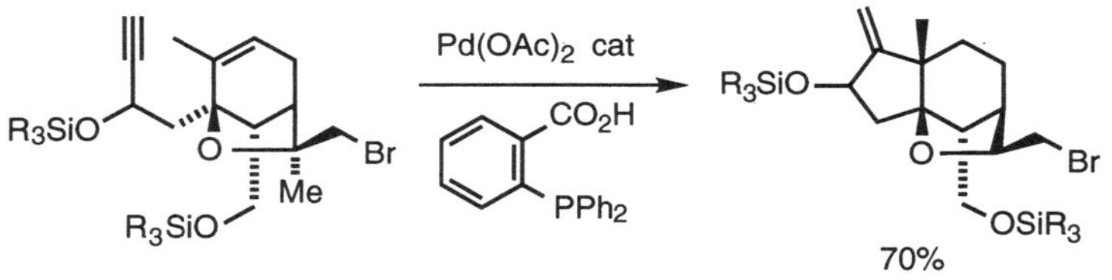

Eq. 4.28

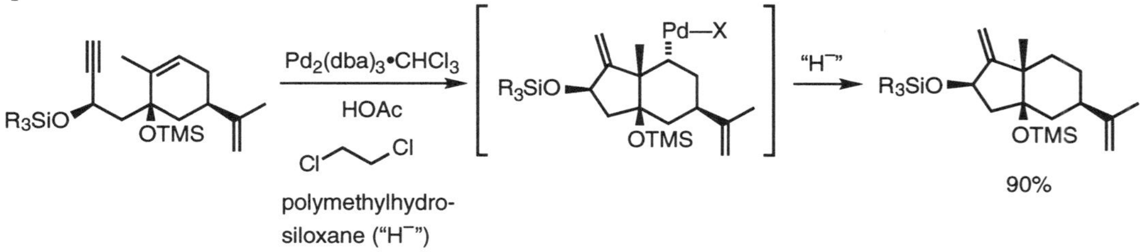

Eq. 4.29

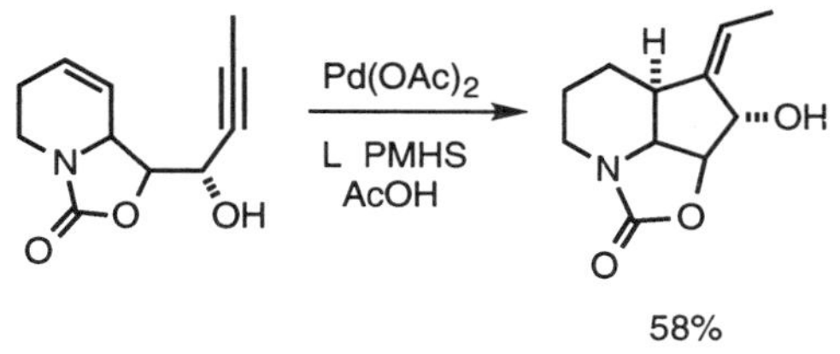

58%

Eq. 4.30

40-80%

X = C(CO₂Bn)₂, NCOPh

R =

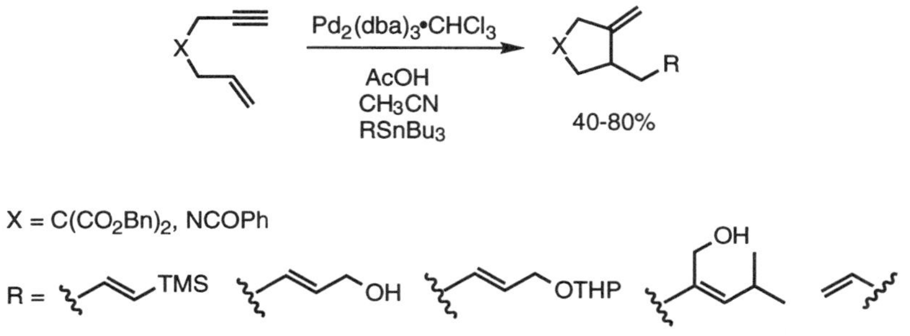

Multiple insertions are possible, provided β-hydrogen elimination is slower than the next insertion, a feature achieved by careful selection of substrate.[39] The example in Eq. 4.31 is illustrative. Insertion is a *cis* process, and β-elimination requires that the metal and hydrogen achieve a *syn* coplanar relationship. In Eq. 4.31, a *cis* insertion leads to σ-alkylpalladium(II) complex which has no β-hydrogen *syn*

Eq. 4.31

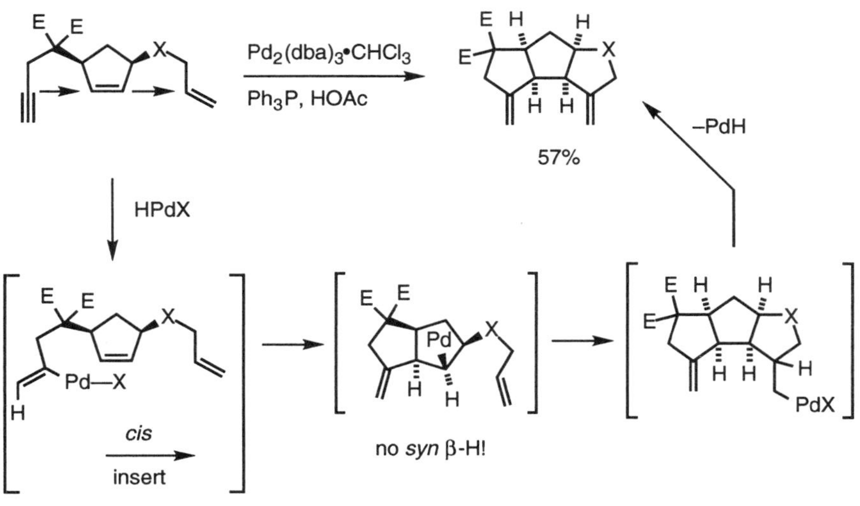

57%

coplanar, making β-elimination disfavored. Instead, a suitably-situated alkene undergoes insertion, and the resulting σ-alkylpalladium(II) complex can then undergo β-elimination. The ultimate example of this is shown in Eq. 4.32 in which all but the final center formed are quaternary, and have no β-hydrogens.[40] This process has been used to generate 1,3,5-hexatrienes which undergo a facile electrocyclic reaction to produce tricyclic material (Eq. 4.33).[41] Note that in none of these cases could σ-carbon-metal complexes be detected, since they were present only in low concentrations because the insertion equilibria lay far to the left.

Eq. 4.32

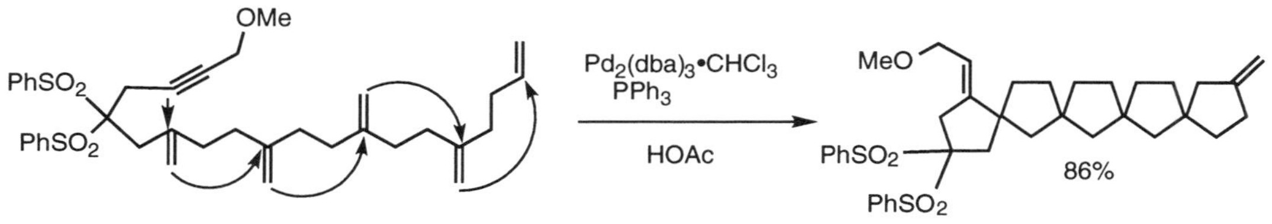

86%

Eq. 4.33

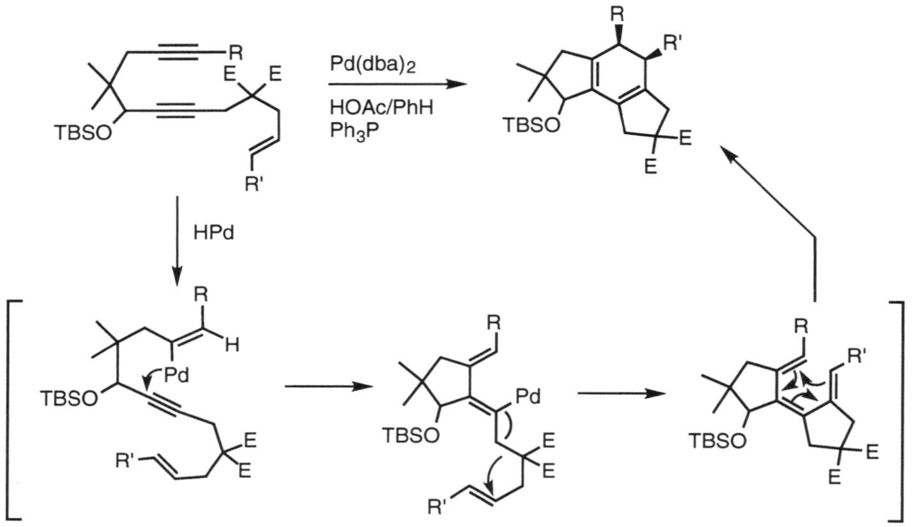

The completely opposite situation obtains for the electron-poor, d°, Zr(IV) complex $Cp_2Zr(H)Cl$, which reacts under mild conditions with a variety of alkenes to give stable, isolable σ-alkyl complexes of the type $Cp_2Zr(R)Cl$ (Eq. 4.34).[42]

Eq. 4.34

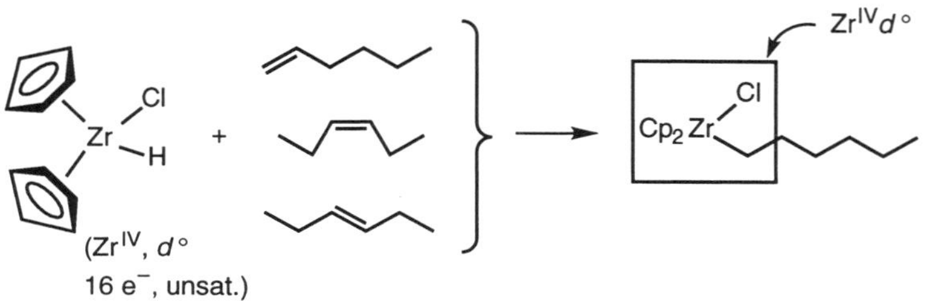

This reaction is remarkable in several respects. The addition occurs under mild conditions, and the resulting alkyl complexes are quite stable. Regardless of the initial position of the double bond in the substrate, the Zr ends up at the sterically least-hindered accessible position of the olefin chain. This rearrangement of the zirconium to the terminal position occurs by a Zr–H elimination, followed by a readdition that places Zr at the less-hindered position of the alkyl chain in each instance (Eq. 4.35). This migration proceeds because of the steric congestion about Zr caused by the two cyclopentadienyl rings. The order of reactivity of various olefins toward hydrozirconation is terminal olefins > *cis* internal olefins > *trans* internal olefins > exocyclic olefins > cyclic olefins and terminal > disubstituted > trisubstituted olefins. Tetrasubstituted olefins and trisubstituted cyclic olefins fail to react. Finally, 1,3-dienes add ZrH to the less-hindered double bond to give γ,δ-unsaturated alkylzirconium complexes.

Eq. 4.35

Alkynes also react with $Cp_2Zr(H)Cl$, adding Zr–H in a *cis* manner. With unsymmetric alkynes, addition of Zr–H gives mixtures of alkenyl Zr complexes with the less-hindered complex predominating. Equilibration of this mixture with a slight excess of $Cp_2Zr(H)Cl$ leads to mixtures greatly enriched in less-hindered complex. The good news is that Cp_2ZrHCl is a metal hydride which readily adds to a wide variety of alkenes and alkynes to give a metal-alkyl species, in principle making alkenes and alkynes sources of alkyl groups in synthesis. The bad news is that Cp_2ZrRCl is Zr(IV), d^0 and undergoes very little useful organic chemistry.

Zirconium(IV) alkyls do undergo two useful reactions. They are readily cleaved by electrophiles such as bromine or peroxides, with retention of configuration, to give straight-chain alkyl halides or alcohols respectively[43] even from mixtures of alkene isomers. (Zirconium(IV) alkenyls undergo electrophilic cleavage with retention of alkene geometry.) In addition, they insert carbon monoxide, giving stable acylzirconium(IV) complex which can be cleaved by acid to give aldehydes.[44] However it was not until the development of efficient transmetallation processes that hydrozirconation achieved real utility in organic synthesis.

4.4 σ-Carbon-Metal Complexes from Transmetallation/Insertion Processes

Transmetallation (transfer of an R group) from main-group metals to transition metals is extensively exploited in organic synthesis, notwithstanding the fact that the process is poorly understood (Chapter 2). Its utility lies in the ability to use main-group organometallic chemistry to "activate" organic substrates (e.g., hydroboration of alkenes, direct mercuration of aromatics), then transfer that substrate to transition metals to take advantage of their unique reactivity (e.g., oxidative addition, insertion). One such useful class of reaction is transmetallation/insertion (Eq. 4.36).

Eq. 4.36

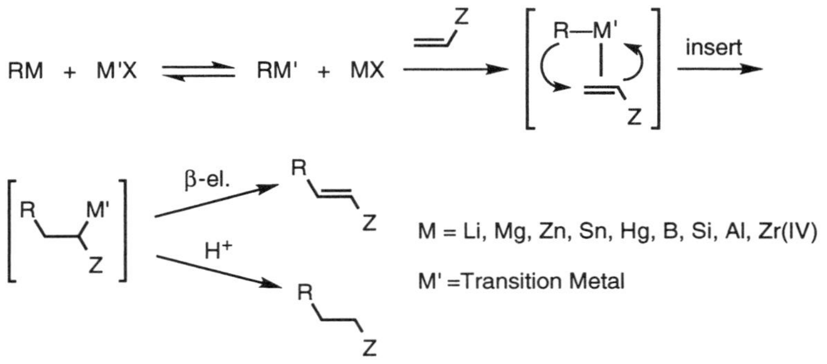

M = Li, Mg, Zn, Sn, Hg, B, Si, Al, Zr(IV)

M' = Transition Metal

As mentioned above, although alkenes and alkynes are readily hydrozirconated, the resulting σ-alkylzirconium(IV) species have only minimal reactivity. However, they will undergo transmetallation to nickel(II) complexes, and the resulting organonickel complexes undergo 1,4-addition to conjugated enones (Eq. 4.37).[45] Multiple transmetallations increase the utility of the process. For example, in

Eq. 4.37

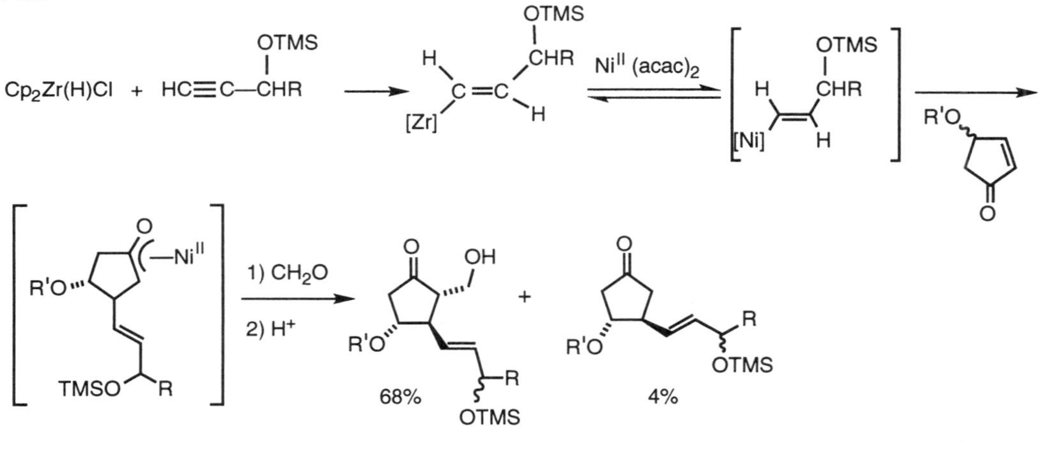

Eq. 4.38,[46] hydrozirconation is used to activate the alkyne, but the σ-alkenylzirconium(IV) species cannot directly transmetallate to copper. However,

organolithium reagents will alkylate the alkenylzirconium species which then transmetallates to copper, which then does 1,4-additions to an enone. This process can be made catalytic in the copper salt by using appropriate conditions (Eq. 4.39).[47] Copper salts also catalyze the conjugate addition of vinyl alanes (from Zr-catalyzed carboalumination of alkynes) to enones (Eq. 4.40),[48] and the addition of *alkyl*zirconium species (from hydrozirconation of alkenes rather than alkynes) to enones.[49]

Eq. 4.38

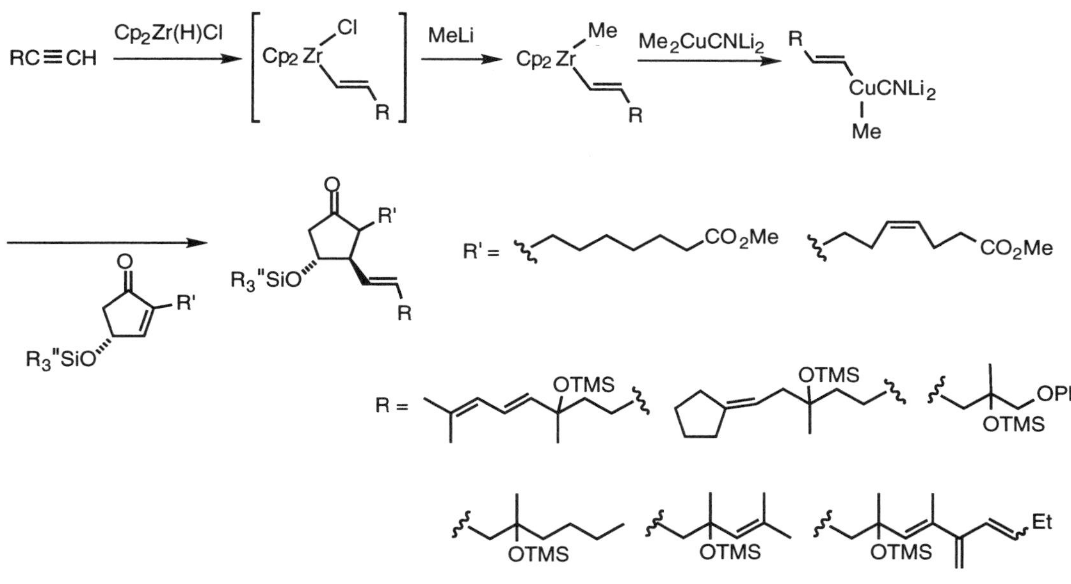

Eq. 4.39

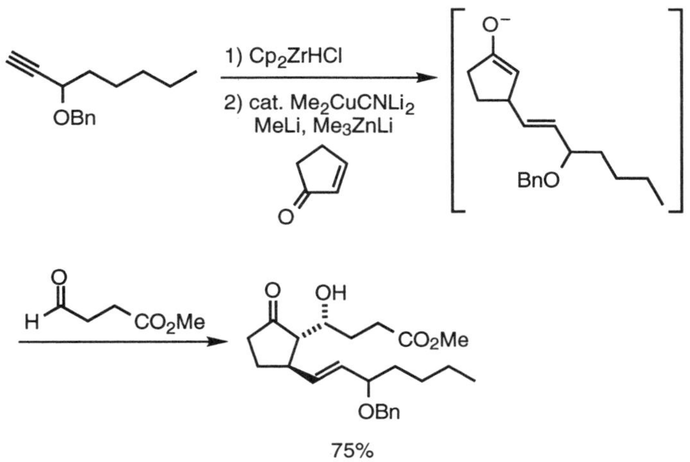

75%

Eq. 4.40

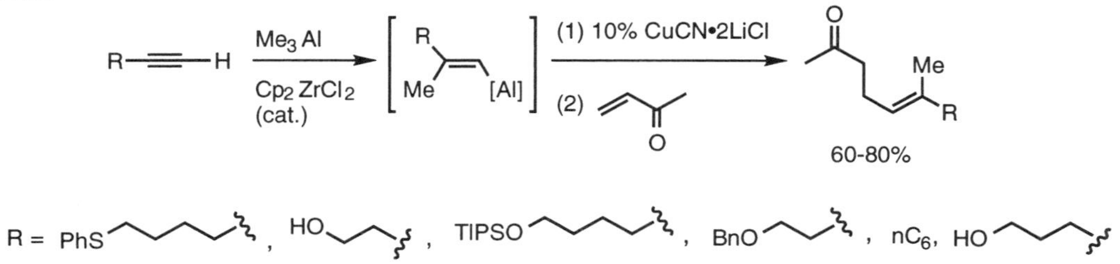

$R = PhS\diagdown\diagdown\diagdown$, $HO\diagdown\diagdown$, $TIPSO\diagdown\diagdown\diagdown$, $BnO\diagdown\diagdown$, nC_6, $HO\diagdown\diagdown\diagdown$

Transmetallation from mercury, thallium and tin to palladium(II) is also a powerful synthetic technique.[50] Aromatic compounds undergo direct, electrophilic mercuration or thallation and, although these main-group organometallics have limited reactivity, they will transfer their aryl groups to palladium(II), which has a very rich organic chemistry. When transmetallation/insertion chemistry is used, the reaction is, in most cases, stoichiometric in palladium, since Pd(II) is required for insertion, and Pd(0) is produced in the last step (Eq. 4.41). Efficient reoxidation has

Eq. 4.41

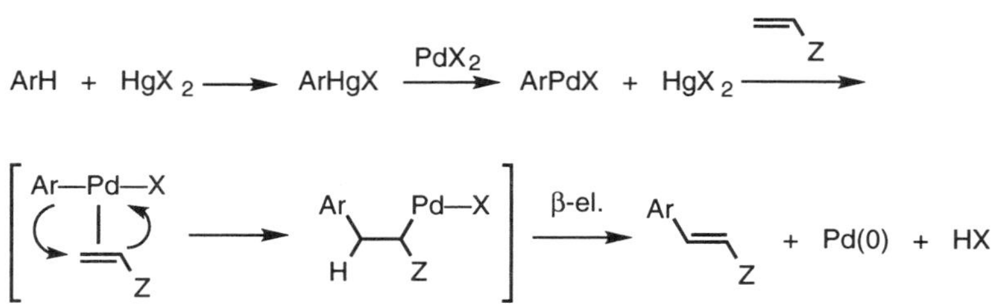

only been achieved in a few cases[50] (Eq. 4.37). However, this can still be a useful process, since Pd(0) can be recovered and reoxidized in a separate step (Eqs. 4.42[51] and 4.43 [52]). In the special case of an allylic chloride as the alkene (Eq. 4.44),[53] the

Eq. 4.42

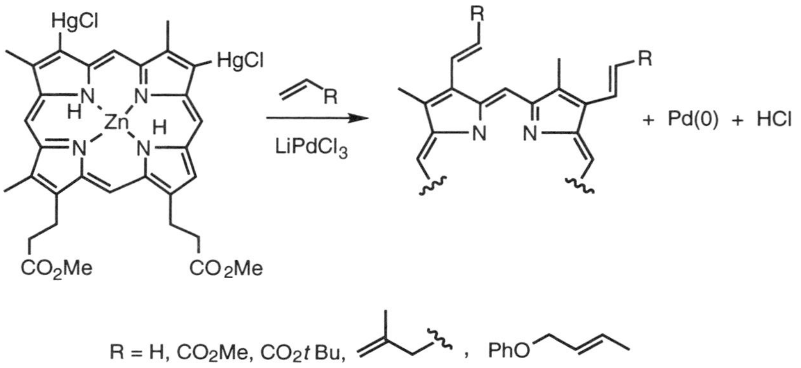

$R = H$, CO_2Me, $CO_2t\,Bu$, $\diagup\diagdown\diagdown$, $PhO\diagdown\diagup\diagdown$

Eq. 4.43

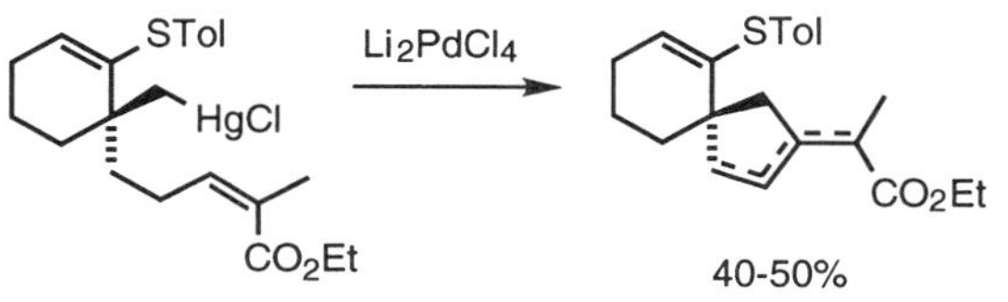

Eq. 4.44

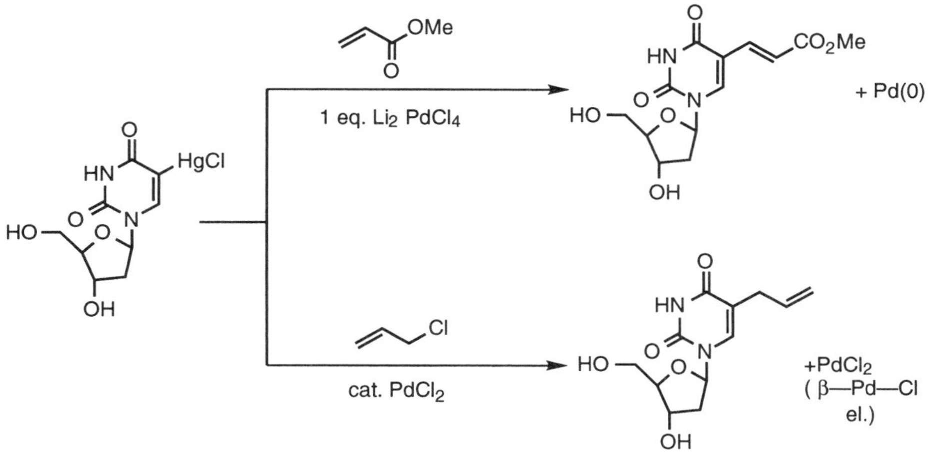

reaction *is* catalytic in Pd(II) since, in this case, β-Cl elimination to give $PdCl_2$, rather than β-hydride elimination to give Pd(0) and HCl, is more facile. Transmetallation from arylthallium(III) complexes is inefficiently catalytic (Eq. 4.45).[54] In this case, Tl(III) oxidizes Pd(0) to Pd(II). Transmetallation from tin has been used to synthesize arylglycosides (Eq. 4.46).[55] Even arylmanganese(I) complexes transmetallate to Pd(II), permitting olefination (Eq. 4.47).[56]

Eq. 4.45

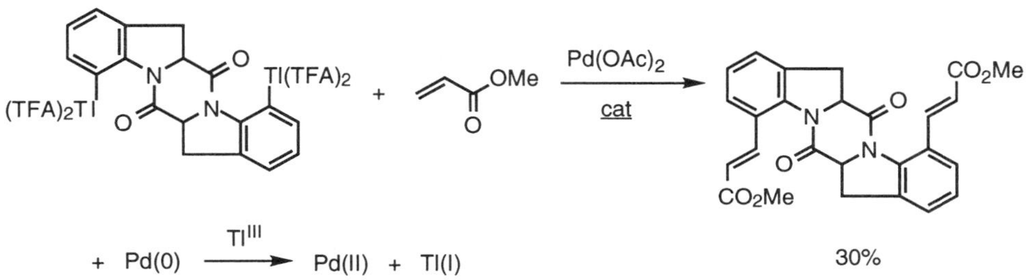

Eq. 4.46

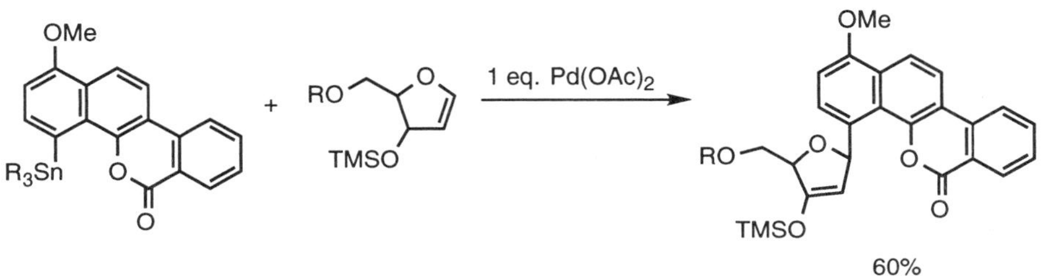

Eq. 4.47

4.5 σ-Carbon-Metal Complexes from Oxidative Addition/Transmetallation Sequences[57]

One of the most general approaches to the formation of σ-carbon metal complexes is the oxidative addition (Chapter 2) of organic halides to low-valent transition metals, most commonly palladium(0) and nickel(0) ((a) in Figure 4.5). The reaction is usually restricted to aryl and alkenyl halides, or at least halides lacking β-hydrogens, since β-hydride elimination is very fast at temperatures above -20°C, limiting what can be done with the σ-carbon metal complex. With alkenyl halides, retention of stereochemistry is observed. The order of reactivity is I > OTf > Br >>> Cl,[58] such that chlorides are rarely reactive enough to participate (although this is changing with new catalysts being developed).[59] From the viewpoint of the substrate, the metal is being oxidized and the substrate reduced. Thus electron-deficient substrates are more reactive than electron-rich substrates. The resulting (unstable) σ-carbon metal halide complex can undergo a variety of useful reactions. Treatment with a main-group organometallic results in transmetallation, (b) producing a diorganometal complex, which undergoes rapid reductive elimination (c) to give coupling (carbon-carbon bond formation) with regeneration of the low valent metal catalyst (Figure 4.5). Because reductive elimination is *faster* than β-hydride elimination, the main-group organometallic *may* contain β-hydrogens, and is not restricted to sp^2 carbon groups. Transmetallation and reductive elimination usually occur with retention of stereochemistry.

Because CO insertion into metal-carbon σ-bonds is facile at temperatures as low as -20°C, carrying out the oxidative addition under an atmosphere of CO results in the capture of the σ-carbon metal complex by CO to produce a σ-acyl species (d). Transmetallation (e) to this complex results in carbonylative coupling (f) (Figure 4.5). Both of these processes have been extensively used in synthesis.

By far the most extensively developed oxidative addition/transmetallation chemistry involves palladium(0) catalysis, and transmetallation from Li, Mg, Zn, Zr, B, Al, Sn, Si, Ge, Hg, Tl, Cu and Ni. There is an overwhelming amount of literature on the subject, but all examples have common features. A very wide range of palladium catalysts or catalyst precursors have been used (Figure 4.6) and, in many cases, catalyst choice is not critical. These catalysts include preformed, stable palladium(0) complexes such as $Pd(PPh_3)_4$ ("tetrakis") or $Pd(dba)_2/Pd_2(dba)_3 \bullet CHCl_3$ plus phosphine[60]; *in situ* generated palladium(0) phosphine complexes, made by

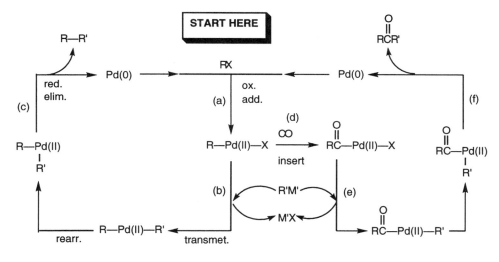

Figure 4.5 Oxidative Addition/Transmetallation

reducing palladium(II) complexes in the presence of phosphines; and "naked" or ligandless palladium(0) species, including, in some cases, palladium on carbon. (It should be noted that palladium(II) is very readily reduced to palladium(0) by alcohols, amines, CO, olefins, phosphines, and main group organometallics.[61])

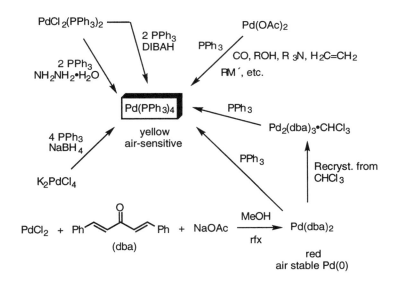

Figure 4.6 Preparation of Palladium(0) Complexes

Phosphine ligands are normally required for oxidative addition of bromides, but they suppress the reaction with aryl iodides, thus discrimination between halides is efficient. In these coupling processes, almost invariably it is the transmetallation step that is rate limiting, and is the least understood. It is important that *both* metals ultimately benefit energetically from the transmetallation step and, depending on the metal, additives are sometimes required.

One of the earliest examples of oxidative addition/transmetallation is the nickel- or palladium-catalyzed Grignard reaction. (Although Ni(II) or Pd(II) complexes are the most common catalyst precursors, the active catalyst species is the metal(0) formed by reduction of the metal(II) complex by the Grignard reagent.) Nickel phosphine complexes catalyze the selective cross-coupling of Grignard reagents with aryl and olefinic halides (Eq. 4.48).[62] The reaction works well with primary and secondary alkyl, aryl, alkenyl, and allylic Grignard reagents. A wide variety of simple, fused, or substituted aryl and alkenyl halides are reactive

Eq. 4.48

R = alkyl, aryl, hetaryl, alkenyl

substrates, but simple alkyl halides fail to react. With simple aryl and alkyl Grignard reagents, the chelating diphosphine $Ph_2P(CH_2)_2PPh_2$ complex of nickel chloride is the most effective catalyst, whereas the $Me_2P(CH_2)_2PMe_2$ complex is most suitable for alkenyl and allylic Grignard reactions, and the *bis* Ph_3P complex is best for reactions of sterically hindered Grignards or substrates.

This process is particularly useful for the coupling of heterocyclic substrates (Eq. 4.49).[63] In this case the Grignard reagent, the halide, or both can be heterocycles.

Eq. 4.49

Configurationally unstable secondary alkyl Grignard reagents can be coupled to a variety of aryl and alkenyl halides with high asymmetric induction using (aminoalkylferrocenyl)phosphines having both carbon-centered and planar chirality (Eq. 4.50).[64] This is a *kinetic resolution* of the racemic (but rapidly equilibrating) Grignard reagent — that is, one enantiomer of the Grignard reagent reacts with the optically active phosphine metal(II) σ-carbon complex faster than does the other enantiomer, and the Grignard reagent equilibrates (inverts) faster than it couples. Aryl triflates also couple efficiently to Grignard reagents. With prochiral bis triflates, optically active biaryls were produced (Eq. 4.51).[65]

Eq. 4.50

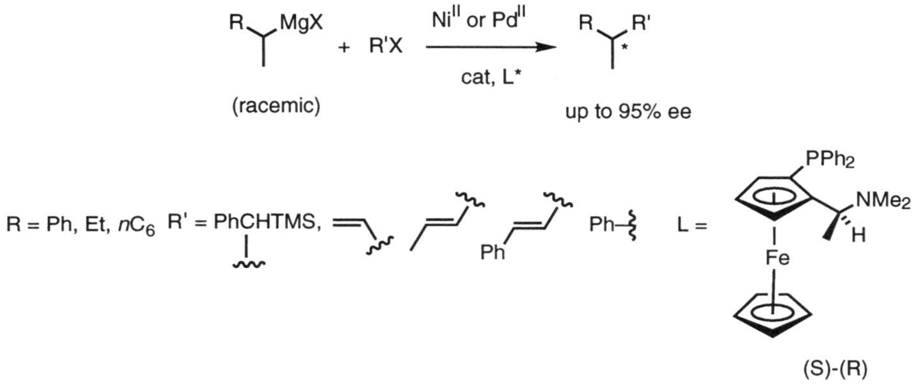

Eq. 4.51

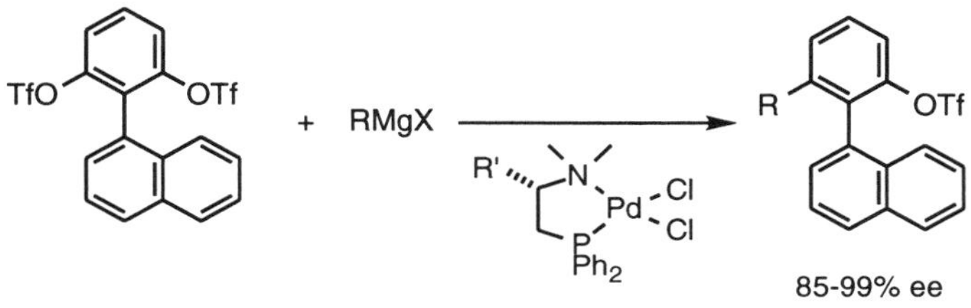

Organozinc halides are among the most efficient main-group organometallics in transmetallation reactions to palladium (Eq. 4.52),[66] while organolithium reagents are among the least. However, treatment of an organolithium reagent with one

Eq. 4.52

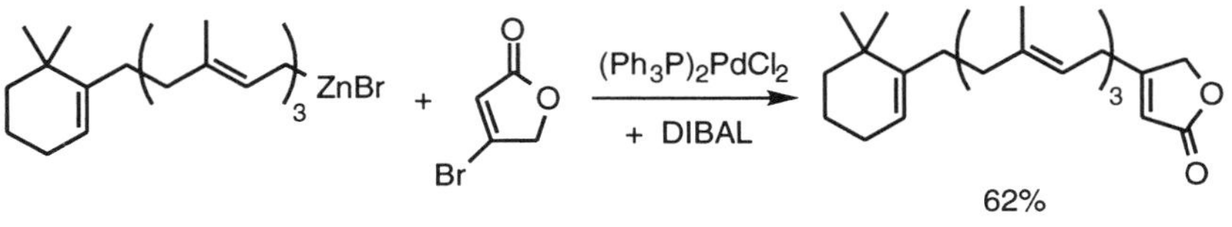

equivalent of anhydrous zinc chloride in THF generates, *in situ*, an organozinc halide which couples effectively (Eq. 4.53[67]). Recent applications of this approach to nonlinear optical materials include the synthesis of donor-acceptor oligothiophenes,[68] and oligofurans (Eq. 4.54).[69] These coupling reactions can also be readily carried out on solid supports and are potentially useful for combinatorial chemistry.[70] This facile Li to ZnCl transfer also allows the use of ligand-directed lithiation to provide coupling partners in this process. Another attractive feature of organozinc halides is that a wide variety of *functionalized* reagents are available directly from functionalized halides and activated zinc (see Eqs. 4.14 to 4.16 and

Eq. 4.53

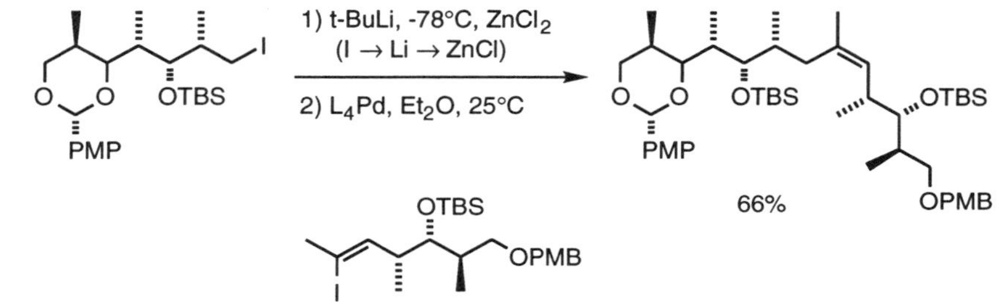

Eq. 4.54

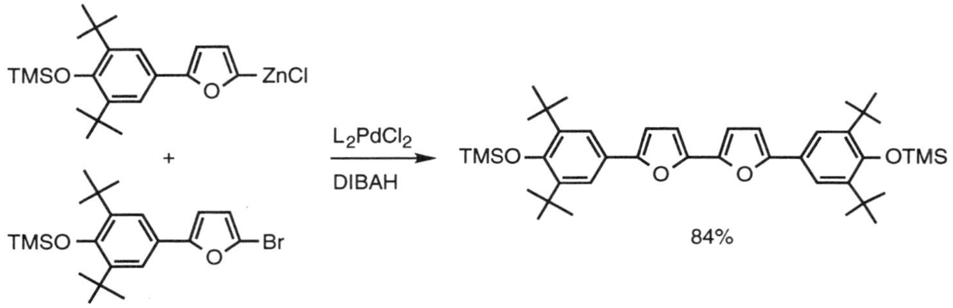

Figure 4.2) permitting the introduction of carbonyl groups, nitriles, halides, and other species into these coupling reactions.

σ-Alkenylzirconium(IV) complexes from hydrozirconation of *terminal* alkynes also transmetallate efficiently to palladium(II)-carbon complexes and are effective coupling partners in Pd(0)-catalyzed oxidative addition/transmetallation processes (Eq. 4.55).[71] However, the corresponding σ-alkenylzirconium(IV) complexes from hydrozirconation of *internal* alkynes fail entirely! Addition of zinc

Eq. 4.55

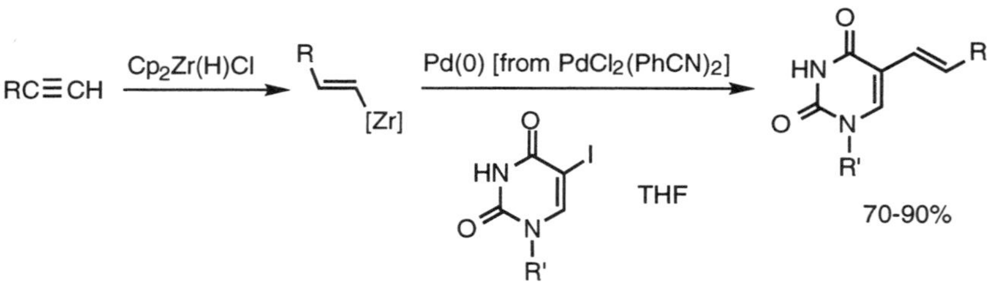

chloride to these reactions promotes this process, probably via a Zr → Zn → Pd transmetallation sequence (Eqs. 4.56[72] and 4.57[73]). This, again, illustrates the efficacy of zinc organometallics in these coupling processes.

Hydroboration is potentially a very attractive way to bring alkynes and alkenes into oxidative addition/transmetallation cycles, since the process has been the subject of long term, intense development and a truly staggering number of cases has been studied. However boron is electrophilic, and attached alkyl groups lack

Eq. 4.56

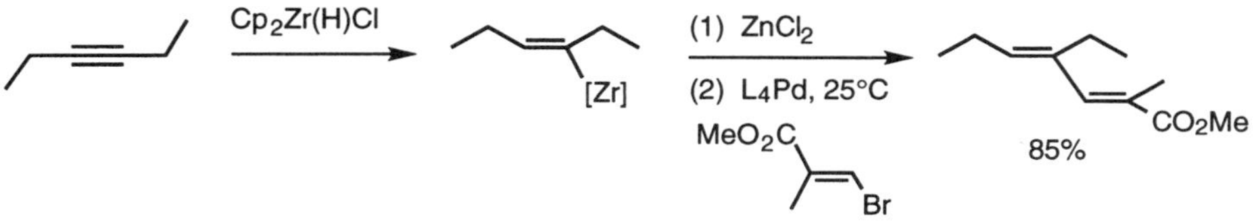

Eq. 4.57

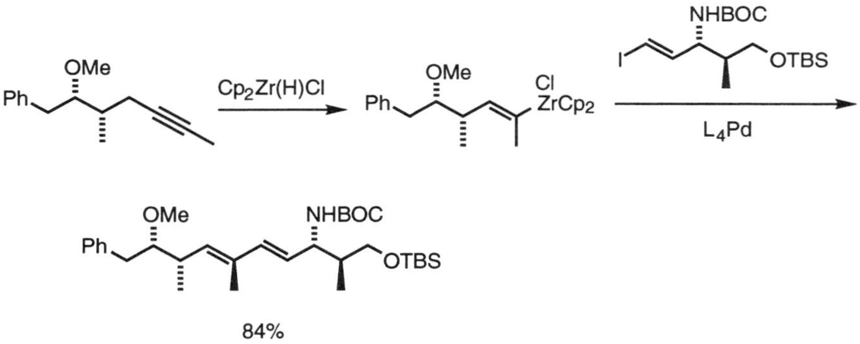

84%

sufficient nucleophilicity to transmetallate to palladium. Addition of anionic bases, such as hydroxide, to neutral organoboranes dramatically increases the nucleophilicity of the organic groups, now attached to a *borate*, and transmetallation from boron to palladium can easily be achieved. This is the basis of the "Suzuki"[74] reaction, which has, to a great extent, supplanted transmetallation from zirconium or aluminum. Thus, alkylboranes (Eq. 4.58)[75] and alkenylboranes (Eqs. 4.59[76] and 4.60[77]) couple efficiently to a wide range of highly functionalized alkenyl halides and triflates. With complex substrates the use of TlOH as base often results in an increase of both the rate and the yield of the process.[76] In the case of base-sensitive substrates, fluoride can be used in place of hydroxide in many cases,.[78,79]

Eq. 4.58

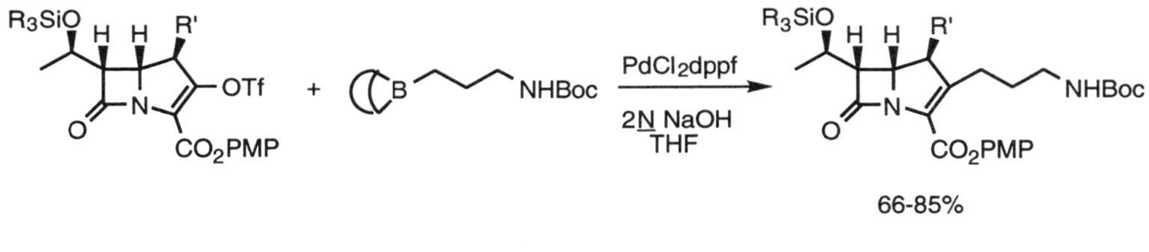

66-85%

Eq. 4.59

82%

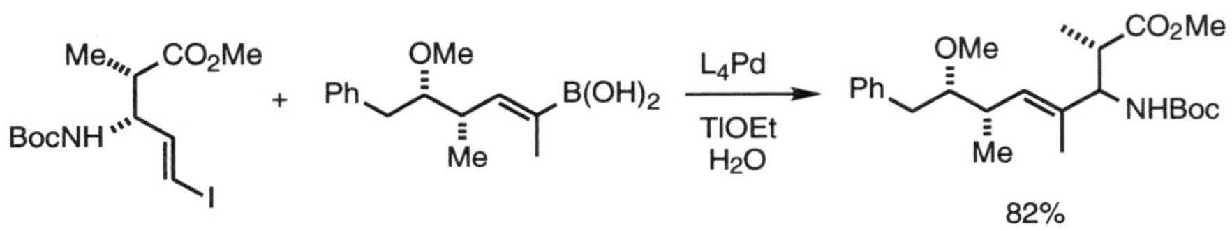

Eq. 4.60

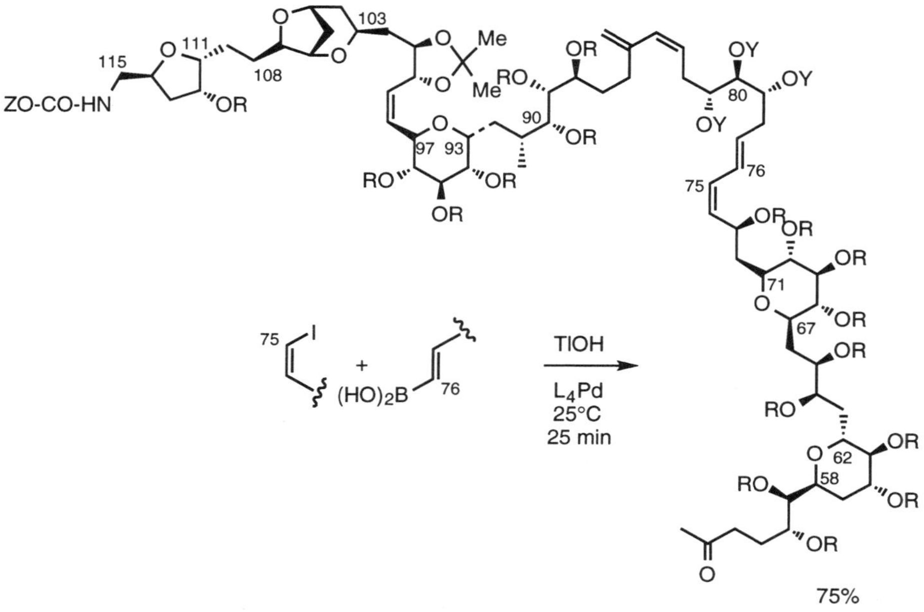

Suzuki coupling has virtually become the method of choice for making biaryls,[57] because arylboronic acids are easy to synthesize, stable to handle, and the boron-containing byproducts are water soluble and non-toxic. Highly hindered, functionalized systems couple efficiently (Eq. 4.61),[80] and the reaction works well

Eq. 4.61

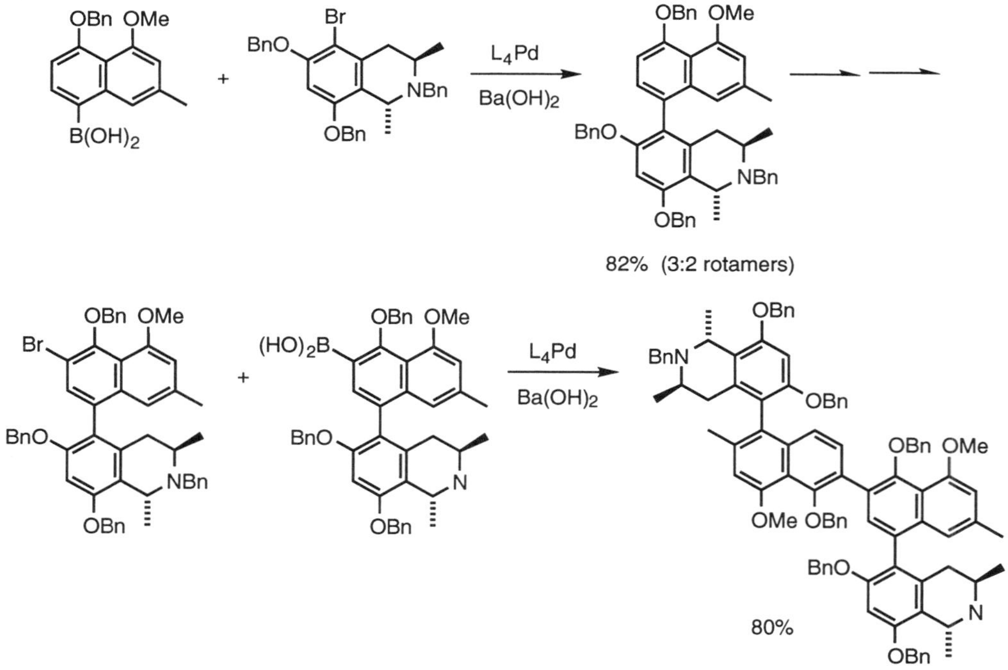

82% (3:2 rotamers)

80%

even on a very large scale (Eq. 4.62).[81] It has become very popular in polymer and materials science and has been used to make AB_2-building blocks for liquid crystal dendrimers,[82] oligophenyls,[83] and macrocyclic oligophenlines,[84] and even covalently-linked organic networks.[85] The reaction proceeds well on polymer supports, and provides the ability to produce combinatorial libraries effectively.[86]

Eq. 4.62

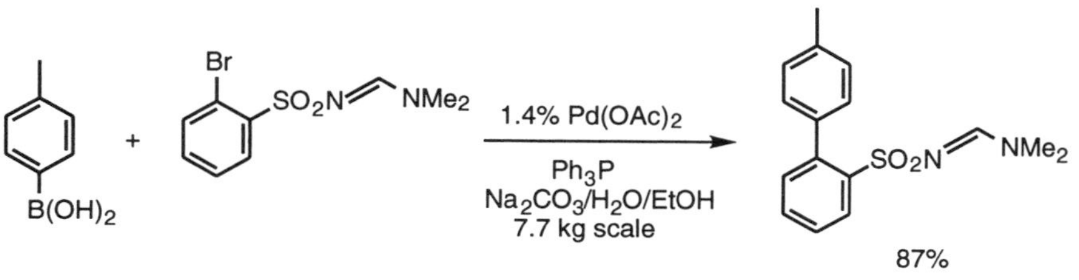

87%

By combining functional-group-directed ortho-lithiation of arenes with Suzuki coupling (Eq. 4.63),[87] a wide range of functionalized arenes, even quite hindered ones can be synthesized (Eq. 4.64)[88] and (Eq. 4.65).[89]

Eq. 4.63

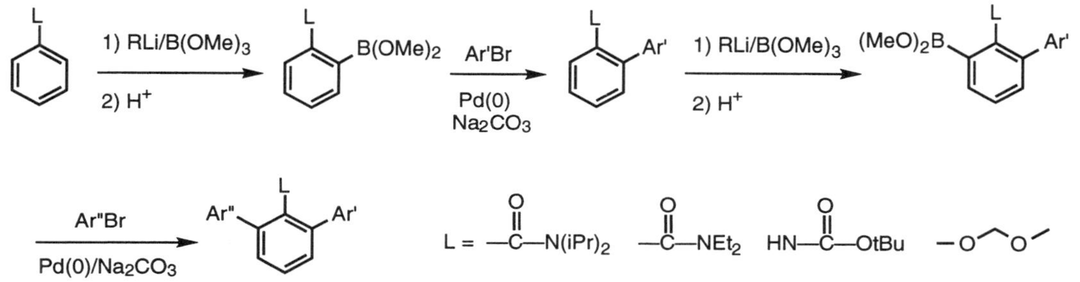

Eq. 4.64

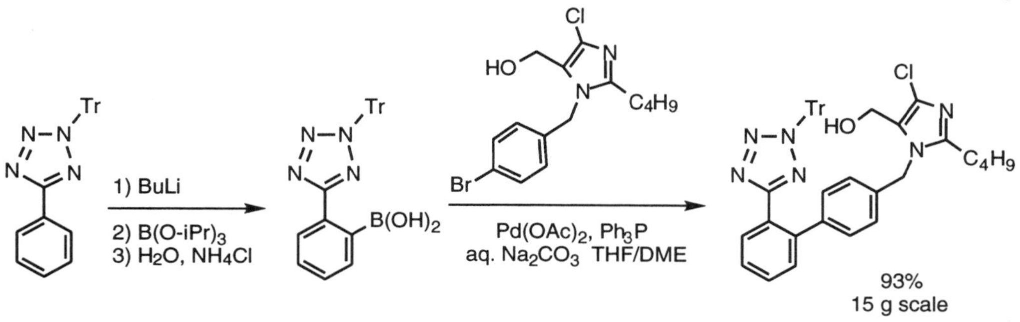

93%
15 g scale

Eq. 4.65

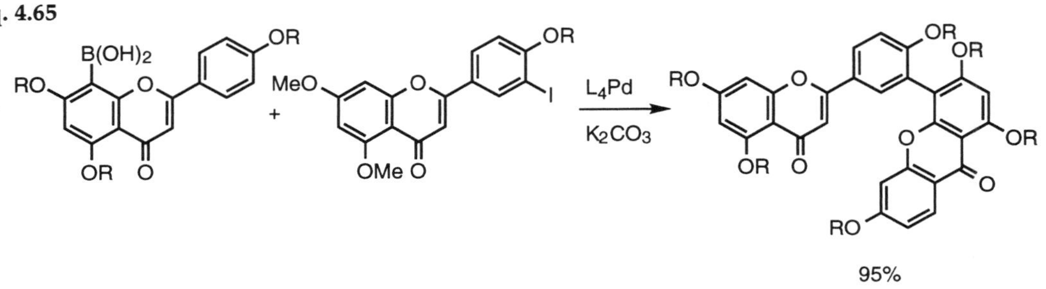

95%

Transmetallation from tin to palladium (The "Stille Reaction") is one of the most extensively utilized and highly developed palladium-catalyzed processes in organic synthesis.[90] A wide variety of structurally-elaborate organotin compounds are easily synthesized, are relatively unreactive toward a wide range of functional groups yet readily transmetallate to palladium, making them ideal reagents for complex oxidative addition/transmetallation sequences. The inherent toxicity of tin compounds and the difficulty in removing tin by-products from reactions have done little to dampen this enthusiasm, and have been somewhat mitigated by the development of tin reagents more readily removed by extraction.[91]

The rate, and ease of transfer from tin to palladium is:

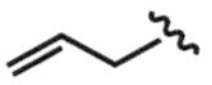

Alkenyl groups transfer with retention of olefin geometry, and allyl groups with allylic transposition. Because simple alkyl groups transfer very slowly, trimethyl or tributylstannyl compounds can be used to transfer a single R group. Transmetallation is almost invariably the rate limiting step in these processes, and relatively unreactive tin reagents often require high (>100°C) reaction temperatures and long reaction times. The use of lower donicity ligands such as Ph_3As or (2-furyl)$_3$P can often enhance transmetallation rates and thus permit the use of milder conditions,[92] as can the use of copper(I) cocatalysts.[93]

In principle, any substrate that will undergo oxidative addition with palladium(0) can be coupled to any organotin group that will transfer. Among the earliest studied substrates were acid chlorides, which are readily converted to ketones (Eq. 4.66).[94] Note that a methyl group can be transferred if it is the only

Eq. 4.66

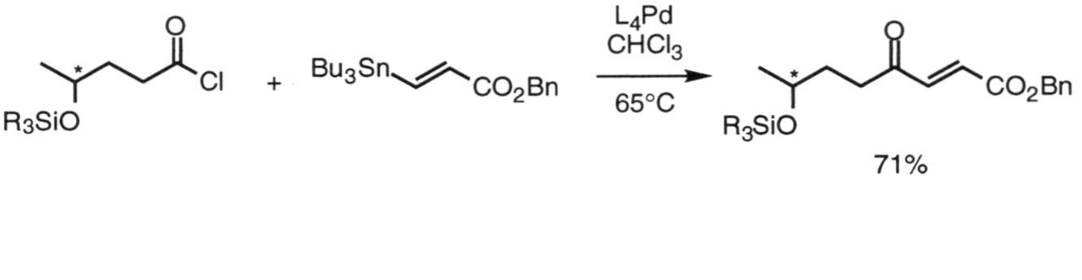

71%

group on tin and conditions are vigorous enough.[95] Macrocycles can be made using an intramolecular version of this process (Eq. 4.67).[96] (A carbon monoxide atmosphere is required to prevent competitive decarbonylation (Chapter 5) of the acid chloride.) Acid chlorides are reduced to aldehydes when tributyltin hydride is used.[97]

Eq. 4.67

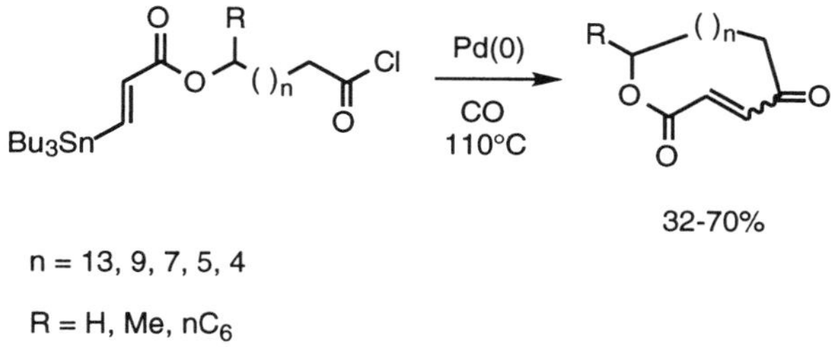

n = 13, 9, 7, 5, 4

R = H, Me, nC_6

Alkenyl halides couple efficiently with acetylenic (Eq. 4.68)[98] and simple alkenyltin compounds,[99] with retention of olefin geometry for both partners (Eq. 4.69).[100] Intramolecular versions are efficient, and a variety of macrocycles have been synthesized using this methodology (Eqs. 4.70[101] and 4.71[102]).

Eq. 4.68

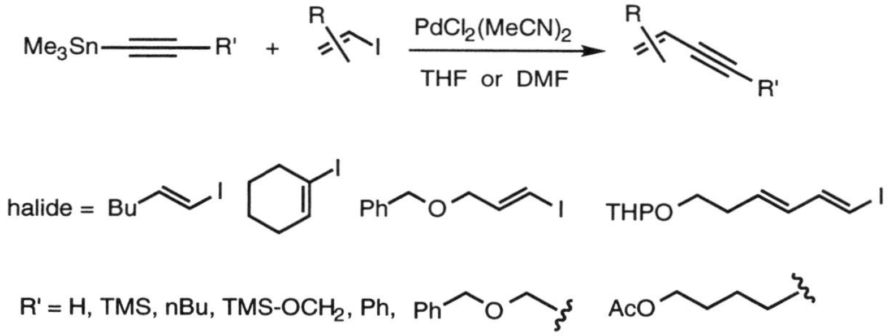

halide =

R' = H, TMS, nBu, TMS-OCH$_2$, Ph,

Eq. 4.69

90%

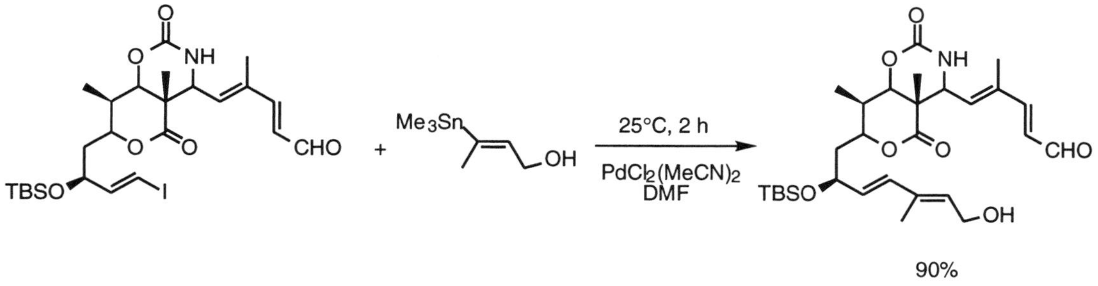

Eq. 4.70

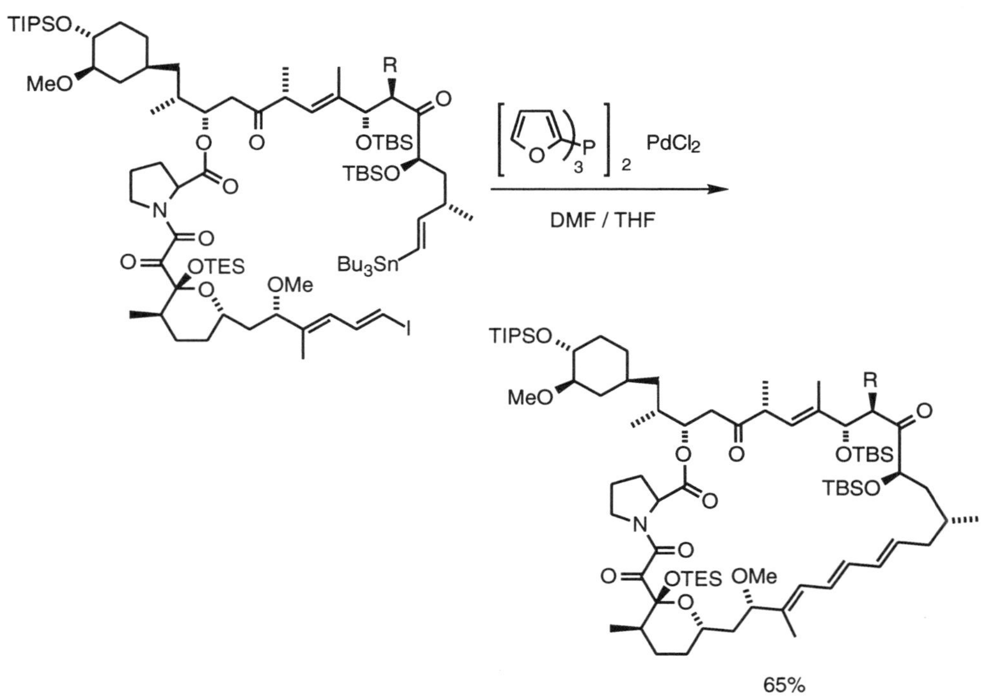

65%

Eq. 4.71

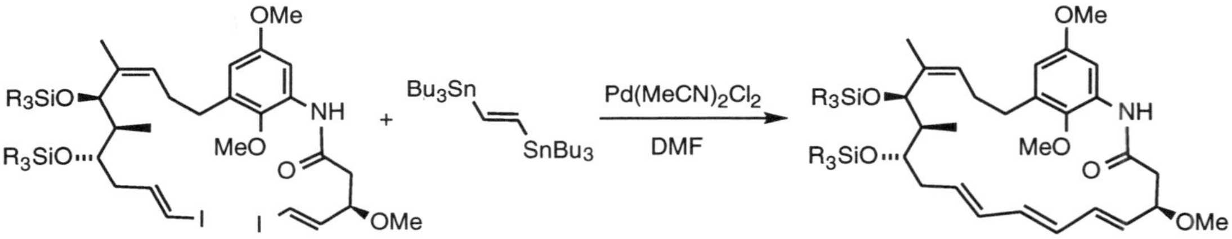

The synthetic utility of this coupling process was dramatically enhanced when conditions were found to make aryl and alkenyl triflates,[103] fluorosulfonates,[104] and phosphonates[105] participate, since this allowed phenols and ketones (via their enol triflates or fluorosulfonates) to serve as coupling substrates. As with other substrates, extensive functionality is tolerated in both partners (Eqs. 4.72[106] and

Eq. 4.72

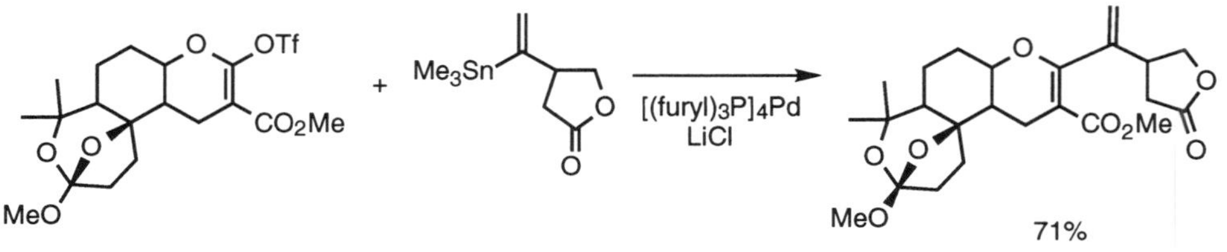

71%

4.73[107]). Triphenylphosphine catalysts often require the addition of LiCl or $ZnCl_2$ to go smoothly but the use of Ph_3As in polar solvents such as DMF obviates the need for additives and brings even unreactive aryl triflates into reaction.[108]

Eq. 4.73

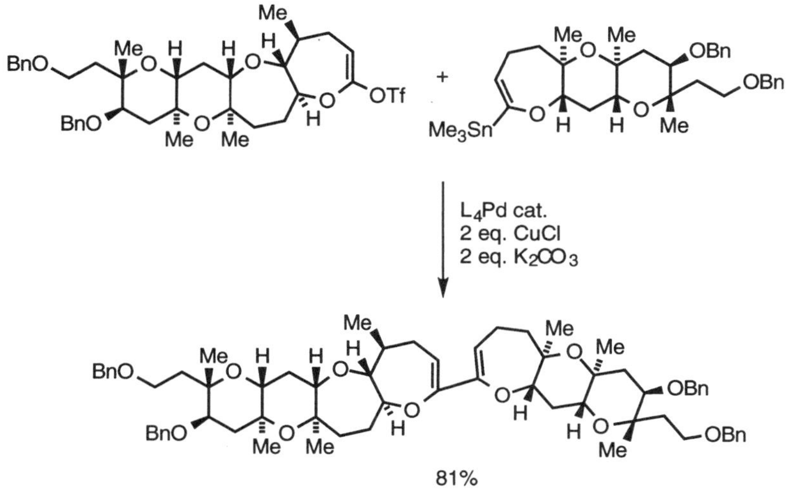

81%

Although rarely used, alkynyl halides also couple efficiently to organostannanes under palladium catalysis (Eq. 4.74).[109]

Eq. 4.74

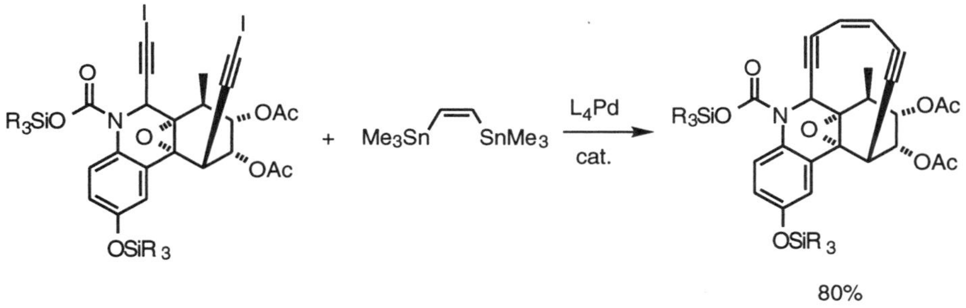

80%

Aryl halides were the first substrates to be coupled with tin reagents using palladium catalysts.[110] Although Suzuki coupling has largely supplanted Stille coupling for the synthesis of biaryls, Stille coupling continues to be extensively used for the synthesis of heteroaromatic systems (Eq. 4.75).[111] By using bis-heteroaryl (eg. thiophene) stannanes and bis aryl-[112] or heteroaryl halides,[113] oligomeric polyarenes can be efficiently synthesized. The development of efficient catalyst systems based on Pd/C[114] and the use of nickel(0) catalysts to bring aryl *chlorides*[115] into reaction has further expanded the utility of Stille coupling.

Palladium-catalyzed amination of aryl halides by stannyl amines had similarly been reported in 1983,[116] but it has recently been shown that stannyl amines are not required.[117] Rather, treatment of aryl bromide, iodides, or triflates with

Eq. 4.75

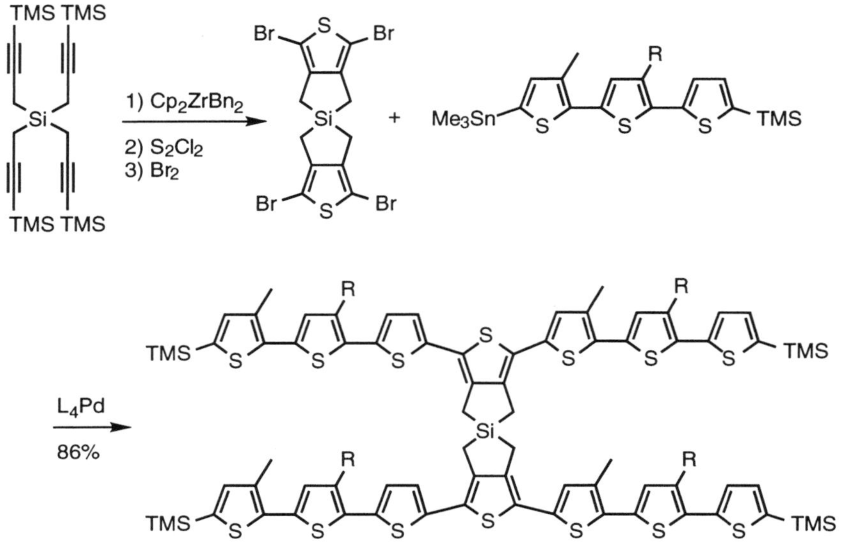

primary or secondary amines in the presence of palladium/phosphine catalysts and sodium t-butoxide[117] or cesium carbonate[118] results in efficient aryl amination (Eq. 4.76). By using activated palladium[119] or nickel[120] catalysts even aryl chlorides are aminated. As is the case with most other palladium catalyzed processes, this amination reaction is finding its way into solid-phase supported combinatorial chemistry[121] and into polymer (polyamine) chemistry by the coupling of diamines to dibromides.[122]

Eq. 4.76

ArBr / ArI / ArOTf + RR'NH $\xrightarrow[\text{or (DPPF)PdCl}_2\text{, t-BuONa, PhCH}_3\text{ 80°C}]{Pd_2(dba)_3/BINAP}$ ArNRR' (60-95%)

Ar = p-NCPh, p-MePh, p-PhPh, p-PhCOPh, p-Et_2NCOPh, o-MeOPh, m-MeOPh, o-Me_2NPh, 3,5-Me_2Ph, m-t-BuO_2CPh, 2,5-Me_2Ph, o-Me, p-MeOPh

Transmetallation from silicon to palladium is not a facile process, and alkenyl silanes can often be carried along as unreactive functional groups in other transmetallation/coupling schemes (e.g., Eq. 4.77[123]). However in the presence of a

Eq. 4.77

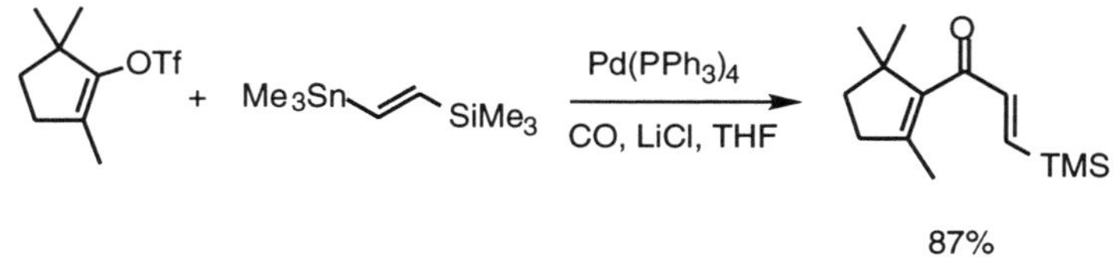

fluoride source, such as tetrabutylammonium fluoride (TBAF), (tris(diethylamino)sulfonium difluorotrimethylsilicate) (TASF),[124] or sodium hydroxide,[125] to aid in Si–C bond cleavage and to provide stable silicon products, oxidative addition/transmetallation from silicon proceeds smoothly with a wide variety of substrates (Eq. 4.78). In contrast to other main-group organometallics, both *cis* and *trans* alkenylsilanes couple with aryl halides to give *trans* products.[126]

Eq. 4.78

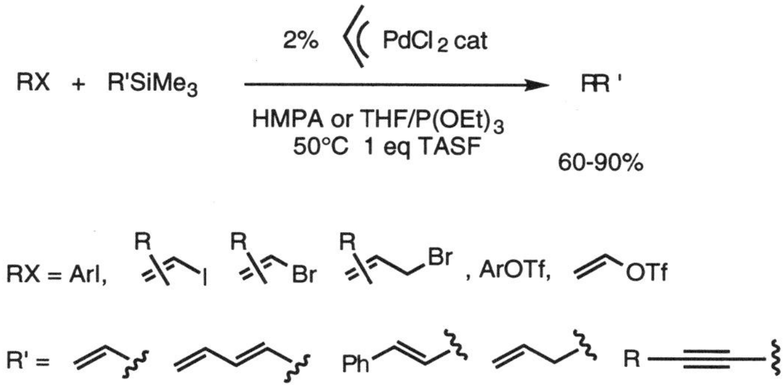

$$RX + R'SiMe_3 \xrightarrow[\substack{HMPA \text{ or } THF/P(OEt)_3 \\ 50°C \ 1 \text{ eq TASF}}]{2\% \ PdCl_2 \text{ cat}} RR'$$

60-90%

RX = ArI, , ArOTf,

R' =

Preformed organocuprates, R_2CuLi, transmetallate to palladium only slowly, and most decompose at temperatures well below that required for transmetallation. However, terminal alkynes couple to aryl and alkenyl halides upon treatment with palladium catalyst, an amine and copper(I) iodide, in a process that almost certainly involves oxidative addition/transmetallation from copper. Although the mechanism has not been studied, a likely one is shown in Eq. 4.79. This reaction has been known for a long time, but only became appreciated with the advent of the ene-diyne antibiotic, antitumor agents such as calicheamycin or esperamycin.[127] It turns out that this is an excellent reaction, which retains the stereochemistry of the alkenyl halide and tolerates functionality (Eq. 4.80[128] and Eq. 4.81[129]).

Eq. 4.79

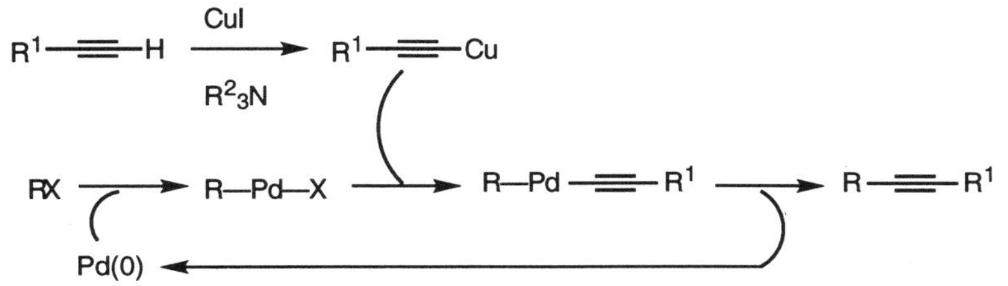

Eq. 4.80

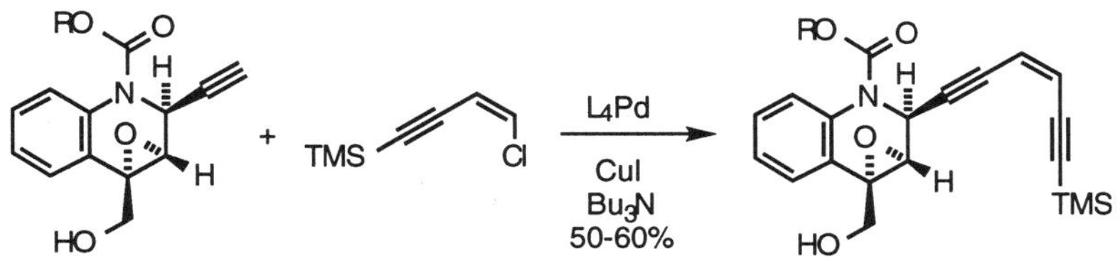

Eq. 4.81

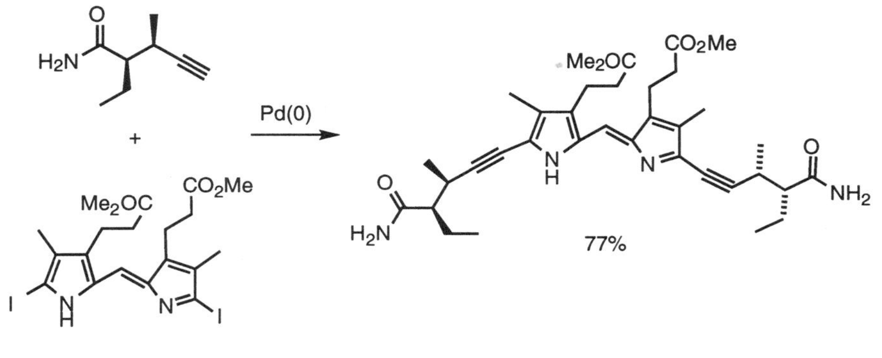

As this process has become more important in the synthesis of highly functionalized organic compounds, modified catalyst conditions have been developed including the use of Pd/C as the palladium catalyst,[130] the use of water as the reaction solvent[131] and the development of catalyst systems that do not require copper co-catalysts. By using tetrabutylammonium iodide as a promoter, aryl triflates can be made to react efficiently.[133]

As with other palladium-catalyzed processes, this one has been put to good use in the area of polymers and materials chemistry.[134] For example, using iterative iodothiophene/ethynylthiophene coupling, oligo (thiophene ethynylenes) were efficiently synthesized (Eq. 4.82).[135] Dendrimers can be made using iodoarene/ethynylbenzene couplings (Eq. 4.83).[136] Other materials applications include the synthesis of photorefractive materials by the coupling of iodocarbazoles with bis ethynylcarbazoles,[137] and the synthesis of fused polycyclic aromatics from tetrahalobenzenes and ethynylarenes.[138]

Eq. 4.82

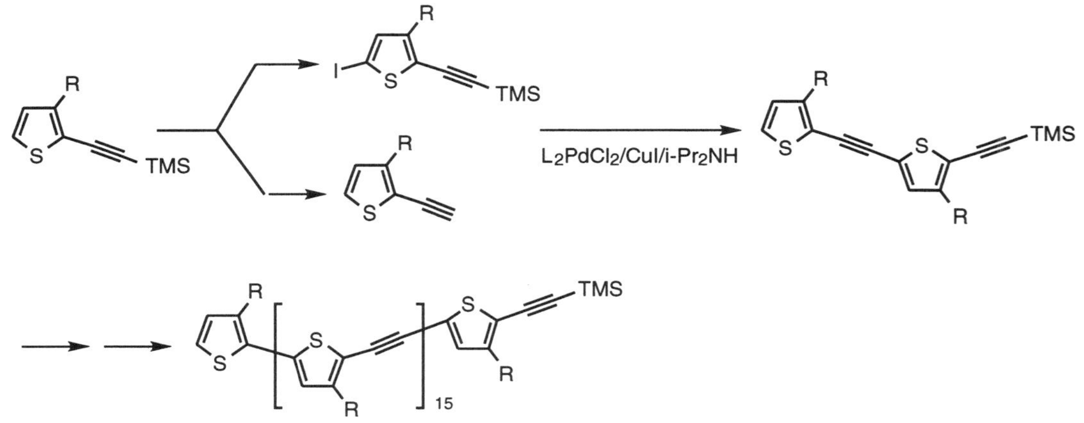

Eq. 4.83

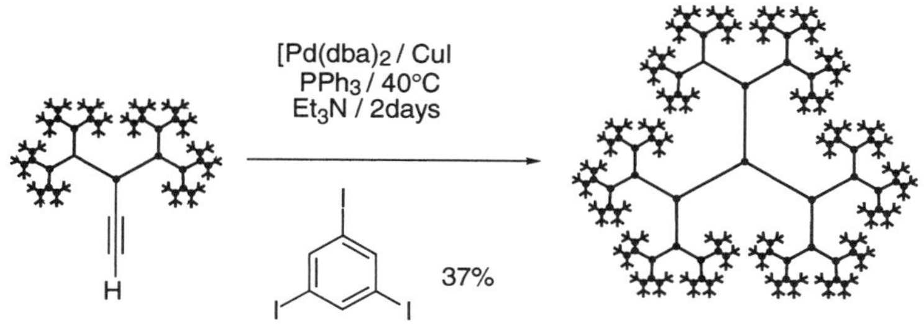

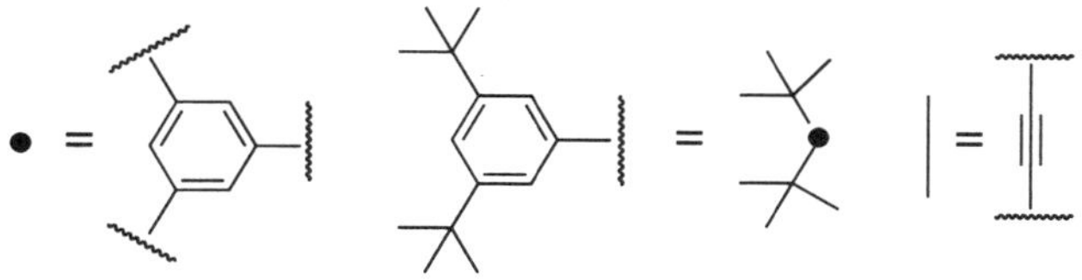

4.6 σ-Carbon-Metal Complexes from Oxidative Addition/Insertion Processes (the Heck Reaction)

Metal-carbon σ-bonds are readily formed by oxidative addition processes, and carbon monoxide and (less easily) alkenes readily insert into metal-carbon σ-bonds. These two processes form the basis of a number of synthetically useful transformations. Because of the very wide range of organic substrates that undergo oxidative addition, carbon monoxide insertion offers one of the best ways to introduce carbonyl groups into organic substrates. The general process is shown in Figure 4.7. As usual, it is limited to substrates lacking β-hydrogens, since β-hydride elimination often occurs under the conditions required for oxidative addition. Thus,

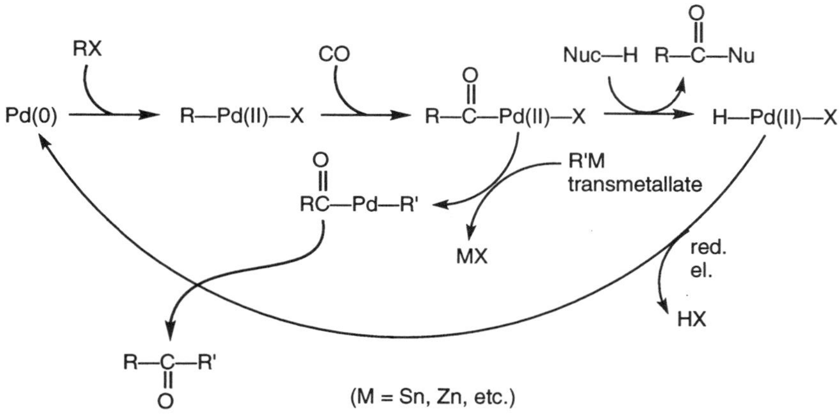

Figure 4.7 Oxidative Addition/CO Insertion

alkenyl, aryl, benzyl, and allyl halides, triflates, and fluorosulfonates[139] are readily carbonylated. Pressure equipment is usually not required, since insertion occurs readily at one atmosphere pressure. A base is required to neutralize the acid produced from the nucleophilic cleavage of the σ-acyl complex. When palladium(II) catalyst precursors are used, rapid reduction to palladium(0) by carbon monoxide ensues. Some ligand, usually triphenylphosphine, is required both to facilitate oxidative addition with bromides and to prevent precipitation of metallic palladium. Alkenyl triflates have been extensively used (Eq. 4.84),[140] and the reaction is efficient even on large scales (Eq. 4.85).[141] Because they are directly available from ketones, this provides a convenient method for the conversion of this common functional group to conjugated esters.

Eq. 4.84

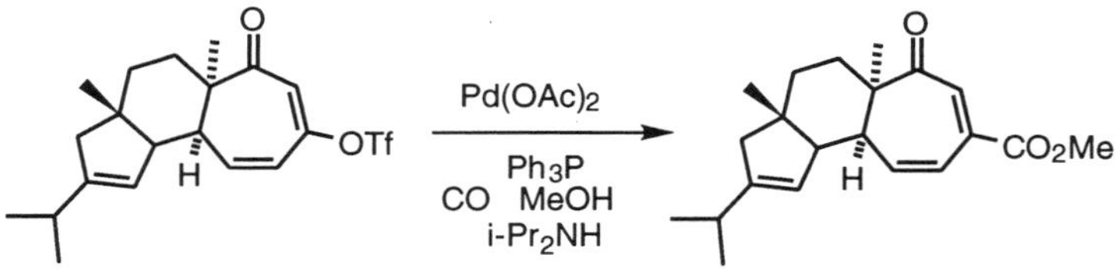

Eq. 4.85

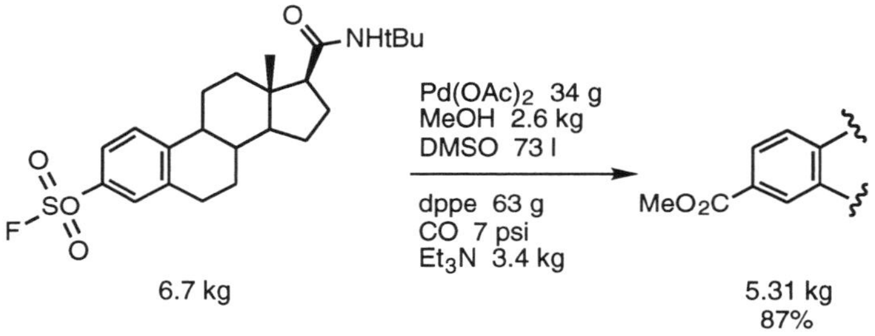

Intramolecular versions are efficient and provide lactones (Eq. 4.86),[142] lactams (Eq. 4.87)[143] and even β-lactams (Eq. 4.88).[144] These carbonylations can also be carried out on a large scale without loss of efficiency (Eq. 4.89).[145] The combination of oxidative addition/CO insertion with transmetallation results in efficient conversion of organic halides and triflates into aldehydes or ketones (Eq. 4.90),[146] (Eq. 4.91),[147]

Eq. 4.86

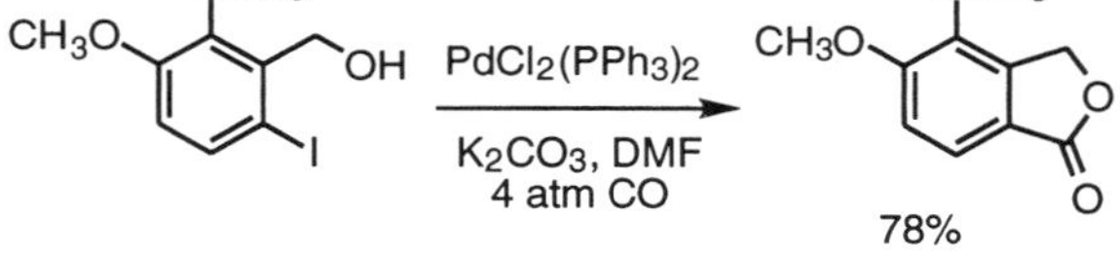

Eq. 4.87

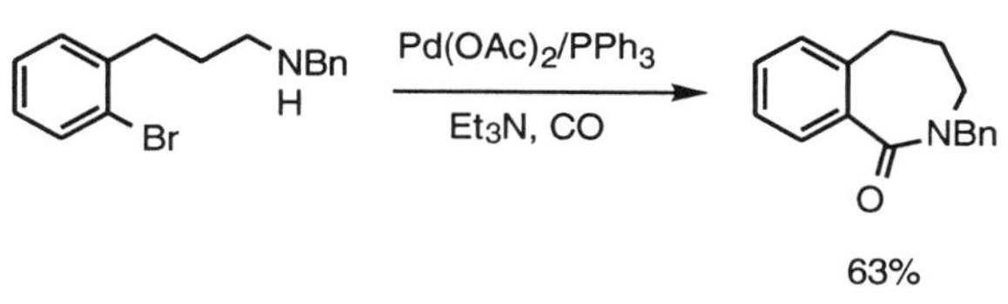

Eq. 4.88

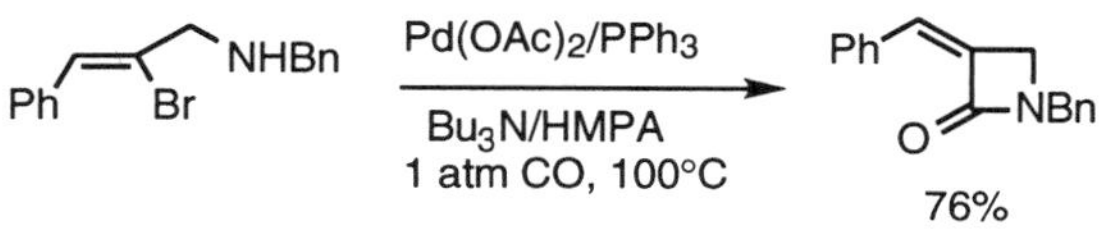

Eq. 4.89

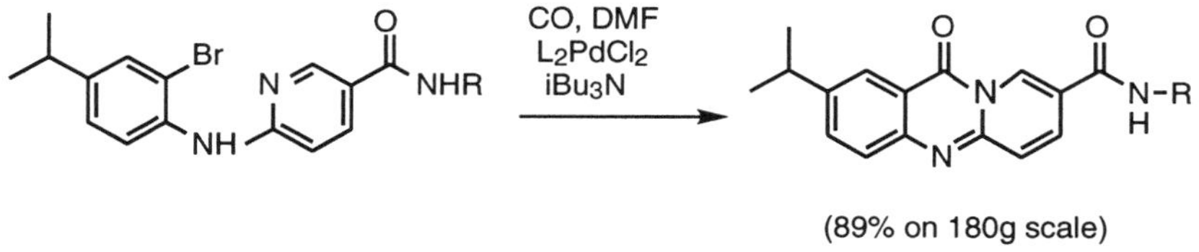

(89% on 180g scale)

Eq. 4.90

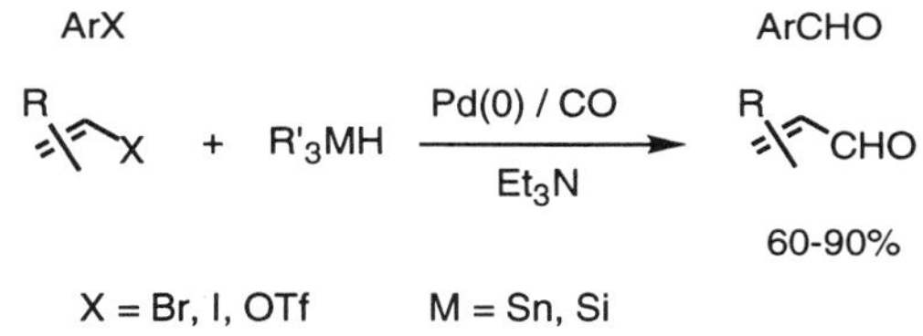

X = Br, I, OTf   M = Sn, Si

Eq. 4.91

65%

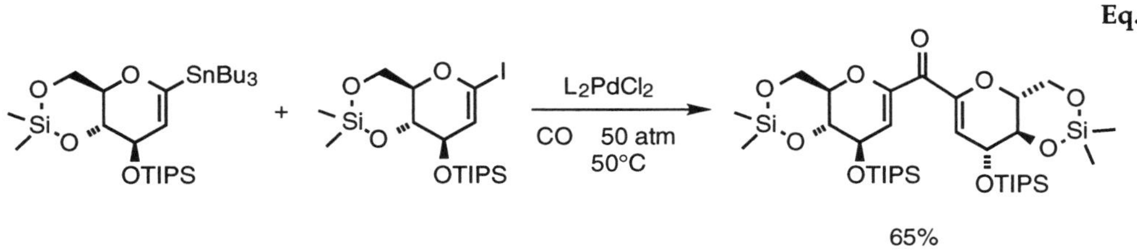

and (Eq. 4.92).[148] The requirement that CO insertion *must* preceed transmetallation can usually be met by adjusting the CO pressure (usually 50-90 psi is required, rather than atmospheric pressure).

Oxidative addition/olefin insertion sequences are even more useful for synthesis, because a greater degree of elaboration is introduced. The general process, known as the Heck reaction,[149] is shown in Figure 4.8. In this case, it is normally the insertion step which is most difficult and it is the structural features on the alkene which impose the most limitations. *All* substrates subject to oxidative addition can,

Eq. 4.92

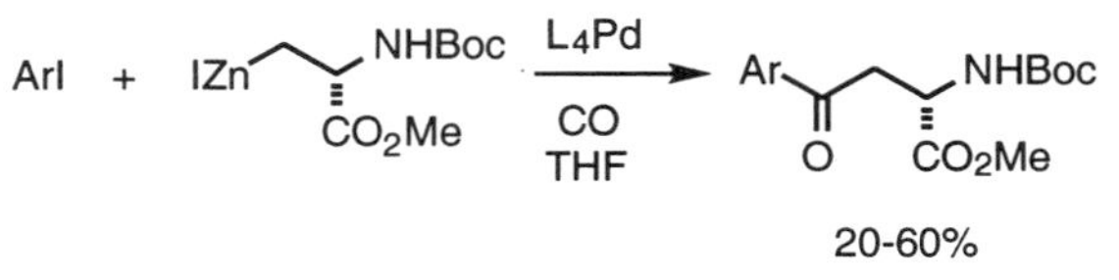

$$ArI + IZn\overset{NHBoc}{\underset{CO_2Me}{\diagdown}} \xrightarrow[\substack{CO \\ THF}]{L_4Pd} Ar\overset{\substack{}}{\underset{\substack{O}}{\diagdown}}\overset{NHBoc}{\underset{CO_2Me}{}}$$

20-60%

Ar = Ph, α-Naphth, p-MePh, o-MeOPh, m-MeOPh, p-MeOPh, o-H$_2$NPh, m-H$_2$NPh, p-FPh, o-FPh, n-FPh, o-O$_2$NPh, m-O$_2$NPh, p-O$_2$NPh

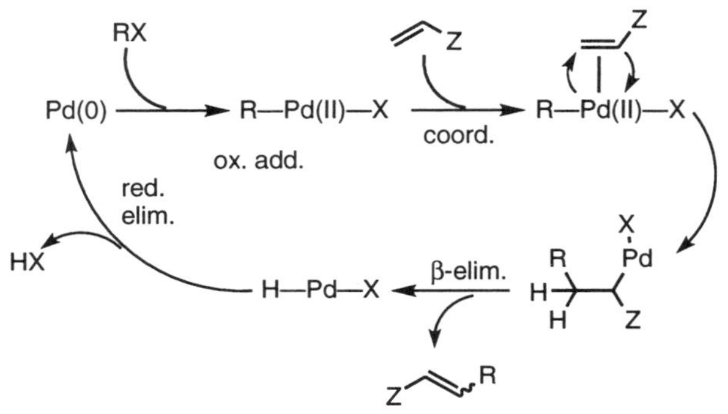

Catalysts: L$_4$Pd, Pd(dba)$_2$, PdCl$_2$L$_2$ + DIBAL,
Pd(OAc)$_2$ + red. agent (CO, CH$_2$=CH$_2$, R$_3$N, R$_3$P, ROH)

Figure 4.8 Oxidative Addition/Olefin Insertion

in principle, participate in this process, although, again, β-hydrogens are not tolerated because β-elimination competes with insertion.

Typically, the Heck reaction is carried out by heating the substrate halide or triflate, the alkene, a catalytic amount of palladium(II) acetate, and an excess of a tertiary amine in acetonitrile. The tertiary amine rapidly reduces Pd(II) to Pd(0) by coordination/β-elimination/reductive elimination (Eq. 4.93).[150] With aryl iodides as substrates, the addition of phosphine is not required, and in fact, inhibits the

Eq. 4.93

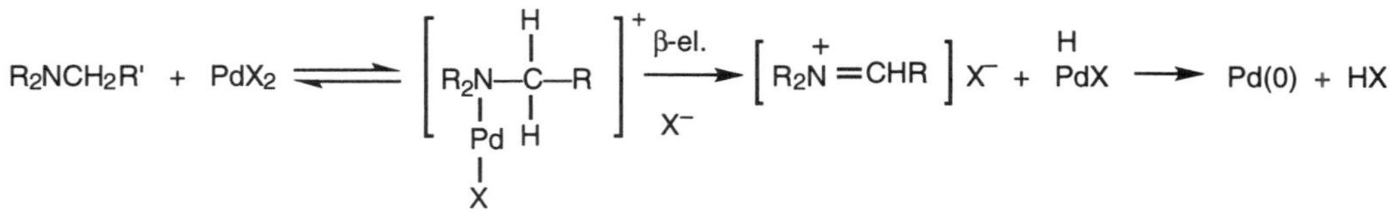

$$R_2NCH_2R' + PdX_2 \rightleftharpoons \left[R_2N\overset{\overset{H}{|}}{\underset{\underset{X}{\overset{|}{Pd}}}{C}}\overset{\overset{}{|}}{\underset{H}{}}R \right]^{+} \xrightarrow[X^-]{\beta\text{-el.}} \left[R_2\overset{+}{N}=CHR \right] X^- + \overset{H}{PdX} \longrightarrow Pd(0) + HX$$

reaction. Even palladium on carbon will catalyze Heck reactions of aryl iodides.[151] In contrast, with bromides as substrates, a hindered tertiary phosphine like tri-*o*-tolylphosphine is required to suppress quaternization. Recently, considerably milder, more efficient conditions for the Heck reaction have been developed. These involve using polar solvents such as DMF with added quaternary ammonium chlorides to promote the reaction.[152] The soluble quaternary ammonium chloride keeps the concentration of soluble chloride high, and halide ions stabilize and activate palladium(0) complexes during these reactions.[153] Quaternary ammonium sulfates also promote mild conditions, and allow the reaction to be carried out under aqueous conditions.[154] For industrial applications, related conditions which result in very high catalyst turnover have been developed.[155]

A very wide range of alkenes will undergo this reaction, including electron-deficient alkenes, like acrylates, conjugated enones and nitriles, unactivated alkenes like ethylene and styrene, and electron-rich alkenes like enol ethers and enamides. With electron-poor and unactivated alkenes the insertion step is sterically controlled and insertion occurs to place the R group at the less-substituted position. The R group is delivered from the same face as the metal (*cis* insertion). Substituents on the β-position of the alkene inhibit insertion and, in intermolecular reactions, β-disubstituted alkenes are unreactive. The β-hydrogen elimination is also a *cis* process (Pd and H must be *syn* coplanar).

With electron-rich alkenes, the regioselectivity of insertion is more complex.[156,157] With cyclic enol ethers and enamides, arylation α- to the heteroatom is always favored, but with acyclic systems mixtures are often obtained. Bidentate phosphine ligands favor α-arylation, as do electron-rich aryl halides, while electron-poor aryl halides tend to result in β-arylation. The regioselectivity of *intra*molecular processes is also difficult to predict, and seems to depend more on ring size than alkene substitution. Even "aromatic" double bonds can insert (see below).

The Heck reaction has become very popular with synthetic chemists and a very broad range of applications to complex molecule syntheses has been reported. The process enjoys a wide tolerance to functionality and normally proceeds with high regio- and stereoselectivity (Eq. 4.94)[158] and (Eq. 4.95).[159] It is efficient with

Eq. 4.94

60-91%

X = Ph, CO$_2$Me, CH$_2$OH, CH(OEt)$_2$, CH$_2$OAc, CH$_2$NHCO$_2$PMP

polymer-supported substrates and is finding increased application in combinatorial chemistry (Eq. 4.96).[160] Polymer and materials scientists have also exploited the Heck reaction, as in the synthesis of poly(pyridylvinylene phenylenevinylenes) (Eq. 4.97),[161] copolymers containing nonlinear optical chromophores having large photorefractive effects,[162] and other polyphenylenevinylene systems.[163]

Eq. 4.95

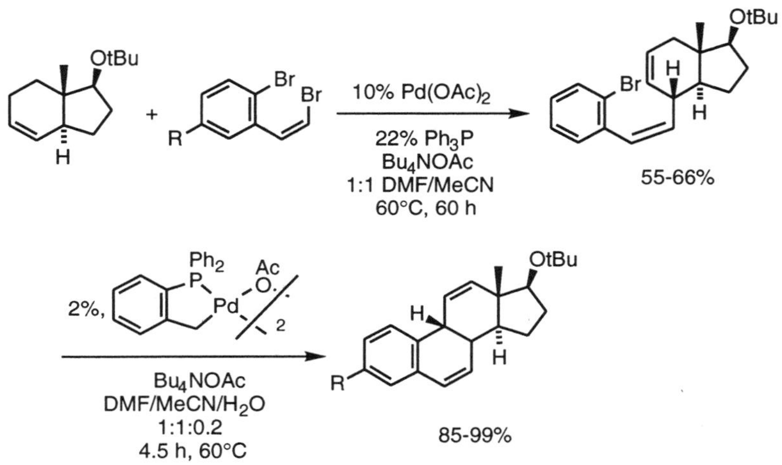

55-66%

85-99%

Eq. 4.96

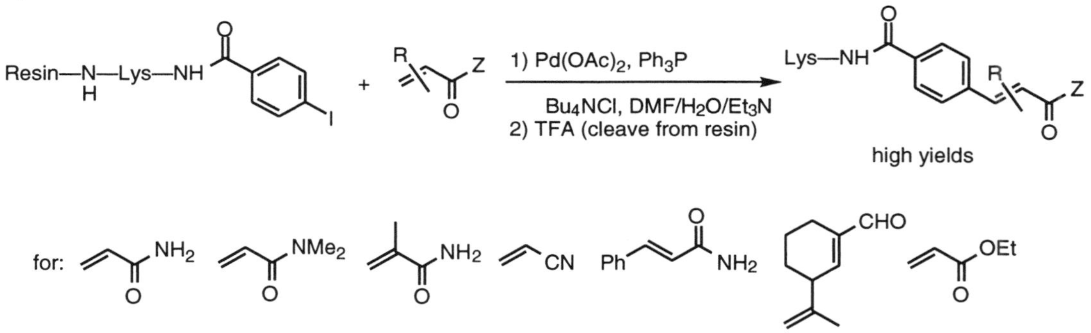

high yields

for:

Eq. 4.97

R = OC$_{16}$H$_{33}$, C$_{12}$H$_{25}$, CO$_2$C$_{12}$H$_{25}$

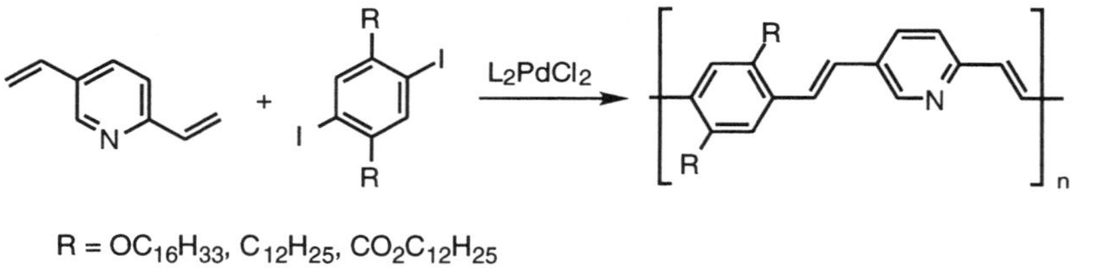

As is often the case with organometallic reactions, intramolecular versions are more efficient than intermolecular versions (Eq. 4.98).[164] Although the

Eq. 4.98

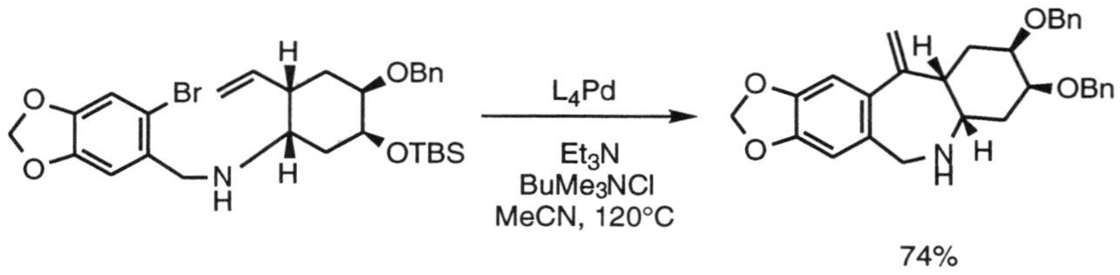

74%

regioselectivity of intramolecular Heck reactions (Eq. 4.99)[165] is usually high, the mode of addition (exo vs endo) is less predictable and is often dictated by ring size (Eq. 4.100).[166] In some cases, the regioselectivity can be altered by the choice of

Eq. 4.99

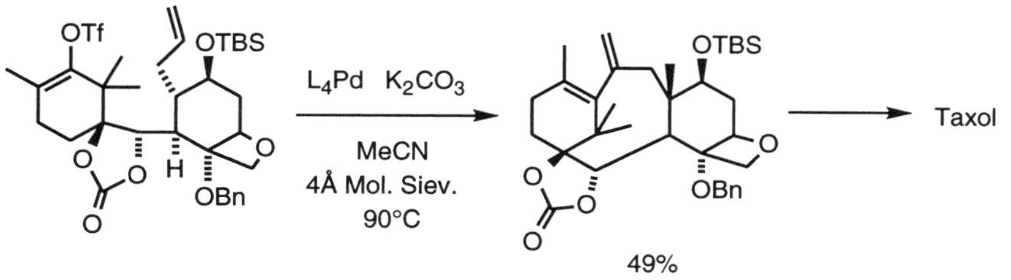

Eq. 4.100

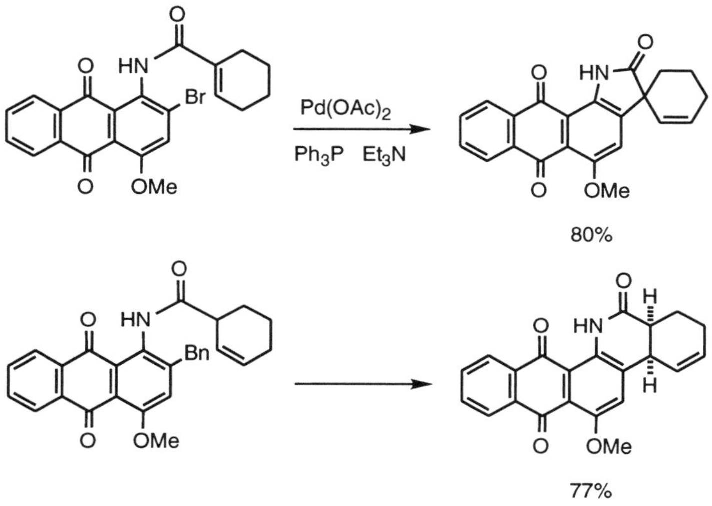

conditions (Eq. 4.101).[167] (The *stereo* chemistry of cyclization depends on the orientation of the olefin relative to the palladium-carbon σ-bond[168]). By use of high dilution or site-isolation by polymer supports, macrocycles are readily available by the Heck reaction (Eq. 4.102).[169]

A major advance came with the ability to induce asymmetry into the Heck reaction.[170] An obvious requirement for asymmetric induction is that the newly-formed stereogenic center be an sp³ center (as in Eqs. 4.95 and 4.100) rather than the more usual sp² centers (Eqs. 4.94, 4.96 and 4.102).[107d] For this to be the case, β-hydride elimination must occur *away* from the newly-formed center (β'), rather than toward it (β) (Eq. 4.103). This restricts asymmetric induction to cyclic systems or to the formation of quaternary centers since it is only in these cases that β-elimination towards the newly-formed center is prevented, either by the inability of that β-hydrogen to achieve the *syn* orientation required for β-elimination, or by the lack of that β-hydrogen. When these conditions are met, excellent enantioselectivity can be achieved (Eq. 4.104),[171] (Eq. 4.105),[172] (Eq. 4.106),[173] and (Eq. 107).[174]

Eq. 4.101

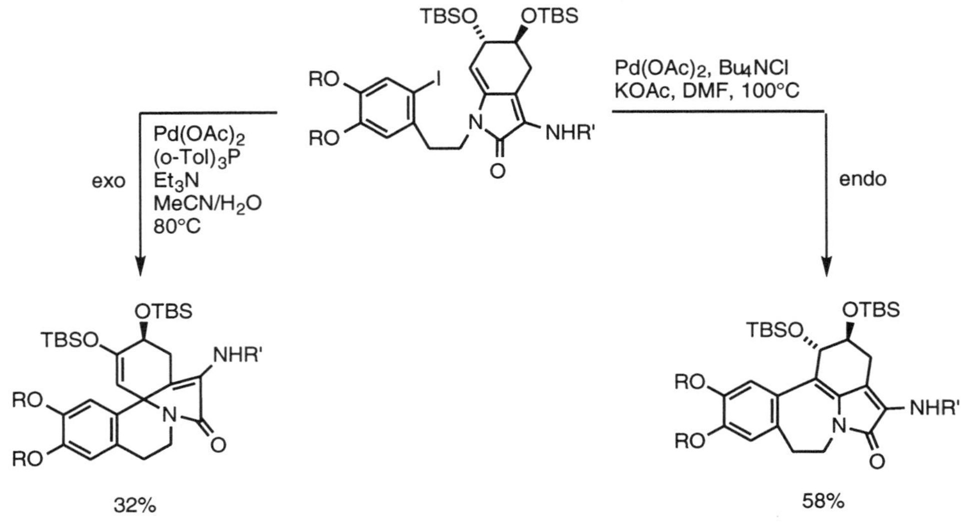

exo

Pd(OAc)₂
(o-Tol)₃P
Et₃N
MeCN/H₂O
80°C

Pd(OAc)₂, Bu₄NCl
KOAc, DMF, 100°C

endo

32%

58%

Eq. 4.102

Pd(OAc)₂

Ph₃P
Bu₄NCl
DMF/H₂O/Et₃N

78%

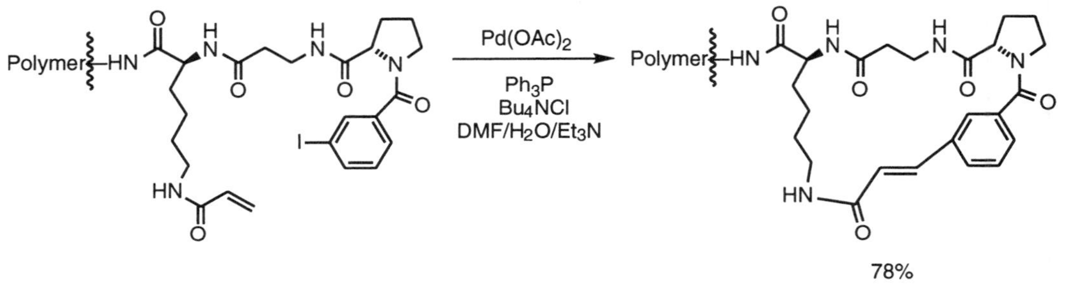

Eq. 4.103

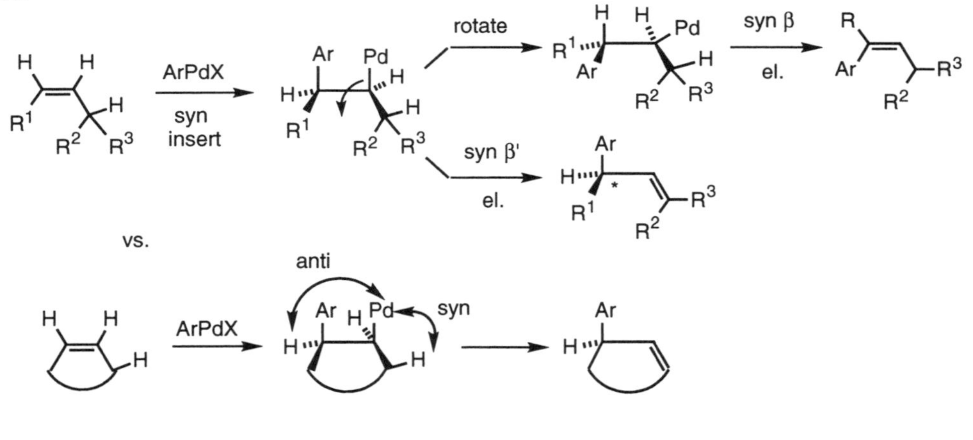

ArPdX
syn
insert

rotate

syn β
el.

syn β'
el.

vs.

anti

ArPdX

syn

100

Eq. 4.104

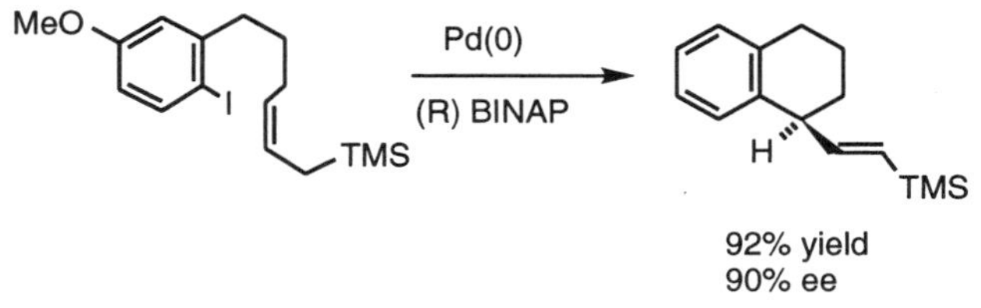

Eq. 4.105

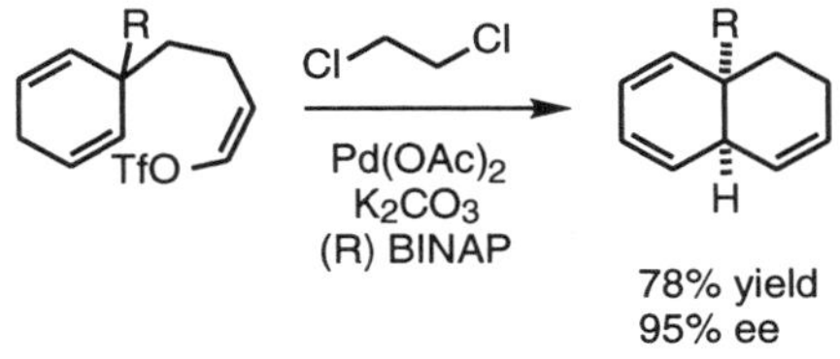

Eq. 4.106

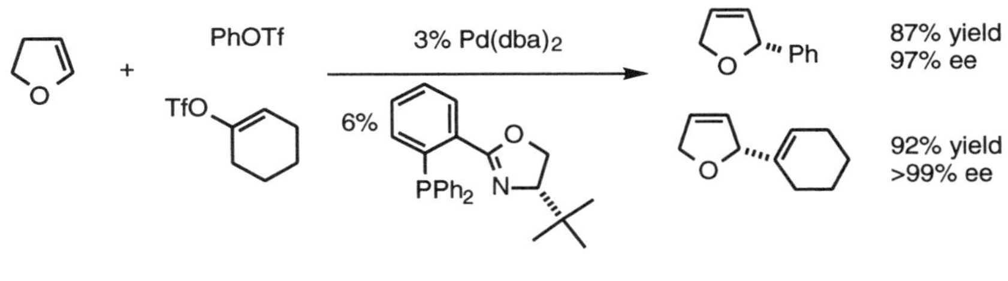

Eq. 4.107

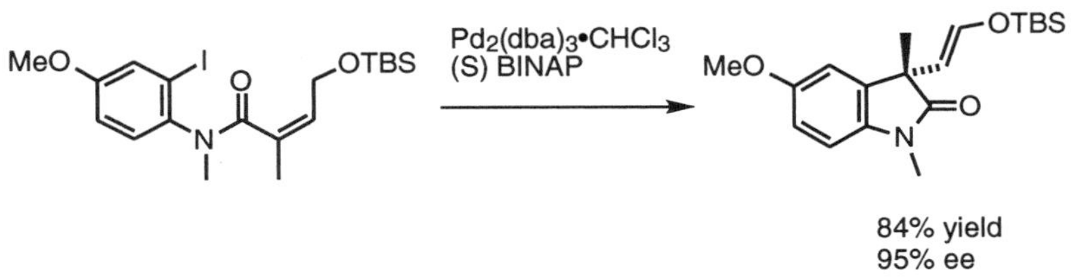

Heck oxidative addition-insertion sequences are not limited to simple olefins. Even aromatic π-systems can insert (Eq. 4.108)[175] and (Eq. 4.109).[176]

Eq. 4.108

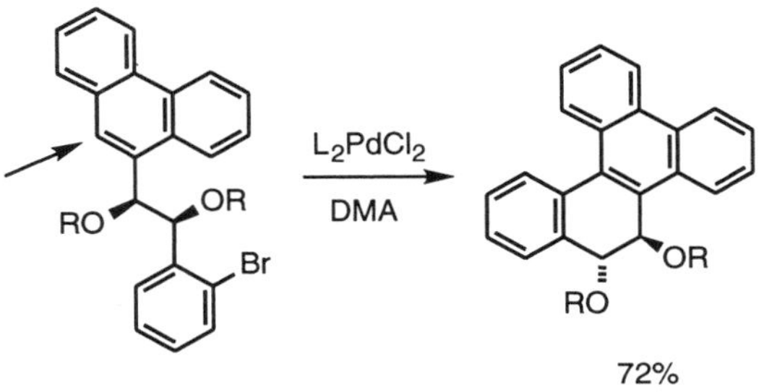

Eq. 4.109

80-90%   X = O, S

The olefin insertion step in the Heck reaction generates a new metal-carbon σ-bond which, in principle, could undergo any of the reactions of metal-carbon σ-bonds (Figure 4.1), if only β-hydrogen elimination did not occur so readily. If β-elimination could be suppressed or if a process even more favored than β-elimination (such as CO insertion) could be arranged, then the metal could be used more than once, and several new bonds could be formed, in sequence, in a single catalytic process. This is, indeed, possible, and these "cascade" reactions are becoming powerful tools in the synthesis of polycyclic compounds.[177]

For example, the reaction in Eq. 4.110[178] involves an oxidative addition/CO insertion/olefin insertion/CO insertion/cleavage process that forms three carbon-

Eq. 4.110

75%

carbon bonds per turn of catalyst. This process works because the initial oxidative adduct has no accessible β-hydrogens, and CO insertion is faster than olefin insertion for σ-alkylpalladium complexes but slower for σ-acyl complexes. The major challenge in the system was to make the final CO insertion into a σ-alkyl-palladium(II) complex, which does have β hydrogens, competitive with β-elimination. This was achieved by using a relatively high pressure of CO. Indeed, under one atmosphere of CO, β elimination was the main course of the reaction. This downhill sequence of events is critical in these multiple cascade reactions, and when met, an impressive number of bonds per catalyst cycle can be achieved (Eq. 4.111)[179] and (Eq. 4.112).[180]

Eq. 4.111

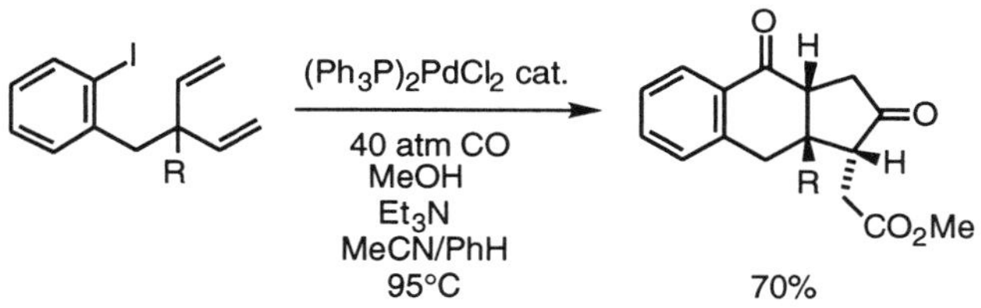

70%

Eq. 4.112

82% yield
68% ee

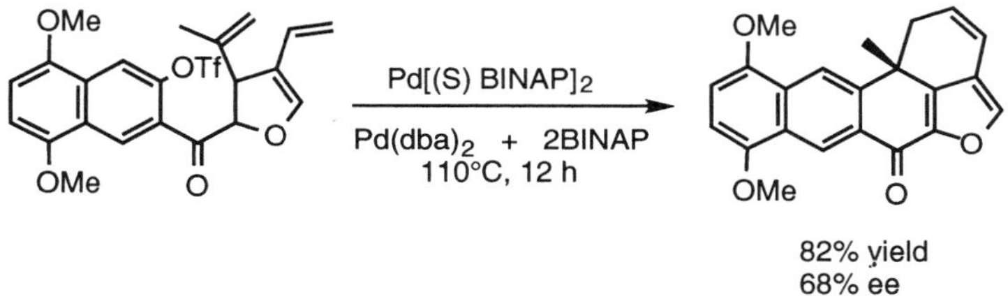

Alkynes insert more readily than alkenes, and a variety of "carbocyclization" procedures[181] involving this functional group have been developed (Eq. 4.113),[182] (Eq. 4.114),[183] and (Eq. 4.115).[184]

Eq. 4.113

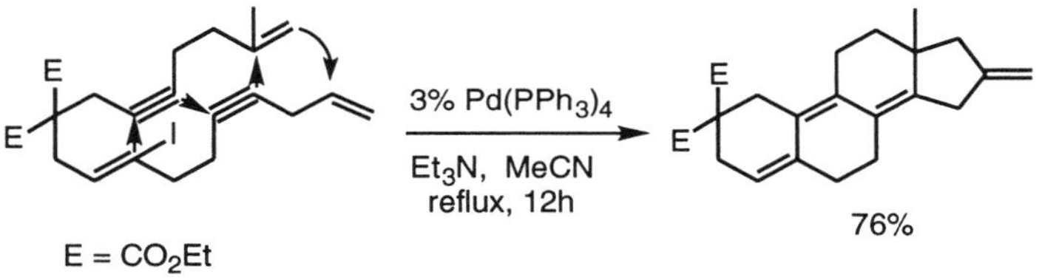

E = CO$_2$Et

76%

Eq. 4.114

71%

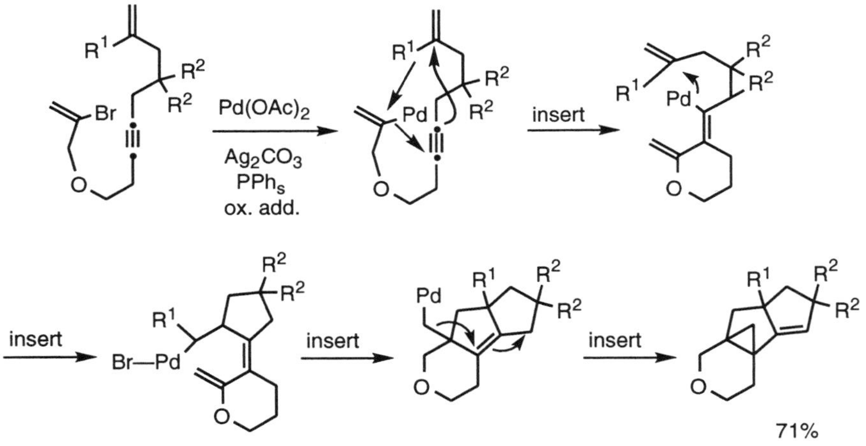

Eq. 4.115

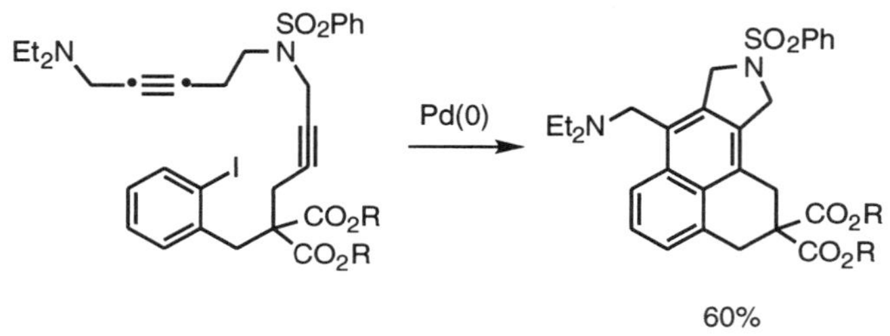

60%

In most of the above cases, the last step of the cascade was a β-hydrogen elimination. However, truncation by transmetallation (Eq. 4.116)[185] or nucleophilic attack (Eq. 4.117)[186] is also possible. Almost endless variations of these themes are possible, and most are likely to be attempted.

Eq. 4.116

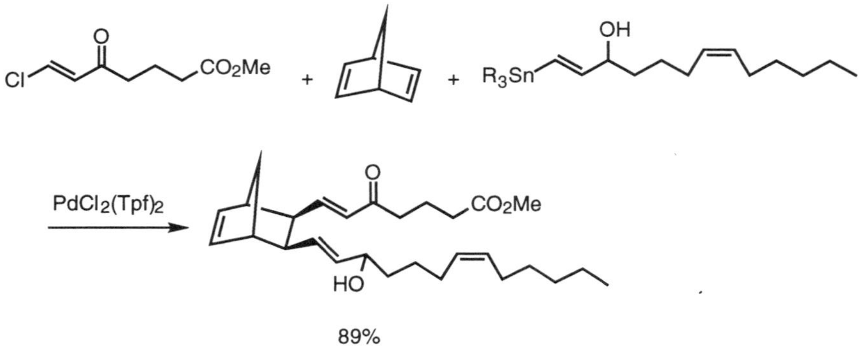

89%

Eq. 4.117

72%

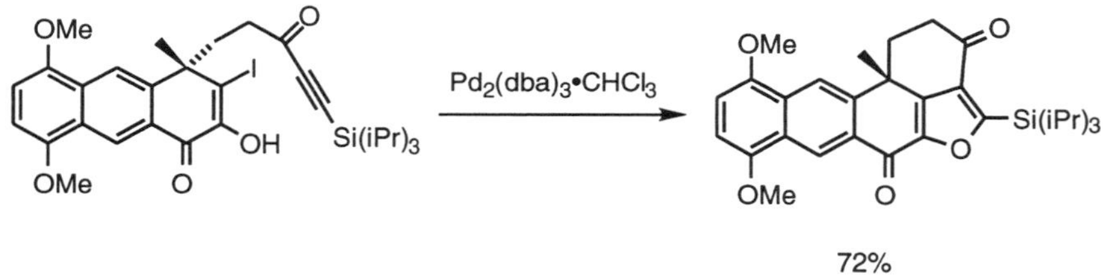

4.7   σ-Carbon-Metal Complexes from Ligand-Directed Cyclometallation Processes

Palladium(II) salts are reasonably electrophilic, and under appropriate conditions (electron-rich arenes, Pd(OAc)$_2$, boiling acetic acid) can directly palladate arenes via an electrophilic aromatic substitution process. However this process is neither general nor efficient, and thus has found little use in organic synthesis. Much more general is the ligand-directed orthopalladation shown in Eq. 4.118.[187] This process is very general, the only requirement being a lone pair of electrons in a benzylic position to precoordinate to the metal, and conditions conducive to electrophilic aromatic substitution (usually HOAc solvent). The site of palladation is always ortho to the directing ligand and, when the two ortho positions are inequivalent, palladation results at the sterically less congested position, regardless of the electronic bias. (This is in direct contrast to *o*-lithiation, which is electronically controlled and will usually occur at the more hindered position on rings having several electron donating groups.) The resulting chelate-stabilized, σ-arylpalladium(II) complexes are almost invariably quite stable, are easily isolated, purified and stored, and are quite unreactive. This has limited their use in synthesis and has prevented successful catalytic processes involving *o*-palladation.

Eq. 4.118

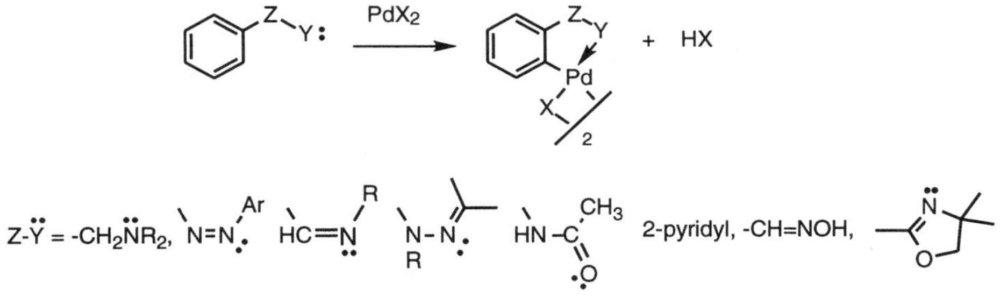

These orthopalladated species *can* be made to react like "normal" σ-aryl metal complexes by decomplexing the ortho ligand, usually accomplished by treating the complex with the substrate and a large excess of triethyl amine, to compete with the ortho ligand. Under these conditions alkene insertion (Eq. 4.119)[188] and CO insertion (Eq. 4.120)[189] are feasible. Although catalysis has not yet been

Eq. 4.119

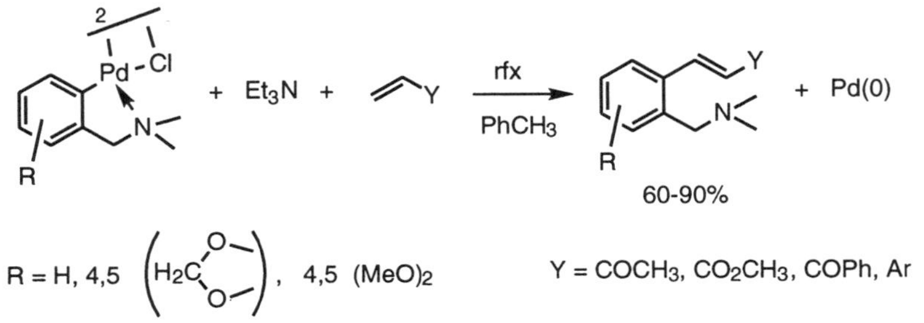

Eq. 4.120

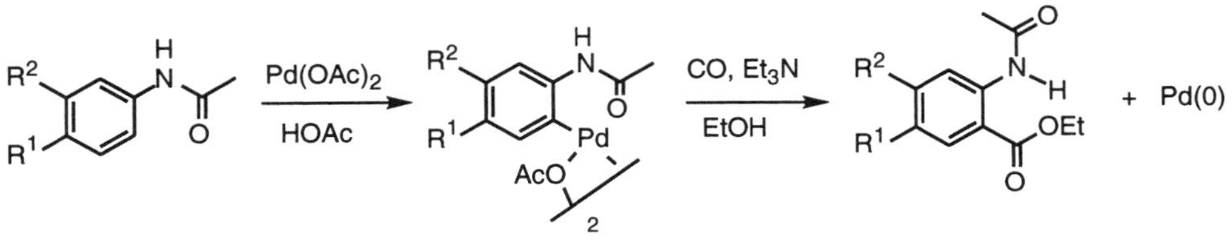

achieved, the potential power of ligand-directed metallation is illustrated in Eq. 4.121,[190] in which an unactivated methyl group was palladated solely because it had the misfortune of being within the range of a ligand-directed palladation.

Eq. 4.121

4.8　σ-Carbon-Metal Complexes from Reductive Cyclodimerization of Alkenes and Alkynes

Low-valent metals from the two extremes of the transition series (Ni, Ti, Zr) undergo reaction with alkenes and alkynes to form five-membered metallacycles in which the unsaturated substrates are formally reductively coupled, and the metal is formally oxidized (Eq. 4.122). These metallacycles can be converted to a number of interesting organic compounds.

Eq. 4.122

The most extensively studied complex to effect this reaction is "zirconocene," "Cp_2Zr," an unstable, unsaturated, d^2 Zr(II) complex.[191] It can be generated in many ways but the most convenient is to treat Cp_2ZrCl_2 with butyllithium and allow the unstable dibutyl complex to warm to above -20°C in the presence of substrate (Figure 4.9). This generates a zirconium-butene complex which can exchange with substrate alkene or alkyne and promote the cyclodimerization. The zirconium butene complex can be considered either as a Zr(II)-alkene complex or a Zr(IV) metallacyclopropane complex. These are the two extreme bonding forms for metal alkene complexes, with the metallacyclopropane form denoting complete two electron transfer from the metal to the olefin. For the zirconium butene complex, the metallacyclopropane form is probably the more accurate representation, based on its reaction chemistry and the isolation of a stable PMe_3 complex. Given that "Cp_2Zr" in fact reductively dimerizes alkenes and alkynes (Eq. 4.122) (2 e⁻ transfer from Zr(II) to alkenes), it is not surprising that the zirconium butene complex also involves extensive electron transfer to the alkene.

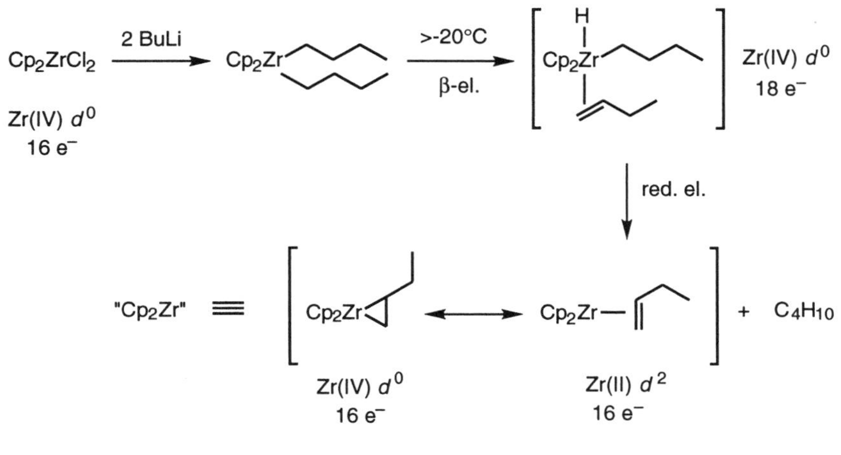

Figure 4.9 Formation of Zirconocene

The five-membered zirconacycles formed by cyclodimerization of alkenes can be transformed into a variety of interesting products. Protolytic cleavage produces the alkane, iodine cleavage the diiodide, and treatment with carbon monoxide and iodine, the cyclopentanone (Eq. 4.123).[192] Cyclodimerization of alkynes produces zirconacyclopentadienes, which undergo electrophilic cleavage to produce a number of unusual heterocycles (Eq. 4.124).[193]

Eq. 4.123

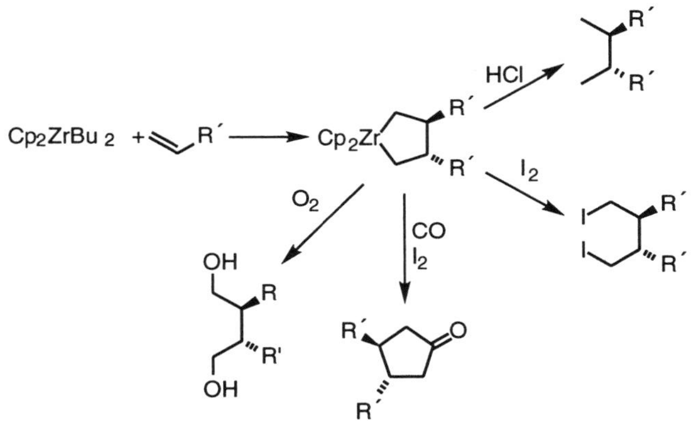

Eq. 4.124

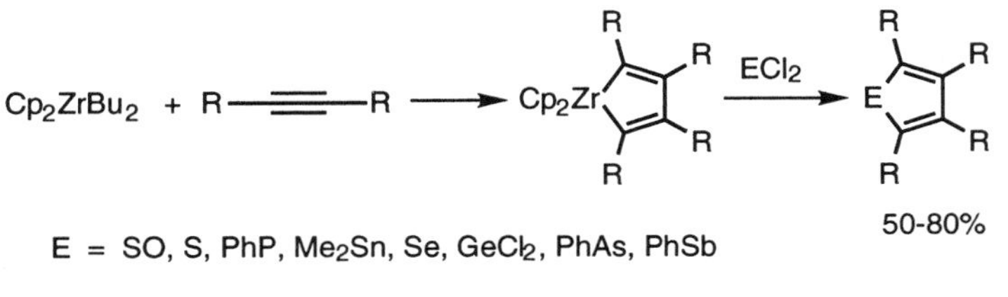

E = SO, S, PhP, Me2Sn, Se, GeCl2, PhAs, PhSb

50–80%

Eq. 4.125

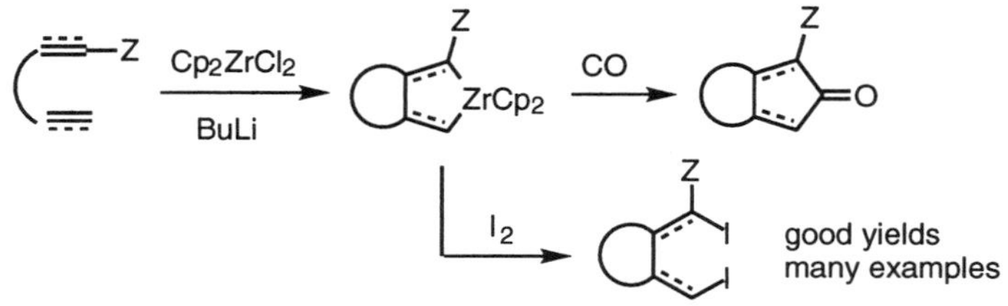

Intramolecular enyne, diyne, and diene cyclizations are also effective, forming mono or bicyclic compounds, depending on the method of cleavage (Eq. 4.125).[194] A range of functional groups is tolerated, and the reaction is highly stereoselective (Eq. 4.126),[195] (Eq. 4.127),[196] and (Eq. 4.128).[197]

Eq. 4.126

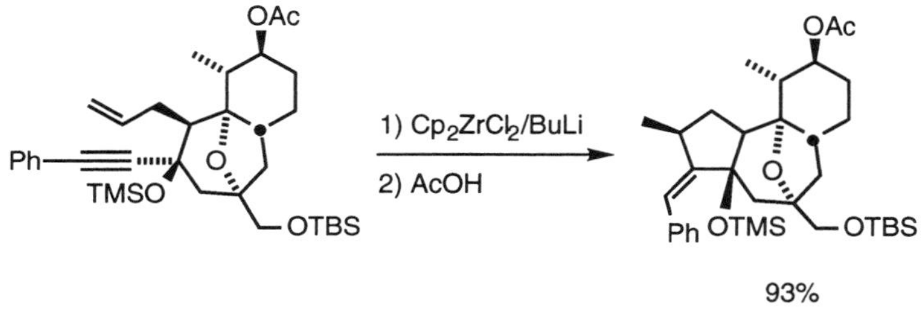

93%

Eq. 4.127

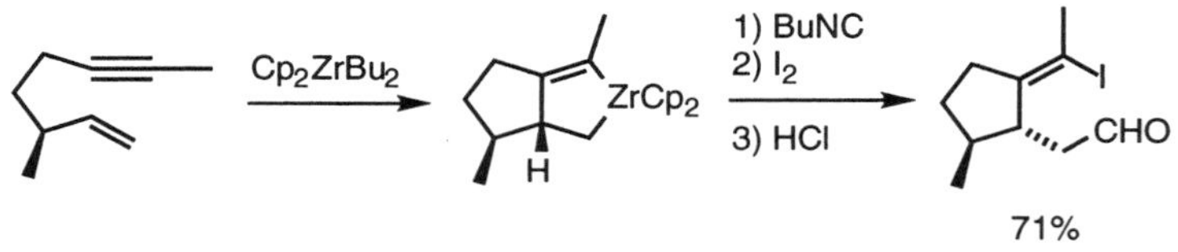

71%

Eq. 4.128

79%

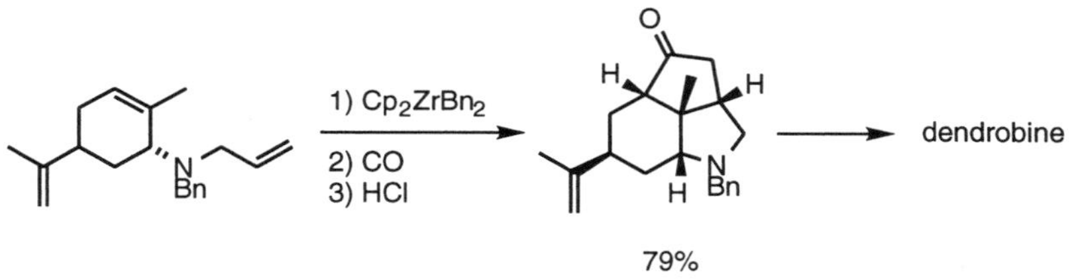

When cleaved with allylic chlorides, η^1-alkyl-η^3-allylzirconium complexes are produced, which undergo a number of useful transformations (Eq. 4.129).[198] By doing this iteratively, bicyclic ketones are produced (Eq. 4.130).[199]

Eq. 4.129

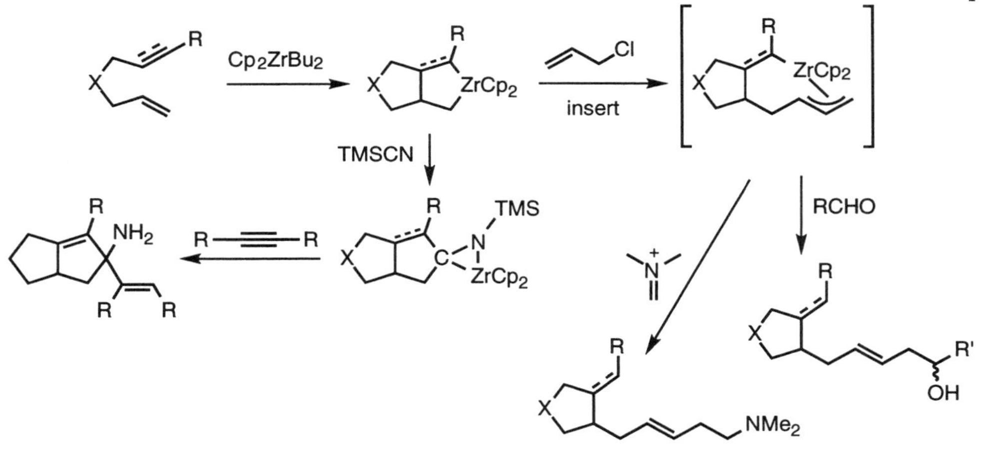

Eq. 4.130

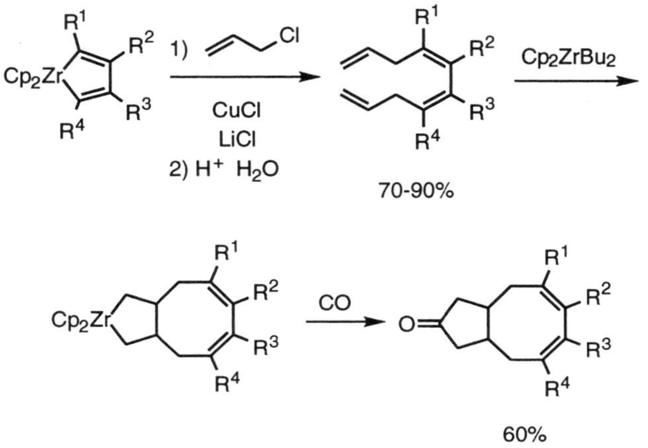

70-90%

60%

Related low-valent titanium complexes similarly "reductively dimerize" enynes or dienes giving metallacyclopentenes and -pentanes which are readily cleaved by electrophiles to produce interesting compounds (Eq. 4.131).[200]

Eq. 4.131

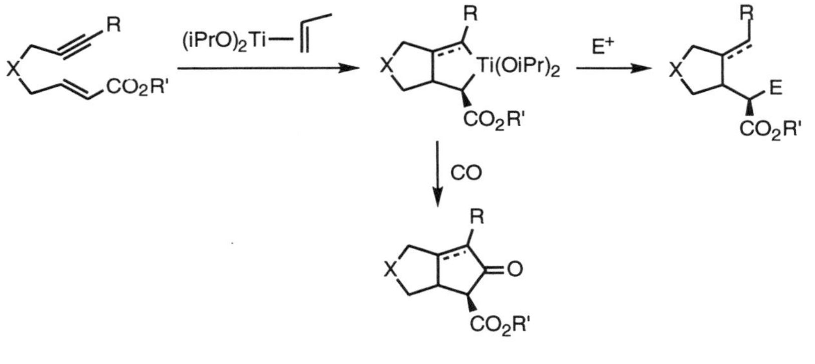

Treatment of $Cp_2Zr(Me)Cl$ with lithiated amines produces unstable zirconium-imine complexes, probably by a sequence of Cl^- displacement, β-hydride elimination and reductive elimination (Eq. 4.132). These insert alkynes to give azazirconacyclopentenes which are cleaved by acid to give allyl amines.[201] By switching over to *CpTiMe₂Cl*, transient imidotitanium species which cycloadd to alkynes are produced, again making metallacycles that undergo useful electrophilic cleavages (Eq. 4.133).[203]

Eq. 4.132

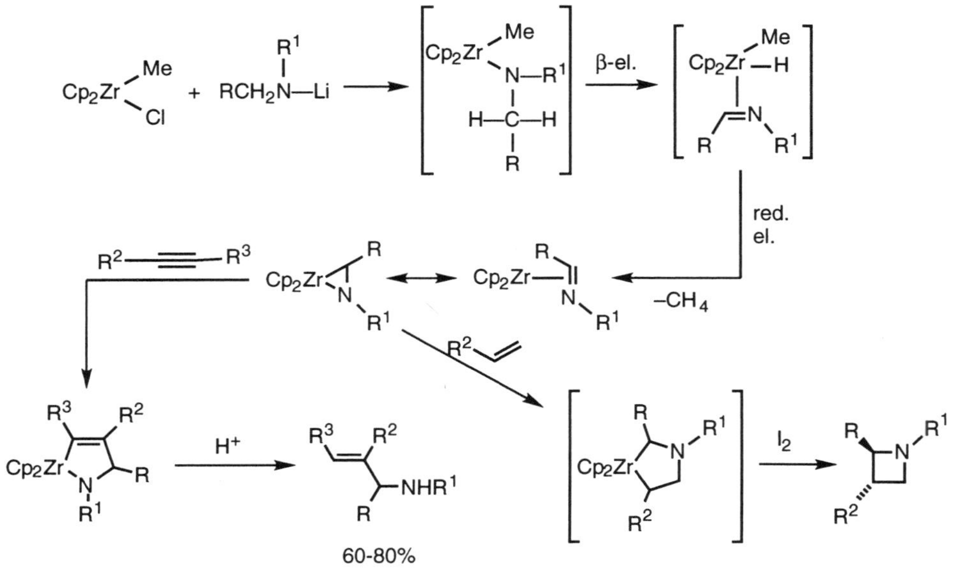

Eq. 4.133

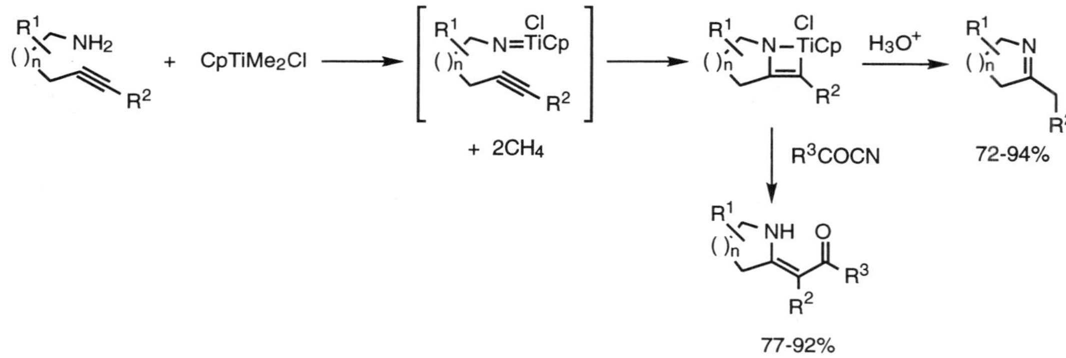

In a closely related process, zirconocene dichloride catalyzes the reaction between alkenes and Grignard reagents to produce new Grignard reagents. The intramolecular version is the easiest to follow (Eq. 4.134).[204] Treatment of α,ω-dienes

Eq. 4.134

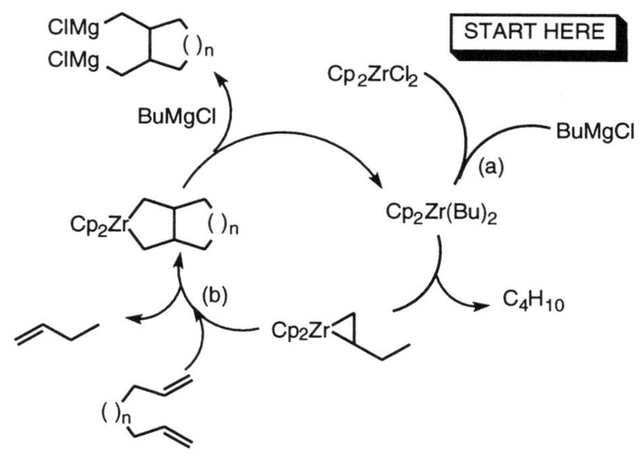

with n-butylmagnesium chloride in the presence of catalytic amounts of Cp_2ZrCl_2 produces cyclized, *bis*-Grignard reagents, which can be further functionalized. The reaction is thought to start as in Figure 4.9. Treatment of zirconocene dichloride with n-butylGrignard reagent produces the reactive zirconacyclopropane/zirconocene butene complex (a). This reacts with the diene to produce the zironacyclopentane (b). Once this occurs, the zirconacyclopentane is cleaved by the Grignard reagent to generate the *bis* Grignard and regenerate the $Cp_2Zr(Bu)_2$. With monoolefins, Cp_2ZrCl_2 catalyzes the addition of EtMgCl to alkenes to give homologated Grignard reagents (Eq. 4.135).[205] The process is similar to that with diolefins, except in this case, the Grignard provides one "alkene", and the cleavage gives a mono- rather than a *bis*-Grignard reagent.

Eq. 4.135

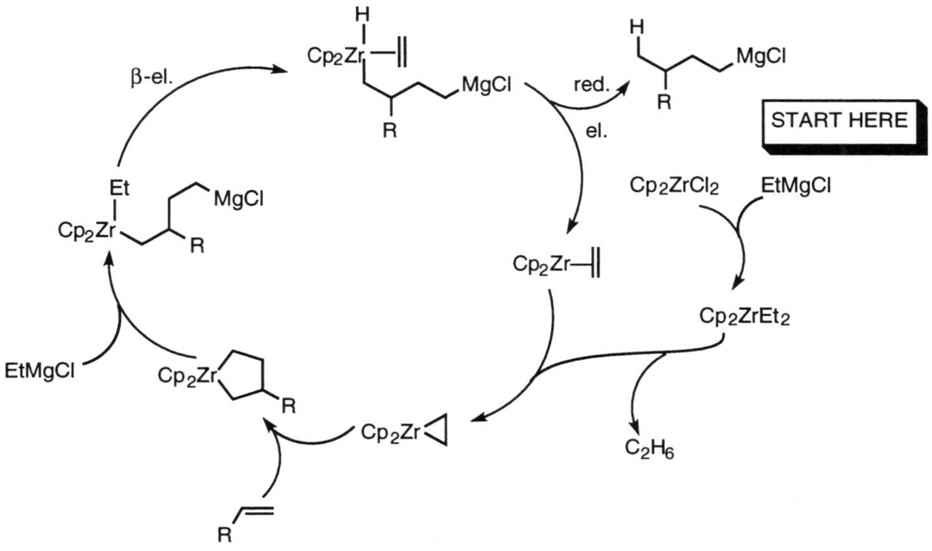

Perhaps most synthetically interesting is the reaction with allylic alcohols and ethers (Eq. 4.136).[206] Here it is clear that the allylic oxygen has both stereochemical and regiochemical consequences, indicating complexation to zirconium along the course of the reaction. With allylic alcohols, the zirconium reagent is delivered to the *same* face as the OH group, while with allylic ethers it comes from the opposite face, implying that OH complexes Zr, but OCH_3 does not. The reaction is also sensitive to steric hindrance (Eq. 4.137). Given a choice between an allylic methyl ether and the substantially larger t-butyldimethylsilyl ether, reaction occurs exclusively at the less hindered alkene.

Eq. 4.136

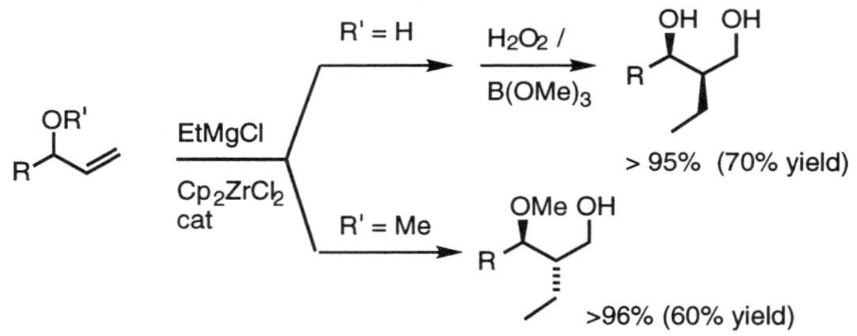

Eq. 4.137

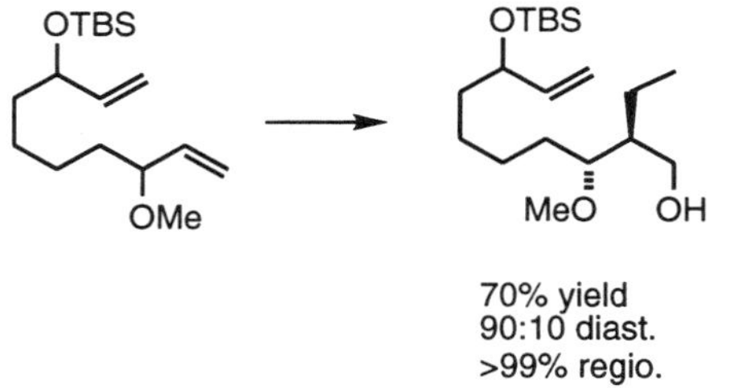

70% yield
90:10 diast.
>99% regio.

By using optically active zirconocene complexes, this ethylmagnesation of olefins has been made asymmetric.[207] The most synthetically-useful reactions involve oxygen or nitrogen heterocycles which are ring-opened in the process, to produce optically active, functionalized terminal olefins (Eq. 4.138). The mechanism by which

Eq. 4.138

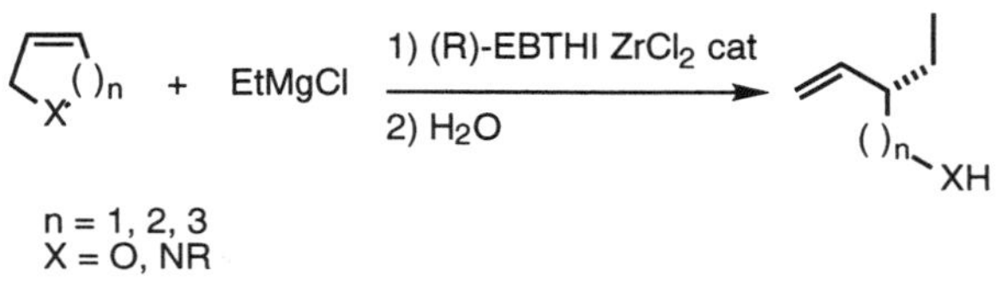

n = 1, 2, 3
X = O, NR

this is thought to proceed is shown in Eq. 4.139 and involves the same steps seen above. Asymmetric induction is thought to result from complexation of the substrate alkene in such a way as to minimize the steric hindrance between the chiral auxilliary and the ring atoms of the substrate. The presence of a β-heteroatom in the final organomagnesium complex leads to production of a terminal alkene by facile elimination. An elegant example of the use of both the allylic alcohol-directed carbomagnesation and the asymmetric ring opening of dihydrofurans is seen in the total synthesis of Fluvivicine (Eq. 4.140).[208]

Eq. 4.139

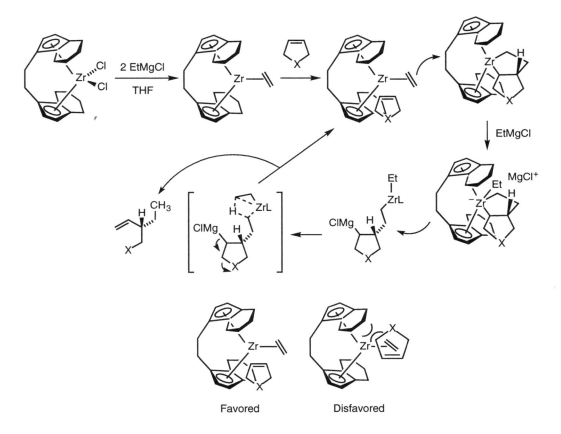

Favored   Disfavored

Because the chiral zirconium catalyst so efficiently discriminates in its reactions with cyclic ethers, it can be used to kinetically resolve racemic dihydrofurans and dihydropyrans.[209] The process takes advantage of the fact that one enantiomer of the dihydropyran undergoes ring opening substantially faster than the other, and is consumed, while the less reactive enantiomer can be recovered unchanged and with very good enantiomeric excess. With dihydrofurans, both react, but give different products (regioisomers) which are separable.

The other class of synthetically useful reductive cyclodimerizations of alkenes and alkynes comes from the opposite end of the periodic chart, being catalyzed by Ni(0) d^{10} complexes.[210] The general process is shown in Eq. 4.141. In this case the metallacycle is not stable, and has not been observed, but is inferred from the products.

Eq. 4.140

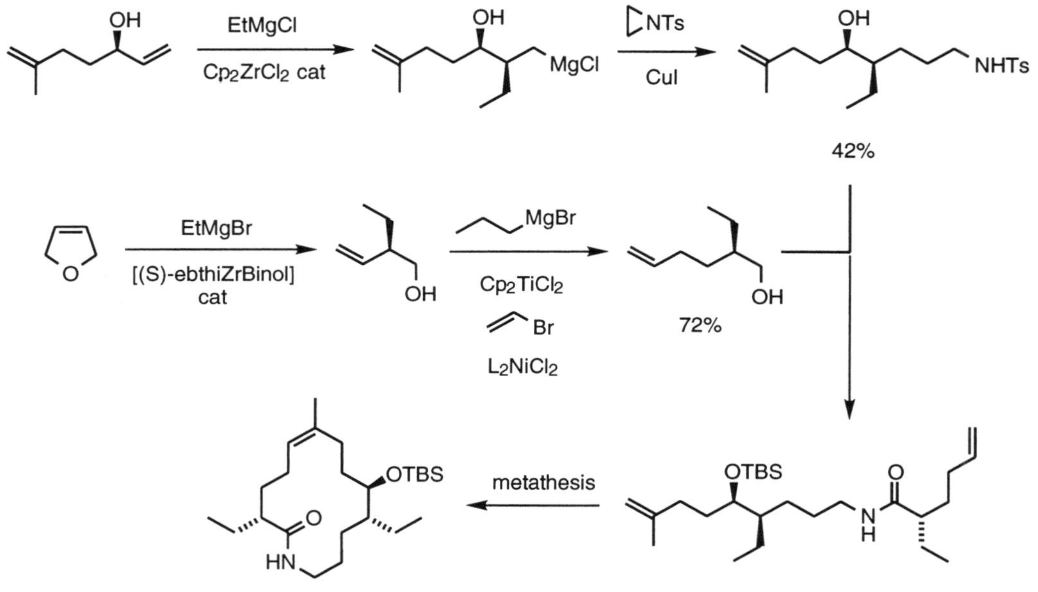

Eq. 4.141

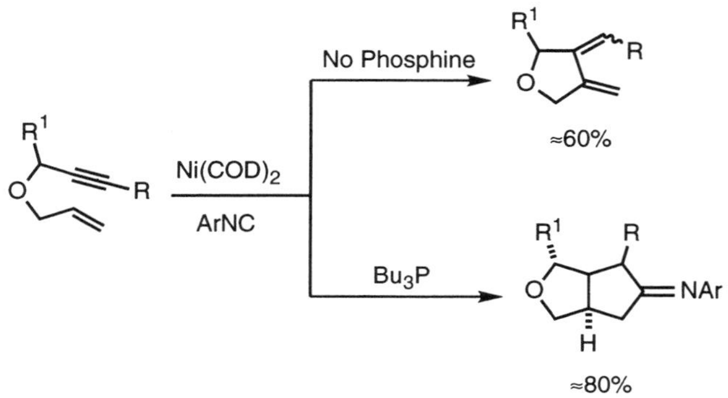

In the absence of added phosphine, simple cyclodimerization, followed by β-elimination/reductive elimination occurs (Eq. 4.142), giving dienes in reasonable yield. With added tributylphosphine, insertion of an isonitrile (which is isoelectronic and isostructural with CO, and inserts readily into metal carbon σ-bonds) occurs, giving the bicyclic imine which can be hydrolyzed to the ketone. Diynes undergo similar cyclizations to give cyclopentadienone imines.

There are a number of closely related reductive cyclizations of unsaturated species which, however, proceed via π-allylmetal intermediates. These are discussed in Chapter 9.

Eq. 4.142

References

1. Kharasch, M.S.; Tawney, P.O. *J. Am. Chem. Soc.* **1941**, *63*, 2308.
2. For early reviews see: (a) Posner, G.H. *Org. React.* **1972**, *19*, 1. (b) **1979**, *22*, 253. (c) *An Introduction to Synthesis Using Organocopper Reagents,* Wiley, New York, 1980.
3. For recent reviews, see: (a) Lipshutz, B.H.; Sengupta, S. *Org. React.* **1992**, *41*, 135. (b) Lipshutz, B.H. "Transition Metal Alkyl Complexes from RLi and CuX" in "Comprehensive Organometallic Chemistry II," Abel, E.W.; Stone, F.G.A.; Wilkinson, G., Eds., Pergamon, Oxford, UK, Vol. 12, pp. 59-130, 1995. (c) Lipshutz, B.H. "Organocopper Reagents and Procedures" in "Organometallics in Synthesis. A Manual II," Schlosser, M., Ed., John Wiley and Sons, Chichester, U.K., 1998.
4. McMurry, J.E.; Scott, W.J. *Tetrahedron Lett.* **1980**, *21*, 4313.
5. For a review dealing with 1,5-, 1,6-, 1,8-, 1,10-, and 1,12-additions see: Krause, N.; Gerold, A. *Angew. Chem. Int. Ed., Engl.* **1997**, *36*, 187.
6. (a) For a review, see: Taylor, J.K. *Synthesis* **1985**, 365. (b) Suzuki, M; Yanagisawa, A.; Noyori, R. *J. Am. Chem. Soc.* **1988**, *110*, 4718.
7. Posner, G.H.; Whitten, C.E.; Sterling, J.J. *J. Am. Chem. Soc.* **1973**, *95*, 7788.
8. Lindstedt, E.-L.; Nilsson, M.; Olsson, T. *J. Organomet. Chem.* **1987**, *334*, 255.
9. (a) Bertz, S.H.; Dabbagh, G.; Villacorta, G.M. *J. Am. Chem. Soc.* **1982**, *104*, 5824. (b) Bertz, S.H.; Dabbagh, G. *J. Org. Chem.* **1984**, *49*, 1119.
10. (a) Martin, S.F.; Fishpaugh, J.R.; Power, J.M.; Giolando, D.M.; Jones, R.A.; Nunn, C.M.; Cowley, A.H. *J. Am. Chem. Soc.* **1988**, *110*, 7226. (b) Bertz, S.H.; Eriksson, M.; Miao, G.; Snyder, J.P. *J. Am. Chem. Soc.* **1996**, *118*, 10906.
11. For reviews see: (a) Lipshutz, B.H.; Wilhelm, R.S.; Kozlowski, T.J. *Tetrahedron* **1984**, *40*, 5005. (b) Lipshutz, B.H. *Syn. Lett.* **1990**, 119. (c) For a different opinion, see: Bertz, S.H. *J. Am. Chem. Soc.* **1990**, *112*, 4031; Snyder, J.P.; Tipsword, G.E.; Spangler, D.P. *J. Am. Chem. Soc.* **1992**, *114*, 1507. For the latest (but not necessarily the last) word see: Huang, H.; Alvarez, K.; Lui, Q.; Barnhart, T.M.; Snyder, J.P.; Penner-Hahn, J.E. *J. Am. Chem. Soc.* **1996**, *118*, 8808. Bertz, S.H.; Miao, G.; Eriksson, M. *J. Chem. Soc., Chem. Comm.* **1996**, 815. Boche, G.; Bosold, F.; Marsch, M.; Harms, K. *Angew. Chem. Int. Ed. Engl.* **1998**, *37*, 1684.
12. (a) Lipshutz, B.H.; Siegmann, K.; Garcia, E. *J. Am. Chem. Soc.* **1991**, *113*, 8161. (b) For a review see: Lipshutz, B.H.; Siegmann, K.; Garcia, E. *Tetrahedron* **1992**, *48*, 2579. (c) Lipshutz, B.H.; Siegmann, K.; Garcia, E.; Kayser, P. *J. Am. Chem. Soc.* **1993**, *115*, 9276.
13. (a) Lipshutz, B.H.; Liu, Z.P.; Kayser, F. *Tetrahedron Lett.* **1994**, *35*, 5567. (b) Rawal, V.H.; Florgancic, A.S.; Singh, S.P. *Tetrahedron Lett.* **1994**, *35*, 8985. (c) Lin, G-Q.; Zhong, M. *Tetrahedron Lett.* **1997**, *38*, 1087.
14. Coleman, R.S.; Grant, E.B. *J. Am. Chem. Soc.* **1995**, *117*, 10889.
15. Maruyama, K.; Yamamoto, Y. *J. Am. Chem. Soc.* **1977**, *99*, 8068.
16. Yamamoto, Y.; Maruyama, K. *J. Am. Chem. Soc.* **1978**, *100*, 3241.
17. Ghribi, A.; Alexakis, A.; Normant, J.F. *Tetrahedron Lett.* **1984**, *25*, 3075.

115

18. (a) For a review see: Alexakis, A.; Mangeney, P.; Ghribi, A.; Marek, I.; Sedrani, R.; Guir, C.; Normant, J.F. *Pure Appl. Chem.* **1988**, *60*, 49. (b) Mangeney, P.; Alexakis, A.; Normant, J.F. *Tetrahedron Lett.* **1987**, *28*, 2363.
19. (a) For reviews see: Oppolzer, W. *Tetrahedron* **1987**, *43*, 1969, 4057. (b) Rossiter, B.E.; Swingle, N.M. *Chem. Rev.* **1992**, *92*, 771. (c) Oppolzer, W.; Kingma, A.J. *Helv. Chim Acta* **1989**, *72*, 1337.
20. (a) Bergdahl, M.; Nilsson, M.; Olsson, T. *J. Organomet. Chem.* **1990**, *391*, C19. (b) Bergdahl, M.; Nilsson, M.; Olsson, T.; Stern, K. *Tetrahedron* **1991**, *47*, 9691. (c) Eriksson, M.; Johansson, A.; Nilsson, M.; Olsson, T. *J. Am. Chem. Soc.* **1996**, *118*, 10904. (d) Urbom, E.; Knühl, G.; Helmchen, G. *Tetrahedron* **1996**, *52*, 971.
21. (a) Tanaka, K.; Ushio, H.; Kawabata, Y.; Suzuki, H. *J. Chem. Soc. Perkins I* **1991**, 1445. (b) For a review: Rossiter, B.E.; Swingle, N.M. *Chem. Rev.* **1992**, *92*, 771. (c) Alexakis, A.; Vastra, J.; Burton, J.; Benhaim, C.; Mangeney, P. *Tetrahedron Lett.* **1998**, *39*, 7869 and references therein.
22. Smith, A.B.; Jerris, P.J. *J. Org. Chem.* **1982**, *47*, 1845.
23. (a) Eis, M.J.; Ganem, B. *Tetrahedron Lett.* **1985**, *26*, 1153. (b) Tian, X.; Hudlicky, T.; Konigsberger, K. *J. Am. Chem. Soc.* **1995**, *117*, 3643.
24. Ghribi, A.; Alexakis, A.; Normant, J.F. *Tetrahedron Lett.* **1984**, *25*, 3083.
25. Rieke, R.D.; Stack, D.E.; Dawson, B.T.; Wu, T-C. *J. Org. Chem.* **1993**, *58*, 2483.
26. Stack, D.E.; Klein, W.R.; Rieke, R.D. *Tetrahedron Lett.* **1993**, *34*, 3063.
27. Ekert, G.W.; Pfennig, D.R.; Suchan, S.D.; Donovan, T.A., Jr.; Aouad, E.; Tehrani, S.F.; Gunnersen, J.N.; Dong, L. *J. Org. Chem.* **1995**, *60*, 2361.
28. For reviews see: (a) Knochel, P. "Zinc and Cadmium" in "Comprehensive Organometallic Chemistry II," Abel, E.W.; Stone, F.G.A.; Wilkinson, G., Eds., Pergamon, Oxford, UK, Vol. 11, pp. 159-183, 1995. (b) Knochel, P. *Synlett* **1995**, 393. (c) Knochel, P.; Singer, R. *Chem. Rev.* **1993**, *93*, 2117.
29. Lipshutz, B.H.; Wood, M.R.; Tirado, R. *J. Am. Chem. Soc.* **1995**, *117*, 6126.
30. Crimmins, M.T.; Nantermet, P.G.; Trotter, B.W.; Vallin, I.M.; Watson, P.S.; McKerlie, L.A.; Reinhold, T.L.; Cheung, A.W-H.; Stetson, K.A.; Dedopoulow, D.; Gray, J.L. *J. Org. Chem.* **1993**, *58*, 1038. Based on Nakamura, E.; Aoki, S.; Sekiya, K.; Oshino, H.; Kuwajima, I. *J. Am. Chem. Soc.* **1987**, *109*, 8056.
31. Crimmins, M.J.; Jung, D.K.; Grail, J.L. *J. Am. Chem. Soc.* **1993**, *115*, 3146.
32. Linderman, R.J.; Griedel, B.D. *J. Org. Chem.* **1990**, *55*, 5428. (b) Linderman, R.J.; McKenzie, J.R. *J. Organomet. Chem.* **1989**, *361*, 31. (c) Linderman, R.J.; Godfrey, A.J. *J. Am. Chem. Soc.* **1988**, *110*, 6249.
33. (a) Alexakis, A.; Commercon, A.; Coulentianos, C.; Normant, J.F. *Pure Appl. Chem.* **1983**, *55*, 1759. (b) Normant, J.F.; Alexakis, A. *Synthesis* **1981**, 841. (c) Gardette, M.; Alexakis, A.; Normant, J.F. *Tetrahedron* **1985**, *41*, 5887.
34. Rao, S.A.; Knochel, P. *J. Am. Chem. Soc.* **1991**, *113*, 5735.
35. (a) Trost, B.M. *Acct. Chem. Res.* **1990**, *23*, 34. (b) Trost, B.M.; Krische, M.J. *Synlett* **1998**, 1.
36. Trost, B.M.; Krische, M.J. *J. Am. Chem Soc.* **1996**, *118*, 233.
37. Yamada, H.; Aoyagi, S.; Kibayashi, C. *Tetrahedron Lett.* **1996**, *37*, 8787.
38. Yamada, H.; Aoyagi, S.; Kibayashi, C. *Tetrahedron Lett.* **1997**, *38*, 3027.
39. For a recent example see: Holzapfel, C.W.; Marcus, L. *Tetrahedron Lett.* **1997**, *38*, 8585.
40. (a) Trost, B.M.; Shi, Y. *J. Am. Chem. Soc.* **1991**, *113*, 701. For full papers see: (b) Trost, B.M.; Tanoury, G.J.; Lautens, M.; Chan, C.; MacPherson, D.T. *J. Am. Chem. Soc.* **1994**, *116*, 4255 and (c) Trost, B.M.; Romero, D.L.; Rise, F. *J. Am. Chem. Soc.* **1994**, *116*, 4268.
41. Trost, B.M.; Shi, Y. *J. Am. Chem. Soc.* **1992**, *114*, 791.
42. Schwartz, J.; Labinger, J.A. *Angew. Chem. Int. Ed.* **1976**, *15*, 333.
43. Gibson, T. *Tetrahedron Lett.* **1982**, *23*, 157.
44. Bertelo, C.A.; Schwartz, J. *J. Am. Chem. Soc.* **1976**, *98*, 262.
45. Loots, M.J.; Schwartz, J. *Tetrahedron Lett.* **1978**, 4381; *J. Am. Chem. Soc.* **1977**, *99*, 8045.
46. (a) Babiak, K.A.; Behling, J.R.; Dygos, J.H.; McLaughlin, K.T.; Ng, J.S.; Kalish, V.J.; Kramer, S.W.; Shone, R.L. *J. Am. Chem. Soc.* **1990**, *112*, 7441. (b) Lipshutz, B.H.; Ellsworth, E.L. *J. Am. Chem. Soc.* **1990**, *112*, 7440. (c) Dygos, J.H.; Adamek, J.P.; Babiak, K.A.; Behling, J.R.; Medich, J.R.; Ng, J.S.; Wieczoerek, J.J. *J. Org. Chem.* **1991**, *56*, 2549. (d) Lipshutz, B.H.; Keil, R. *J. Am. Chem .Soc.* **1992**, *114*, 7919.
47. (a) Lipshutz, B.H.; Wood, M.R. *J. Am. Chem. Soc.* **1993**, *115*, 12625. (b) Lipshutz, B.H.; Wood, M.R. *J. Am. Chem. Soc.* **1994**, *116*, 11689.
48. (a) VanHorne, D.E.; Negishi, E-I. *J. Am. Chem. Soc.* **1978**, *100*, 2252. (b) Lipshutz, B.H.; Dimock, S.H. *J. Org. Chem.* **1991**, *56*, 5761.
49. Wipf, P.; Xu, W.J.; Smitrovich, J.H.; Lehman, R.; Venanzi, L.M. *Tetrahedron* **1994**, *50*, 1935.
50. For a review see: Leong, W.W.; Larock, R.C., "Transition Metal Alkyl Complexes: Main Group Transmetallation/Insertion Chemistry" in "Comprehensive Organometallic Chemistry II," Abel, E.W.; Stone, F.G.A.; Wilkinson, G., Eds., Pergamon, Oxford, UK, Vol. 12, pp. 131-160, 1995.

116

51. Morris, I.K.; Snow, K.M.; Smith, N.W.; Smith, K.M. *J. Org. Chem.* **1990**, *55*, 1231.
52. Takemoto, Y.; Kuraoka, S.; Ohra, T.; Yonetoku, Y.; Iwata, C. *Tetrahedron* **1997**, *53*, 603.
53. Ruth, J.L.; Bergstrom, D.E. *J. Org. Chem.* **1978**, *43*, 2870; Bergstrom, D.E.; Inoue, H.; Reddy, P.A. *J. Org. Chem.* **1982**, *47*, 2174; Hacksell, U.; Daves, G.D., Jr. *J. Org. Chem.* **1983**, *48*, 2870.
54. Somei, M.; Kawesaki, T. *Chem. Pharm. Bull* **1989**, *37*, 3426.
55. Outten, R.A.; Daves, Jr., J.D. *J. Org. Chem.* **1987**, *52*, 5064.
56. Cambie, R.C.; Metzler, M.R.; Rutledge, P.S.; Woodgate, P.D. *J. Organomet. Chem.* **1992**, *429*, 59.
57. For a review see: Farina, V. "Transition Metal Alkyl Complexes: Oxidative Additions and Transmetallations" in "Comprehensive Organometallic Chemistry II," Abel, E.W.; Stone, F.G.A.; Wilkinson, G., Eds., Pergamon, Oxford, UK, Vol. 12, pp. 161-240, 1995. For a review on its use in biaryl synthesis see: Stanforth, S.P. *Tetrahedron* **1998**, *54*, 263.
58. Jutland, A.; Mosleh, A. *Organometallics* **1995**, *14*, 1810.
59. Herrmann, W.A.; Brossmer, C.; Reisinger, C-P.; Priermeier, T.; Beller, M.; Fischer, H. *Angew. Chem. Int. Ed. Engl.* **1995**, *34*, 1844.
60. The order of reactivity for phosphine/Pd(dba)$_2$-generated catalysts is Pd(dba)$_2$ + 2Ph$_3$P > 1 DIOP > 1 dppf > 1 BINAP; addition of 2 equiv. of a diphosphine gives an inactive catalyst: Amatore, C.; Broeka, G.; Jutland, A.; Khalil, F. *J. Am. Chem. Soc.* **1997**, *119*, 5176. For a review see: Amatore, C.; Jutland, A. *Coord. Chem. Rev.* **1998**, *178-180*, 511.
61. Amatore, C.; Carre, E.; Jutland, A.; M'Barke, M.A. *Organometallics* **1995**, *14*, 1818.
62. Tamao, K.; Sumitani, K.; Kiso, Y.; Zembayashi, M.; Fujioka, H.; Kodama, S-I.; Nakajima, I.; Minato, A.; Kumada, M. *Bull Chem. Soc. Jpn.* **1976**, *49*, 1958.
63. For a review see: Kalinin, V.N. *Synthesis* **1992**, 413.
64. (a) Hayashi, T.; Konishi, M.; Ito, H.; Kumada, M. *J. Am. Chem. Soc.* **1982**, *104*, 4962. (b) Hayashi, T.; Konishi, M.; Fukushima, M.; Kanehira, K.; Hioki, T.; Kumada, M. *J. Org. Chem.* **1983**, *48*, 2195. (c) For a review see: Sawamura, M.; Ito, Y. *Chem. Rev.* **1992**, *92*, 857.
65. (a) Hayashi, T.; Niizuma, S.; Kamikawa, T.; Suzuki, N.; Hozumi, Y. *J. Am. Chem. Soc.* **1995**, *117*, 9101. (b) Kamikawa, T.; Nozumi, Y.; Hayashi, T. *Tetrahedron Lett.* **1996**, *37*, 3161. (c) Kamikawa, T.; Hayashi, T. *Synlett* **1997**, 163.
66. (a) Kobayashi, M.; Negishi, E-I. *J. Org. Chem.* **1980**, *45*, 5223. (b) Negishi, E-I.; Owczarczyk, Z. *Tetrahedron Lett.* **1991**, *32*, 6683.
67. Smith, A.B., III; Qiu, Y.; Jones, D.R.; Kobayashi, K. *J. Am. Chem. Soc.* **1995**, *117*, 12011.
68. (a) Effenberger, F.; Würthner, F.; Steyke, F. *J. Org. Chem.* **1995**, *60*, 2082. (b) Würthner, F.; Vollmer, M.; Effenberger, F.; Emele, P.; Meyer, D.U.; Port, H.; Wolf, H.C. *J. Am. Chem. Soc.* **1995**, *117*, 8090. (c) Wu, X.; Rieke, R. *J. Org. Chem.* **1995**, *60*, 6658. (d) Yu, D.; Gharavi, A.; Yu, L. *J. Am. Chem. Soc.* **1995**, *117*, 11680. (e) Nakayama, J.; Ting, Y.; Sugihara, Y.; Ishii, A. *Heterocycles* **1997**, *44*, 75.
69. Takahashi, K.; Gunjii, A.; Yanagi, K.; Miki, M. *Tetrahedron Lett.* **1995**, *36*, 8055.
70. (a) Rottlander, M.; Knochel, P. *Synlett* **1997**, 1084. For the synthesis of related functionalized aromatic oligomers see: (b) Bruno, J.M.; Chang, M.N.; Choi-Sledeski, Y.M.; Green, D.M.; McGarry, D.G.; Regan, J.R.; Volz, F.A. *J. Org. Chem.* **1997**, *62*, 5174.
71. (a) Vincent, P.; Beaucourt, J.P.; Pichat, L. *Tetrahedron Lett.* **1982**, *23*, 63. (b) Okukado, N.; Van Horn, D.E.; Klima, D.L.; Negishi, E-I. *Tetrahedron Lett.* **1978**, 1027.
72. (a) For a review see: Negishi, E-I. *Acct. Chem. Res.* **1982**, *15*, 340. (b) Negishi, E-I.; Okukado, N.; King, A.O.; Van Horn, D.E.; Spiegel, B.I. *J. Am. Chem. Soc.* **1978**, *100*, 2254.
73. (a) Panek, J.S.; Hu, T. *J. Org. Chem.* **1997**, *62*, 4912. (b) Panek, J.S.; Hu, T. *J. Org. Chem.* **1997**, *62*, 4914.
74. (a) For a review see: Miyaura, N.; Suzuki, A. *Chem. Rev.* **1995**, *95*, 2457. (b) For recent mechanistic studies which show transmetallation in the Suzuki coupling goes with retention see: Ridgeway, B.H.; Woerpel, K.D. *J. Org. Chem.* **1998**, *63*, 458; Matos, K.; Soderquist, J.A. *J. Org. Chem.* **1998**, *63*, 461.
75. Narukawa, Y.; Nishi, K.; Groue, H. *Tetrahedron* **1997**, *53*, 539.
76. Humphrey, J.M.; Aggen, J.B.; Chamberlin, A.R. *J. Am. Chem. Soc.* **1996**, *118*, 11759.
77. Uenishi, J-i.; Bean, J.M.; Armstrong, R.W.; Kishi, Y. *J. Am. Chem. Soc.* **1987**, *109*, 4756.
78. Shen, W. *Tetrahedron Lett.* **1997**, *38*, 5575.
79. (a) Wright, S.W.; Hageman, D.L.; McClure, L.D. *J. Org. Chem.* **1994**, *59*, 6095.
80. (a) Hobbs, P.D.; Upender, V.; Liu, J.; Pollart, D.J.; Thomas, P.W.; Dawson, M.I. *J. Chem. Soc., Chem. Comm.* **1996**, 923. (b) Hobbs, P.D.; Upender, V.; Dawson, M.I. *Synlett* **1997**, 965.
81. Jendralla, H.; Wagner, A.; Mobrath, M.; Wunner, J. *Liebigs Ann. Chem.* **1995**, 1253.
82. Percec, V.; Chu, P.; Ungar, G.; Zhow, J. *J. Am. Chem. Soc.* **1995**, *117*, 11441.
83. (a) Galda, P.; Rahalin, M. *Synthesis* **1996**, 614. (b) Keegstra, M.; DeFeyter, S.; DeSchryver, F.C.; Müllen, M. *Angew. Chem. Int. Ed. Engl.* **1996**, *35*, 774. (c) Kowitz, C.; Wegner, G. *Tetrahedron* **1997**, *53*, 15553. (d) Lambda, J.S.; Tour, J.M. *J. Am. Chem. Soc.* **1994**, *116*, 11723.

84. Hensel, V.; Lützow, K.; Jacob, J.; Gessler, K.; Saenger, W.; Schlüter, A.D. *Angew. Chem. Int. Ed. Engl.* **1997**, *36*, 2654.
85. Feldman, K.S.; Campbell, R.F.; Saunders, J.C.; Ahn, C.; Masters, K.M. *J. Org. Chem.* **1997**, *62*, 8814.
86. (a) Guiles, J.W.; Johnsen, S.G.; Murray, W.V. *J. Org. Chem.* **1996**, *61*, 5169. (b) Brown, S.D.; Armstrong, R.W. *J. Am. Chem. Soc.* **1996**, *118*, 6331.
87. For reviews see: (a) Snieckus, V. *Chem. Rev.* **1990**, *90*, 879. (b) Snieckus, V. *Pure Appl. Chem.* **1994**, *66*, 2155.
88. Larsen, R.D.; King, A.O.; Chen, C.Y.; Corley, E.G.; Foster, B.S.; Roberts, F.E.; Yang, C.; Lieberman, D.R.; Reamer, R.A.; Tschaen, D.M.; Verhoeven, T.R.; Reider, P.J.; Lo, Y.S.; Rossano, L.T.; Brooks, A.S.; Meloni, D.; Moore, J.R.; Arnett, J.F. *J. Org. Chem.* **1994**, *59*, 6391. For a mechanistic study of this process see: Smith, G.B.; Dezeny, G.C.; Hughes, D.L.; King, A.O.; Verhoeven, T.R. *J. Org. Chem.* **1994**, *59*, 8151.
89. Muller, D.; Fleury, J-P. *Tetrahedron Lett.* **1991**, *32*, 2229.
90. For a very thorough review of the subject see: (a) Varina, V.; Krishnamurthy, V.; Scott, W.J. *Org. React.* **1997**, *50*, 1. (b) Farina, V.; Roth, G.P. in "Advances in Metal-Organic Chemistry," Liebeskind, L.S. Ed., JAI Press, Greenwich, CT, 1995, 5, 1. For older reviews (c) Michell, T.N. *Synthesis* **1992**, 803. (d) Stille, J.K. *Angew. Chem. Int. Ed. Engl.* **1986**, *25*, 508.
91. Hoshino, M.; Degenkolb, P.; Curran, D.P. *J. Org. Chem.* **1997**, *62*, 8341.
92. Farina, V.; Krishnan, B. *J. Am. Chem. Soc.* **1991**, *113*, 9585.
93. (a) Liebeskind, L.S.; Fengl, R.W. *J. Org. Chem.* **1990**, *55*, 5359. (b) Farina, V.; Kapadia, S.; Krishnan, B.; Wang, C.; Liebeskind, L. *J. Org. Chem.* **1994**, *59*, 5905.
94. (a) Labadie, J.W.; Teuting, D.; Stille, J.K. *J. Org. Chem.* **1983**, *48*, 4634. (b) Labadie, J.W.; Stille, J.K. *J. Am. Chem. Soc.* **1983**, *109*, 669.
95. Kende, A.S.; Roth, B.; Sanfillipo, P.J.; Blacklock, T.J. *J. Am. Chem. Soc.* **1982**, *104*, 5808.
96. Baldwin, J.E.; Adlington, R.M.; Ramcharitar, S.H. *Tetrahedron* **1992**, *48*, 2957.
97. Four, P.; Guibe, F. *J. Org. Chem.* **1981**, *46*, 4439.
98. Stille, J.K.; Simpson, J.H. *J. Am. Chem. Soc.* **1987**, *109*, 2138.
99. Stille, J.K.; Groh, B.L. *J. Am. Chem. Soc.* **1987**, *109*, 813.
100. (a) Kende, A.S.; Liu, K.; Kalder, I.; Dorey, G.; Koch, K. *J. Am. Chem. Soc.* **1995**, *117*, 8258. For use in the synthesis of tetraenes by the coupling of trienyl tin reagents see: (b) Andrus, M.P.; Lepose, S.D.; Turner, T.M. *J. Am. Chem. Soc.* **1997**, *119*, 12159.
101. Smith, A.B., III; Condon, S.M.; McCauley, J.A.; Leazer, J.L., Jr.; Leaky, J.W.; Maleczka, R.E., Jr. *J. Am. Chem. Soc.* **1997**, *119*, 962.
102. Panek, J.S.; Masse, C.E. *J. Org. Chem.* **1997**, *62*, 8290.
103. (a) For a review on enol triflates see: Scott, W.J.; McMurray, J.E. *Acc. Chem. Res.* **1988**, *21*, 47. (b) Echavarren, A.M.; Stille, J.K. *J. Am. Chem. Soc.* **1987**, *109*, 5478.
104. Roth, G.P.; Fuller, C.E. *J. Org. Chem.* **1991**, *56*, 3493.
105. Nicolaou, K.C.; Shi, G-Q.; Gunzer, J.L.; Gärtner, P.; Yang, Z. *J. Am. Chem. Soc.* **1997**, *119*, 5467.
106. Paquette, L.A.; Wang, T-Z.; Swik, M.R. *J. Am. Chem. Soc.* **1994**, *116*, 11323.
107. Nicolaou, K.C.; Sato, M.; Miller, N.D.; Gunzer, J.L.; Renaud, J.; Unterstetter, E. *Angew. Chem. Int. Ed. Engl.* **1996**, *39*, 887.
108. Farina, V.; Krishnan, B.; Marshall, D.R.; Roth, G.P. *J. Org. Chem.* **1993**, *59*, 5434.
109. Shair, M.D.; Yoon, T.; Danishefsky, S.J. *J. Org. Chem.* **1994**, *59*, 3755.
110. Kosugi, M.; Shimizu, Y.; Migita, T. *Chem. Lett.* **1977**, 1423.
111. Wu, R.; Schumm, J-S.; Pearson, D.L.; Tour, J.M. *J. Org. Chem.* **1996**, *61*, 6906.
112. (a) Zhang, Q.T.; Tour, J.M. *J. Am. Chem. Soc.* **1997**, *119*, 9624. (b) Pelter, A.; Jenkins, I.; Jones, D.E. *Tetrahedron* **1997**, *53*, 10357.
113. Zhang, Q.T.; Tour, J.M. *J. Am. Chem. Soc.* **1997**, *119*, 5065.
114. Roth, G.P.; Farina, V.; Liebeskind, L.S.; Peña-Cabrera, E. *Tetrahedron Lett.* **1995**, *36*, 2191.
115. Shirakawa, E.; Yamasaki, K.; Hiyama, T. *J. Chem. Soc. Perkin I* **1997**, 2449.
116. Kosugi, M.; Kamezawa, M.; Migita, T. *Chem. Lett.* **1983**, 927.
117. (a) Wolfe, J.P.; Wagaw, S.; Buchwald, S.L. *J. Am. Chem. Soc.* **1996**, *118*, 7215. (b) Wolfe, J.P.; Buchwald, S.L. *J. Org. Chem.* **1996**, *61*, 1133. (c) For triflates see: Wolfe, J.P.; Buchwald, S.L. *J. Org. Chem.* **1997**, *62*, 1264. (d) Driver, M.J.; Hartwig, J.F. *J. Am. Chem. Soc.* **1996**, *118*, 7217. (e) For triflates see: Louis, J.; Driver, M.S.; Hamann, B.C.; Hartwig, J.F. *J. Org. Chem.* **1997**, *62*, 1268. (f) Hartwig, J.F. *Synlett* **1997**, 329. (g) Wolfe, J.P.; Buchwald, S.L. *J. Org. Chem.* **1997**, *62*, 6066. (h) Wolfe, J.P.; Wagaw, S.; Marcoux, J.-F.; Buchwald, S.L. *Acc. Chem. Res.*, **1998**, *31*, 805.
118. Åhman, J; Buchwald, S.L. *Tetrahedron Lett.* **1997**, *38*, 6363.
119. (a) Reddy, N.P.; Tanaka, M. *Tetrahedron Lett.* **1997**, *38*, 4807. (b) Beller, M.; Riermeier, T.H.; Reisinger, C-P.; Hermann, W.A. *Tetrahedron Lett.* **1997**, *38*, 2073.
120. Wolfe, J.P.; Buchwald, S.L. *J. Am. Chem. Soc.* **1997**, *119*, 6054.

121. Ward, Y.D.; Farina, V. *Tetrahedron Lett.* **1996**, *37*, 6993.
122. (a) Kambara, T.; Izumi, K.; Nakadani, Y.; Narise, T.; Hasegawa, K. *Chem. Lett.* **1997**, 1185. (b) Louie, J.; Hartwig, J.F. *J. Am. Chem. Soc.* **1997**, *119*, 11695.
123. Crisp, G.T.; Scott, W.J.; Stille, J.K. *J. Am. Chem. Soc.* **1984**, *106*, 7500.
124. Hiyama, T. *Syn. Lett.* **1991**, 845.
125. Hagiwara, E.; Gowda, K-i.; Hatanaka, Y.; Hiyama, T. *Tetrahedron Lett.* **1997**, *38*, 439.
126. Rossi, R.; Carpita, A.; Messeri, T. *Gazz Chem. Ital.* **1992**, *122*, 65.
127. For total syntheses of these compounds utilizing this reaction see: Magnus, P.; Carter, P.; Elliot, J.; Lewis, R.; Harling, J.; Pitterns, T.; Batwa, W.E.; Fort, S. *J. Am. Chem. Soc.* **1992**, *114*, 2544. For a review see: Maier, M.E. *Synlett* **1995**, 13.
128. (a) Wender, P.A.; Beckham, S.; O'Leary, J-G. *Synthesis* **1995**, 1279. (b) Nishikawa, T.; Yoshikai, M.; Kawai, T.; Unno, R.; Jomori, T.; Isobe, M. *Tetrahedron* **1995**, *51*, 9339.
129. (a) Jacobi, P.A.; Guo, J.; Rajeswari, S.; Zheng, W. *J. Org. Chem.* **1997**, *62*, 2907. (b) Jacobi, P.A.; Guo, J. *Tetrahedron Lett.* **1995**, *36*, 2717.
130. (a) De la Rosa, M.A.; Velardi, E.; Grizman, A. *Synth. Comm.* **1990**, *20*, 2059. (b) Bleicher, L.; Cosford, N.P.P. *Synlett* **1995**, 1115.
131. Bumagin, N.; Sukholminova, L.I.; Luzckova, E.V.; Tolstaya, T.P.; Beletskaya, I.P. *Tetrahedron Lett.* **1996**, *37*, 897.
132. Nguefack, J.F.; Bolitt, V.; Sinou, D. *Tetrahedron Lett.* **1996**, *37*, 5527.
133. Powell, N.A.; Rychnovsky, S.D. *Tetrahedron Lett.* **1996**, *37*, 7901.
134. For a review see: Tour, J.M. *Chem. Rev.* **1996**, *96*, 537.
135. (a) Pearson, D.L.; Tour, J.M. *J. Org. Chem.* **1997**, *62*, 1376. (b) Jones, L., II; Pearson, D.L.; Schumm, J.S.; Tour, J.M. *Pure Appl. Chem.* **1996**, *68*, 145.
136. Moore, J. *Angew. Chem. Int. Ed. Engl.* **1993**, *32*, 246. See also Zeng, F.; Zimmerman, S.C. *J. Am. Chem. Soc.* **1996**, *118*, 5326.
137. Zhang, Y.; Wada, T.; Wang, L.; Sasake, H. *Tetrahedron Lett.* **1997**, *38*, 1785.
138. Goldfinger, M.B.; Crawford, K.B.; Swager, T.M. *J. Am. Chem. Soc.* **1997**, *119*, 4578.
139. (a) Roth, G.P.; Thomas, J.A. *Tetrahedron Lett.* **1992**, *33*, 1959. (b) Cacchi, S.; Ciattini, P.G.; Moreta, E.; Ortar, G. *Tetrahedron* **1986**, *27*, 3931. (c) Dolle, R.E.; Schmidt, S.J.; Kruse, L.I. *J. Chem. Soc., Chem. Comm.* **1987**, 904.
140. Snider, B.B.; Vo, N.H.; O'Neil, S.V.; Foxman, B.M. *J. Am. Chem. Soc.* **1996**, *118*, 7644.
141. McGuire, M.A.; Sorenson, E.; Ouring, F.W.; Resnick, T.M.; Fox, M.; Baine, N.H. *J. Org. Chem.* **1994**, *59*, 6683.
142. (a) Cowell, A.; Stille, J.K. *J. Am. Chem. Soc.* **1980**, *102*, 4193. (b) Martin, L.D.; Stille, J.K. *J. Org. Chem.* **1982**, *47*, 3630.
143. Mori, M.; Chiba, K.; Ban, Y. *J. Org. Chem.* **1978**, *43*, 1684.
144. (a) Mori, M.; Chiba, K.; Okita, K.; Ban, Y. *Chem. Comm.* **1979**, 698. (b) *Tetrahedron* **1985**, *41*, 387.
145. Tilley, J.W.; Coffen, D.L.; Schaer, B.H.; Lind, J. *J. Org. Chem.* **1987**, *52*, 2469.
146. Kotsuki, H.; Dalta, P.K.; Suenaga, H. *Synthesis* **1996**, 470.
147. Jeanneret, V.; Meerpoel, L.; Vogel, P. *Tetrahedron Lett.* **1997**, *38*, 543.
148. Jackson, R.F.W.; Turner, D.; Block, M.H. *J. Chem. Soc., Perkin Trans I* **1997**, 865.
149. For reviews see: (a) Heck, R.F. *Org. React.* **1982**, *27*, 345. (b) Heck, R.F. "Palladium Reagents in Organic Synthesis," Academic Press, London, UK, 1985. (c) Söderberg, B.C. "Transition Metal Alkyl Complexes: Oxidative Addition and Insertion," in "Comprehensive Organometallic Chemistry II," Abel, E.W.; Stone, F.G.A.; Wilkinson, G., Eds., Pergamon, Oxford, UK, Vol. 12, pp. 241-297, 1995. (d) Overman, L.E. *Pure Appl. Chem.* **1994**, *66*, 1423. (e) de Meijere, E.; Meyer, F.E. *Angew. Chem. Int. Ed. Engl.* **1994**, *33*, 2379. (f) Cabri, W.; Candiani, I. *Acc. Chem. Res.* **1995**, *28*, 2.
150. McCrindle, R.; Ferguson, G.; Arsenault, G.J.; McAlees, A.J.; Stephenson, D.K. *J. Chem. Res. Synop.* **1984**, 360.
151. Andersson, C.M.; Karabelas, K.; Hallberg, A.; Andersson, C. *J. Org. Chem.* **1985**, *50*, 3891.
152. (a) Jeffry, T. *Tetrahedron Lett.* **1985**, *26*, 2667. (b) Larock, R.C.; Baker, B.E. *Tetrahedron Lett.* **1988**, *29*, 905. (c) Jeffrey, T. *Chem. Comm.* **1984**, 1287. (d) Jeffry, T. *Synthesis* **1987**, 70.
153. Amatore, C.; Azzabi, M.; Jutand, A. *J. Am. Chem. Soc.* **1991**, *113*, 8375.
154. (a) Jeffery, T. *Tetrahedron* **1996**, *52*, 10113. (b) Casalnuovo, W.L.; Calabrese, J.C. *J. Am. Chem. Soc.* **1990**, *112*, 4324. (c) Genet, J.P.; Blast, E.; Savignac, M. *Synlett* **1992**, 1715. (d) Jeffery, T. *Tetrahedron Lett.* **1994**, *35*, 3501. (e) Dibowski, H.; Schmidchen, F.P. *Tetrahedron* **1995**, *51*, 2325.
155. (a) Spencer, A. *J. Organomet. Chem.* **1983**, *258*, 101; **1984**, *265*, 323; **1984**, *270*, 115. See also (b) Ohff, M.; Ohff, A.; van der Boom, M.E.; Milstein, D. *J. Am. Chem. Soc.* **1997**, *119*, 11687.
156. (a) Daves, G.D., Jr.; Hallberg, A. *Chem. Rev.* **1989**, *89*, 1433. (b) Daves, G.D. *Acct. Chem. Res.* **1990**, *23*, 201.

157. (a) Cabri, W.; Candiani, I.; Bedeschi, A.; Penco, S. *J. Org. Chem.* **1992**, *57*, 1481. (b) Cabri, W.; Candiani, I.; Bedeschi, A.; Santi, R. *J. Org. Chem.* **1992**, *57*, 3558.
158. Nishi, K.; Narukawa, Y.; Onoue, H. *Tetrahedron Lett.* **1996**, *37*, 2987.
159. Tietze, L.F.; Nöbel, T.; Speacha, M. *Angew. Chem. Int. Ed. Engl.* **1996**, *35*, 2259.
160. Hiroshige, M.; Hauske, J.R.; Zho, P. *Tetrahedron Lett.* **1995**, *36*, 4567.
161. Fu, D.-K.; Xu, B.; Swager, T.M. *Tetrahedron* **1997**, *53*, 15487.
162. Peng, Z.; Charavi, A.R.; Yu, L. *J. Am. Chem. Soc.* **1997**, *119*, 4622.
163. Maddix, T.; Li, W.; Yu, L. *J. Am. Chem. Soc.* **1997**, *119*, 844.
164. Jin, J.; Weinreb, S.M. *J. Am. Chem. Soc.* **1997**, *119*, 2050. For a review on intramolecular Heck reactions see: Grigg, R.; Sridharan, V.; Santhakumar, V.; Thornton-Pett, M.; Bridge, A.W. *Tetrahedron* **1993**, *49*, 5177.
165. (a) Masters, J.J.; Link, J.T.; Snyder, L.B.; Young, W.B.; Danishefsky, S.J. *Angew. Chem. Int. Ed. Engl.* **1995**, *34*, 1723. (b) Young, W.B.; Masters, J.J.; Danishefsky, S.J. *J. Am. Chem. Soc.* **1995**, *117*, 5228.
166. Okita, T.; Isobe, M. *Tetrahedron* **1994**, *50*, 11143.
167. (a) Rigby, J.H.; Hughes, R.J.; Heeg, M.J. *J. Am. Chem. Soc.* **1995**, *117*, 7834. (b) Bombrun, A.; Sageot, O. *Tetrahedron Lett.* **1997**, *38*, 1057.
168. Overman, L.E.; Abelman, M.M.; Kucera, D.J.; Tran, V.D.; Ricca, D.J. *Pure Appl. Chem.* **1992**, *64*, 813.
169. Hiroshige, M.; Hauske, J.R.; Zhou, R. *J. Am. Chem. Soc.* **1995**, *117*, 11590.
170. For a review see: Shibasaki, M.; Boden, C.D.J.; Kojima, A. *Tetrahedron* **1997**, *53*, 7371.
171. Tietze, L.F.; Raschile, T. *Synlett* **1995**, 597.
172. Kojima, A.; Boden, C.J.; Shibasaki, M. *Tetrahedron Lett.* **1997**, *38*, 3459.
173. (a) Loiseleur, O.; Meier, P.; Pfaltz, A. *Angew. Chem. Int. Ed. Engl.* **1996**, *35*, 200. (b) Loiseleur, O.; Hayashi, M.; Sames, N.; Pfaltz, A. *Synthesis* **1997**, 1338. See also (c) Traebesinger, G.; Albinati, A.; Feiben, N.; Kunz, R.W.; Pregosin, P.S.; Tschoener, M. *J. Am. Chem. Soc.* **1997**, *119*, 6315.
174. Ashimori, A.; Matsuura, T.; Overman, L.E.; Poon, D. *J. Org. Chem.* **1993**, *58*, 6949.
175. Tang, X-Q.; Harvey, R.G. *Tetrahedron Lett.* **1995**, *36*, 6037.
176. Burwood, M.; Davies, B.; Diaz, I.; Grigg, R.; Molina, P.; Sridharan, V.; Hughes, M. *Tetrahedron Lett.* **1995**, *36*, 9053.
177. For reviews see: (a) Heumann, A.; Reglier, M. *Tetrahedron* **1996**, *52*, 9289. (b) Grigg, R.; Sridharan, V. in "Comprehensive Organometallic Chemistry II," Abel, E.W.; Stone, F.G.A.; Wilkinson, G., Eds., Pergamon, Oxford, UK, 1995, Vol. 12, pp. 299-321.
178. Negishi, E-I.; Sawada, H.; Tour, J.M.; Wei, Y. *J. Org. Chem.* **1988**, *53*, 913.
179. (a) Coperet, C.; Ma, S.; Negishi, E-i. *Angew. Chem. Int. Ed. Engl.* **1996**, *35*, 2125. For full papers on the subject see: (b) Coperet, C.; Ma, S.; Sugihara, T.; Negishi, E-i. *Tetrahedron* **1996**, *52*, 11529. (c) Negishi, E-i.; Ma, S.; Amanfu, J.; Coperet, C.; Miller, J.A.; Tour, J.M. *J. Am. Chem. Soc.* **1996**, *118*, 5919.
180. (a) Maddaford, S.P.; Andersen, N.G.; Cristofoli, W.A.; Keays, B.A. *J. Am. Chem. Soc.* **1996**, *118*, 10766. (b) Keay, B.A.; Maddaford, S.P.; Cristofoli, W.A.; Andersen, N.G.; Passafaro, M.W.; Wilson, N.S.; Nieman, J.A. *Can. J. Chem.* **1997**, *75*, 1163.
181. For reviews see: (a) Malacria, M. *Chem. Rev.* **1996**, *96*, 289. (b) Ojima, I.; Tzamarioudalsi, M.; Li, Z.; Donovan, R.J. *Chem. Rev.* **1996**, *96*, 635. (c) Negishi, E-i.; Copret, C.; Ma, S.; Liou, S-Y.; Liu, F. *Chem. Rev.* **1996**, *96*, 365.
182. Zhang, Y.; Wu, G.; Agnel, G.; Negishi, E-I. *J. Am. Chem. Soc.* **1990**, *112*, 8590.
183. Meyer, F.E.; Parsons, P.J.; deMeijere, A. *J. Org. Chem.* **1991**, *56*, 6487.
184. Grigg, R.; Logananthan, V.; Sridharan, V. *Tetrahedron Lett.* **1996**, *37*, 3399.
185. Oda, H.; Kobayashi, T.; Kosugi, M.; Migita, T. *Tetrahedron* **1995**, *51*, 695.
186. Kojima, A.; Takemoto, T.; Sodeoka, M.; Shibasaki, M. *J. Org. Chem.* **1996**, *61*, 4876.
187. For reviews see: (a) Ryabov, A.D. *Synthesis* **1985**, 233. (b) Omae, I. *Chem. Rev.* **1979**, *79*, 287. (c) Pfeffer, M.; Dehand, P. *Coord. Chem. Rev.* **1976**, *18*, 327.
188. Brisdon, B.J.; Nair, P.; Dyke, S.F. *Tetrahedron* **1981**, *37*, 173.
189. Horino, H.; Inoue, N. *J. Org. Chem.* **1981**, *46*, 4416.
190. Carr, K.; Sutherland, J.K. *J. Chem. Soc., Chem. Comm.* **1984**, 1227.
191. For reviews see: (a) Broene, R.D. "Reductive Dimerization of Alkenes and Alkynes" in "Comprehensive Organometallic Chemistry II," Abel, E.W.; Stone, F.G.A.; Wilkinson, G., Eds., Pergamon, Oxford, UK, Vol. 12, pp. 326-347, 1995. (b) Hanzawa, Y.; Ito, H.; Taguchi, T. *Synlett* **1995**, 299. (c) Negishi, E.; Takahashi, T. *Acc. Chem. Res.* **1994**, *27*, 124.
192. Swanson, D.R.; Rousset, C.J.; Negishi, E-I.; Takahashi, T.; Seki, T.; Saburi, M.; Uchida, Y. *J. Org. Chem.* **1989**, *54*, 3521.
193. (a) Fagan, P.J.; Nugent, W.A. *J. Am. Chem. Soc.* **1988**, *110*, 2310. See also (b) Hara, R.; Nishihara, Y.; Landre, P.D.; Takahashi, T. *Tetrahedron Lett.* **1997**, *38*, 447.

120

194. (a) Nugent, W.A.; Taber, D.F. *J. Am. Chem. Soc.* **1989**, *111*, 6435. (b) Negishi, E-I.; Holmes, S.J.; Tour, J.M.; Miller, J.A.; Cederbaum, F.E.; Swanson, D.R.; Takahashi, T. *J. Am. Chem. Soc.* **1989**, *111*, 3336. (c) Negishi, E-I.; Miller, S.R. *J. Org. Chem.* **1989**, *54*, 6014.
195. Wender, P.A.; Rice, K.D.; Schmute, M.E. *J. Am. Chem. Soc.* **1997**, *119*, 7897.
196. Negishi, E-i.; Ma, S.; Sugihara, T.; Noda, Y. *J. Org. Chem.* **1997**, *62*, 1920.
197. (a) Mori, M.; Saitoh, F.; Uesaka, N.; Okamura, K.; Date, T. *J. Org. Chem.* **1994**, *59*, 4993. (b) Uesaka, N.; Saitoh, F.; Mori, M.; Shibasaki, M.; Okamura, K.; Date, T. *J. Org. Chem.* **1994**, *59*, 5633.
198. (a) Luker, T.; Whitby, R.J. *Tetrahedron Lett.* **1995**, *36*, 4109. (b) Probert, G.D.; Whitby, R.J.; Coote, S.J. *Tetrahedron Lett.* **1995**, *36*, 4113. (c) Gordon, C.J.; Whitby, R.J. *Synlett* **1995**, 77. (d) For related insertion of propargyl chloride see: Gordon, G.J.; Whitby, R.J. *J. Chem. Soc., Chem. Comm.* **1997**, 1045.
199. Takahashi, T.; Kotora, M.; Kasai, K.; Suzuki, N. *Organometallics* **1994**, *13*, 4183.
200. (a) Urabe, H.; Suzuki, K.; Sato, F. *J. Am. Chem. Soc.* **1997**, *119*, 10014. (b) See also Urabe, H.; Takeda, T.; Hideura, D.; Sato, F. *J. Am. Chem. Soc.* **1997**, *119*, 11295. (c) Garcia, A.M.; Maseareñas, J.L.; Castedo, L.; Mouriño, A. *J. Org. Chem.* **1997**, *62*, 6353.
201. (a) Buchwald, S.L.; Watson, B.T.; Wannamaker, M.W.; Dewan, J.C. *J. Am. Chem. Soc.* **1989**, *111*, 4486. (b) Coles, N.; Whitby, R.J.; Blagg, J. *Syn. Lett.* **1992**, 143.
202. Barluenga, J.; Sanz, R.; Fañanas, F. *J. Org. Chem.* **1997**, *62*, 5953.
203. Fairfax, D.; Stein, M.; Livinghouse, T.; Jensen, M. *Organometallics* **1997**, *16*, 1523.
204. Knight, K.S.; Waymouth, R.M. *J. Am. Chem. Soc.* **1991**, *113*, 6268.
205. Takahashi, T.; Seki, T.; Nitto, Y.; Saburi, M.; Rousset, C.J.; Negishi, E-I. *J. Am. Chem. Soc.* **1991**, *113*, 6266.
206. (a) Hoveyda, A.H.; Xu, Z. *J. Am. Chem. Soc.* **1991**, *113*, 5079. (b) Hoveyda, A.H.; Xu, Z,.; Morken, J.P.; Houri, A.F. *J. Am. Chem. Soc.* **1991**, *113*, 8950. (c) Hoveyda, A.H.; Morken, J.P.; Houri, A.F.; Xu, Z. *J. Am. Chem. Soc.* **1992**, *114*, 6692.
207. For a review see Hoveyda, A.H.; Morken, J.P. *Angew. Chem. Int. Ed. Engl.* **1996**, *35*, 1262.
208. Xu, Z.; Johannes, C.W.; Houri, A.F.; La, D.S.; Cogan, D.A.; Hofichina, G.E.; Hoveyda, A.H. *J. Am. Chem. Soc.* **1997**, *119*, 10302.
209. Visser, M.S.; Heron, N.M.; Didiuk, M.T.; Segal, J.F.; Hoveyda, A.H. *J. Am. Chem. Soc.* **1996**, *118*, 4291 and references therein.
210. For a review see: Tamao, K.; Kobayashi, K.; Ito, Y. *Syn. Lett.* **1992**, 539.

CHAPTER 5

Synthetic Applications of Transition Metal Carbonyl Complexes

5.1   Introduction

Carbon monoxide coordinates to virtually all transition metals, and an enormous number of complexes with a bewildering array of structures are known. However, most synthetically useful reactions of metal carbonyl complexes, those in which the CO ligand is more than a spectator ligand, involve the first row, homoleptic ("all CO") carbonyl complexes shown in Figure 5.1. Carbon monoxide is among the best π-acceptor ligands, stabilizing low oxidation states and high electron density on the metal. Many neutral M° carbonyl complexes can be reduced to very low *formal* oxidation states (-I through -III), and the resulting anions are powerful

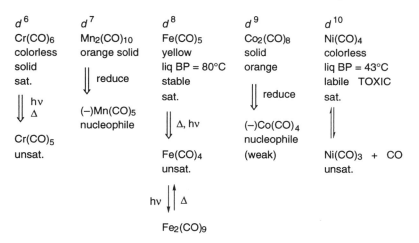

Figure 5.1   Homoleptic Metal Carbonyls

nucleophiles. The ease of migratory insertion of CO into metal-carbon σ-bonds makes metal carbonyls attractive catalysts for the introduction of carbonyl groups into organic substrates, and the reversibility of this process is central to *decarbonylation* processes. The general reactions that metal carbonyls undergo are shown in Figure 5.2. Specific examples will be discussed in the following sections.

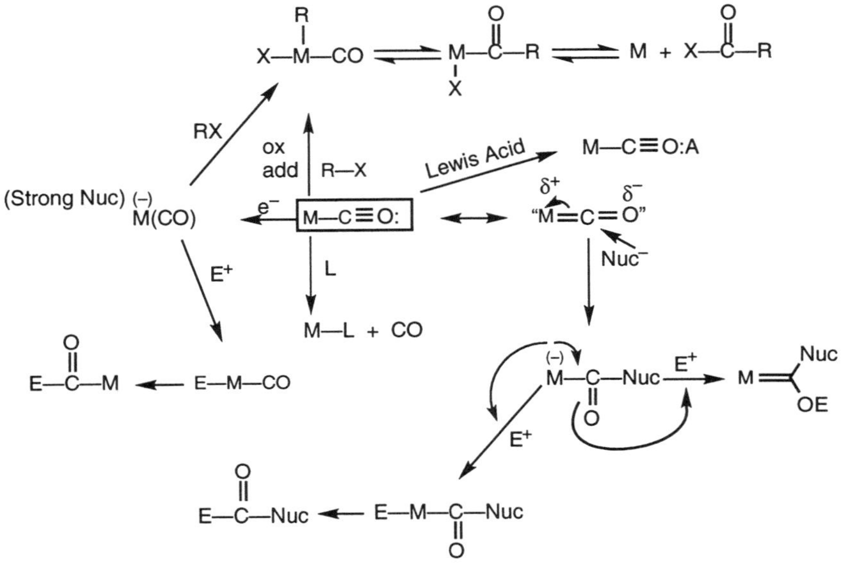

Figure 5.2 Reactions of Metal Carbonyls

5.2   Coupling Reactions of Metal Carbonyls

Allylic halides undergo clean coupling at the less-substituted terminus when treated with nickel carbonyl in polar solvents such as DMF or THF (Eq. 5.1).[2] The reaction is reasonably well-understood, and involves dissociation of one CO from the coordinatively saturated, very labile, nickel carbonyl (a) followed by oxidative

Eq. 5.1

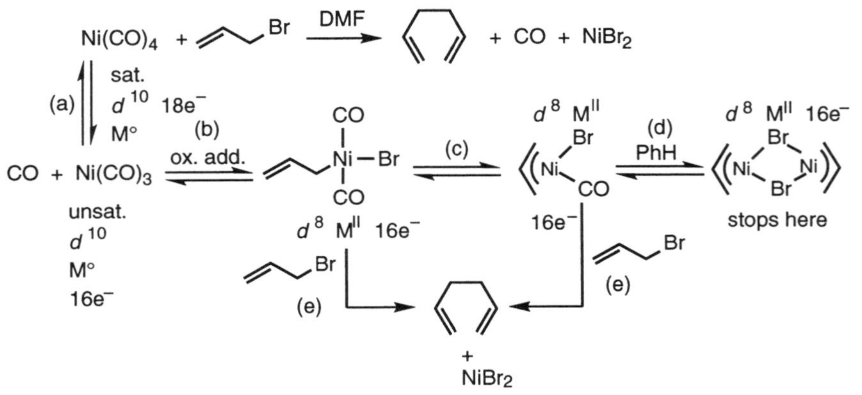

addition of the allylic halide to the unsaturated $Ni(CO)_3$ fragment (b) giving ultimately the η^3-allylnickel carbonyl intermediate (c). This complex has been detected spectroscopically (ν_{CO} 2060 cm^{-1}) and can be generated independently from CO treatment of η^3-allylnickel bromide dimer. When the reaction is run in apolar solvents such as benzene, coupling does not occur; rather the η^3-allylnickel bromide dimer can be isolated in excellent yield (d) (Chapter 9). However, in DMF both the dimer and the η^3-allylnickel carbonyl bromide monomer react quickly with excess allylic bromide to give the coupled product, biallyl. All steps save the last one appear to be reversible, since simple treatment of η^3-allylnickel bromide dimer with carbon monoxide in DMF results in efficient coupling, and both allyl bromide and $Ni(CO)_4$ can be detected in solution. Under high dilution conditions α,ω-bis-allyl halides are cyclized (Eq. 5.2).[3] In this way simple 12, 14, and 18 membered rings,[4] humulene[5] and macrocyclic lactones[6] have been prepared. This potentially useful coupling has been utilized very little, primarily because nickel carbonyl is a volatile (BP 43°C), highly toxic, colorless liquid that is difficult to handle and to dispose.

Eq. 5.2

In contrast, iron pentacarbonyl is considerably less toxic, less volatile, and easier to handle, and a number of synthetically useful procedures involving it have been developed. It is, however, not very labile, and since it is coordinatively saturated, one CO must be lost before reactions can take place. As a consequence, reactions involving $Fe(CO)_5$ frequently require heating, sonication, or photolysis to generate the coordinatively unsaturated $Fe(CO)_4$ fragment. The species $Fe(CO)_4$ can be generated under much milder conditions from the gold/orange dimer, $Fe_2(CO)_9$, prepared by irradiation of $Fe(CO)_5$ in acetic acid. Very gentle heating of solutions (suspensions) of $Fe_2(CO)_9$ releases $Fe(CO)_4$ for reactions in which substrates or products are thermally or photochemically unstable.

Reaction of α,α'-disubstituted-α,α'-dibromoketones or bis(sulfonyl)ketones[8] with $Fe(CO)_4$ generates an iron(II) oxallyl cation (Eq. 5.3) which undergoes a variety of synthetically interesting [3+4] cycloadditions to give seven-membered rings (Eq. 5.4). Generation of the iron-oxallyl cation is usually done *in situ*, and is likely to involve oxidative addition (nucleophilic attack) of the bromoketone to the electron-rich iron(0) species, to give an iron(II) C-enolate. Rearrangement to the O-enolate followed by loss of Br$^-$ gives the reactive oxallyl cation. A variety of dienes undergo cycloaddition, including five membered aromatic heterocycles, to give bridge bicyclic systems (Eq. 5.5). α-Substitution on the bromoketone is required to stabilize positive charge in the complex, so dibromoacetone itself does not react. However, tetrabro-

Eq. 5.3

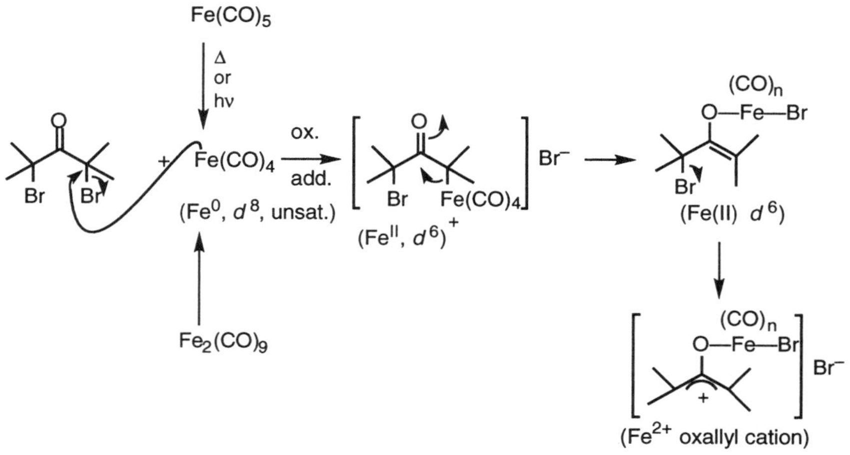

Eq. 5.4

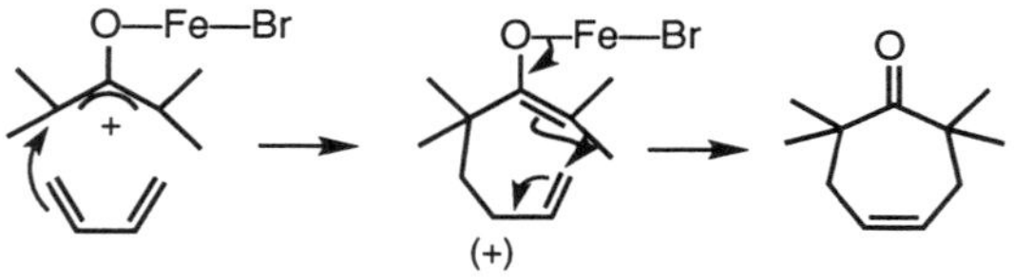

Eq. 5.5

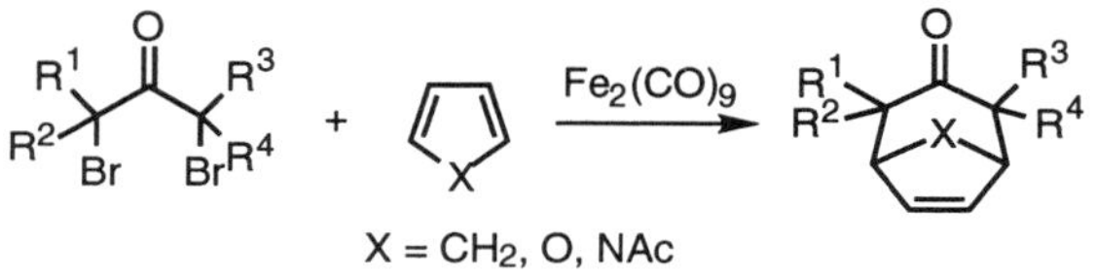

X = CH$_2$, O, NAc

moacetone is reactive, permitting the introduction of the unsubstituted acetone fragment, after a subsequent reductive debromination (Eq. 5.6).[9] This strategy has been used to synthesize tropane alkaloids.[10] Cycloheptatrienone undergoes cyclo-

Eq. 5.6

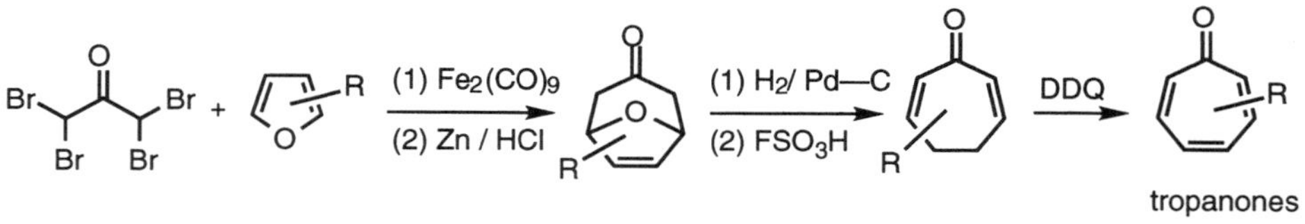

tropanones

addition to the carbonyl group instead of the olefin, probably because of the highly stabilized cation resulting from initial attack at oxygen (Eq. 5.7).[11] Dioxofulvenes undergo a very unusual 3+3 cycloaddition (Eq. 5.8).[12]

Iron oxallyl cation complexes also undergo cycloaddition reactions with electron-rich alkenes, such as styrene or enamines, to produce cyclopentanones (Eq. 5.9).[13] Even relatively nonnucleophilic olefins such as N-tosyl enamines react cleanly (Eq. 5.10).[14]

Eq. 5.7

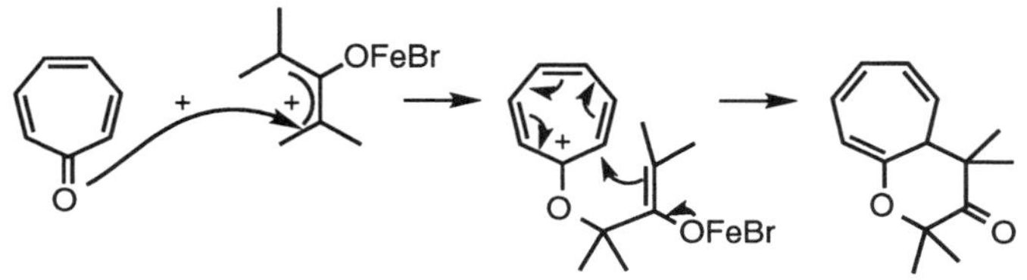

Eq. 5.8

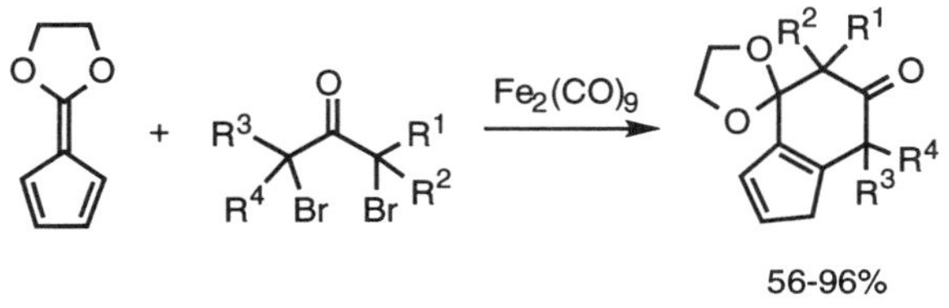

56-96%

Eq. 5.9

Y = Ph, NR₂

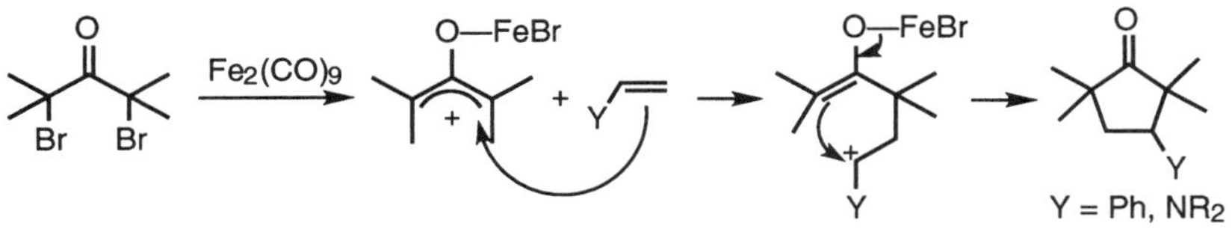

Eq. 5.10

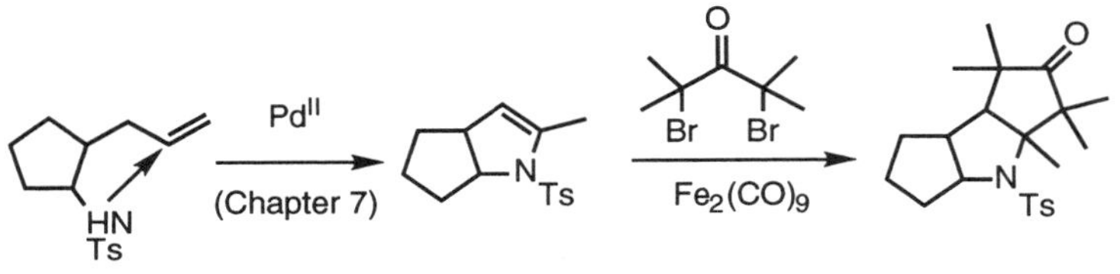

5.3   Carbonylation Reactions[15]

By far the most common synthetic reactions of transition metal carbonyls are those which introduce a carbonyl group into the product. A large number of industrial processes, including hydroformylation (oxo), hydrocarboxylation, and the Monsanto acetic acid process, are predicated on the facility with which CO inserts into metal carbon bonds. Carbonylation reactions are also useful in the synthesis of fine chemicals.

Although nickel carbonyl couples allylic halides in polar solvents, in less polar solvent in the presence of methanol they are converted to β,γ-unsaturated esters (Eq. 5.11). Under these conditions CO insertion/methanol cleavage is faster than reaction with allyl halides (coupling) which requires very polar solvents.

Eq. 5.11

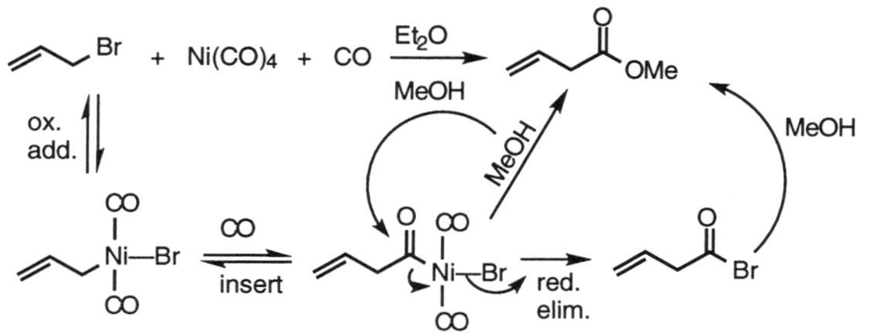

Iron carbonyl promotes a very unusual cyclocarbonylation of allyl epoxides to produce β-lactones or γ-lactones, depending on conditions.[16] The process is somewhat complex, and involves treatment of vinyl epoxides with $Fe(CO)_4$, generated from $Fe(CO)_5$ by heat, sonication or irradiation, or from $Fe_2(CO)_9$ under milder conditions (Eq. 5.12). $Fe(CO)_4$ is a good nucleophile, and can attack (oxidative addition) the vinyl epoxide in an S_N2' manner to give a cationic σ-alkyl-

Eq. 5.12

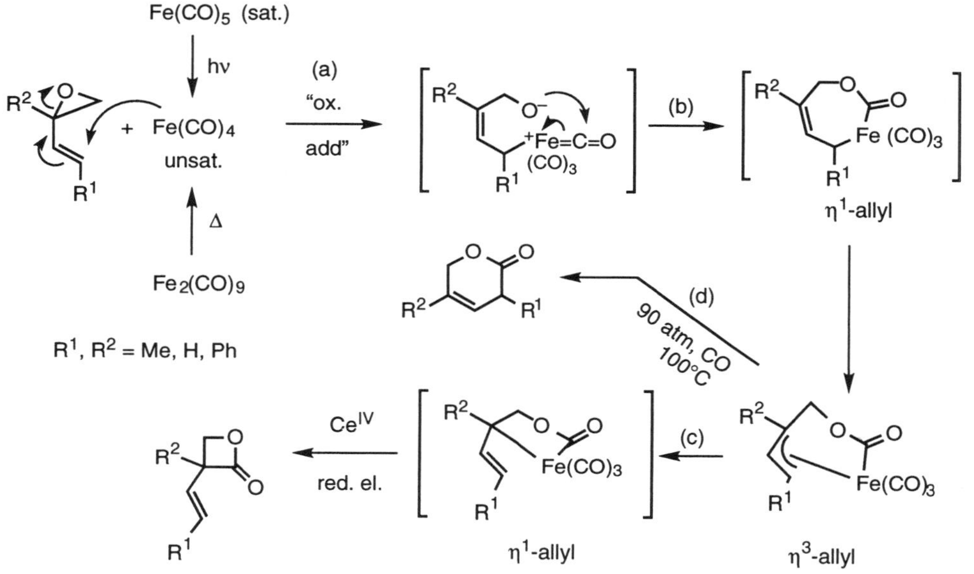

iron(II) carbonyl complex (a). Cationic metal carbonyl complexes are prone to nucleophilic attack at a carbonyl group, in this case producing a "ferrilactone" complex (b), best represented as the η^3-allyl complex, which is stable and isolable. Oxidation of this complex with cerium(IV) leads to the β-lactone (c), presumably by an "oxidatively driven reductive elimination" from the five membered η^1-allyl complex. In contrast, decomposition of the η^3-allyl complex by treatment with high pressures of carbon monoxide at elevated temperatures produces the δ-lactone, formally by reductive elimination from the *seven* membered η^1-allyl complex. This chemistry has been used in the total synthesis of a variety of natural products (Eq. 5.13[17] and Eq. 5.14[18]).

Eq. 5.13

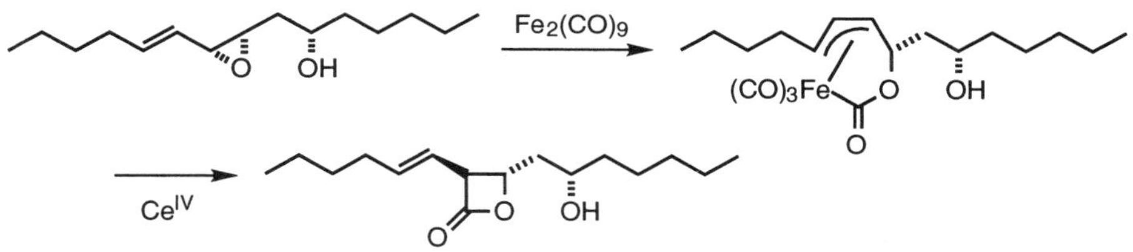

Eq. 5.14

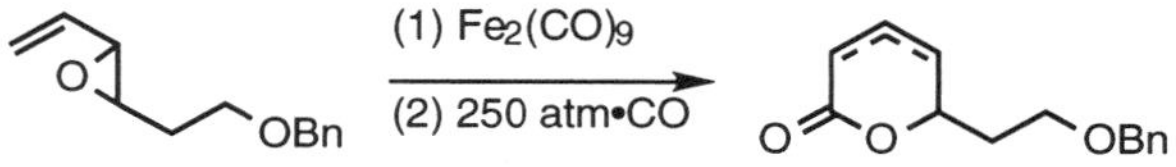

These η^3-allyliron lactone complexes can also be used to synthesize β-lactams. In the presence of Lewis acids, amines will attack the allyl complex in an S_N2' manner to produce a rearranged η^3-allyliron lactam complex. Oxidation of this complex produces β-lactams (Eq. 5.15).[19] The same η^3-allyliron lactam complexes are available directly from the amino alcohol, and it is likely that the conversion of the lactone complex to the lactam complex proceeds through the amino alcohol.

Eq. 5.15

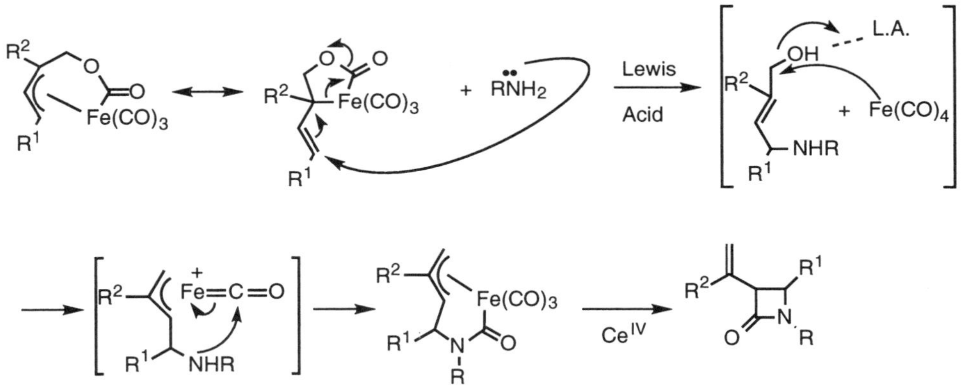

In the above carbonylations, the first step is nucleophilic attack of the zero-valent metal on the substrate (S_N2-like oxidative addition, Chapter 2). Anionic metal carbonyl complexes are even stronger nucleophiles, and effect a number of synthetically useful carbonylation reactions.

Probably the most intensively-studied and highly-developed (for organic synthesis) anionic metal carbonyl species is disodium tetracarbonylferrate, $Na_2Fe(CO)_4$, Collman's reagent.[20] This d^{10} Fe(-II) complex is easily prepared by reduction of the commercially available iron pentacarbonyl by sodium-benzophenone ketyl in dioxane at reflux. (This complex is also available commercially.)

This complex is useful in synthesis specifically because of its high nucleophilicity and the ease of the CO-insertion reaction in this system. Figure 5.3 summarizes this chemistry. Organic halides and tosylates react with $Na_2Fe(CO)_4$

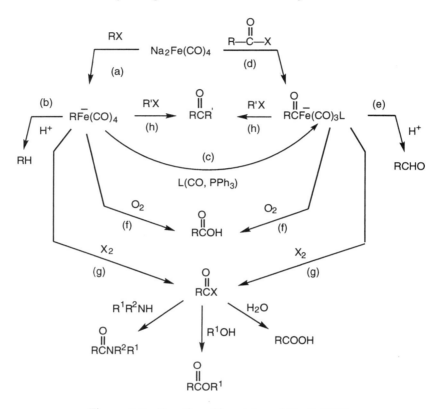

Figure 5.3   Reaction Chemistry of $Na_2Fe(CO)_4$

with typical S_N2 kinetics (second-order), stereochemistry (inversion), and order of reactivity ($CH_3 > RCH_2 > R'RCH$; $RI > RBr > ROTs > RCl$; vinyl, aryl, inert) to produce coordinatively saturated anionic d^8 alkyliron(0) complexes (a). These undergo protolytic cleavage to produce the corresponding hydrocarbon, an overall reduction of the halide (b). In the presence of excess carbon monoxide or added triphenylphosphine, migratory insertion takes place to form the acyl complex (c), accessible directly by the reaction of $Na_2Fe(CO)_4$ with acid halides (d). This acyl complex reacts with acetic acid to produce aldehydes, providing a high-yield conversion of alkyl and acyl halides to aldehydes (e). Oxidative cleavage of either the acyl or alkyliron(0) complex by oxygen (f) or halogen (g) produces carboxylic acid derivatives. Finally, the acyliron(0) complex itself is still sufficiently nucleophilic to react with reactive alkyl iodides to give unsymmetric ketones in excellent yield (h). Interestingly, the alkyliron(0) complex reacts in a similar fashion with alkyl iodides to produce unsymmetric ketones inserting CO along the way (h).

Thus this chemistry provides routes from alkyl and acid halides to alkanes, aldehydes, ketones, and carboxylic acid derivatives.

The reagent is quite specific for halides. Ester, ketone, nitrile, and olefin functionality is tolerated elsewhere in the molecule. In mixed halides (i.e., chlorobromo compounds) the more reactive halide is the exclusive site of reaction. The major limitation is the high basicity (pK_b about that of OH^-) of $Na_2Fe(CO)_4$, which results in competing elimination reactions with tertiary and secondary substrates. Additionally, allylic halides having alkyl groups δ to the halide fail as substrates, since stable 1,3-diene iron complexes form preferentially. Finally, since the migratory insertion fails when the R group contains adjacent electronegative groups, such as halogen or alkoxy groups, the syntheses involving insertion are limited to simple primary or secondary substrates.

The chemistry described above has been subjected to very close experimental scrutiny, and as a consequence, the mechanistic features are understood in detail. Both the alkyliron(0) complex and the acyliron(0) complex have been isolated as airstable, crystalline $[(Ph_3P)_2N]^+$ salts, fully characterized by elemental analysis and IR and NMR spectra. These salts undergo the individual reactions detailed in Figure 5.3. Careful kinetic studies indicate that ion-pairing effects dominate the reactions of $Na_2Fe(CO)_4$, and account for the observed 2×10^4 rate increase (tight ion pair $[NaFe(CO)_4]^-$ versus solvent-separated ion pair $[Na^+:S:Fe(CO)_4]^-$ as the kinetically active species) in going from THF to N-methylpyrrolidone. The rate law, substrate order of reactivity, stereochemistry at carbon, and activation parameters (particularly the large negative entropy of activation) are all consistent with an S_N2-type oxidative-addition, with no competing one-electron mechanism. The migratory-insertion step is also subject to ion-pairing effects, and is accelerated by tight ion pairing $[Li^+ > Na^+ > Na\text{-}crown^+ > (Ph_3P)_2N^+]$. The insertion reaction is overall second-order, first-order in both $NaRFe(CO)_4$ and added ligand, with about a twenty-fold difference in rate over the range of ligands Ph_3P (slowest) to Me_3P (fastest).[21]

An interesting variation of this chemistry[22] results in the production of ethyl ketones from organic halides, $Na_2Fe(CO)_4$, and ethylene (Eq. 5.16). The reaction is

Eq. 5.16

$$ RX + Fe(CO)_4^{2-} \xrightarrow{L} RCFe(CO)_3L^- + CH_2 = CH_2 \longrightarrow RCCH_2CH_2Fe(CO)_r^- $$

$$ \longrightarrow RCCHCH_3 \xrightarrow{H^+} RCCH_2CH_3 $$
$$ \underset{Fe(CO)_n}{|^-} $$

thought to proceed via the acyl iron(0) complex, which inserts ethylene into the acyl-metal bond and rapidly rearranges to the α-metalloketone, which then cannot further insert ethylene since the R group now contains an adjacent electronegative acyl group. Intramolecular versions of this can be used to synthesize cyclic ketones (Eq. 5.17),[23] but in some cases the process is not regioselective (Eq. 5.18).[24]

Eq. 5.17

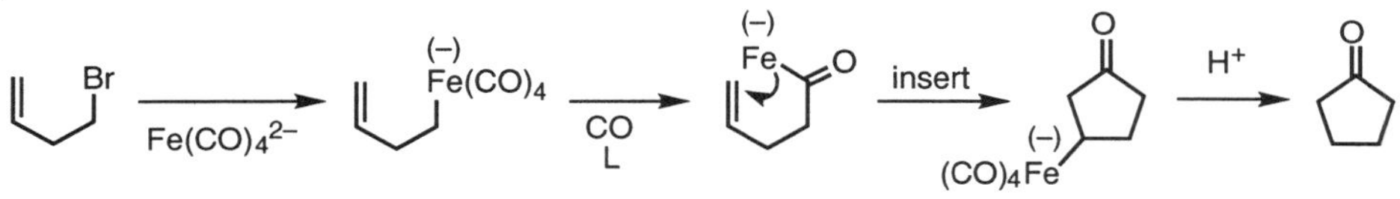

Eq. 5.18

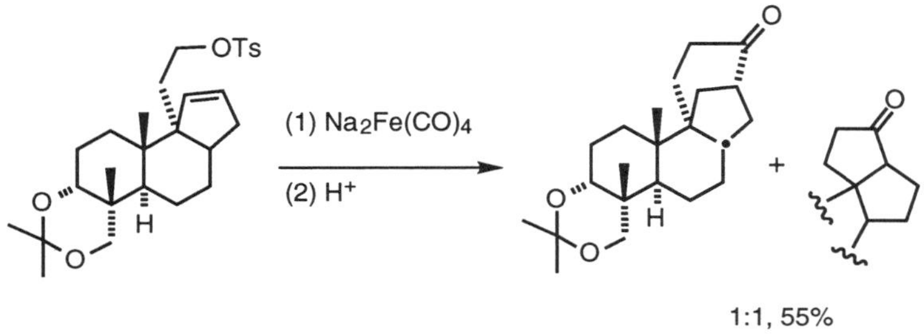

1:1, 55%

Anionic nickel acyl complexes, generated *in situ* by the reaction of organolithium reagents with the toxic, volatile nickel carbonyl, effect a number of synthetically useful transformations including the acylation of allylic halides, 1,4-acylation of conjugated enones, and 1,2-addition to quinones (Eq. 5.19).[25] Because of

Eq. 5.19

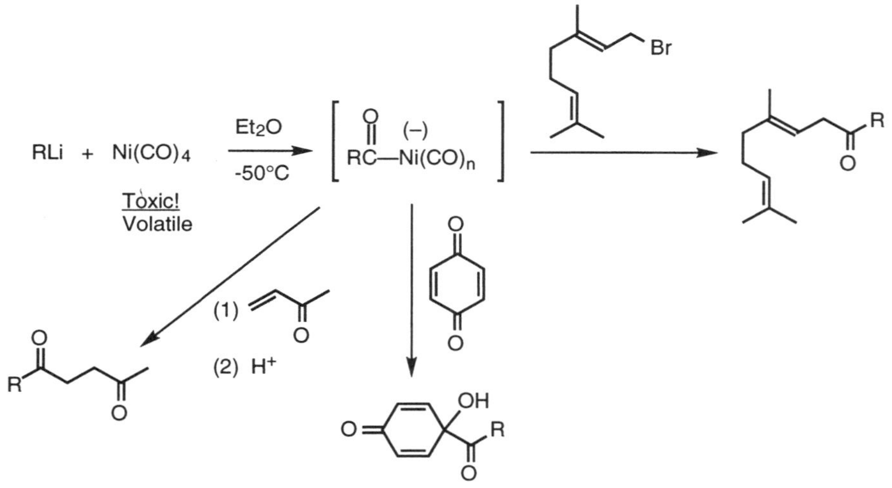

the toxicity of the reagent, few groups have had the courage to use it, although one nice synthetic application has been reported (Eq. 5.20).[26] The same general reaction chemistry can be achieved replacing nickel carbonyl with the orange, air stable, non-volatile crystalline solid $(PPh_3)(CO)_2(NO)Co$, a complex much more easily handled.[27]

No one has yet seen fit to use this in synthesis, however. (This complex is isoelectronic with $Ni(CO)_4$, a nickel(0), d^{10} complex, since the nitrosyl ligand is formally a cation, making this a cobalt (-1), d^{10} complex.)

Eq. 5.20

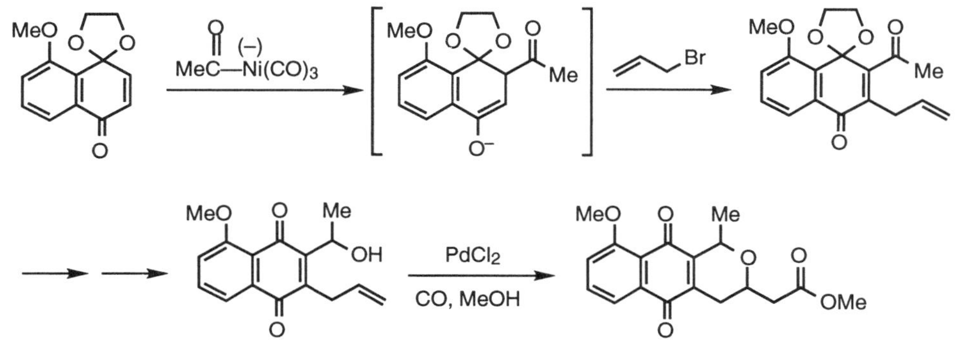

Alkenyl and aryl iodides are converted to esters by treatment with nickel carbonyl and sodium methoxide in methanol (Eq. 5.21).[28] The mechanism of this process is not known but is likely to involve an anionic nickel acyl species. Intramolecular versions have also been developed (Eq. 5.22),[29] including multiple CO insertions (Eq. 5.23).[30]

Eq. 5.21

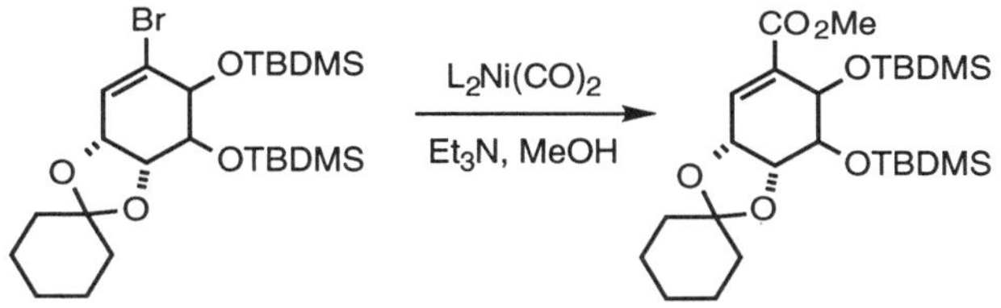

Eq. 5.22

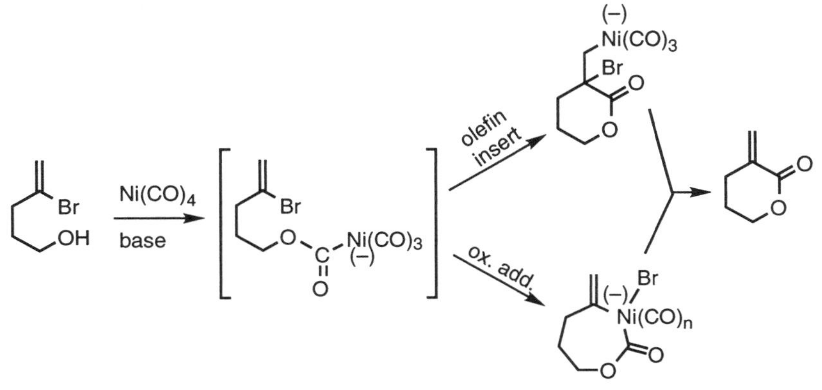

Eq. 5.23

5.4   Decarbonylation Reactions[31]

Much of this chapter has been devoted to consideration of methods for the introduction of carbonyl groups into organic substrates via a "migratory insertion" reaction, in which a metal alkyl or aryl group migrates to an adjacent, coordinated CO to produce a σ-acyl complex, which is then cleaved to produce the organic carbonyl compound. The reverse process, in which organic carbonyl compounds (specifically aldehydes and acid chlorides) are decarbonylated, is also possible, and often quite useful in organic synthesis. Although a number of transition metal complexes will function as decarbonylation agents, by far the most efficient is $(Ph_3P)_3RhCl$, the Wilkinson hydrogenation catalyst. This complex reacts with alkyl, aryl, and alkenyl aldehydes under mild conditions to produce the corresponding hydrocarbon and the very stable $(Ph_3P)_2Rh(Cl)CO$, as in Eq. 5.24. The course of this stoichiometric decarbonylation can easily be monitored by the change in color of the solution, from the deep red of $(Ph_3P)_3RhCl$ to the canary yellow of $(Ph_3P)_3Rh(Cl)CO$. *trans*-α-Alkylcinnamaldehydes are decarbonylated with retention of olefin geometry, producing *cis*-substituted styrenes.[32] More impressively, chiral aldehydes are converted to chiral hydrocarbons with overall retention and with a high degree of stereoselectivity (Eq. 5.25).[33] The mechanism proposed for this decarbonylation is

Eq. 5.24

Eq. 5.25

that in Eq. 5.26. It involves, as a first step, oxidative addition of the aldehyde to the rhodium complex. The next step is the reverse of the carbonyl insertion reaction, that is, migration of the alkyl group from carbonyl to metal. It is well-known that migratory insertion is a reversible process, and can be driven in either direction by the appropriate choice of reaction conditions. In addition, it has been shown in other cases that this reversible transformation proceeds with retention of stereochemistry of the migrating alkyl group; thus the retention of stereochemistry in the decarbonylation of chiral aldehydes. The last step is the reductive elimination of the alkane (RH) and the production of $RhCl(CO)L_2$. This last step is irreversible (alkanes do not oxidatively add to $RhCl(CO)L_2$), and drives the entire process to completion. Since $RhCl(CO)L_2$ is much less reactive in oxidative addition reactions than $RhClL_3$ (CO is a π acceptor and withdraws electron density from the metal), it does not react with aldehydes under these mild conditions, and the decarbonylation described above is stoichiometric. When the reaction is carried out at temperatures in excess of 200°C, both $RhClL_3$ and $RhCl(CO)L_2$ function as decarbonylation catalysts, presumably because $RhCl(CO)L_2$ will oxidatively add aldehydes under these severe conditions.

Eq. 5.26

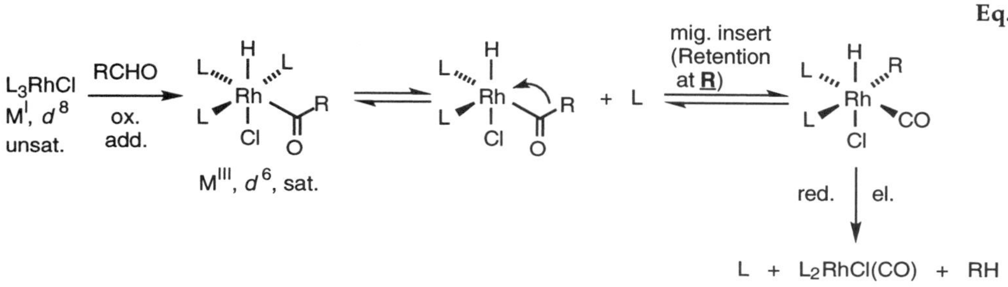

Quite complex aldehydes can be decarbonylated without problems (Eq. 5.27[34] and Eq. 5.28[35]).

Eq. 5.27

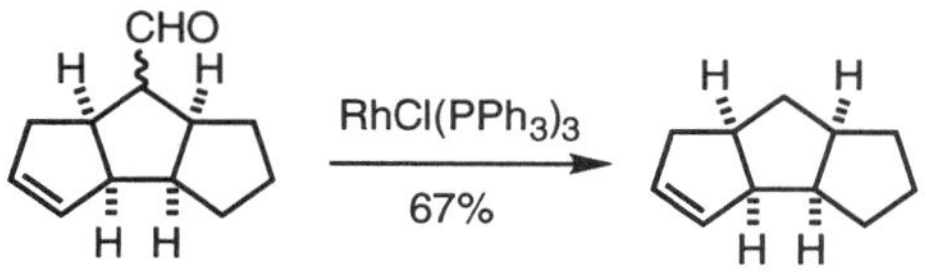

Eq. 5.28

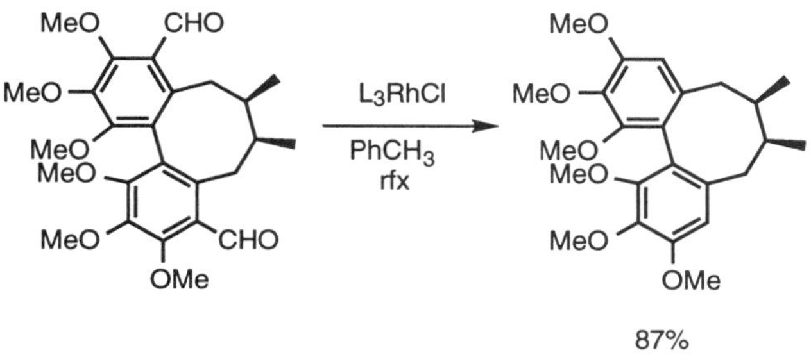

87%

Acid chlorides also undergo decarbonylation upon treatment with RhCl(PPh₃)₃, but the reaction suffers several complications not encountered with aldehydes. The reaction of acid chlorides is most straightforward with substrates that have no β-hydrogens. In these cases, decarbonylation occurs smoothly to give the alkyl chloride. In contrast to aldehydes, the initial oxidative adduct is quite stable, and has been isolated and fully characterized in several instances. Heating this adduct leads to decarbonylation, via a mechanism thought to be strictly analogous to that involved in aldehyde decarbonylation. However, the decarbonylation of acid halides differs from that of aldehydes in several respects. Decarbonylation of optically active acid chlorides with RhCl(PPh₃)₃ produces racemic alkyl chlorides. This is in direct contrast to aldehydes, which undergo decarbonylation with a high degree of retention.

Acid halides with β-hydrogens undergo decarbonylation by RhCl(PPh₃)₃ to produce primarily olefins resulting from β-hydride elimination from the σ-alkyl intermediate, rather than reductive elimination (Eq. 5.29). Branched acid chlorides that can undergo β-elimination in several directions give mixtures of products with the most substituted olefins predominating.

Eq. 5.29

Contrary to older literature reports, aroyl chlorides and α,β-unsaturated acid chlorides do *not* undergo decarbonylation to give the corresponding unsaturated chloride, but rather produce stable aryl metal complexes or quaternary phosphonium salts.[36]

5.5   Metal Acyl Enolates[37]

Many of the metal-acyl species discussed above are quite stable, and can be isolated and handled easily. One complex, CpFe(CO)(PPh₃)(COCH₃), has been developed into a useful synthetic reagent. This complex has several interesting characteristics: (1) it is easy to prepare and handle, (2) it is chiral at iron, and can be resolved, and (3) the protons α to the acyl group are acidic and the corresponding metal acyl enolate undergoes reaction with a variety of electrophiles (Eq. 5.30). More interesting, because the complex is chiral at iron and one face is hindered by the triphenylphosphine ligand, reactions of these acyliron enolates occur with very high stereoselectivity (Eq. 5.31). The observed absolute stereochemistry results from alkylation of the E-enolate with the carbonyl group and O⁻ *anti*, from the less hindered face. Oxidative cleavage of the resulting complex produces the carboxylic acid derivative in good yield with high ee.[38]

Eq. 5.30

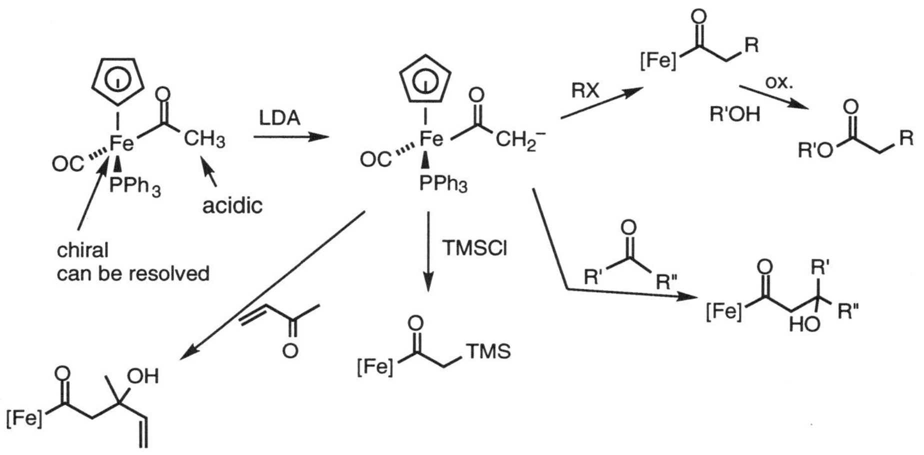

Eq. 5.31

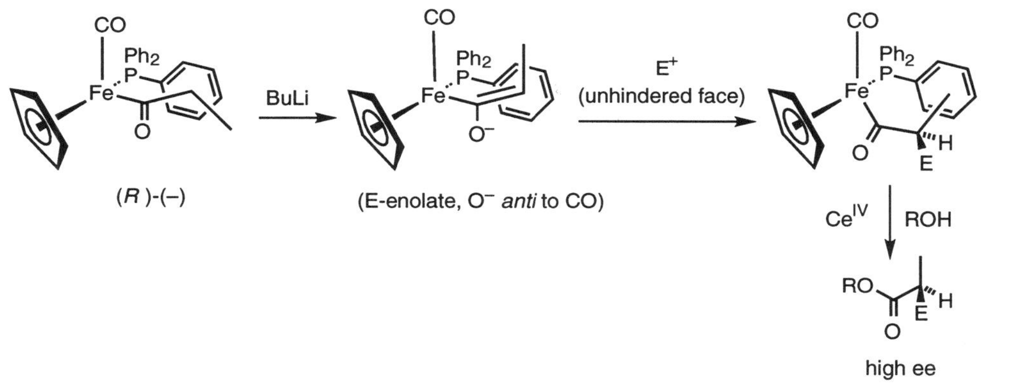

This chemistry has been used extensively in the synthesis of optically active compounds. Two such cases are shown in Eq. 5.32[39] and Eq. 5.33.[40]

As might be expected, α,β-unsaturated iron acyl complexes also undergo reactions in a highly diastereoselective manner. Both conjugate addition/enolate trapping (Eq. 5.34)[41] and γ-deprotonation-α-alkylation (Eq. 5.35)[42] enjoy a high degree of stereocontrol. Vinylogous iron acyl complexes also undergo a variety of alkylation reactions (Eq. 5.36)[43] although these have not yet been used in total synthesis.

Other metal acyl enolate complexes, particularly those of molybdenum and tungsten, have been prepared and their reaction chemistry studied (for example Eq.

Eq. 5.32

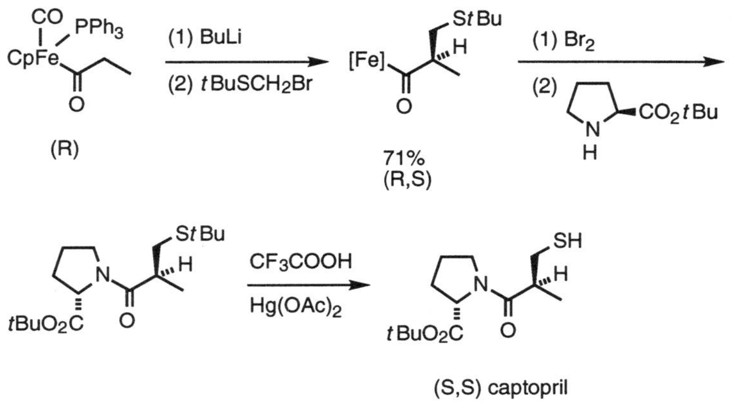

71%
(R,S)

(S,S) captopril

Eq. 5.33

56%

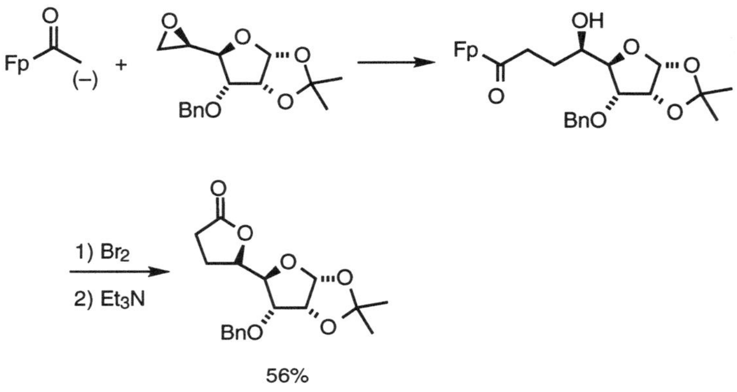

Eq. 5.34

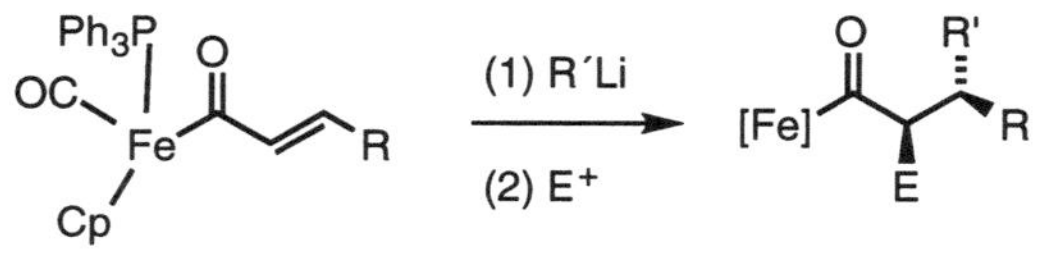

Eq. 5.35

γ-deprotonation   α-alkylation

138

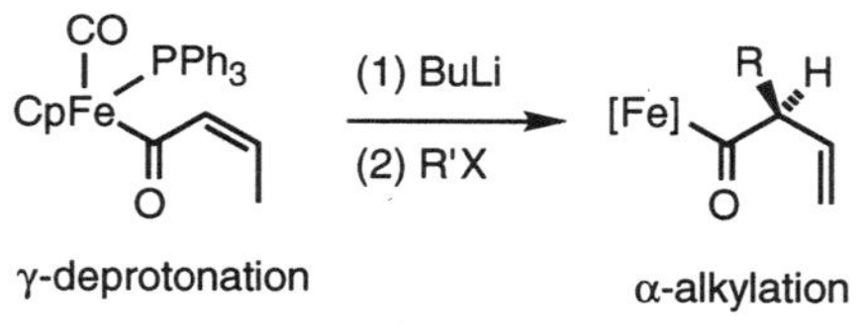

Eq. 5.36

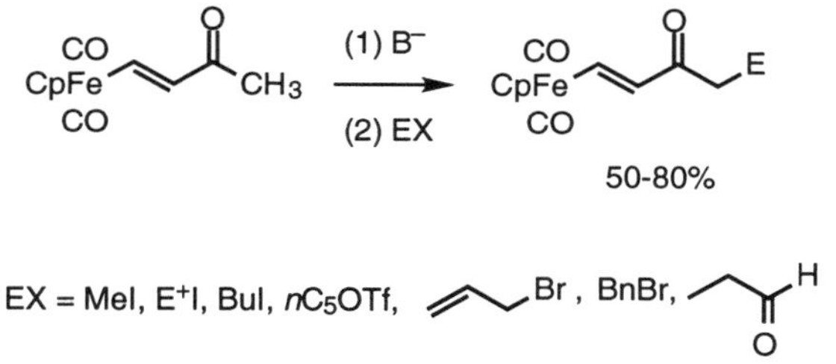

$$EX = MeI, E^+I, BuI, nC_5OTf, \quad \text{(allyl)}Br, BnBr, \quad \text{(acetaldehyde)}$$

5.37[44]). However the focus of these studies has been more organometallic/mechanistic than synthetic, so the potential of these systems in the synthesis of complex molecules is yet to be realized.

Eq. 5.37

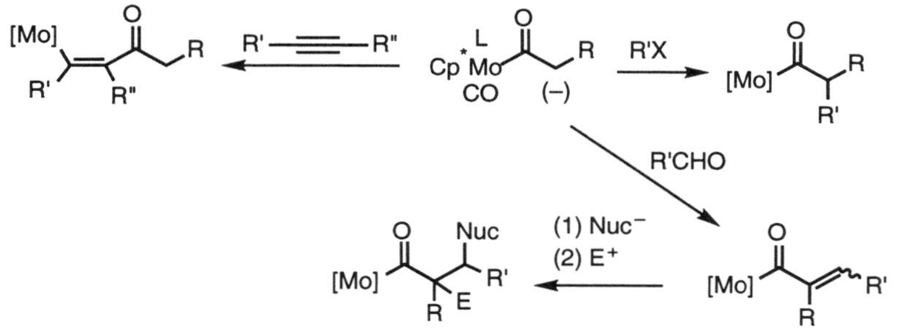

5.6   Bridging Acyl Complexes

Reaction of triiron dodecacarbonyl with thiols, followed by conjugated acid halides produces diiron bridged acyl complexes (Eq. 5.38).[45] These complexes undergo highly exo-selective Diels Alder reactions with dienes under very mild conditions. Decomplexation by oxidative cleavage of the bridging acyl species produces the corresponding thioester (Eq. 5.38).[46] Complexation to iron clearly

Eq. 5.38

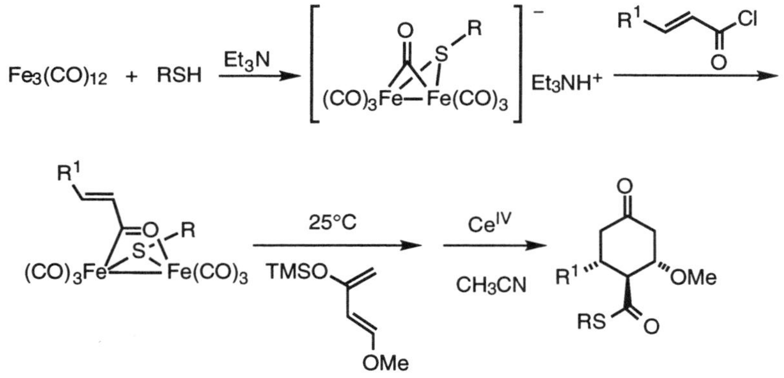

activates the enone towards Diels Alder addition since methyl crotonate undergoes reaction with the same dienes only at much higher temperatures .and with lower selectivity. In contrast, nitrones undergo *endo* selective 1,3-dipolar cycloaddition to these bridging acyl species.[47] By using a chiral thiol to prepare the bridging acyl species, reasonable asymmetric induction has been achieved (Eq. 5.39).[48]

Eq. 5.39

60-96%
>20:1

References

1. For a review see: Bates, R.W., "Transition Metal Carbonyl Compounds," in "Comprehensive Organometallic Chemistry II", Abel, E.W.; Stone, F.G.A.; Wilkinson, G., Eds., Elsevier Science Ltd., Oxford, UK, 1995, Vol. 12, pp. 349-386.
2. Corey, E.J.; Semmelhack, M.F.; Hegedus, L.S. *J. Am. Chem. Soc.* **1968**, *90*, 2416.
3. Dauben, W.G.; Beasley, G.H.; Broadhurst, M.D.; Muller, B.; Peppard, D.J.; Pesnelle, P.; Suter, C. *J. Am. Chem. Soc.* **1974**, *96*, 4724.
4. Corey, E.J.; Wat, E. *J. Am. Chem. Soc.* **1967**, *89*, 2757.
5. Corey, E.J.; Hamanaka, E. *J. Am. Chem. Soc.* **1967**, *89*, 2758.
6. Corey, E.J.; Kirst, H.A. *J. Am. Chem. Soc.* **1972**, *94*, 667.
7. (a) For a review, see: Noyori, R. *Acct. Chem. Res.* **1979**, *12*, 61. (b) Takaya, H.; Makino, S.; Hayakawa, Y.; Noyori, R. *J. Am. Chem. Soc.* **1978**, *100*, 1765, 1778.
8. Hardinger, S.A.; Bayne, C.; Kantorowski, E.; McClellan, R.; Larres, L.; Nuesse, M.-A. *J. Org. Chem.* **1995**, *60*, 1104.
9. Noyori, R.; Sato, T.; Hayakawa, Y. *J. Am. Chem. Soc.* **1978**, *100*, 2561.
10. Hayakawa, Y.; Baba, Y.; Makino, S.; Noyori, R. *J. Am. Chem. Soc.* **1978**, *100*, 1786.
11. Ishizu, T.; Harano, K.; Yasuda, M.; Kanematsu, K. *J. Org. Chem.* **1981**, *46*, 3630.
12. Hong, B-C.; Sun, S-S. *Tetrahedron Lett.* **1996**, *37*, 659.
13. Hayakawa, Y.; Yokoyama, K.; Noyori, R. *J. Am. Chem. Soc.* **1978**, *100*, 1791, 1799.
14. Hegedus, L.S.; Holden, M. *J. Org. Chem.* **1985**, *50*, 3920.
15. Colquhoun, H.M.; Thompson, D.J.; Twigg, M.V. "Carbonylation", Plenum Press, New York, 1991.
16. For a review see: Ley, S.V.; Cox, L.R.; Meek, G. *Chem. Rev.*, **1996**, *96*, 423.
17. Bates, R.W.; Fernandez-Moro, R.; Ley, S.V. *Tetrahedron Lett.* **1991**, *32*, 2651.
18. Kotecha, N.R.; Ley, S.V.; Mantegani, S. *Synlett* **1992**, 395.
19. (a) Annis, G.D.; Hebblethwaite, E.M.; Hodgson, S.T.; Hollingshead, D.M.; Ley, S.V. *J. Chem. Soc., Perkin I* **1983**, 2851. (b) Horton, A.M.; Hollinshead, D.M.; Ley, S.V. *Tetrahedron* **1984**, *40*, 1737.
20. Collman, J.P. *Acct. Chem. Res.* **1975**, *8*, 342.
21. Collman, J.P.; Finke, R.G.; Cawse, J.N.; Brauman, J.I. *J. Am. Chem. Soc.* **1977**, *99*, 2515; *J. Am. Chem. Soc.* **1978**, *100*, 4766.
22. Cooke, M.P., Jr.; Parlman, R.M. *J. Am. Chem. Soc.* **1975**, *97*, 6863.
23. Merour, J.Y.; Roustan, J.L.; Charrier, C.; Collin, J.; Benaim, J. *J. Organomet. Chem.* **1973**, *51*, C24.
24. (a) McMurry, J.E.; Andrus, A.; Ksander, G.M.; Musser, J.H.; Johnson, M.A. *J. Am. Chem. Soc.* **1979**, *101*, 1330. (b) *Tetrahedron* **1981**, *27 Supplement*, 319.
25. (a) Corey, E.J.; Hegedus, L.S. *J. Am. Chem. Soc.* **1969**, *91*, 4926. (b) See also Hermanson, J.R.; Gunther, M.L.; Belletire, J.L.; Pinhas, A.R. *J. Org. Chem.* **1996**, *60*, 1900.
26. Semmelhack, M.F.; Keller, L.; Sato, T.; Spiess, E.J.; Wulff, W. *J. Org. Chem.* **1985**, *50*, 5566.
27. Hegedus, L.S.; Perry, R.J. *J. Org. Chem.* **1985**, *50*, 4955.
28. Blacker, A.J.; Booth, R.J.; Davies, G.M.; Sutherland, J.K. *J. Chem. Soc., Perkin Trans I* **1995**, 2861.

29. (a) Semmelhack, M.F.; Brickner, S.J. *J. Org. Chem.* **1981**, *46*, 1723. (b) Llebaria, A.; Delgado, A.; Camps, F.; Moreto, J.M. *Organometallics* **1993**, *12*, 2825.
30. Llebaria, A.; Camps, F.; Moreto, J.N. *Tetrahedron* **1993**, *49*, 1283.
31. For a review see: Tsuji, J. in *Organic Syntheses via Metal Carbonyls* vol. 2, Wender, I.; Pino, P., eds., Wiley, New York, 1977, pp. 595-654. Catalytic decarbonylation of aldehydes has recently been achieved using Wilkinsons catalyst and $(PhO)_2P(O)N_3$: O'Connor, J.; Ma, J. *J. Org. Chem.* **1992**, *57*, 5075.
32. Tsuji, J.; Ohno, K. *Tetrahedron Lett.* **1967**, 2173.
33. (a) Walborsky, H.M.; Allen, L.E. *Tetrahedron Lett.* **1970**, 823. (b) Walborsky, A.M.; Allen, L.E. *J. Am. Chem. Soc.* **1971**, *93*, 5465.
34. Jung, M.E.; Rayle, H.L. *J. Org. Chem.* **1997**, *62*, 4601.
35. Tanaka, M.; Ohshima, T.; Mitsuhashi, H.; Maruno, M.; Wakamatsu, T. *Tetrahedron* **1995**, *51*, 11693.
36. (a) Kampmeier, J.A.; Rodehorst, R.M.; Philip, Jr., J.B. *J. Am. Chem. Soc.* **1981**, *103*, 1847. (b) Kampmeier, J.A.; Mahalingam, S. *Organometallics* **1984**, *3*, 489. (c) Kampmeier, J.A.; Harris, S.H.; Rodehorst, R.M. *J. Am. Chem. Soc.* **1981**, *103*, 1478.
37. For reviews see: (a) Davies, S.G. *Pure Appl. Chem.* **1988**, *60*, 13. (b) Blacksburn, B.K.; Davies, S.G.; Sutton, K.H.; Whittaker, M. *Chem. Soc. Rev.* **1988**, *17*, 147. (c) Ref. 1.
38. For a review, see: Davies, S.G.; Bashiardes, G.; Beckett, R.P.; Coote, S.J.; Dordor-Hedgecock, I.M.; Goodfellow, C.L.; Gravatt, G.L.; McNally, J.P.; Whittaker, M. *Phil. Trans R. Soc. Lond. A* **1988**, *326*, 619.
39. Bashiardes, G.; Davies, S.G. *Tetrahedron Lett.* **1987**, *28*, 5563.
40. Davies, S.G.; Kellie, H.M.; Polywoka, R. *Tetrahedron:Asymmetry* **1994**, *5*, 2563.
41. Davies, S.G.; Dordor-Hedgecock, I.M.; Easton, R.J.C.; Preston, S.C.; Sutton, K.H.; Walker, J.C. *Bull Soc. Chim. Fr.* **1987**, 608.
42. Davies, S.G.; Easton, R.J.C.; Gonzalez, A.; Preston, S.C.; Sutton, K.H.; Walker, J.C. *Tetrahedron* **1986**, *42*, 3987.
43. Mattson, M.N.; Helquist, P. *Organometallics* **1992**, *11*, 4.
44. Rusik, C.A.; Collins, M.A.; Gamble, A.S.; Tonker, T.L.; Templeton, J.L. *J. Am. Chem. Soc.* **1989**, *111*, 2550.
45. Seyferth, D.; Archer, C.M.; Ruschke, D.P. *Organometallics*, **1991**, *10*, 3363 and references therein.
46. Gilbertson, S.R.; Zhao, X.; Dawson, D.P.; Marshall, K.L. *J. Am. Chem. Soc.*, **1993**, *115*, 8517.
47. Gilbertson, S.R.; Dawson, D.P.; Lopez, O.D.; Marshall, K.L. *J. Am. Chem. Soc.*, **1995**, *117*, 4431.
48. Gilbertson, S.R.; Lopez, O.D. *J. Am. Chem. Soc.*, **1997**, *119*, 3399.

CHAPTER 6

Synthetic Applications of Transition Metal Carbene Complexes

6.1   Introduction

Carbene complexes — complexes having formal metal-to-carbon double bonds — are known for metals across the entire transition series, although relatively few have been developed as useful reagents for organic synthesis. The two bonding extremes for carbene complexes are represented by the *electrophilic*, heteroatom, stabilized "Fischer" carbenes and the *nucleophilic*, methylene or alkylidene, "Schrock" carbenes. Between these two extremes lie the "Grubbs-" (and also different "Schrock-") type carbenes, which are particularly useful for alkene- and alkyne-metathesis reactions. Although they share common structural and reactivity features, their chemistry is sufficiently different to be discussed separately.

6.2   Electrophilic Heteroatom Stabilized "Fischer" Carbene Complexes[1]

The most intensively developed carbene complexes for use in organic synthesis are the electrophilic Fischer carbene complexes of the Group 6 metals, Cr, Mo, W. These are readily synthesized by the reaction of the air-stable, crystalline metal hexacarbonyl with a range of organolithium reagents (Eq. 6.1). One of the six equivalent CO groups undergoes attack to produce the stable, anionic lithium acyl "ate" complex, in which the negative charge is stabilized and extensively delocalized into the remaining five, π-accepting, electron-withdrawing CO groups. These "ate" complexes are normally isolated as the stable tetramethylammonium salt, which can be prepared on a large scale and stored for months without substantial decomposition. Treatment with hard alkylating agents such as methyl triflate or

Eq. 6.1

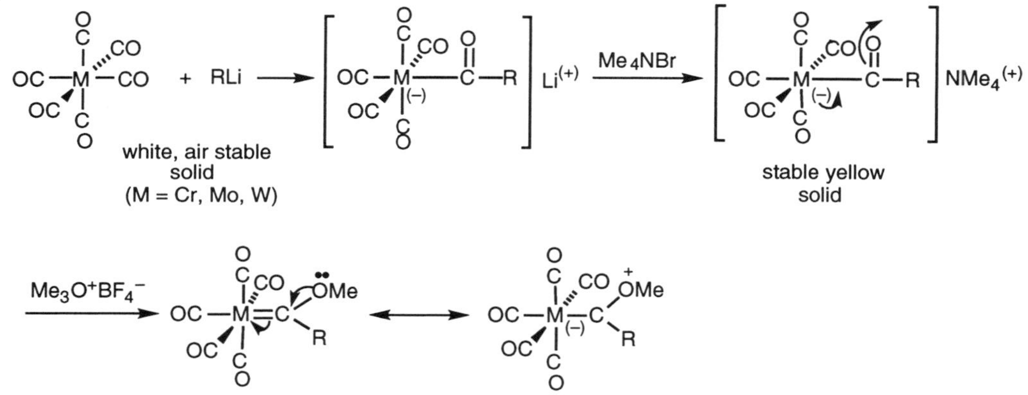

trimethyloxonium salts (methyl Meerwein's reagent), with alkyl halides under phase transfer conditions,[2] or with dimethyl sulfate,[3] results in alkylation at oxygen, producing the alkoxycarbene complex in excellent yield.

Although synthesis via organolithium reagents is the most common approach to carbene complexes, they can be generated by any reaction that produces anionic acyl complexes. Chromium hexacarbonyl is easily reduced to the dianion by sodium naphthalenide or potassium/graphite intercolate. Treatment with acid chlorides generates the anionic acyl complex which again undergoes o-alkylation to produce carbene complexes.[4] Cyclic alkoxycarbene complexes result from intramolecular O-alkylation,[5] while aminocarbene complexes are produced from amides,[6] utilizing TMSCl to assist in elimination of oxygen (Figure 6.1).

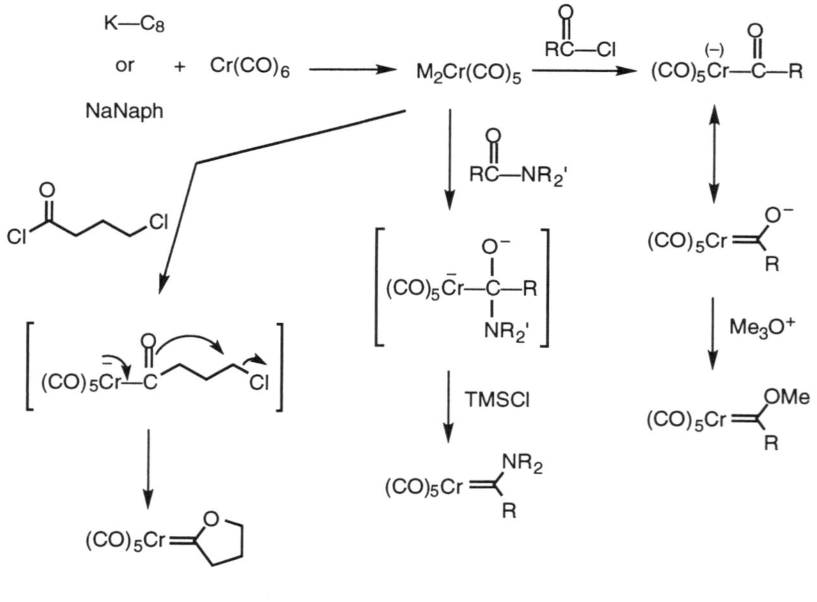

Figure 6.1 Synthesis of Chromium Carbene Complexes from $M_2Cr(CO)_5$

Other, less general, routes to carbene complexes involve the reaction of functionalized organozinc reagents with active $Cr(CO)_5$ species (Eq. 6.2)[7] and the reaction of acetylenic alcohols with active $Cr(CO)_5$ species (Eq. 6.3)[8], (Eq. 6.4),[9] and (Eq. 6.5).[10]

These group 6 carbene complexes are yellow to red crystalline solids, and are easily purified by crystallization or chromatography on silica gel. As solids they are quite air-stable and easy to handle. In solution, they are slightly air-sensitive, particularly in the presence of light, and reactions are best carried out under an inert atmosphere. The heteroatom is required for stability (the diphenyl carbene complex

Eq. 6.2

$$FGR \overset{}{\underset{2}{\diagdown}} Zn \; + \; Cr(CO)_5(THF) \longrightarrow \left[(CO)_5\overset{(-)}{Cr} \diagdown\diagup RFG \right]$$

$$FG = Cl(CH_2)_n-, \; Br(CH_2)_n-, \; AcO(CH_2)_n-$$

CO insert

$$(CO)_5Cr = \overset{OMe}{\underset{RFG}{\diagup}} \xleftarrow{Me_3O^+} \left[(CO)_5\bar{Cr}-\overset{O}{\overset{\|}{C}} \diagdown\diagup RFG \right]$$

Eq. 6.3

$$Cr(CO)_6 \xrightarrow[Et_2O]{h\nu} (CO)_5Cr(OEt_2) \longrightarrow (CO)_5Cr = \overset{O}{\underset{}{\diagdown}}\overset{R^2}{\underset{R^1}{}}$$

with acetylenic alcohol $R^2, R^1 OH$

via

$$\left[(CO)_5Cr - \overset{H}{\underset{R^2 \; HO \; R^1}{\|}} \right] \longrightarrow \left[(CO)_5Cr = \bullet = \overset{H}{\underset{R^1}{\diagdown}} \overset{R^2}{\underset{OH}{}} \right]$$

Eq. 6.4

$$Cr(CO)_6 \; + \; \overset{Li}{\underset{}{}} = \bullet = \overset{OMe}{\underset{Ph}{}} \longrightarrow \left[(CO)_5Cr = \overset{OLi}{\underset{OMe}{}} \overset{Ph}{} \right] \longrightarrow$$

$$\left[(CO)_5Cr = \overset{O}{\underset{Li}{}} \overset{OMe}{\underset{Ph}{}} \right] \xrightarrow[Si \; gel]{H_2O} (CO)_5Cr = \overset{O}{\underset{}{}} \overset{OMe}{\underset{Ph}{}}$$

Eq. 6.5

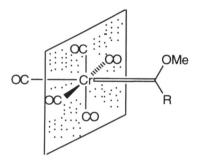

can be made, but decomposes above –20°C), and this stability results from extensive delocalization of the lone pair on the heteroatom into the strongly electron-withdrawing metal pentacarbonyl fragment. This delocalization is evidenced by restricted rotation (14-25 kcal/mol) about the heteroatom-carbene carbon bond, and can be correlated to ^{53}Cr NMR chemical shifts.[11] For delocalization the orbital of the lone pair must be colinear with the p-orbital of the carbene carbon; because the chromium pentacarbonyl fragment presents the carbene-carbon with a "wall" of CO's, α-branching of the substituents on the carbene-carbon results in steric congestion which may prevent efficient overlap and compromise the stability of the carbene complex (Figure 6.2). A practical consequence of this is the difficulty experienced in preparing branched carbene complexes for use in the synthesis of complex molecules.

Figure 6.2   The "CO Wall" in Group 6 Carbene Complexes

Electrophilic carbene complexes have a very rich chemistry, and undergo reaction at several sites. Fischer carbene complexes are coordinatively saturated, metal(0) d^6 complexes which undergo ligand exchange (CO loss) by a dissociative process (Eq. 6.6). Since loss of CO is a prerequisite for substrate coordination and subsequent reaction, this exchange process is central to most synthetically-useful reactions. It can be driven thermally ($T_{1/2}$ for CO exchange = 5 minutes at 140°C) or photochemically, and most organic reactions of carbene complexes involve one of these modes of activation.

Eq. 6.6

$$(CO)_5Cr=\overset{OMe}{\underset{R}{\big<}} \quad \overset{\Delta \text{ or } h\nu}{\rightleftharpoons} \quad (CO)_4Cr=\overset{OMe}{\underset{R}{\big<}} \; + \; CO \quad \overset{L}{\longrightarrow} \quad (CO)_4Cr=\overset{OMe}{\underset{\underset{L}{|}}{\big<}}R$$

$(M^0, d^6, 18e^-, \text{sat})$

Because CO groups are strongly electron-withdrawing, the metal-carbene carbon bond is polarized such that the carbene-carbon is electrophilic, and generally subject to attack by a wide range of sterically unencumbered nucleophiles (Eq. 6.7). In most cases the resulting tetrahedral intermediate is unstable, and the alkoxy group is ejected, producing a new carbene complex. This is one of the best ways to prepare Fischer carbene complexes containing heteroatoms other than oxygen, and the process is quite analogous to "transesterification" of organic esters. In fact, in many of their reactions the analogy between alkoxycarbene complexes and organic esters is remarkable.

Eq. 6.7

$$(CO)_5Cr \underset{\delta^-}{\overset{OMe}{\big<}}\underset{\delta^+ R}{} \quad \overset{Nuc^-}{\longrightarrow} \quad \left[(CO)_5Cr \overset{OMe}{\underset{Nuc}{\overset{|}{\underset{|}{C}}}}{-}R \right] \quad \longrightarrow \quad (CO)_5Cr=\overset{Nuc}{\underset{R}{\big<}} \; + \; MeO^-$$

$Nuc = R'O^-, NH_3, RNH_2{}^{12}$
<u>Small</u> R_2NH, RSH

Protons α- to the carbene-carbon are quite acidic ($pK_a \approx 12$) and can be removed by a variety of bases to give an "enolate" anion stabilized by delocalization into the metal carbonyl fragment (Eq. 6.8).[13] These anions are only weakly nucleophilic, but in the presence of Lewis acid catalysts they react with epoxides (Eq. 6.9)[14] and aldehydes (Eq. 6.10)[15] to produce homologated carbene complexes.

Eq. 6.8

$$(CO)_5Cr=\overset{OMe}{\underset{CH_3}{\big<}} \quad \overset{B^-}{\longrightarrow} \quad (CO)_5Cr=\overset{OMe}{\underset{\underset{(-)}{CH_2}}{\big<}} \quad \longleftrightarrow \quad (CO)_5Cr^-\overset{OMe}{\underset{CH_2}{\big<}}$$

Eq. 6.9

$$(CO)_5Cr=\overset{OMe}{\underset{CH_3}{\big<}} \quad \overset{BuLi}{\longrightarrow} \quad (CO)_5Cr=\overset{OMe}{\underset{CH_2^-}{\big<}} \quad \overset{\overset{O}{\triangle}}{\underset{BF_3 \cdot OEt_2}{\longrightarrow}}$$

$$\left[(CO)_5Cr=\overset{OMe}{\big<} \cdots O^- \right] \quad \longrightarrow \quad (CO)_5Cr=\overset{O}{\big<}$$

Eq. 6.10

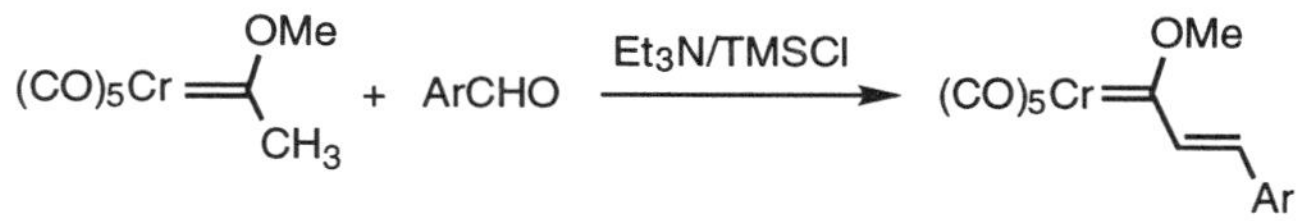

$$(CO)_5Cr = \overset{OMe}{\underset{CH_3}{\big|}} \quad + \quad ArCHO \quad \xrightarrow{\quad Et_3N/TMSCl \quad} \quad (CO)_5Cr = \overset{OMe}{\underset{Ar}{\big|}}$$

These reactions are important, since they allow elaboration of carbene fragments for ultimate incorporation into organic substrates. Alkylation of these anions with active halides often results in messy polyalkylation. Use of triflates mitigates this problem, as does replacing one of the acceptor CO ligands with a donor phosphine ligand.[16] This makes the chromium fragment a poorer acceptor, decreases the acidity of the α-protons and increases the reactivity of the anion. Similarly, aminocarbene complexes produce more reactive α-anions than do alkoxycarbenes (compare the α-proton acidity of amides vs. esters, N being a stronger donor than O), and are more easily α-alkylated (below). α-Anions of aminocarbene complexes also alkylate aldehydes (Eq. 6.11[17]) and add 1,4- to conjugated enones (Eq. 6.12).[18] With optically active amino groups, high enantioselectivity can be achieved.[18d]

Eq. 6.11

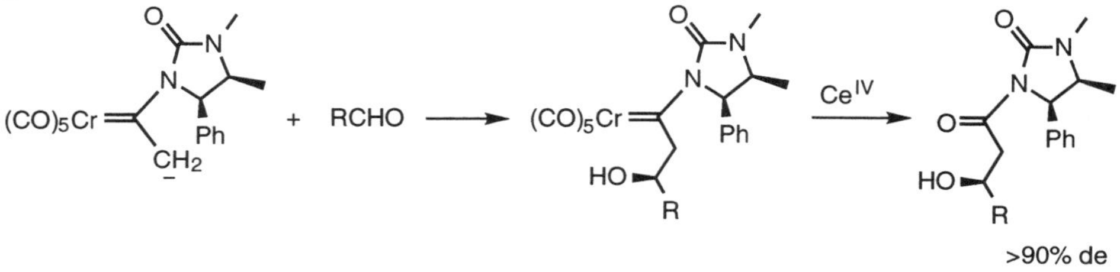

>90% de

Eq. 6.12

>95% ee

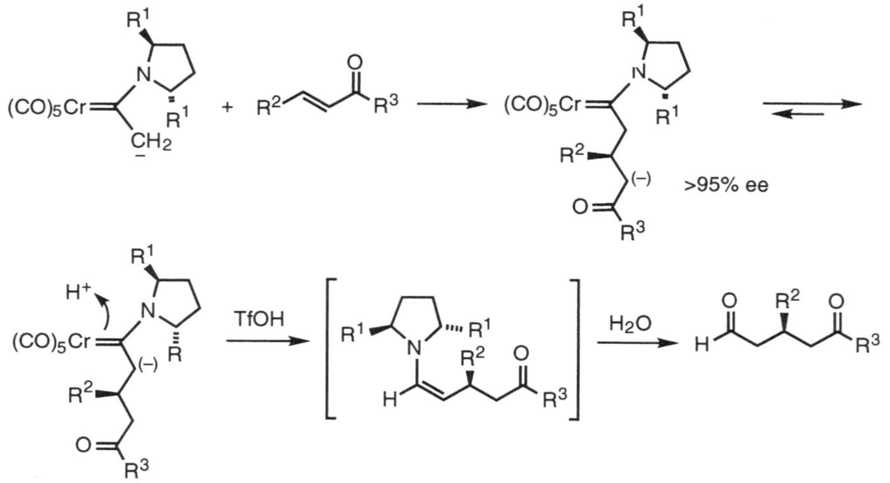

Strong bases which irreversibly deprotonate carbene complexes are often required. Weaker bases such as pyridine catalyze the decomposition of alkoxycarbene complexes into enol ethers, by reversible α-deprotonation, followed by reprotonation on the metal and reductive-elimination (Eq. 6.13).

Eq. 6.13

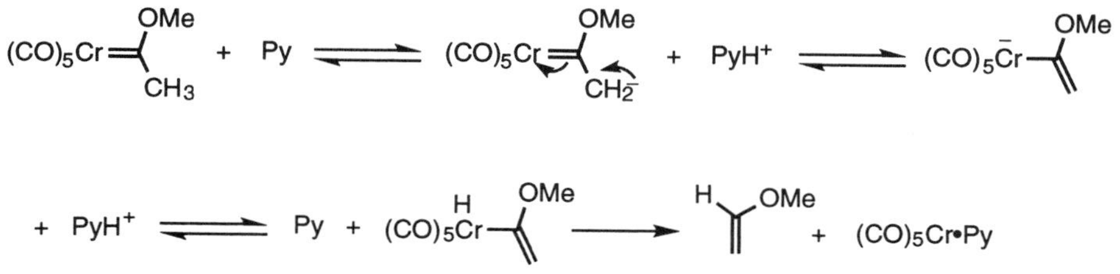

Although originally viewed as a nuisance, this process is quite useful for the synthesis of complex enol ethers from alkoxycarbene complexes (Eq. 6.14).[19] When combined with the formation of cyclic alkoxycarbene complexes from alkynols (Eq. 6.3)[8b] quite an efficient synthesis of cyclic enol ethers results (Eq. 6.15[8b] and Eq. 6.16).[20] The reaction even works with alkynyl amines (Eq. 6.17)[21] and has been made modestly catalytic (Eq. 6.18).[22]

Eq. 6.14

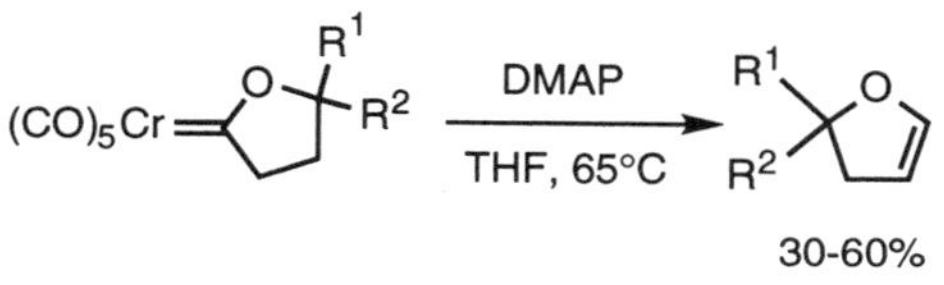

30-60%

Eq. 6.15

44%

55%

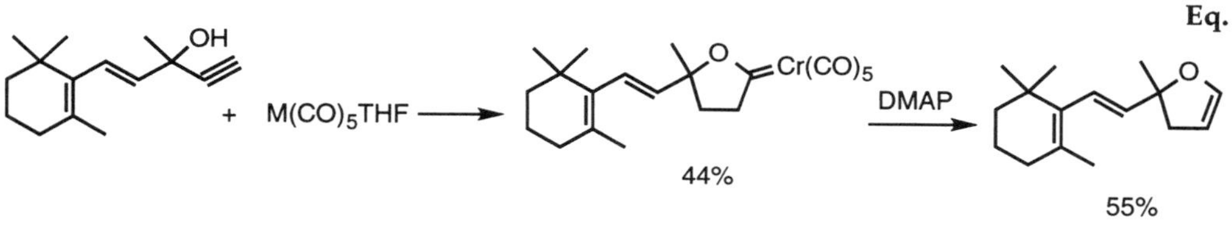

Eq. 6.16

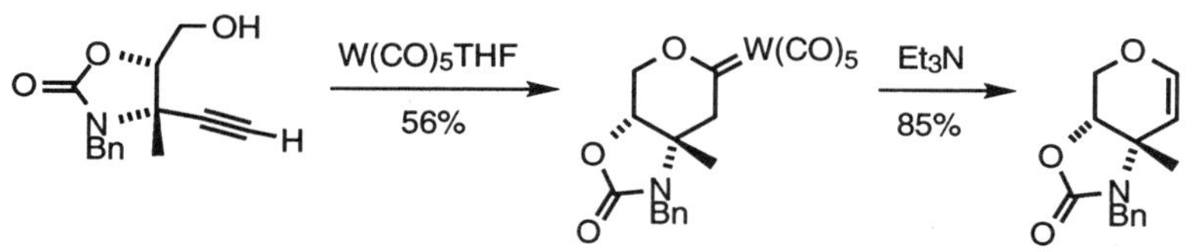

Eq. 6.17

n = 1, 68%
n = 2, 20%

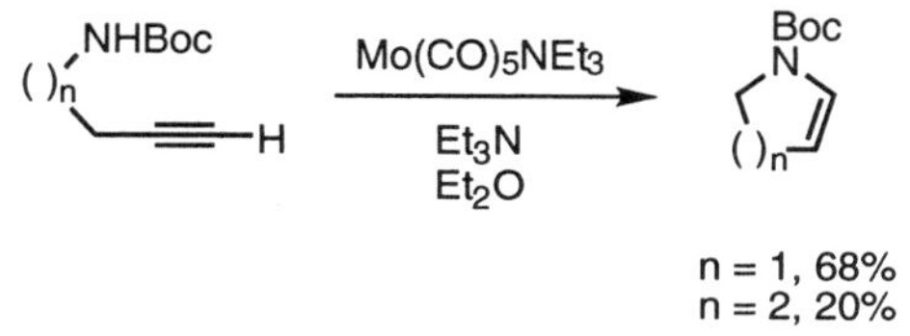

Eq. 6.18

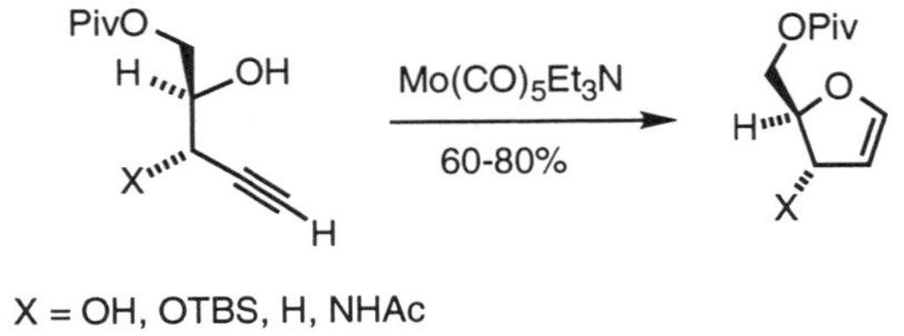

X = OH, OTBS, H, NHAc

α,β-Unsaturated carbene complexes, prepared by aldol condensation (Eq. 6.10) or from vinyllithium reagents, undergo many reactions common to unsaturated esters but are considerably more reactive. For instance, Diels-Alder reactions are 2 × 10^4 faster with carbene complexes than with esters, permitting very facile elaboration of carbene complexes (Eq. 6.19).[23] This, again, is a consequence of the electron-

Eq. 6.19

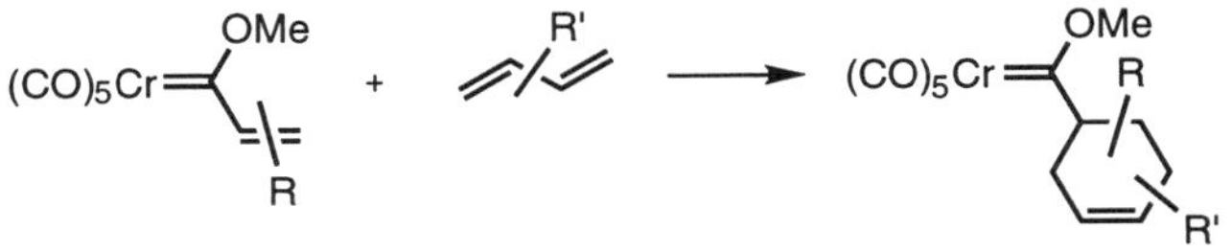

withdrawing ability of the $Cr(CO)_5$ fragment.[24] Intramolecular version are facile (Eq. 6.20),[25] and stereoselectivity is often high (Eq. 6.21).[26] Alkynyl carbene complexes are also highly reactive towards Diels-Alder cycloaddition, and highly functionalized complexes result when heteroatom dienes are used (Eq. 6.22).[27] Unsaturated carbene complexes also undergo 1,3-dipolar cycloadditions to nitrones[28] and diazoalkenes.[29]

Eq. 6.20

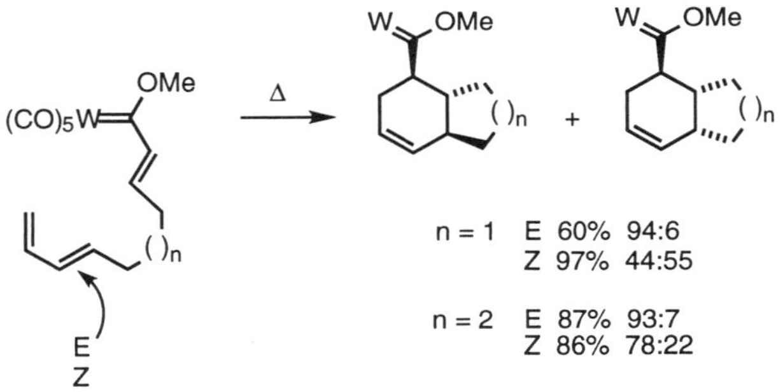

n = 1 E 60% 94:6
      Z 97% 44:55

n = 2 E 87% 93:7
      Z 86% 78:22

Eq. 6.21

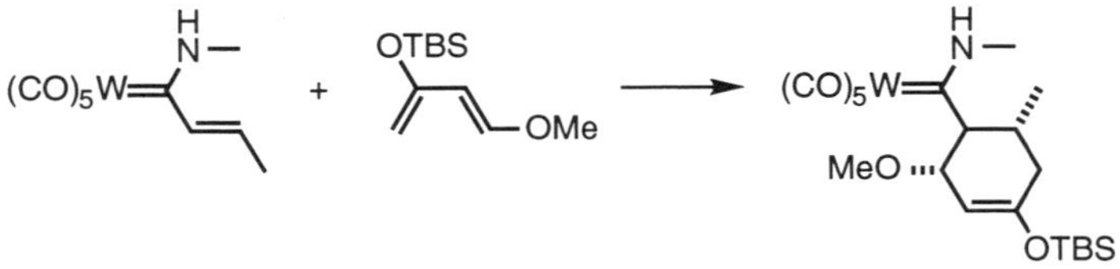

Michael addition reactions to conjugated carbene complexes are also efficient. In fact, in the exchange of the alkoxy group by amines in α,β-unsaturated

Eq. 6.22

carbene complexes, Michael addition of the amine is the major competing pathway. Michael addition provides a very useful route for elaborating carbene complexes, for ultimate use in synthesis. Stabilized enolates add readily, and the resulting carbene anion is inert to further attack by nucleophiles, permitting further functionalization of the newly-introduced carbonyl group (Eq. 6.23).[30] The reaction of alkynylcarbene

Eq. 6.23

complexes with β-dicarbonyl enolates results in Michael addition followed by displacement of the alkoxy group by the enolate oxygen. Reaction of these complexes with enamines results in a remarkable cyclization (Eq. 6.24).[31]

Once elaborated, the carbene fragment can be converted to organic products in a number of ways. Oxidation with Ce(IV) or dimethyldioxirane (Eq. 6.25)[32] generates the carboxylic acid derivative, as does oxidation of the tetramethylammonium salt in the presence of a nucleophile.[13] However, the metal-carbon double bond can be, and has been, put to better use.

One of the first organic reactions of Group 6 carbene complexes studied was the cyclopropanation of electrophilic olefins (Eq. 6.26).[34] The reaction is carried out by simply heating mixtures of the carbene complex and olefin, although often at elevated temperatures. Moderate to good yields of the cyclopropane are obtained. With 1,2-disubstituted alkenes, the stereochemistry of the alkene is maintained in the product, but mixtures of stereoisomers at the (former) carbene carbon are obtained. Dienes, if they are not too highly substituted, are monocyclopropanated (Eq. 6.27).[35] 1,3-Dienoic esters cycloproponate exclusively at the double bond remote (γ,δ) from the ester.[36] In many cases, in addition to the cyclopropane product, varying amounts of a by-product corresponding to vinyl C–H insertion are obtained (Eq. 6.28).[37]

Eq. 6.24

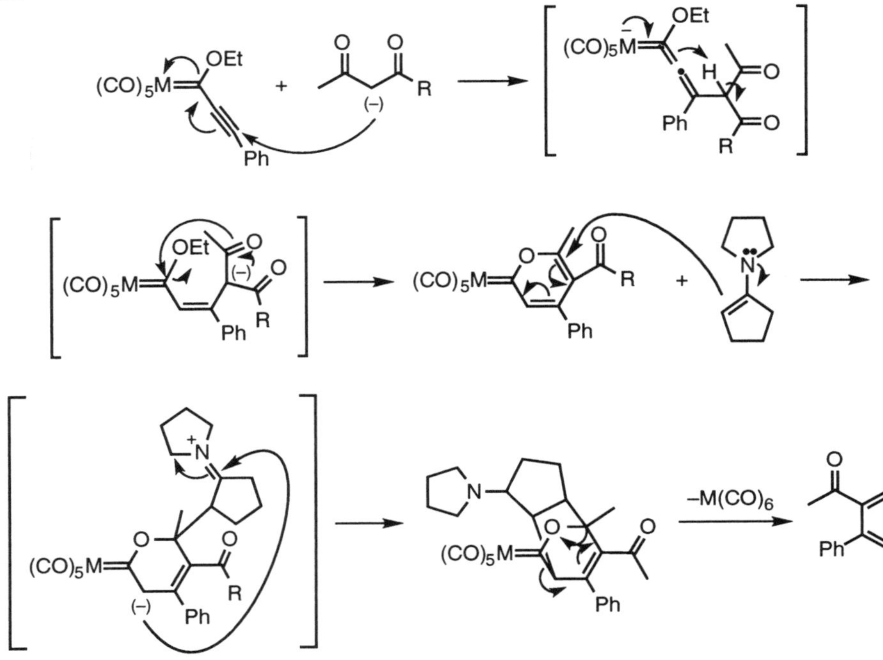

(CO)5M
OEt
Ph
R
(–)
H
–M(CO)6

Eq. 6.25

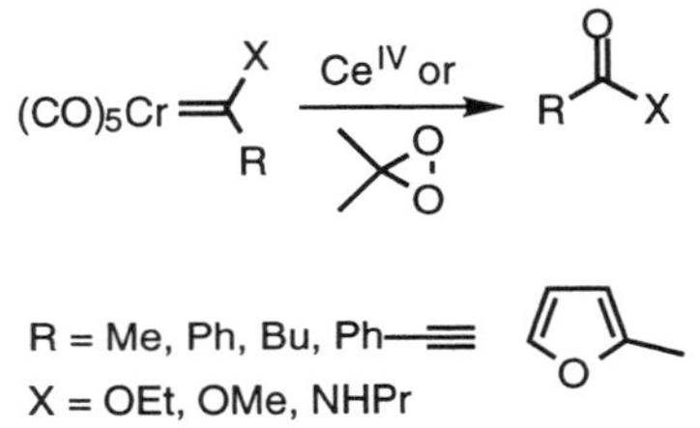

(CO)5Cr
X
R
Ce^IV or
R = Me, Ph, Bu, Ph—≡
X = OEt, OMe, NHPr

Eq. 6.26

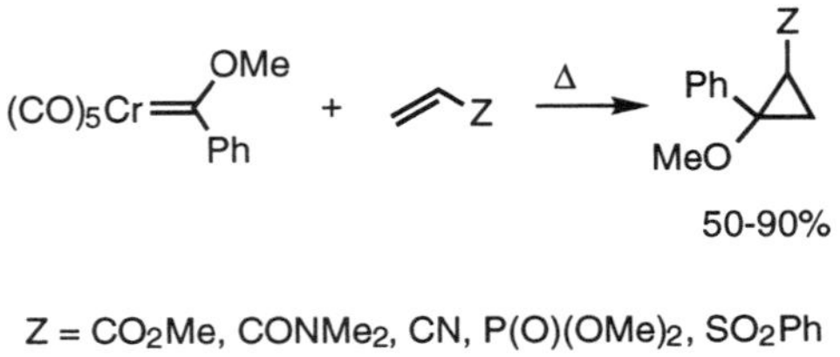

(CO)5Cr
OMe
Ph
Z
Δ
MeO
50-90%
Z = CO2Me, CONMe2, CN, P(O)(OMe)2, SO2Ph

Although cyclopropanes are the expected product of the reaction of free organic carbenes and olefins, there is ample evidence that free organic carbenes are not involved in the cyclopropanation reactions of transition metal carbene complexes. Rather the route in Figure 6.3 is followed. Cyclopropanation is suppressed by CO pressure, indicating that loss of CO to generate a vacant coordination site is required. Coordination of the alkene (a), followed by formal 2+2 cycloaddition to the metal-carbon double bond (b), would generate a

Eq. 6.27

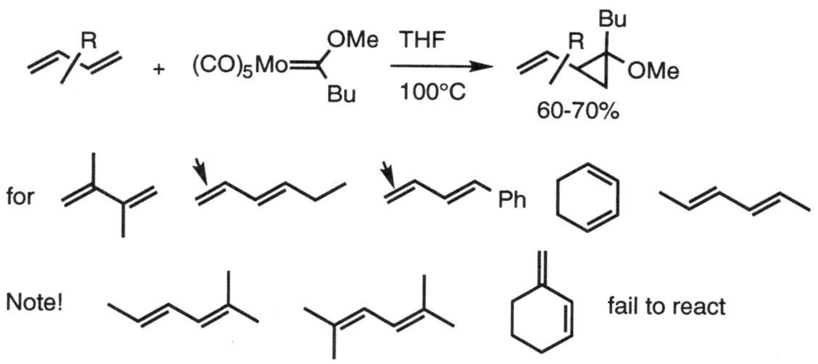

Eq. 6.28

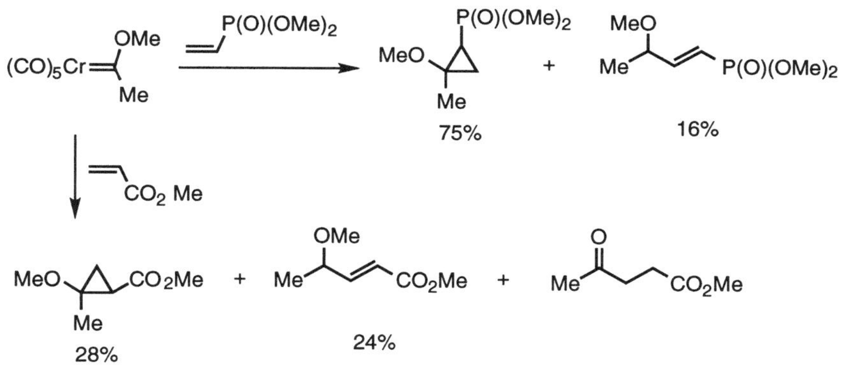

metallocyclobutane (c). Depending on the regioselectivity of this cycloaddition step, the alkene substituent may end up either α or β to the metal. Reductive elimination (d) gives the cyclopropane. The observed "C–H" insertion by-product could be formed by β-hydride elimination (e) then reductive elimination (f) from the metallacyclobutane.

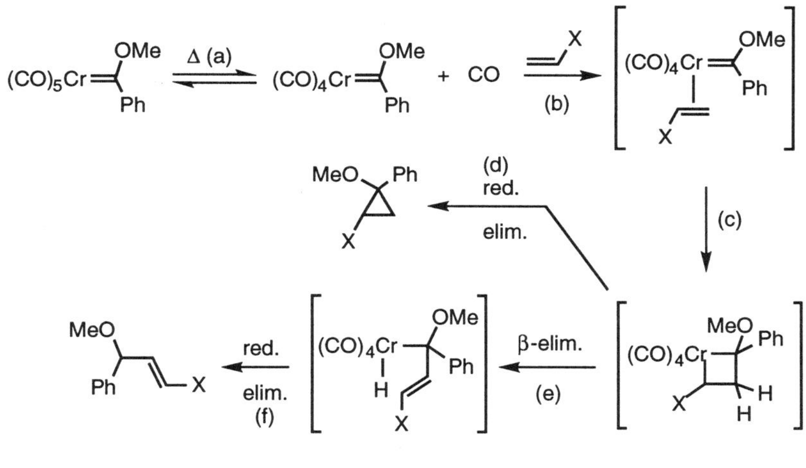

Figure 6.3   Mechanism of Cyclopropanation

Evidence for metallacyclobutane intermediates comes from attempted cyclopropanation of electron-rich alkenes. Under normal reaction conditions, cyclopropanation does not result; rather a new alkene resulting from metathesis (alkene group interchange) is obtained (Eq. 6.29). Olefin metathesis is a well-known

Eq. 6.29

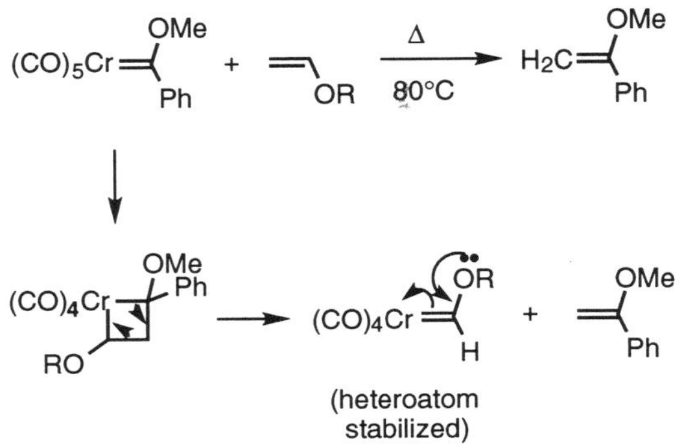

process for carbene complexes, and has found many uses in synthesis (See 6.5). It is thought to occur by the cycloaddition/cycloreversion process shown. Because metathesis of electron-rich olefins produces heteroatom-stabilized carbene complexes, whereas electron-poor olefins would produce unstabilized carbene complexes, metathesis is the favored pathway with electron-rich substrates. Electron-rich alkenes can be cyclopropanated by unstable acyloxycarbene complexes, generated *in situ* by O-acylation of tetramethyl ammonium (but not lithium, because of tight ion pairing) "ate" complexes (Eq. 6.30),[38] or under high (>100 atm) pressures of CO, which seems to suppress metathesis.

Eq. 6.30

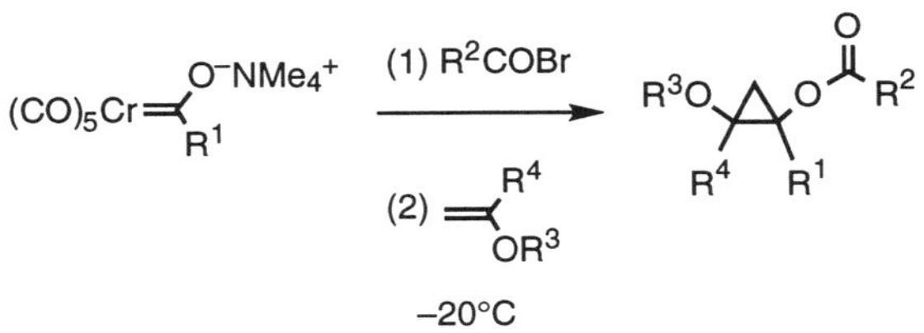

In contrast to electron-poor and electron-rich alkenes, simple aliphatic alkenes do not undergo intermolecular cyclopropanation by chromium alkoxycarbenes under any conditions. However, in general, intramolecular reactions are considerably more favorable, and aryl-*alken*oxycarbene complexes undergo extremely facile intramolecular cyclopropanation of the unactivated alkene (see below).

Carbene complexes with structurally complex alkoxy groups are not directly available by the standard synthesis (Figure 6.1) because the corresponding alkyl triflates or oxonium salts are difficult to synthesize. Further, they cannot be made by alkoxy-alkoxy interchange, since these reactions (in contrast to alkoxy-amine exchange) are slow, inefficient, and beset with side reactions. However, acyloxycarbene complexes, generated *in situ* by reaction of acid halides with ammonium acylate complexes, react cleanly, even with structurally complex alcohols, to produce alkoxycarbene complexes in excellent yield (Eq. 6.31).[39] When homoallylic alcohols were exchanged into (aryl)(acyloxy) carbenes, the carbene

Eq. 6.31

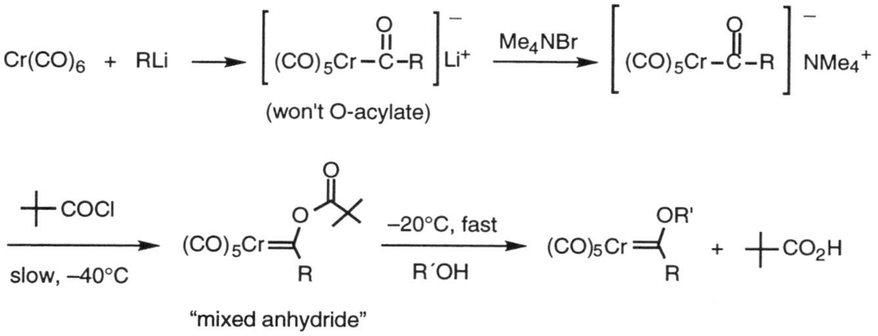

$$Cr(CO)_6 \; + \; RLi \;\longrightarrow\; \left[(CO)_5Cr-\overset{\overset{\displaystyle O}{\|}}{C}-R \right]^{-} Li^{+} \xrightarrow{\;Me_4NBr\;} \left[(CO)_5Cr-\overset{\overset{\displaystyle O}{\|}}{C}-R \right]^{-} NMe_4^{+}$$

(won't O-acylate)

$$\xrightarrow[\text{slow, }-40°C]{\text{$+$COCl}} \quad (CO)_5Cr=\langle \overset{O}{\underset{R}{\overset{\|}{\cdots}}} \quad \xrightarrow[\text{R}'OH]{-20°C,\text{ fast}} \quad (CO)_5Cr=\langle \overset{OR'}{\underset{R}{}} \;+\; +CO_2H$$

"mixed anhydride"

complex could not be isolated. Instead, spontaneous, intramolecular cyclopropanation occurred, even under modest pressures of carbon monoxide (Eq. 6.32).[40] This is a general phenomenon, and a variety of cyclopropanated tetrahydrofurans, pyrans

Eq. 6.32

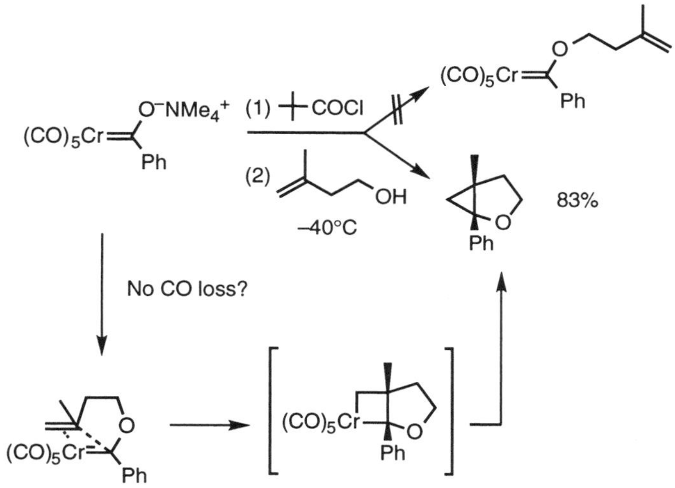

and pyrrolidines are available by this route. This increase in reactivity is clearly entropic; when the alkene is more than three carbons removed from the oxygen, cyclopropanation does not occur under conditions sufficiently severe to decompose the complex. However, more remote alkenes *will* cyclopropanate if they are conformationally held in proximity to the carbene system (Eq. 6.33).[41]

Eq. 6.33

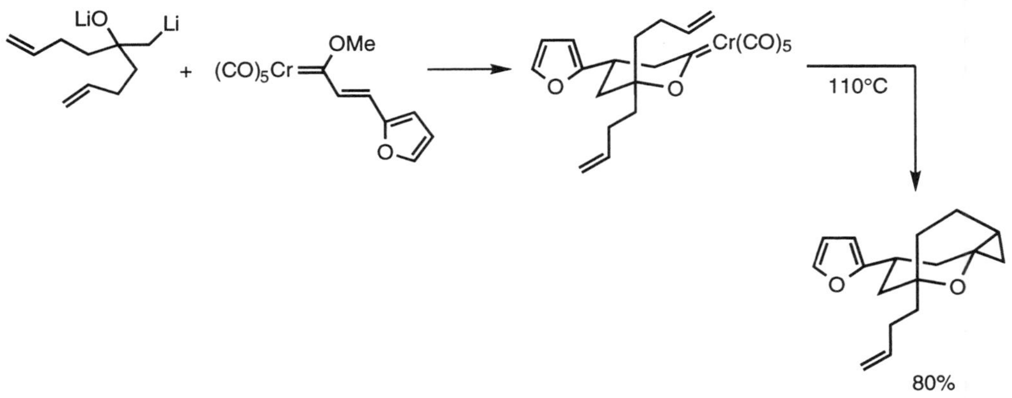

80%

Alkynes cycloadd to carbene complexes more readily than do alkenes. Treatment of enynes with chromium carbene complexes produces bicyclic compounds in excellent yield, through a process of alkyne cycloaddition/metathesis/alkene cycloaddition/reductive elimination (Eq. 6.34).[42]

Eq. 6.34

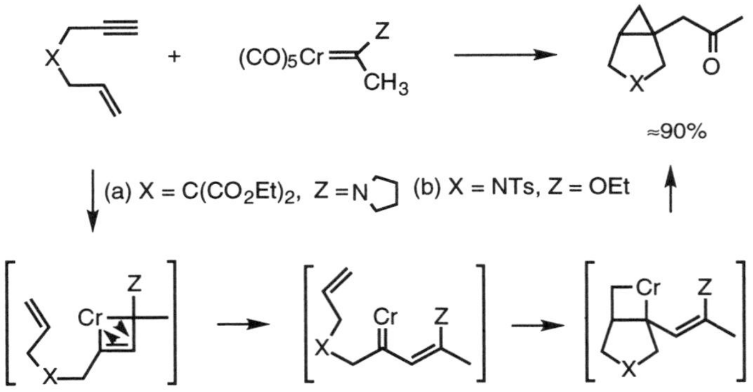

≈90%

(a) X = C(CO₂Et)₂, Z = N⟩   (b) X = NTs, Z = OEt

Many variations of this theme are possible and have been tried, including tethering the alkene to the carbene[43] (Eq. 6.35),[44] the alkyne to the carbene (Eq. 6.36),[45] both the

Eq. 6.35

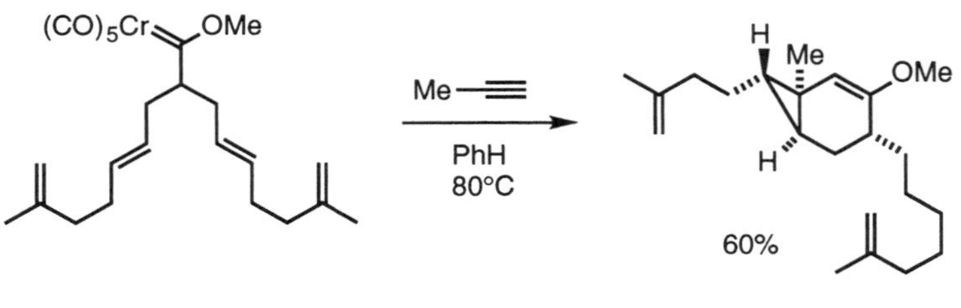

60%

Eq. 6.36

60%

alkyne and alkene tethered to the carbene,[46] and the alkyne tethered to a 1,3 diene (Eq. 6.37).[47] All lead to a remarkable increase in complexity in a single step.

Eq. 6.37

87%

Another thermal reaction of carbene complexes of great use in synthesis is the Dötz reaction, the reaction of unsaturated alkoxycarbene complexes with alkynes, to produce hydroquinone derivatives. The general transformation is shown in Eq. 6.38.[48] Although the overall sequence is complex, the mechanism is reasonably well-understood, and involves many of the standard processes of organometallic chemistry. Again, the reaction begins with a thermal loss of one CO, to generate a vacant coordination site. Coordination of the alkyne followed by cycloaddition generates the metallacyclobutene, which inserts CO to give a metallacyclopentenone. Fragmentation of this metallacycle generates a metal-bound vinyl ketene, for which there is excellent experimental evidence (X-ray in one case, isolation of the vinylketene with sterically hindered alkynes[49]). Cyclization of this vinyl ketene followed by enolization produces the hydroquinone derivative, as its $Cr(CO)_3$ complex (Chapter 10). The free ligand is obtained by oxidative removal of chromium by exposure to air and light. Depending on the metal, and the solvent, and in particular, the heteroatom, the CO insertion step may be supplanted by a metathesis step, producing indenes rather than hydroquinones. This is particularly common for aminocarbene complexes.[50]

Eq. 6.38

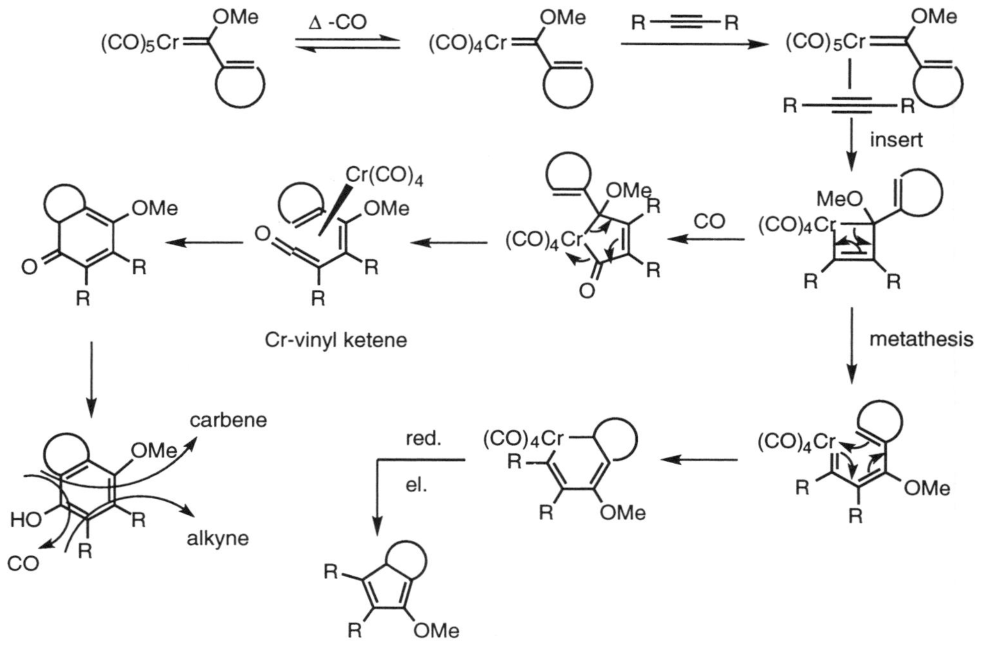

Because of the stability and ease of handling of Group 6 carbene complexes, and the many ways they can be elaborated, highly functionalized, structurally complex carbene complexes are relatively easily prepared, making their use in organic synthesis appealing. The Dötz reaction has been extensively utilized to make quinone derivatives. Selected examples are shown in Eq. 6.39,[51] Eq. 6.40,[52] and Eq.

Eq. 6.39

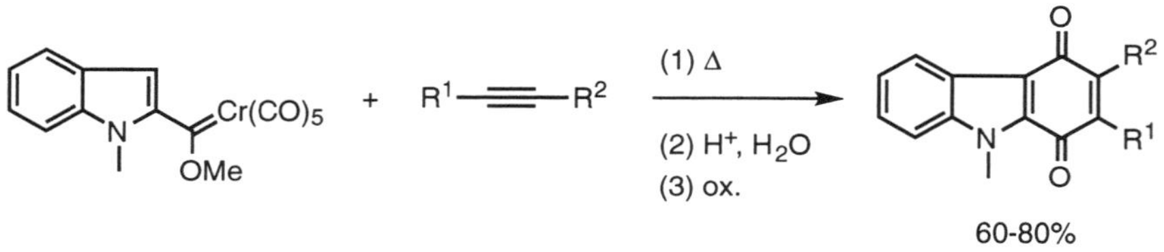

Eq. 6.40

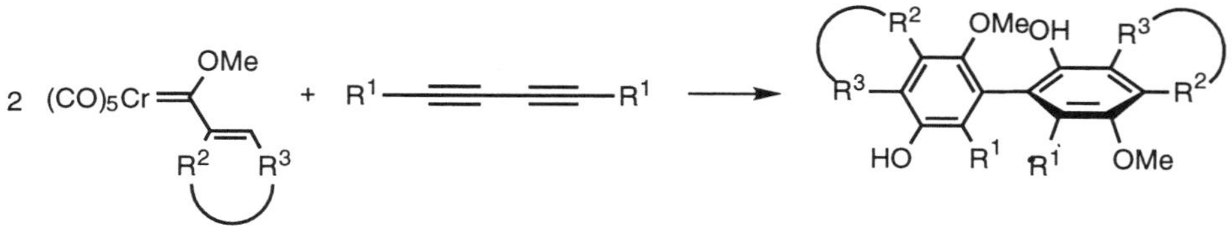

6.41.[53] As is the case with cyclopropanation, a cascade of different reactions can be arranged so as to make many bonds in a single reaction sequence. Such a case is seen in Eq. 6.42,[54] in which a Diels-Alder cycloaddition is followed by an alkyne metathesis, and then a Dötz reaction, all in a one-pot reaction.

Eq. 6.41

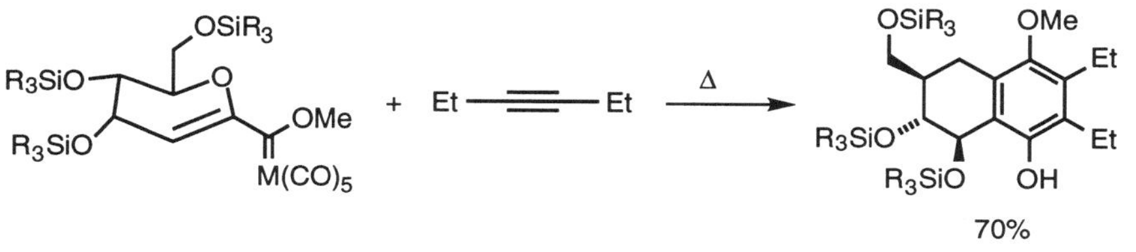

70%

Eq. 6.42

62%

via
Diels Alder

Product

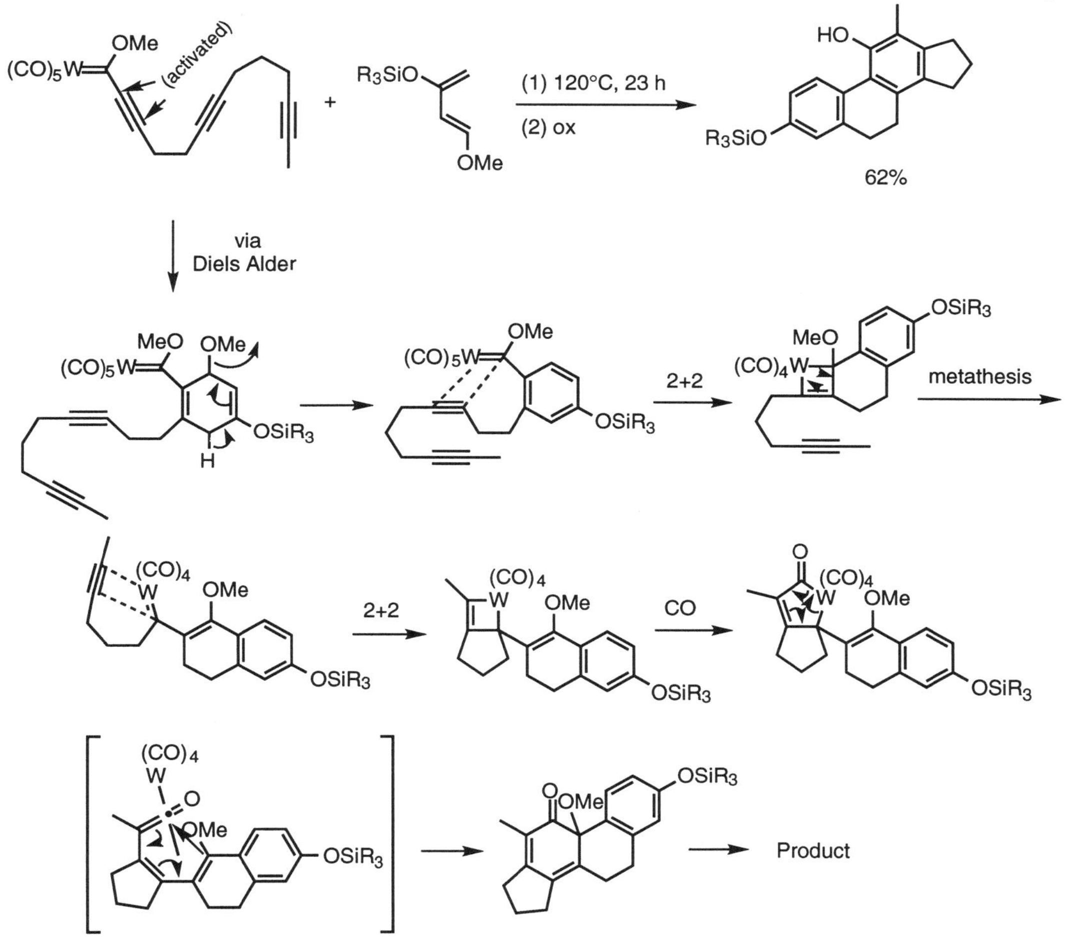

The observation that photolysis of chromium carbene complexes with visible light resulted in insertion of CO into the metal-carbon double bond to generate species with ketene-like reactivity (see below)[55] led to the development of a very clever alternative to the Dötz benzannulation process (Eq. 6.43).[56] By synthesizing *cis*-dienyl carbene complexes, the key vinylketene intermediate in the Dötz reactions could be generated by photochemically-driven CO insertion, resulting in an efficient

Eq. 6.43

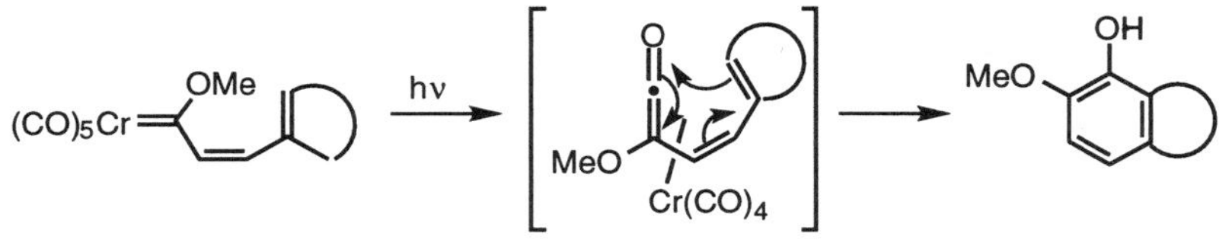

benzannulation process. Thermal or photochemical insertions of *isonitriles* into the metal-carbon double bond similarly produces vinylketeneimine intermediates, which also cyclize readily (Eq. 6.44).[57] A very nice application of both of these processes is seen in Eq. 6.45.[58]

Eq. 6.44

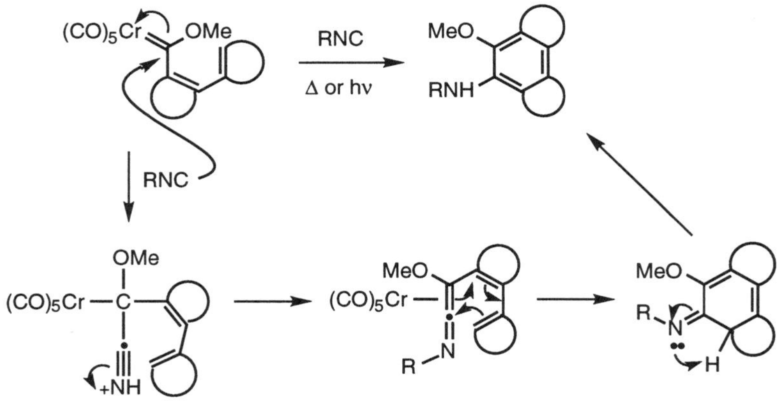

Eq. 6.45

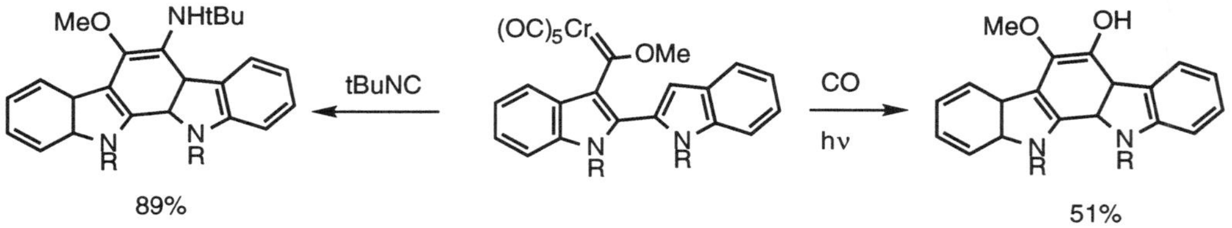

The Dötz reaction is also quite sensitive to the nature of the heteroatom on the carbene complex. Although alkoxycarbene complexes usually give hydroquinone derivatives, aminocarbene complexes fail to insert CO, and give indane derivatives instead (Eq. 6.38).[59] When carried out intramolecularly, the reaction is remarkably diastereoselective (Eq. 6.46).[60] When combined with Diels-Alder cycloaddition, complex systems can be produced efficiently (Eq. 6.47).[61]

As can be seen, the Dötz reaction has many steps, each of which can be diverted along other pathways, leading to other products. The main problem in the development of this chemistry was to develop conditions which led to a single major product, since the reaction is quite sensitive to both the reaction conditions and structural features of the reactants.

Eq. 6.46

Eq. 6.47

For example, when cyclopropylcarbene complexes are subjected to the typical Dötz reaction, totally unexpected products result (Eq. 6.48).[62] Although the

Eq. 6.48

mechanism is not well-understood, it is likely that things proceed as usual to the metallacyclopentenone stage (b), at which point, the "sp^2" like cyclopropyl group inserts as if it were an olefin (cyclopropyl groups are readily cleaved by a variety of transition metals) to give a metallacyclooctadienone (c). This then reinserts to give a bicyclic system (d)(shown as occurring in two steps, for clarity), which fragments to give the cyclopentadienone and an alkene (e). The dienone is reduced by the low-valent chromium fragment in the presence of water. Intramolecular versions also work.[63] When the metal is changed from chromium to the larger tungsten, seven-membered rings form, apparently from direct reductive elimination from the metallacyclooctadienone (Eq. 6.49).[64] (Note that isomerization of the double bonds via 1,5-hydride shifts to give the most stable product occurs under the relatively severe reaction conditions.)

Eq. 6.49

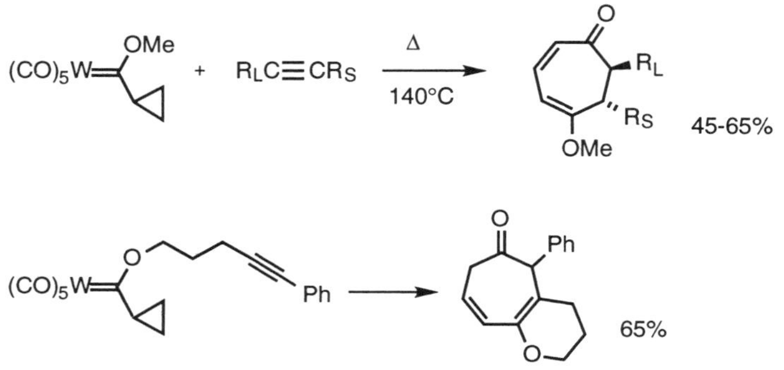

Iron carbene complexes analogous to the Group 6 Fischer carbenes are easily prepared, either by the Fischer route (Eq. 6.1) or from Collman's reagent ($Na_2Fe(CO)_4$) (Eq. 6.50). In the thermal reaction with alkenes, no cyclopropenes are formed; rather, C–H insertion products predominate (Eq. 6.51).[65] With alkynes, iron alkoxycarbenes react by a completely different route from Group 6 analogs, inserting *two* CO's and giving a mixture of equilibrating pyrones (Eq. 6.52)![66]

Eq. 6.50

Eq. 6.51

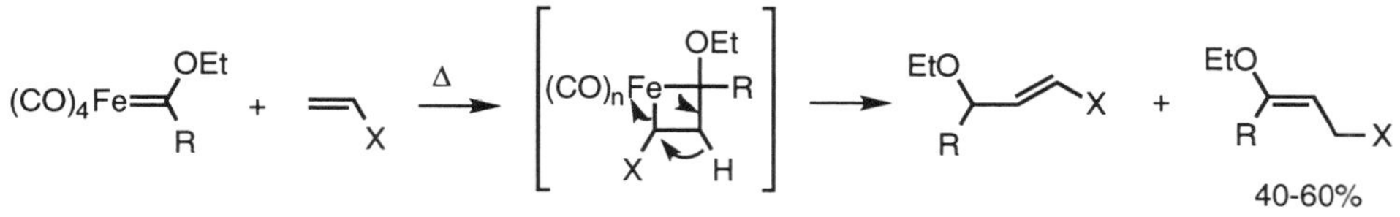

Eq. 6.52

In addition to thermal reactions, chromium heteroatom carbene complexes have a very rich photochemistry which results in a number of synthetically useful processes. All of these complexes are colored and have absorptions in the 350-450 nm range. This visible absorption has been assigned as an allowed metal-to-ligand charge transfer (MLCT) band.[67] From the molecular orbital diagram of the complex, the HOMO is metal-d-orbital centered, and the LUMO is carbene-carbon p-orbital centered. Thus, photolysis into the MLCT results in a *formal*, reversible, one electron oxidation of the metal, since, in the absorption process, an electron is removed from the d-centered HOMO to the p-centered LUMO.[68] One of the best ways to drive insertion of carbon monoxide into metal-carbon bonds is to oxidize the metal (Chapters 2 and 4) and, indeed, photolysis of chromium carbene complexes with visible light appears to drive a reversible insertion of CO into the metal carbon double bond, producing a metallacyclopropanone, best represented as a metal-bound ketene (Eq. 6.53).[69] Since excited states of transition metal complexes are usually too short-lived to undergo intermolecular reactions, this is almost surely a ground-state complex which, if not trapped, deinserts CO, regenerating the carbene complex. Thus, many carbene complexes can be recovered unchanged after extended periods of photolysis in the absence of reagents which react with ketenes, but are completely consumed in a few hours when photolyzed in the presence of such reagents.

Eq. 6.53

From a synthetic point of view,[70] this is very useful, since it permits the generation of unusual, electron-rich alkoxy-, amino-, and thioketenes not easily available by conventional routes, under very mild conditions (visible light, any solvent, Pyrex vessels, no additional reagents). The ketene is metal-bound, and generated in low concentrations, thus normal ketene side reactions such as dimerization, or incorporation of more than one ketene unit into the product, are not observed. However, the species thus generated undergo most of the normal reactions of ketenes, and herein lies their utility.

The reaction of ketenes with imines to produce β-lactams (the Staudinger reaction) has been extensively developed for the synthesis of this important class of antibiotics. Most biologically active β-lactams are optically active, and have an amino group α- to the lactam carbonyl. To prepare these from chromium carbene complex chemistry, an optically active complex having a hydrogen on the carbene carbon is necessary. These are not available from conventional Fischer synthesis, since anionic metal formyl complexes are strong reducing agents and transfer hydrides to electrophiles. The requisite optically active aminocarbene complex can be made efficiently by an alternative route involving the reaction of $K_2Cr(CO)_5$ with the appropriate amide, followed by addition of TMSCl to promote the displacement of the oxygen (Figure 6.1).[71]

This optically active aminocarbene complex undergoes efficient reaction with a very wide range of imines to give optically active β-lactams in excellent chemical yield, and with very high diastereoselectivity (Eq. 6.54).[72] The absolute

Eq. 6.54

stereochemistry of the α-position is determined by the chiral auxiliary on nitrogen, which transfers its absolute configuration to this position (R → R, S → S), while the relative (cis-trans) stereochemistry between the two new chiral centers is determined by the imine.[73] When the reaction is carried out under a modest (60 psi) pressure of CO, chromium hexacarbonyl can be recovered and reused. The chiral auxiliary is easily removed by hydrolysis of the acetonide, followed by either reductive (H₂, Pd/C) or oxidative (IO₄⁻) cleavage of the amino alcohol. This chemistry has been used in a formal synthesis of (+) 1-carbacephalothin (Eq. 6.55).[74]

Eq. 6.55

Imidazolines are particularly interesting substrates for this β-lactam forming process, since they fail to form β-lactams when allowed to react with ketenes, or with ester enolates. However, photochemical reaction with chromium carbene complexes produce azapenams, a new class of β-lactam, in good yield (Eq. 6.56).[75,76] This is one of several cases for which chromium carbene photochemistry and classical ketene chemistry differ in detail.

Eq. 6.56

70%   >99% ee

Ketenes also undergo facile 2+2 cycloaddition reactions with olefins, producing cyclobutanones, with a high degree of stereoselectivity, favoring the more hindered cyclobutanone ("masochistic stereoinduction"[77]). Photolysis of chromium alkoxycarbene complexes in the presence of a wide variety of alkenes produces cyclobutanones in high yield, with roughly the same stereoselectivity observed using conventionally generated ketenes (Eq. 6.57).[78] As expected, the reaction is restricted to relatively electron-rich olefins, and electron-deficient olefins fail altogether. It is also restricted to alkoxycarbene complexes or aminocarbene complexes having aryl groups on nitrogen to decrease the nucleophilicity of the ketene.[79] Intramolecular versions also work well, giving access to a variety of bicyclic systems. Even unconventional cycloadditions proceed in remarkably good yield (Eq. 6.58).[80] With optically active alkenes, high diastereoselectivity is observed (Eq. 6.59).[81] The resulting optically active cyclobutanones undergo Baeyer-Villager oxidation/elimination to produce butenolides, useful synthetic intermediates.

Eq. 6.57

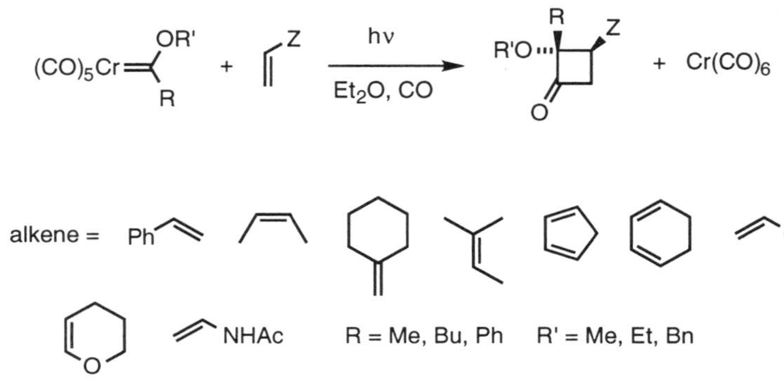

Eq. 6.58

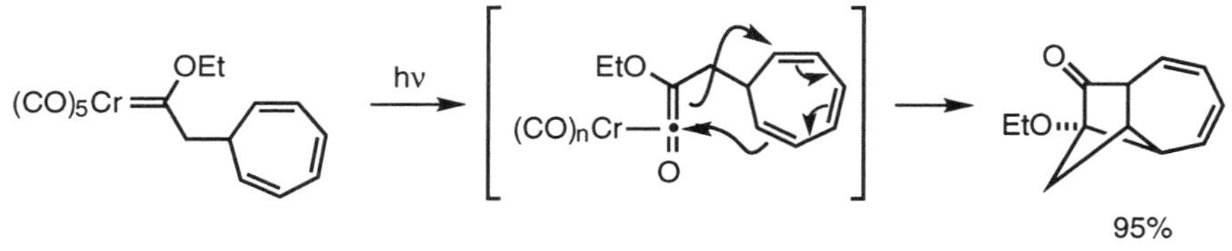

95%

Eq. 6.59

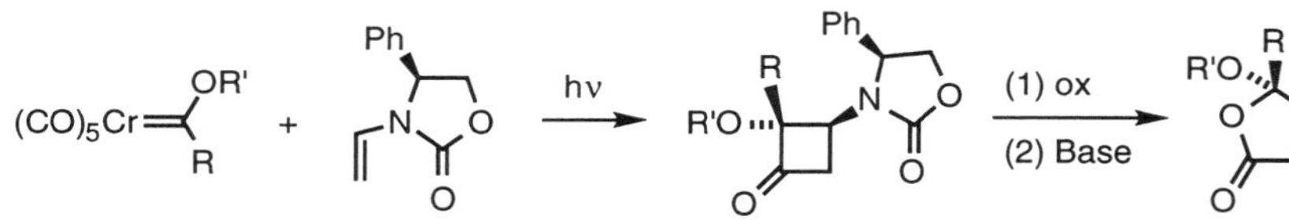

Ketenes react with alcohols or amines to give carboxylic acid derivatives. Similarly, photolysis of chromium carbene complexes in the presence of alcohols produces esters in excellent yield. Use of aminocarbene complexes, prepared from amides (Figure 6.1) provides a very direct synthesis of α-amino acid derivatives from amides, a transformation not easily achieved using conventional methodology.[82] Lactams, including β-lactams, undergo this transformation as well, producing cyclic amino acids.

Optically active α-amino acids are an exceptionally important class of compounds, not only because of their central biological role as the fundamental units of peptides and proteins, but also because of their role in the development of new pharmaceutical derivatives. By synthesizing optically active aminocarbene complexes and taking advantage of their α-carbon reactivity, a variety of natural (S) and unnatural (R) amino acids can be synthesized (Eq. 6.60).[83] With aldehydes as electrophiles, homoserines could be efficiently synthesized.[84]

Eq. 6.60

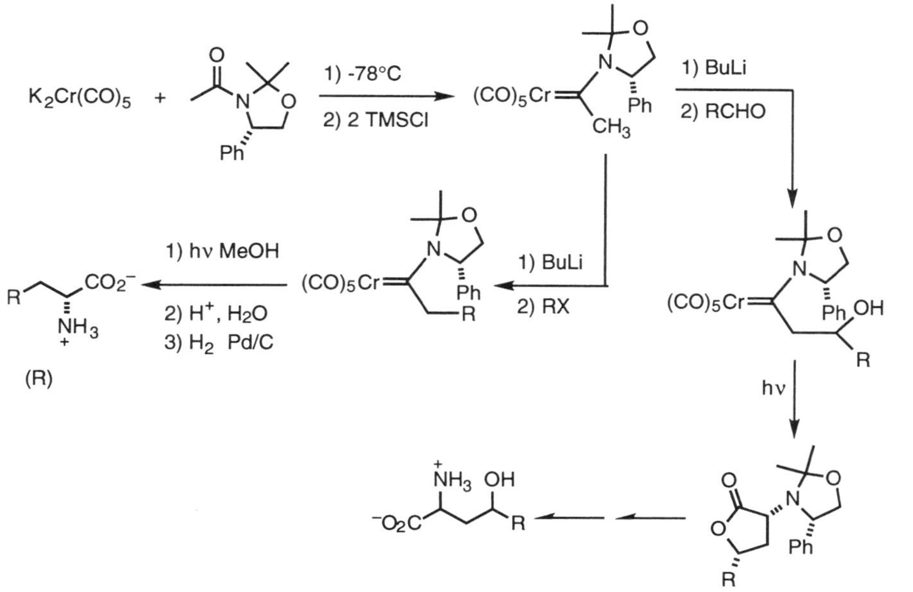

By replacing the alcohol with an optically active α-amino acid ester, dipeptides can be synthesized in a process that forms the peptide bond and the new stereogenic center on the carbene-complex-derived amino acid fragment in a single step (Eq. 6.61).[85] This permits the direct introduction of natural (or unnatural) amino acid fragments into peptides utilizing visible light as the coupling agent. In this case, "double diastereoselection" is observed with the (R,S,S) dipeptide being the matched and the (S,R,S) dipeptide being the mismatched pair. Even sterically hindered α,α-dialkyl- and N-alkyl-α,α-dialkylamino acid esters couple in good yield,[86] as do polystyrene bound-,[87] and soluble PEG-bound[88] amino acid residues.

Eq. 6.61

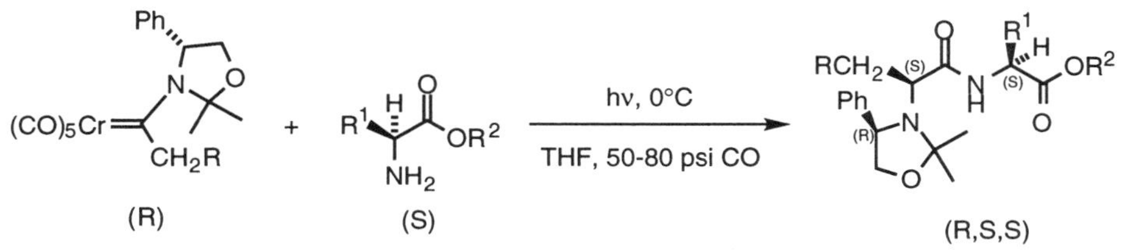

6.3   Electrophilic, Nonstabilized Carbene Complexes[89]

Group 6 carbene complexes lacking a heteroatom on the carbene carbon are unstable and difficult to prepare. As a consequence, they have not been extensively utilized in synthesis (although they do cyclopropanate olefins[90]). In contrast, a number of stable, easily prepared iron carbonyl complexes are precursors to unstabilized electrophilic carbene complexes. When these precursor complexes are treated with appropriate reagents in the presence of alkenes, cyclopropanation results. Among the most convenient of these carbene precursors are the methylthiomethyl complexes, easily prepared on large scale by the reaction of $CpFe(CO)_2Na$ with CH_3SCH_2Cl.[91] S-Methylation followed by warming in the presence of an excess of electron-rich alkene produces the unstable cationic carbene complex which then cyclopropanates the alkene. The corresponding methoxymethyl complex behaves similarly (Eq. 6.62). Perhaps the most versatile approach relies upon the reaction of the strongly nucleophilic $CpFe(CO)_2Na$ with aldehydes[92] or

Eq. 6.62

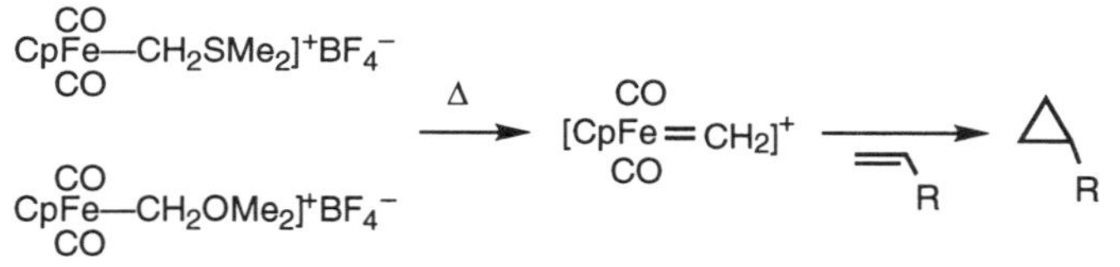

ketals[93] to produce the requisite α-oxoiron complexes. These undergo α-elimination to produce cationic carbene complexes when treated with alkylating or silylating agents (Eq. 6.63).

Eq. 6.63

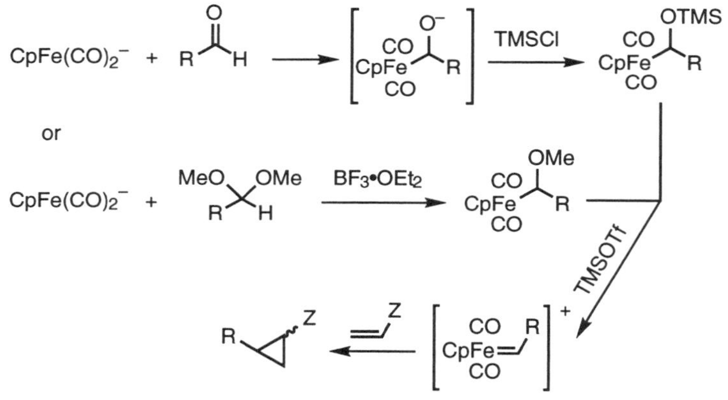

This cyclopropanation is somewhat limited. Only relatively unfunctionalized, electron-rich alkenes undergo the reaction, and relatively few carbene groups — "CH_2", "$CHCH_3$", "$C(CH_3)_2$", "$CHPh$", "CHC_3H_5" — can be transferred. With substituted carbene complexes, mixtures of *cis* and *trans* disubstituted cyclopropanes are obtained.

By replacing one of the carbon monoxide groups with a phosphine, the iron carbene complex precursors become chiral at iron, and can be resolved. Additional asymmetry can be introduced by using an optically active phosphine, giving complexes chiral both at iron and phosphorus. These optically active carbene complexes cyclopropanate alkenes with quite good enantioselectivity, but only modest *cis/trans* selectivity (Eq. 6.64). The mechanism of this asymmetric cyclopropanation has been extensively studied.[94] Because chiral metal complexes induce asymmetry in the products, "free" carbenes cannot be involved. Unexpectedly perhaps, metallacyclobutanes also are not involved. From careful stereochemical and labeling studies, the mechanism shown in Eq. 6.64 has been established. It involves attack of the alkene on the electrophilic carbene complex to generate an electrophilic center at the γ-carbon. If this carbon bears an electron-donating group, this intermediate is sufficiently long-lived to permit rotation about the β-γ-bond, leading to loss of the original olefin stereochemistry. This γ-cationic center is then attacked by the Fe-C_α bond from the back side resulting in inversion at this center.

Eq. 6.64

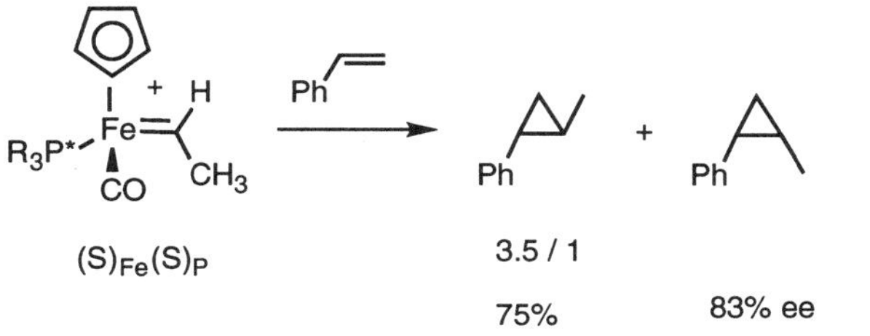

$(S)_{Fe}(S)_P$ 3.5 / 1

75%   83% ee

Eq. 6.65

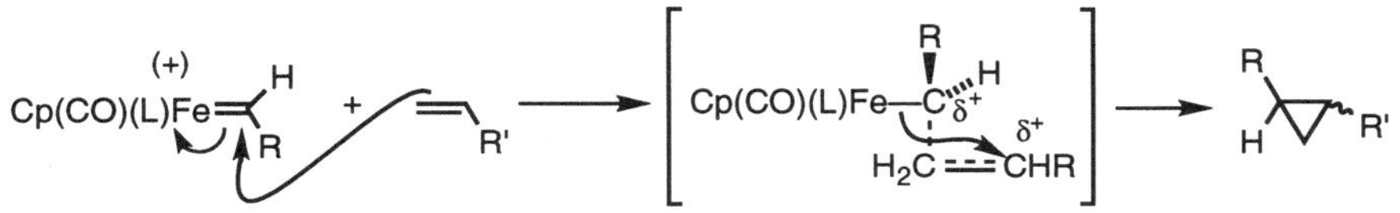

Consistent with this mechanism, iron-carbene intermediates can effect cationic cyclization of polyenes (Eq. 6.66)[95] as well as carbene-like C-H insertion reactions (Eq. 6.67).[96]

Eq. 6.66

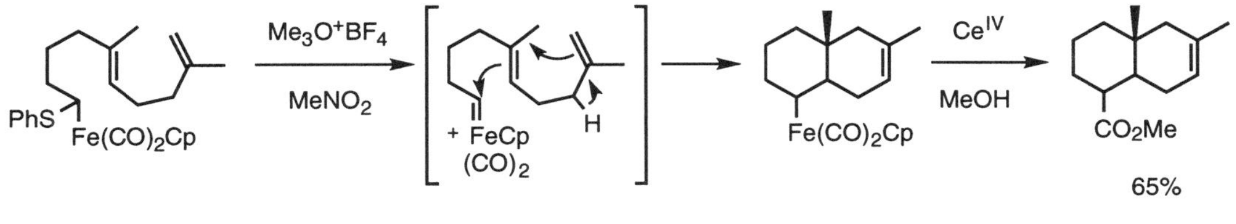

65%

Eq. 6.67

60-90%

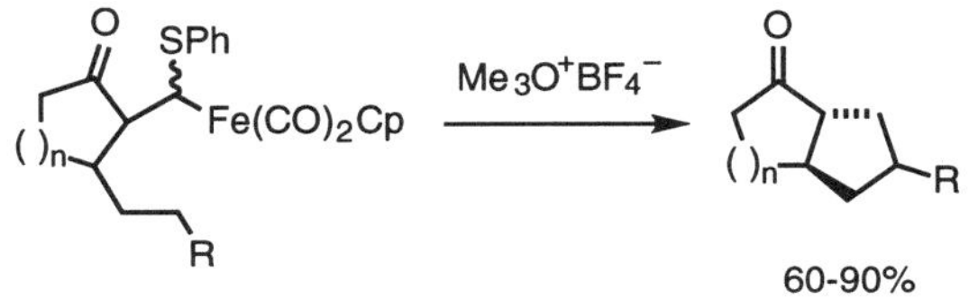

6.4 Metal-Catalyzed Decomposition of Diazo Compounds Proceeding Through Unstabilized Electrophilic Carbene Intermediates

Transition metals catalyze a number of important reactions of organic diazo compounds, among which cyclopropanation of olefins is one of the most synthetically useful (Eq. 6.68).[97] A wide variety of transition-metal complexes are efficient catalysts, among which copper(I) triflate, palladium(II) salts, rhodium(II)

Eq. 6.68

$ML_n = Rh_2(OAc)_4, CuCl\cdot P(OiPr)_3, Rh_6(CO)_{16}, PdCl_2\cdot 2PhCN$

$X = BrCH_2, ClCH_2, PhO, nBu, OAc, OEt, O\text{-}nBu, iPr, tBu, CH_2=CPh$

$CH_2=CMe, CH_2=C\text{-}tBu, CH=CHOMe, CH=CHCl, CH=CHPh, CH=CHMe$

acetate, and $Rh_6(CO)_{16}$ are most commonly used. The reaction is most effective when relatively stable α-diazocarbonyl compounds are used, since the major competing reaction is decomposition of the diazo compound itself. Olefins ranging from electron-rich enol ethers to electron-poor α,β-unsaturated esters undergo catalytic cyclopropanation with diazo compounds, although alkenes with electron-withdrawing substituents are likely to react by a mechanism different from that of other olefins, and β-substitution drastically suppresses the reaction. The order of reactivity for olefins is electron rich > "neutral" >> electron poor and *cis* > *trans*. Olefin geometry is maintained, but there is little stereoselectivity for the diazo-derived group. Dienes and alkynes are also cyclopropanated. Detailed mechanistic studies have implicated the intermediacy of metal carbene complexes formed by electrophilic addition of the metal catalyst to the diazo compound in this cyclopropanation.[98]

From a synthetic point of view, rhodium(II) acetate is by far the most-used catalyst, because it is commercially available, easy to handle, and reacts quickly and efficiently. Many functional groups are tolerated making elaborate cyclopropanes available (for example Eq. 6.69).[99] Intramolecular versions are quite efficient,

Eq. 6.69

W = Me, H, OSiR₃
X = Me, H
Y = H, OAc, OMe, OSiR₃
Z = H, OAc

41 - 94%

producing polycyclic ring systems (Eq. 6.70).[100] By using rhodium(II) complexes of optically active carboxylates or carboxamides, high asymmetric induction can be achieved (Eq. 6.71)[101] and (Eq. 6.72).[102] Note that with polyenes, the choice of catalyst effects regioselectivity (Eq. 6.73).[103]

Eq. 6.70

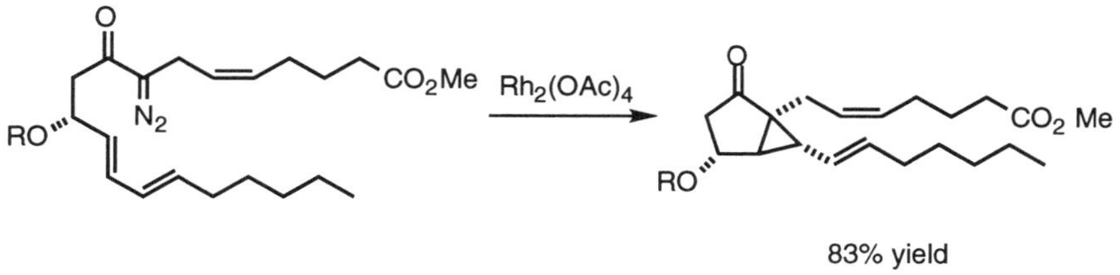

83% yield

Eq. 6.71

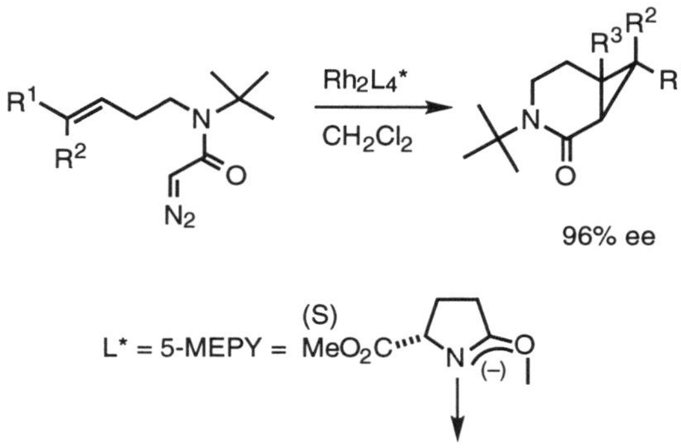

96% ee

$$L^* = 5\text{-MEPY} = $$

Eq. 6.72

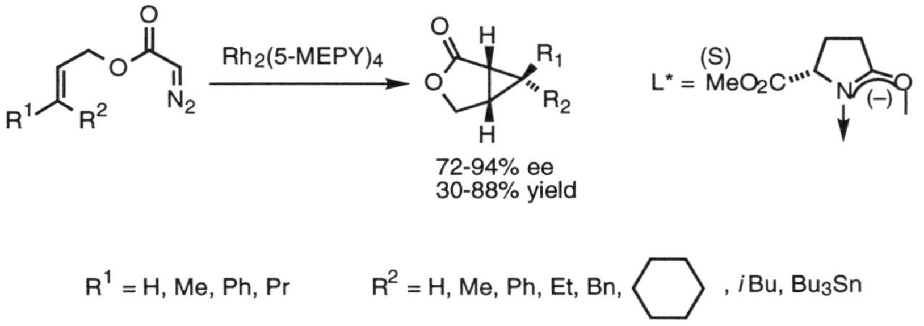

72-94% ee
30-88% yield

R^1 = H, Me, Ph, Pr R^2 = H, Me, Ph, Et, Bn, ⬡ , i Bu, Bu$_3$Sn

Eq. 6.73

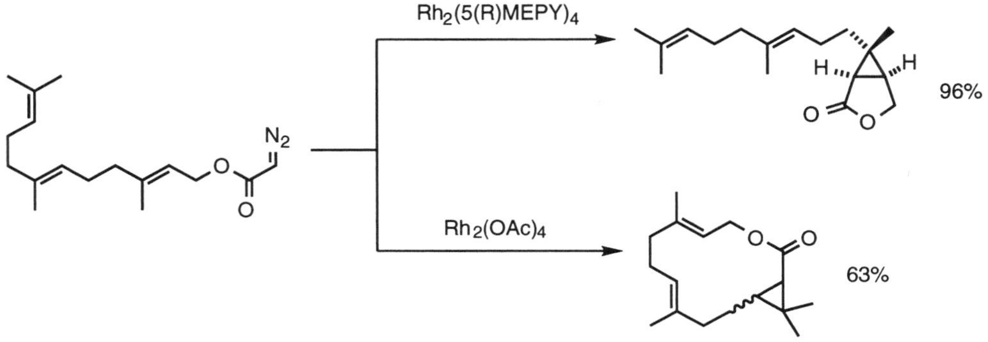

96%

63%

When substrates have additional unsaturation appropriately situated, metathesis "cascade" cyclizations ensue (Eq. 6.74),[67] much like the corresponding reactions of chromium carbene complexes (Eqs. 6.34-6.37). This provides further evidence for the intermediacy of metal carbene complexes in these cyclopropanations, since such metathesis-derived products are unlikely for other types of carbenoid species.

Eq. 6.74

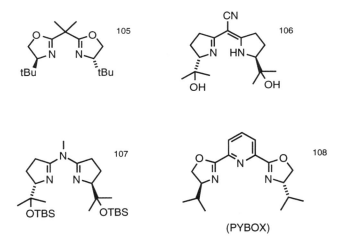

As stated above, a very large number of transition metals catalyze the cyclopropanation of alkenes by diazocompounds and a large number of chiral ligands have been developed. Some, normally used in conjunction with Cu^{2+} or Ru^{2+} are shown in Figure 6.4.

Figure 6.4   Chiral Ligands for Asymmetric Cyclopropanation

Cyclopropanation of alkenes by metal-catalyzed decomposition of diazoalkanes has many parallels to related reactions of stable carbene complexes, and the two systems are clearly mechanistically related. However, two reactions not common for metal carbene complexes — X–H insertion and ylide formation — are observed with metal-catalyzed decomposition of diazoalkanes, and both are synthetically useful.

Rhodium(II) catalyzes the insertion of diazoalkanes into C–H, O–H, N–H, S–H, and Si–H bonds in a process that is very unlikely to involve free carbenes.[109] Detailed mechanisms involving metal carbene complexes have not yet been elucidated, but are likely to involve electrophilic attack of the metal-bound carbene carbon on the X–H bond. The general order of reactivity for C–H insertion is 2° > 1° ≈ 3° but is catalyst dependent.[110] As expected for electrophilic attack, electron-withdrawing groups deactivate C–H bonds and donating groups activate them. Intramolecular insertions favor the formation of five-membered rings, and fluorinated carboxylate ligands on rhodium favor insertion with aromatic C–H bonds.[111] Very high asymmetric induction can be observed when chiral metal complex catalysts are used (Eq. 6.75),[112] strongly implicating an intimate association of the metal with the

Eq. 6.75

n = 1, 2, 3

97% ee

carbene fragment during bond formation. Other synthetically useful C–H insertions are shown in Eq. 6.76[113] and Eq. 6.77.[111] Rhodium(II) salts also catalyze N–H (Eq. 6.78)[114] and O–H (Eq. 6.79)[115] insertions which are synthetically useful.

Finally, transition metals catalyze the reaction between diazo compounds and heteroatom lone pairs to produce ylides, which have a very rich chemistry in their own right (Eq. 6.80).[116] The specific reactivity depends on the nature of the ylide formed. For example sulfur ylides undergo facile [2,3]-rearrangements (Eq. 6.81)[117] or Stevens rearrangement (Eq. 6.82)[118] as do nitrogen ylides (Eq. 6.83)[119] and (Eq. 6.84).[120]

Eq. 6.76

70-80%

88-98% ee

Eq. 6.77

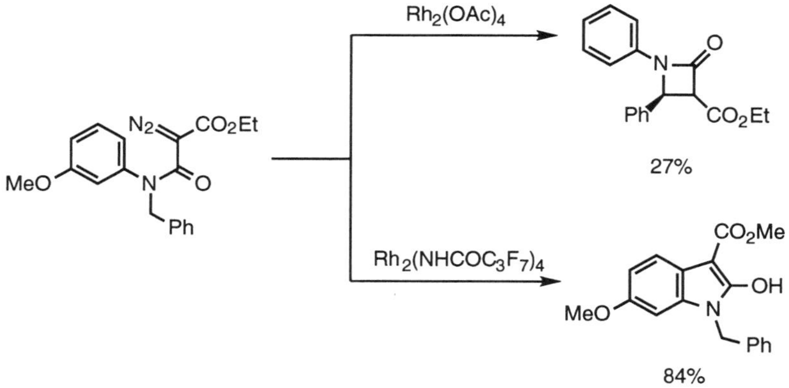

27%

84%

Eq. 6.78

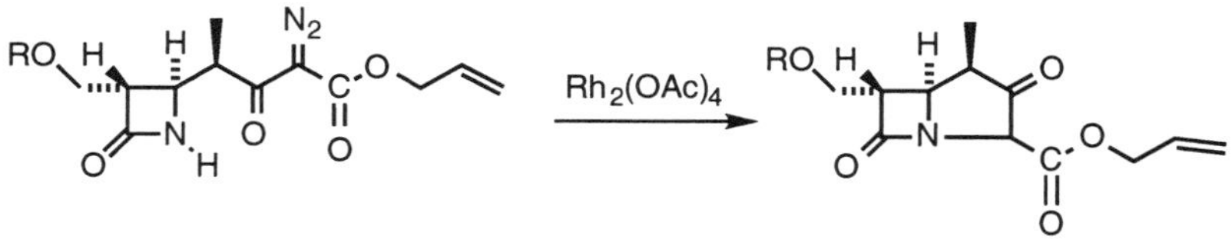

Eq. 6.79

96%

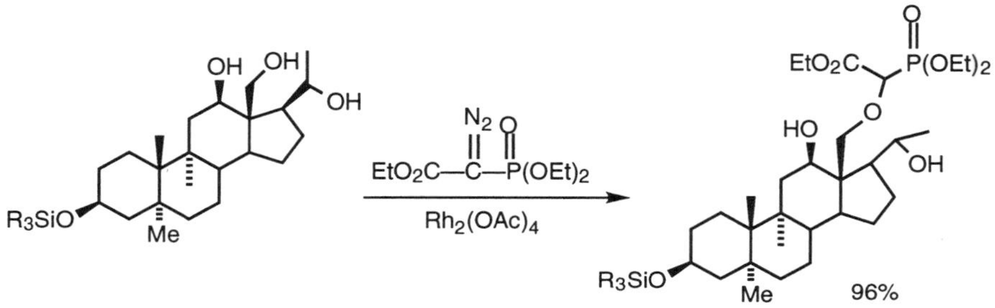

Eq. 6.80

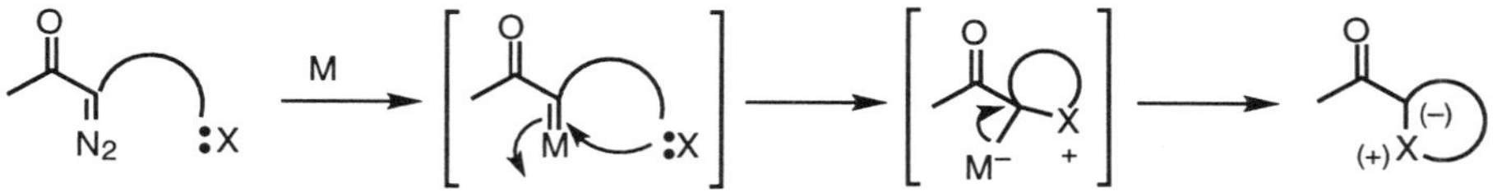

Eq. 6.81

77%

174

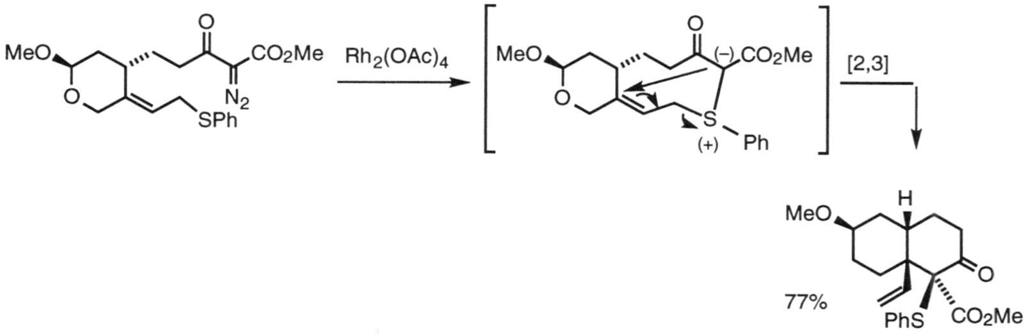

Eq. 6.82

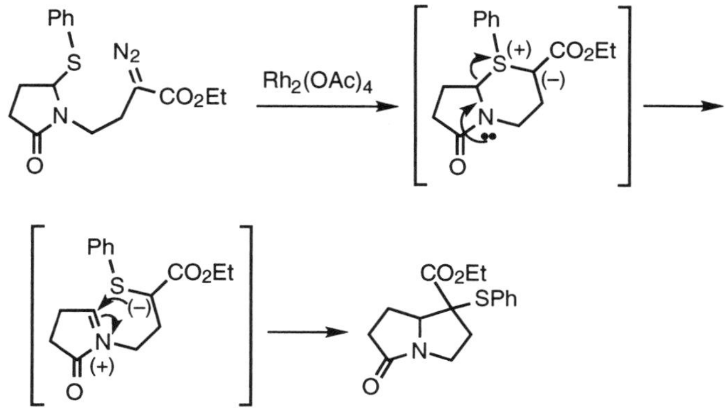

Eq. 6.83

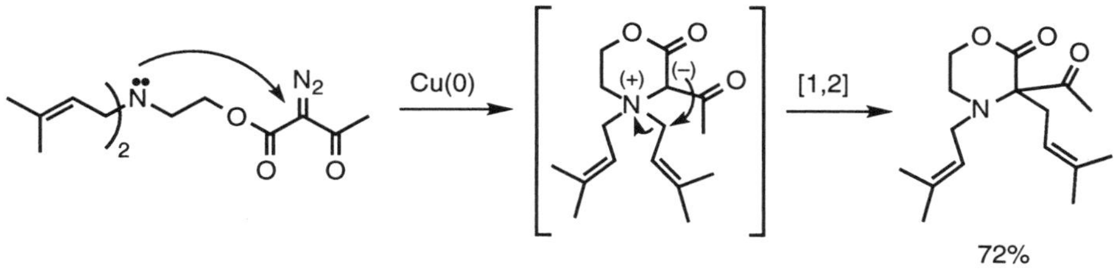

72%

Eq. 6.84

95%

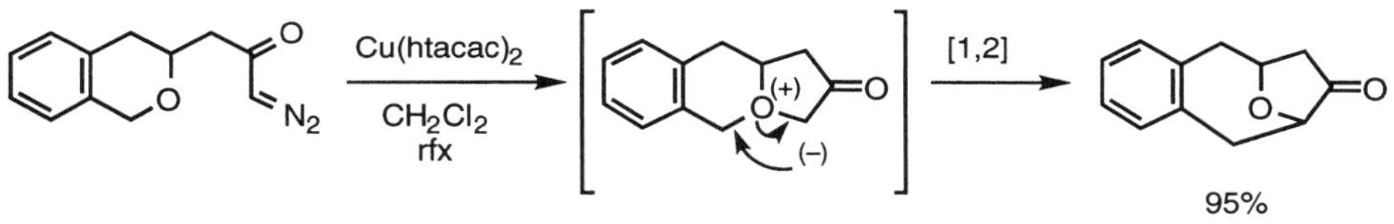

Most impressive from a synthetic point of view is the carbonyl ylide chemistry[121] accessible by metal-catalyzed diazo decomposition, primarily because carbonyl ylides have a rich dipolar cycloaddition chemistry. This allows the construction of very complex ring systems directly and efficiently, as illustrated by the synthesis of the aspidosperma alkaloid ring system in a single step (Eq. 6.85).[122] When coupled with metathesis chemistry (Eq. 6.74)[121] remarkable transformations occur in a single pot (Eq. 6.86).[123]

Eq. 6.85

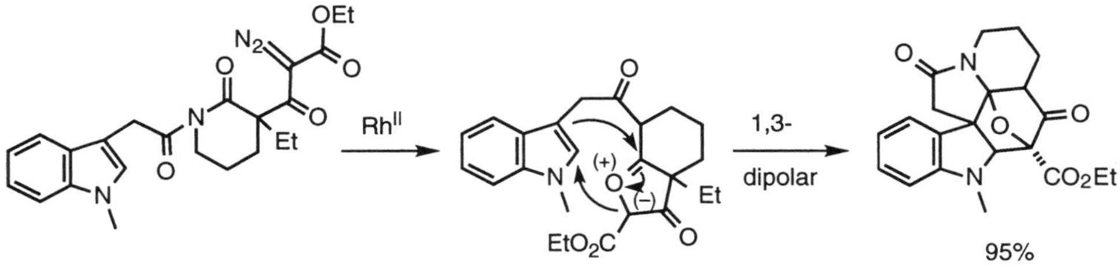

95%

Eq. 6.86

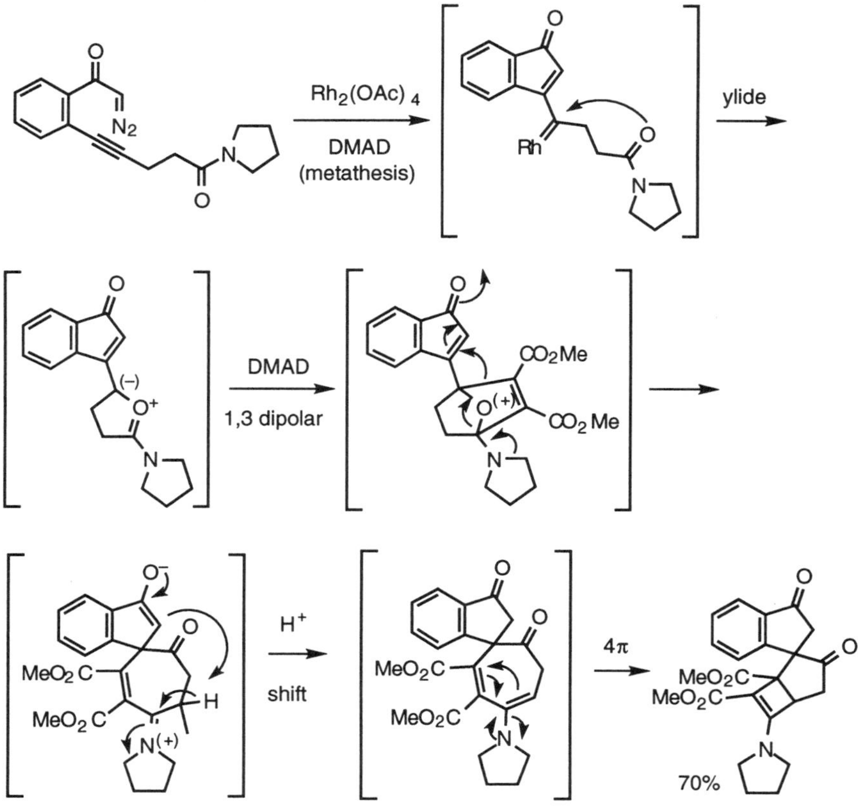

6.5 Metathesis Processes of Electrophilic Carbene Complexes

Olefin metathesis[126] is a process by which metal-carbene complexes catalyze a methylene group interchange via a cycloaddition/cycloreversion proces (Eq. 6.87).

Eq. 6.87

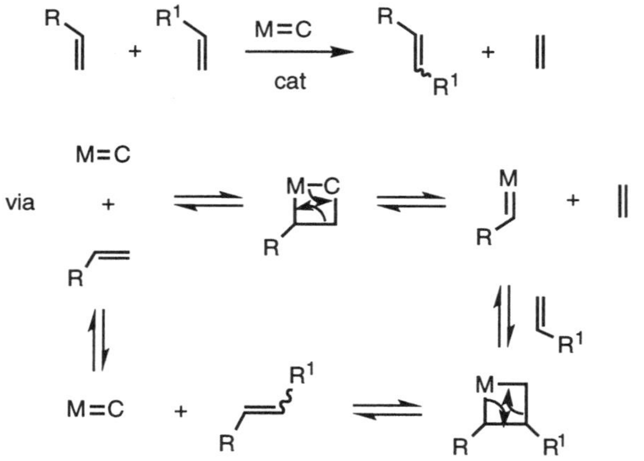

Metathesis has long been of importance industrially, but the high reactivity of the catalysts coupled with their relative intolerance to functional groups drastically limited their use in organic synthesis. However the recent development of two new catalysts that are relatively easy to handle and which perform metathesis efficiently in the presence of a wide array of functional groups has brought the use of this process in complex syntheses.

The catalysts of most use in organic synthesis are the Schrock molybdenum metathesis catalysis **1**[125] and the Grubbs ruthenium catalysts **2**[126] and **3**.[127] They share many common features, and their reaction chemistry overlaps considerably. However, in some cases there is a clear preference for one over the other. The Grubbs catalysts are, in general, easier to prepare and handle, tolerate a wider array of polar functional groups, and are more stable towards air and moisture. The Schrock catalyst has significantly higher metathesis activity, and often is more efficient with sterically-hindered olefins.[128] Together they have had a *major* impact on synthetic methodology by permitting the use of the olefin functional group as a carbon-carbon bond forming moiety.

1

2 R = Ph

3 R = CH = CPh$_2$

One of the earliest applications of olefin metathesis was to polymer synthesis, and initially these new metathesis catalysts were also applied to polymers in ring opening polymerizations (ROMP) (Eq. 6.88).[129] Their tolerance for functionality permitted the synthesis of highly functionalized polymers, even under aqueous conditions (Eq. 6.89).[130] Metathesis is also finding increased application to materials synthesis (Eq. 6.90).[131]

Eq. 6.88

Eq. 6.89

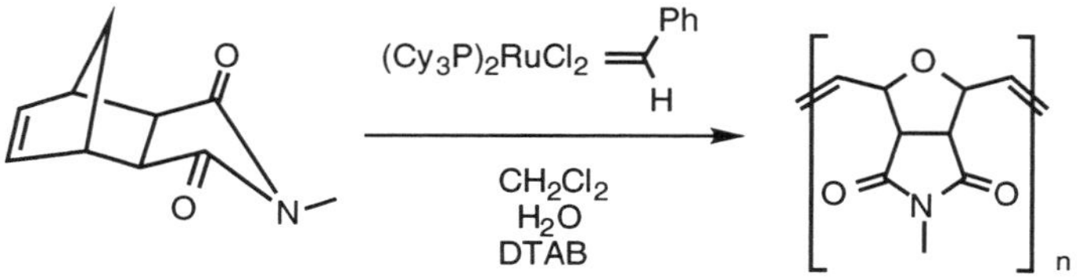

Eq. 6.90

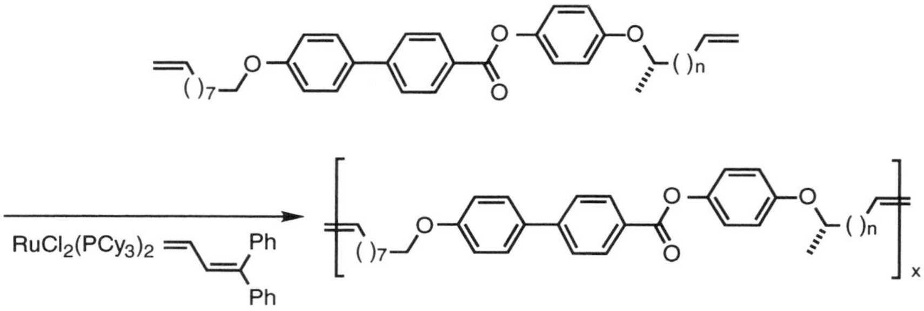

However, it is the application of *ring-closing* metathesis that has had the greatest impact on organic synthesis. Although the synthetic community was slow to appreciate the power of ring closing metathesis, and much of the early work was done by the developers of the catalysts themselves, once it caught on, the use of ring closing metathesis exploded. It can be used to make small rings (Eq. 6.91)[132] and (Eq. 6.92),[133] medium rings (Eq. 6.93),[134] and macrocycles (Eq. 6.94).[135] It is efficient with polymer-bound substrates as illustrated by the combinatorial synthesis of sixty epothilones, shown in Eq. 6.95.[136] It has been used to make catenanes[137,138] and to cap calixarenes,[139] and its use is growing logarithmically. Very recently[140] asymmetric versions utilizing chiral Schrock type catalysts have been introduced.

Eq. 6.91

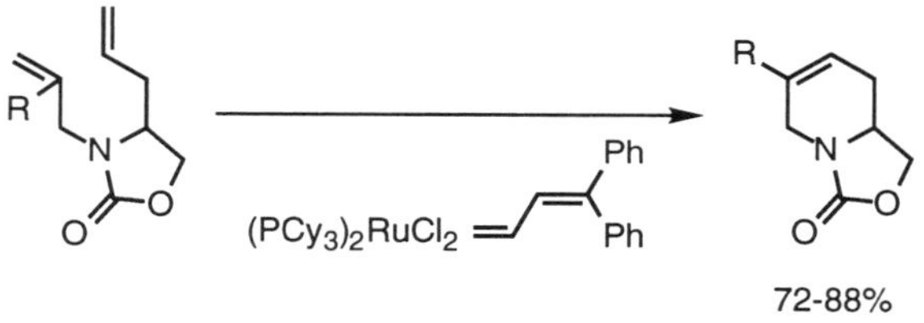

72-88%

Eq. 6.92

70%

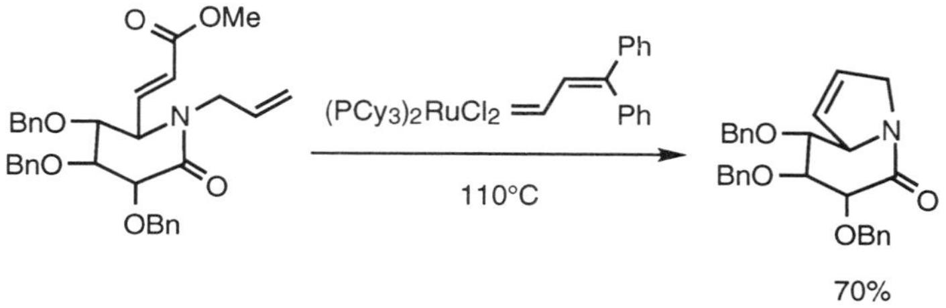

Eq. 6.93

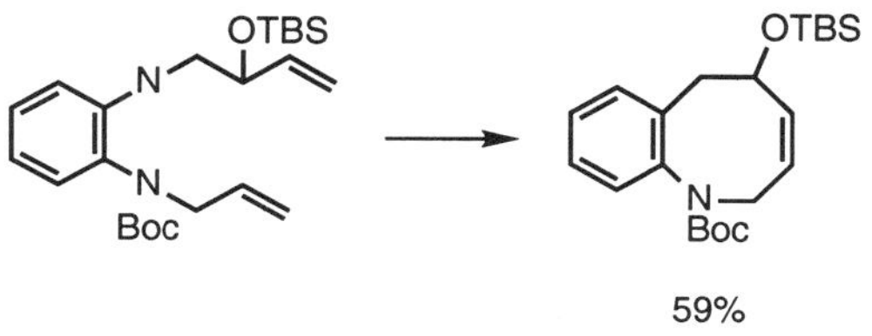

59%

Eq. 6.94

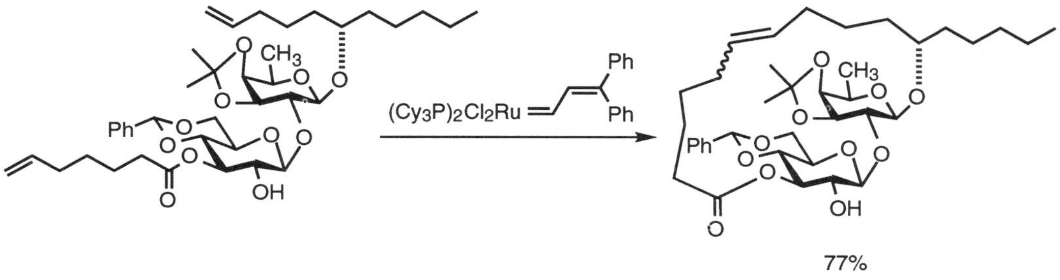

77%

Eq. 6.95

(60 compounds)

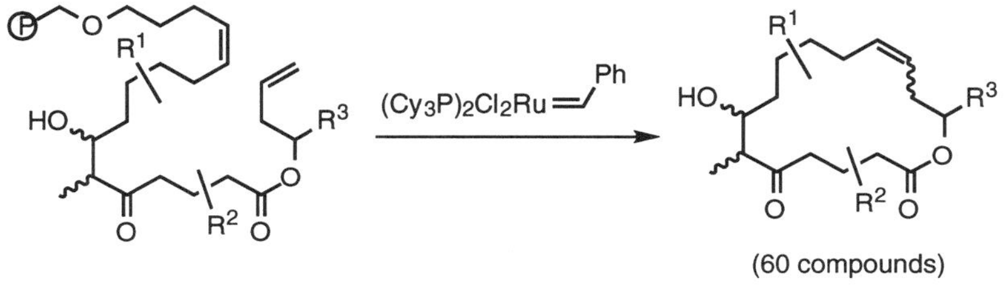

6.6   Nucleophilic "Schrock" Carbene Complexes[141]

Nucleophilic carbene complexes represent the opposite extreme of reactivity and bonding from electrophilic carbene complexes. They are usually formed by metals at the far left of the transition series, in high oxidation states (usually d^0) with strong donor ligands, such as alkyl or cyclopentadienyl, and no acceptor ligands. The carbene ligand is usually simply the $=CH_2$ group. These are so different from electrophilic carbene complexes that the carbene ligand in these complexes is formally considered a four electron, dinegative ligand, $[H_2C::]^{2-}$. These complexes often have Wittig-reagent-like reactivity and are often represented as $M^+-CH_2^- \leftrightarrow M=CH_2$, much like ylides. However, nucleophilic carbene complexes also form metallacycles, and both of these aspects of reactivity are useful in synthesis.

The most extensively studied nucleophilic carbene complex is "Tebbe's" reagent, prepared by the reaction of trimethyl aluminum with titanocene dichloride (Eq. 6.96).[142] In the presence of pyridine this complex is synthetically equivalent to "Cp_2TiCH_2", and it is very efficient in converting carbonyl groups to methylenes (Eq.

6.97).[143] The reaction is thought to proceed via an oxometallacycle which can fragment to give the alkene and a very stable titanium(IV) oxo species (Eq. 6.98). The

Eq. 6.96

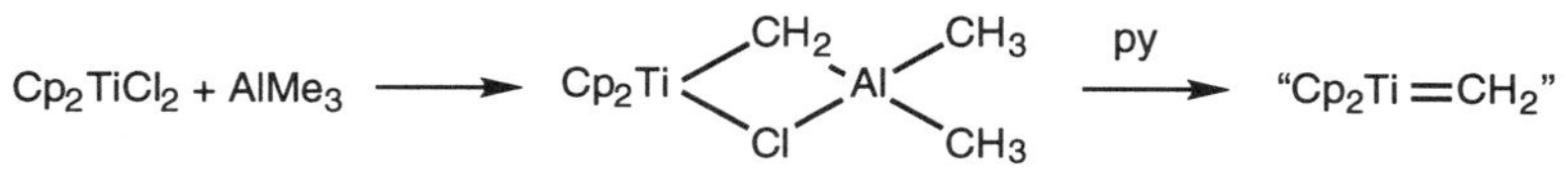

$$Cp_2TiCl_2 + AlMe_3 \longrightarrow Cp_2Ti\big\langle\substack{CH_2 \\ Cl}\big\rangle Al\big\langle\substack{CH_3 \\ CH_3} \xrightarrow{py} \text{``}Cp_2Ti=CH_2\text{''}$$

Eq. 6.97

$$\text{``}Cp_2Ti=CH_2\text{''} + \underset{\underset{X}{\overset{O}{\|}}}{RC} \longrightarrow R-\underset{X}{\overset{CH_2}{\overset{\|}{C}}}$$

$$X = R', H, OR', NR'_2$$

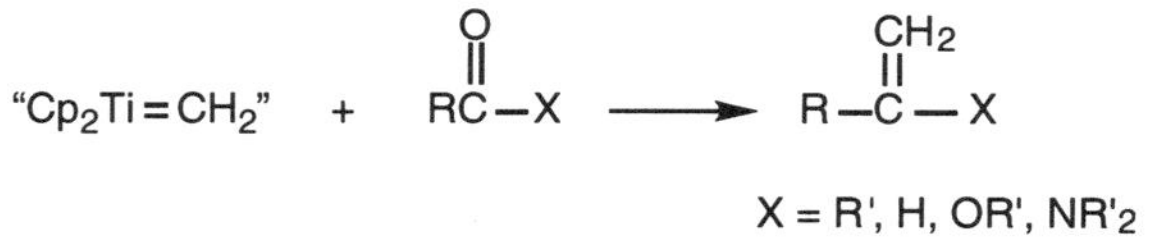

Eq. 6.98

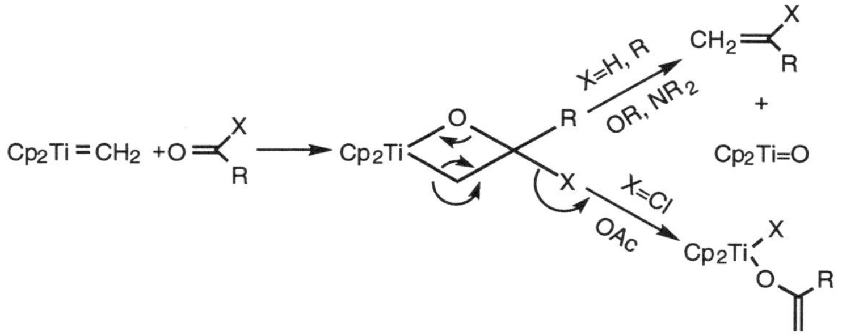

reaction not only works well with aldehydes and ketones, but it also converts esters to enol ethers and amides to enamines! With acid halides and anhydrides, titanium enolates are produced instead. Tebbe's reagent does not enolize ketones and is particularly efficient for methylenating sterically hindered, enolizable ketones which are not methylenated efficiently by Wittig reagents.[144] The reagent tolerates a very high degree of functionality (Eq. 6.99)[145] and (Eq. 6.100).[146]

In addition to methylenating carbonyl compounds, Tebbe's reagent reacts with olefins to form titanacycles. These can be cleaved by a variety of reagents (Eq. 6.101), or can be used to synthesize allenes by a cycloaddition/metathesis sequence

Eq. 6.99

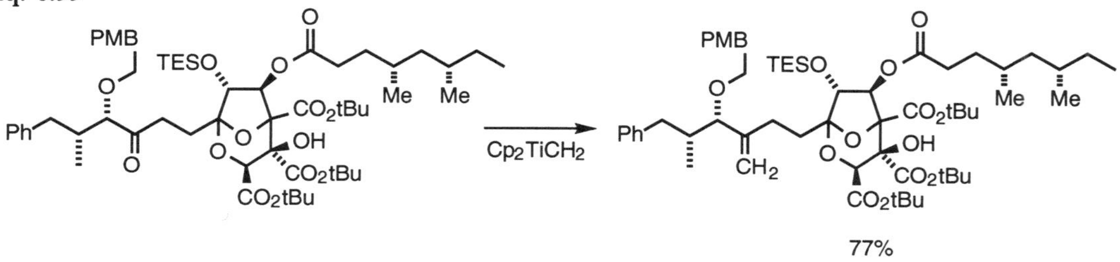

77%

Eq. 6.100

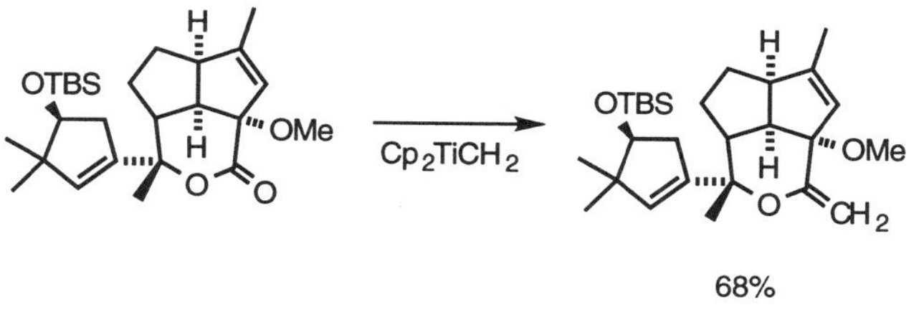

68%

Eq. 6.101

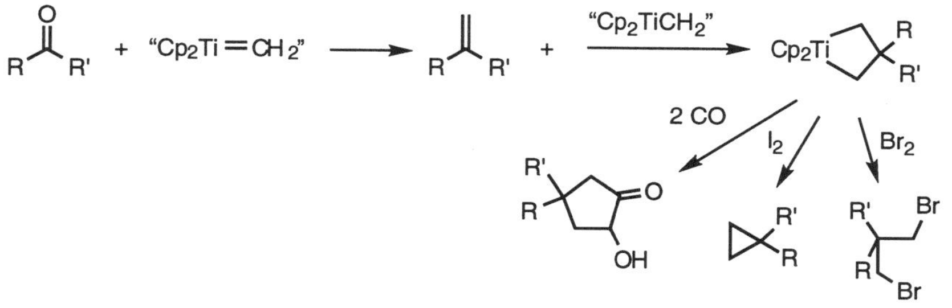

(Eq. 6.102).[147] Methylenation combined with metathesis can be a powerful synthetic tool (Eq. 6.103).[148]

Eq. 6.102

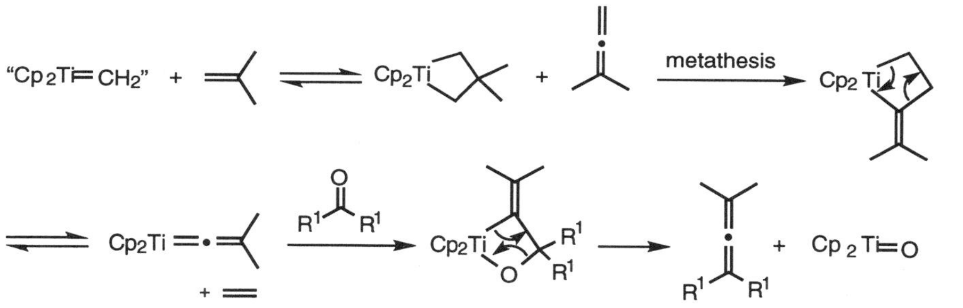

An even more easily prepared methylenating reagent is dimethyl titanocene, Cp_2TiMe_2, generated by the simple treatment of titanocene dichloride with methyllithium. Gentle heating of this complex (~60°C) in the presence of ketones, aldehydes, or esters results in methylation with yields and selectivities similar to those observed with Tebbe's reagent.[86] In some instances, it has advantages over Tebbe's reagent. In the system in Eq. 6.104, Tebbe's reagent fragmented the substrate while Cp_2TiMe_2 cleanly methylenated it. In addition, benzylidene (PhCH=),[151] trimethylsilylmethylene (TMSCH=),[152] allenelidene (CH$_2$=C=),[153] and methylenecyclopropane[154] can be transferred, while Tebbe's reagent is restricted to simple methylene groups. Other systems that efficiently methylenate (4, 5, and 6) are shown below.

Eq. 6.103

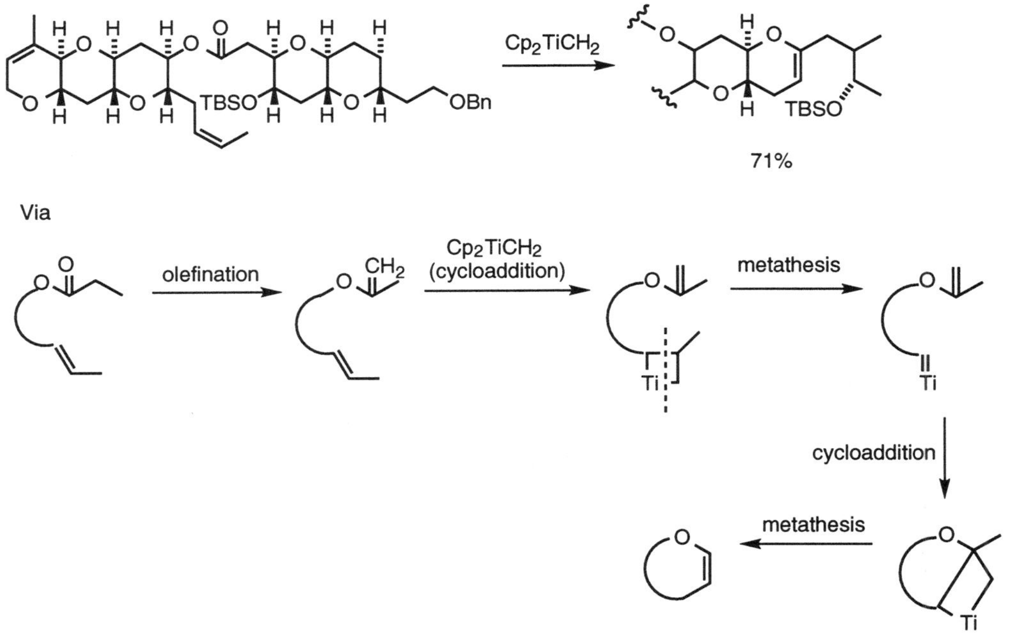

71%

Via

Eq. 6.104

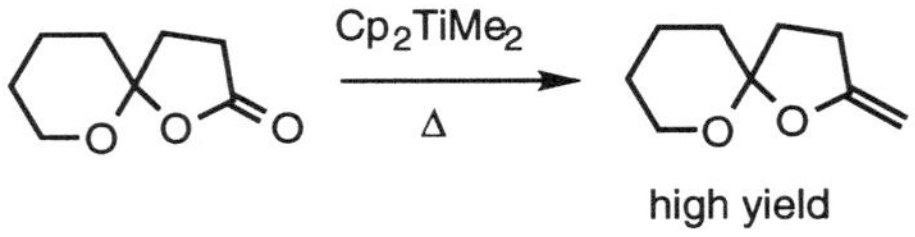

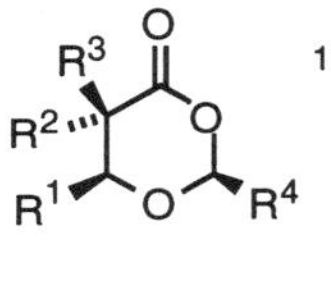

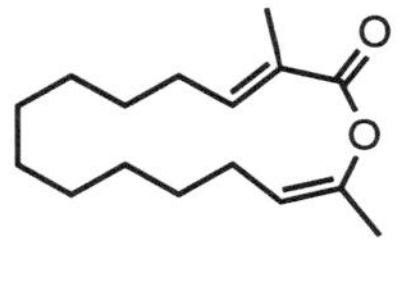

high yield

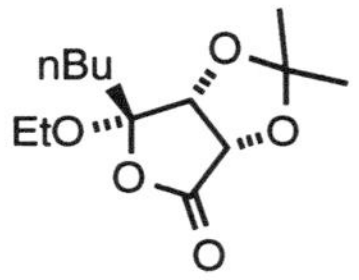

4   **5**   **6**

References

1. Seyferth, D., ed. *Transition Metal Carbene Complexes,* Verlag Chemie, 1983.
2. Hoye, T.R.; Chen, K.; Vyvyan, J.R. *Organometallics* **1993**, *12*, 2806.
3. Bao, J.; Wulff, W.D.; Dominy, J.B.; Fumo, M.J.; Grant, E.B.; Rob, A.C.; Whitcomb, M.C.; Yeung, S-M.; Ostrander, R.L.; Rheingold, A.L. *J. Am. Chem. Soc.* **1996**, *118*, 3392.
4. Semmelhack, M.F.; Lee, G.R. *Organometallics* **1987**, *6*, 1839.
5. Casey, C.P.; Brunswald, W.P. *J. Organomet. Chem.* **1976**, *118*, 309.
6. (a) Hegedus, L.S.; Schwindt, M.A.; DeLombaert, S.; Imwinkelried, R. *J. Am. Chem. Soc.* **1990**, *112*, 2264. (b) Schwindt, M.A.; Lejon, T.; Hegedus, L.S. *Organometallics* **1990**, *9*, 2814.
7. Stadtmüller, H.; Knochel, P. *Organometallics* **1995**, *14*, 3863.
8. (a) Dötz, K.H.; Sturm, W.; Alt, H.G. *Organometallics* **1987**, *6*, 1424. (b) Schmidt, B.; Kocienski, P.; Reid, G. *Tetrahedron* **1996**, *52*, 1617.
9. Christoffers, J.; Dötz, K.H. *Chem. Ber.* **1995**, *128*, 157.

10. Cosset, C.; Del Rio, I.; Le Bozec, H. *Organometallics* **1995**, *14*, 1938.
11. Hafner, A.; Hegedus, L.S.; deWeck, G.; Hawkins, B.; Dötz, K.H. *J. Am. Chem. Soc.* **1988**, *110*, 8413.
12. Merlic, C.A.; Xu, D.; Gladstone, B.G. *J. Org. Chem.* **1993**, *58*, 538.
13. (a) Casey, C.P., "Metal-Carbene Complexes in Organic Synthesis", in *Transition Metal Organometallics in Organic Synthesis*, vol. 1, Alper, H. ed. Academic Press, New York, 1976, pp. 190-23. (b) Bernasconi, C.F.; Sun, W. *J. Am. Chem. Soc.* **1993**, *115*, 12526. (c) For a review on the physical organic chemistry of Fischer carbene complexes see: Bernasconi, C.F. *Chem. Soc. Rev.* **1997**, *26*, 299.
14. Lattuada, L.; Licandro, E.; Maiorana, S.; Molinari, H.; Papagni, A. *Organometallics* **1991**, *10*, 807.
15. Aumann, R.; Heinen, H. *Chem. Ber.* **1987**, *120*, 537.
16. (a) Xu, Y-C.; Wulff, W.D. *J. Org. Chem.* **1987**, *52*, 3263. (b) Armin, S.R.; Sarkar, A. *Organometallics* **1995**, *14*, 547.
17. Powers, T.S.; Shi, Y.; Wilson, K.J.; Wulff, W.D. *J. Org. Chem.* **1994**, *59*, 6882.
18. (a) Anderson, B.A.; Wulff, W.D.; Rahm, A. *J. Am. Chem. Soc.* **1993**, *115*, 4602. (b) Baldoli, C.; Del Butero, P.; Licandro, E.; Maiorana, S.; Papagni, A.; Zannoti-Gerosa, A. *J. Organomet. Chem.* **1995**, *486*, 1995. (c) Shi, Y.; Wulff, W.D.; Yap, G.P.A.; Rheingold, A. *J. Chem. Soc., Chem. Comm.* **1996**, 2600. (d) For a review on asymmetric synthesis with Fischer complexes see: Wulff, W.D. *Organometallics* **1998**, *17*, 3116.
19. (a) McDonald, F.E.; Connolly, C.B.; Gleason, M.M.; Towne, T.B.; Treiber, K.D. *J. Org. Chem.* **1993**, *58*, 6952. (b) McDonald, F.E.; Schultz, C.C. *J. Am. Chem. Soc.* **1994**, *116*, 9363.
20. McDonald, F.E.; Zhu, H-Y. *Tetrahedron* **1997**, *53*, 11061.
21. McDonald, F.E.; Chatterjee, A.K. *Tetrahedron Lett.* **1997**, *38*, 7687.
22. McDonald, F.E.; Gleason, M.M. *J. Am. Chem. Soc.* **1996**, *118*, 6648.
23. Wang, S.L.B.; Wulff, W.D. *J. Am. Chem. Soc.* **1990**, *112*, 4550. Wulff, W.D.; Bauta, W.E.; Kaesler, R.W.; Lankford, P.J.; Miller, R.A.; Murray, C.K.; Yang, D.C. *J. Am. Chem. Soc.* **1990**, *112*, 3642.
24. Alkenyl carbene complexes were estimated to have reactivity comparable to maleic anhydride. See Adam, H.; Albrecht, T.; Sauer, J. *Tetrahedron Lett.* **1994**, *35*, 557.
25. Wulff, W.D.; Power, T.S. *J. Org. Chem.* **1993**, *58*, 2381.
26. Anderson, B.; Wulff, W.D.; Power, T.S. *J. Am. Chem. Soc.* **1992**, *114*, 10784.
27. Barluenga, J.; Tomas, M.; Ballesteros, A.; Santamaria, J.; Swarez-Sobrino, A. *J. Org. Chem.* **1997**, *62*, 9229.
28. Loft, M.S.; Mowlem, T.J.; Widdowson, D.A. *J. Chem. Soc. Perkin Trans I* **1995**, 97.
29. Barluenga, J.; Fernandez-Mari, F.; Viado, A.L.; Aguilar, E.; Olano, B. *J. Chem. Soc. Perkin Trans I* **1997**, 2267.
30. (a) Aoki, S.; Fujimura, T.; Nakamura, E. *J. Am. Chem. Soc.* **1992**, *114*, 2985. (b) Nakamura, E.; Tanaka, K.; Fujimura, T.; Aoki, S.; Williard, P.G. *J. Am. Chem. Soc.* **1993**, *115*, 9015.
31. Aumann, R.; Meyer, A.G.; Fröhlich, R. *J. Am. Chem. Soc.* **1996**, *118*, 10853.
32. Lluch, A.M.; Jordi, L.; Sanchez-Baeza, F.; Ricart, S.; Camps, F.; Messeguer, A.; Moreto, J.M. *Tetrahedron Lett.* **1992**, *33*, 3021.
33. Soderberg, B.C.; Bowden, B.A. *Organometallics* **1992**, *11*, 2220.
34. For reviews see: (a) Dötz, K.H. *Angew. Chem. Int. Ed. Engl.* **1984**, *23*, 587. (b) Reissig, H-U. *Organomet. in Synthesis* **1989**, *2*, 311. For other studies see: Reissig, H-U. *Organometallics* **1990**, *9*, 3133. (c) Doyle, M.P., "Transition Metal Carbene Complexes: Cyclopropanation" in "Comprehensive Organometallic Chemistry II", Abel, E.W.; Stone, F.G.A.; Wilkinson, G., Eds., Pergamon, Oxford, UK, 1995, Vol. 12, pp. 387-395.
35. Harvey, D.F.; Lund, K.P. *J. Am. Chem. Soc.* **1991**, *113*, 8916.
36. Buchert, M.; Hoffmann, M. Reissig, H-U. *Chem. Ber.* **1995**, *128*, 605.
37. Wienand, A.; Reissig, H.U. *Angew. Chem. Int. Ed. Engl.* **1990**, *29*, 1129.
38. Murray, C.K.; Yang, D.C.; Wulff, W.D. *J. Am. Chem. Soc.* **1990**, *112*, 5660.
39. Semmelhack, M.F.; Bozell, J.J. *Tetrahedron Lett.* **1982**, *23*, 2931.
40. Soderberg, B.C.; Hegedus, L.S. *Organometallics* **1990**, *9*, 3113.
41. Barluenga, J.; Montserrat, J.M.; Flerez, J. *J. Chem. Soc., Chem. Comm.* **1993**, 1068.
42. (a) For a review see: Harvey, D.F.; Sigano, D.M. *Chem. Rev.* **1996**, *96*, 271. (b) Hoye, T.R.; Suriano, J.A. *Organometallics* **1992**, *11*, 2044; **1989**, *8*, 2670. (c) Mori, M.; Wantanuki, S. *J. Chem. Soc., Chem. Comm.* **1992**, 1082. (d) Katz, T.J. *Tetrahedron Lett.* **1991**, *32*, 5895.
43. Alvarez, C.; Parlier, A.; Rudler, H.; Yefsah, R.; Daran, J.C.; Knobler, C. *Organometallics* **1989**, *8*, 2253.
44. Hoye, T.R.; Vyuyan, J.R. *J. Org. Chem.* **1995**, *60*, 4184.
45. Harvey, D.F.; Brown, M.F. *J. Am. Chem. Soc.* **1990**, *112*, 7806. *Tetrahedron Lett.* **1991**, *32*, 2871. *Tetrahedron Lett.* **1991**, 5223, 6311.

46. (a) Harvey, D.F.; Brown, M.F. *J. Org. Chem.* **1992**, *57*, 5559. (b) Harvey, D.F.; Lund, K.P.; Neil, D.A. *J. Am. Chem. Soc.* **1992**, *114*, 8424.
47. (a) Harvey, D.F.; Lund, K.P. *J. Am. Chem. Soc.* **1991**, *113*, 5066. (b) Harvey, D.F.; Grenzer, E.M. *J. Org. Chem.* **1996**, *61*, 159.
48. Dötz, K.H. *Angew. Chem. Int. Ed. Engl.* **1984**, *23*, 587. For a review see: Wulff, W.D., "Transition Metal Carbene Complexes: Alkyne and Vinyl Ketene Chemistry" in "Comprehensive Organometallic Chemistry II", Abel, E.W.; Stone, F.G.A.; Wilkinson, G., Eds., Pergamon, Oxford, UK, 1995, Vol. 12, pp. 469-548.
49. Dötz, K.H.; Muhlemeier, J. *Angew. Chem. Int. Ed. Engl.* **1982**, *21*, 929.
50. (a) Wulff, W.D.; Bax, B.M.; Branwold, T.A.; Chan, K.S.; Gilbert, A.M.; Hsung, R.P. *Organometallics* **1994**, *13*, 102. (b) Wulff, W.D.; Gilbert, A.M.; Hsung, R.P.; Rahm, A. *J. Org. Chem.* **1995**, *60*, 4566.
51. Bauta, W.E.; Wulff, W.D.; Pavkovic, S.F.; Zaluzec, E.J. *J. Org. Chem.* **1989**, *54*, 3249.
52. Bao, J.; Wulff, W.D.; Doming, J.B.; Fumo, M.J.; Grant, E.B.; Rob, A.C.; Whitcomb, M.C.; Yeung, S-M.; Ostrander, R.L.; Rheingold, A.L. *J. Am. Chem. Soc.* **1996**, *118*, 3392.
53. Dötz, K.H.; Ehlenz, R.; Paetsch, D. *Angew. Chem. Int. Ed. Engl.* **1997**, *36*, 2376.
54. (a) Bao, J.; Dragisich, V.; Wenglowsky, S.; Wulff, W.D. *J. Am. Chem. Soc.* **1991**, *113*, 9873. (b) Bao, J.; Wulff, W.D.; Dragisch, V.; Wenglowski, S.; Ball, R.G. *J. Am. Chem. Soc.* **1994**, *116*, 7616.
55. Hegedus, L.S.; deWeck, G.; D'Andrea, S. *J. Am. Chem. Soc.* **1988**, *110*, 2122.
56. (a) Merlic, C.A.; Xu, D. *J. Am. Chem. Soc.* **1991**, *113*, 7418. (b) Merlic, C.A.; Xu, D.; Gladstone, B.G. *J. Org. Chem.* **1993**, *58*, 538.
57. (a) Merlic, C.A.; Burns, E.E.; Xu, D.; Chen, S.Y. *J. Am. Chem. Soc.* **1992**, *114*, 8722.
58. Merlic, C.A.; McInnes, D.M.; You, Y. *Tetrahedron Lett.* **1997**, *38*, 6787.
59. (a) Yamashita, A. *Tetrahedron Lett.* **1986**, *27*, 5915. For reviews on aminocarbene complex chemistry see: (b) Grotjahn, D.B.; Dötz, K.H. *Synlett* **1991**, 381. (c) Schwindt, M.P.; Miller, J.R.; Hegedus, L.S. *J. Organomet. Chem.* **1991**, *413*, 143.
60. Dötz, K.H.; Schäfer, T.O.; Harms, K. *Synthesis* **1992**, 146; *Angew. Chem. Int. Ed. Engl.* **1990**, *29*, 176.
61. Barluenga, J.; Aznar, F.; Barluenga, S. *J. Chem. Soc., Chem. Comm.* **1995**, 1973.
62. (a) Turner, S.U.; Senz, U.; Herndon, J.W.; McMullen, L.A. *J. Am. Chem. Soc.* **1992**, *114*, 8394. (b) Hill, D.K.; Herndon, J.W. *Tetrahedron Lett.* **1996**, *37*, 1359. (c) Herndon, J.W.; Patel, P.P. *Tetrahedron Lett.* **1997**, *38*, 59.
63. Herndon, J.W.; Matasi, J.J. *J. Org. Chem.* **1990**, *55*, 786.
64. Herndon, J.W.; Zora, M.; Patel, P.P. *Tetrahedron* **1993**, *49*, 5507.
65. Semmelhack, M.F.; Tamura, R. *J. Am. Chem. Soc.* **1983**, *105*, 4099; 6750.
66. Semmelhack, M.F.; Tamura, R.; Schnatter, W.; Springer, J. *J. Am. Chem. Soc.* **1984**, *100*, 5363.
67. Foley, H.C.; Strubinger, L.M.; Targos, T.S.; Geoffroy, G.L. *J. Am. Chem. Soc.* **1983**, *105*, 3064.
68. Geoffroy, G.L. *Adv. Organomet. Chem.* **1985**, *24*, 249.
69. (a) Hegedus, L.S.; DeWeck, G.; D'Andrea, S. *J. Am. Chem. Soc.* **1988**, *110*, 2122. (b) For a review on metal ketene complexes see: Geoffroy, G.L.; Bassner, S.L. *Adv. Organomet. Chem.* **1988**, *28*, 1.
70. For a review see: Hegedus, L.S. *Tetrahedron* **1997**, *53*, 4105.
71. (a) Imwinkelried, R.; Hegedus, L.S. *Organometallics* **1988**, *7*, 702. (b) Schwindt, M.A.; Lejon, T.; Hegedus, L.S. *Organometallics* **1990**, *9*, 2814.
72. Hegedus, L.S.; Imwinkelried, R.; Alarid-Sargent, M.; Dvorak, D.; Satoh, Y. *J. Am. Chem. Soc.* **1990**, *112*, 1109.
73. For a discussion of stereoselectivity in these reactions see: Hegedus, L.S.; Montgomery, J.; Narukawa, Y.; Snustad, D.C. *J. Am. Chem. Soc.* **1991**, *113*, 5784.
74. Narukawa, Y.; Juneau, K.N.; Snustad, D.C.; Miller, D.B.; Hegedus, L.S. *J. Org. Chem.* **1992**, *57*, 5453.
75. (a) Betschart, C.; Hegedus, L.S. *J. Am. Chem. Soc.* **1992**, *114*, 5010. (b) Es-Sayed, M.; Heiner, T.; de Meijere, A. *Synlett* **1993**, 57.
76. Hsiao, Y.; Hegedus, L.S. *J. Org. Chem.* **1997**, *62*, 3586.
77. Valenti, E.; Pericas, M.A.; Moyano, A. *J. Org. Chem.* **1990**, *55*, 3582.
78. Soderberg, B.C.; Hegedus, L.S.; Sierra, M.A. *J. Am. Chem. Soc.* **1990**, *112*, 4364.
79. Soderberg, B.C.; Hegedus, L.S. *J. Org. Chem.* **1991**, *56*, 2209.
80. Aumann, R.; Krüger, C.; Goddard, R. *Chem. Ber.* **1992**, *125*, 1627.
81. (a) Hegedus. L.S.; Bates, R.W.; Soderberg, B.C. *J. Am. Chem. Soc.* **1991**, *113*, 923. (b) Reed, A.D.; hegedus, L.S. *J. Org. Chem.* **1995**, *60*, 3717. (c) Reed, A.D.; Hegedus, L.S. *Organometallics* **1997**, *16*, 2313.
82. For a review see: Hegedus, L.S. *Acc. Chem. Res.* **1995**, *28*, 299.
83. Hegedus, L.S.; Schwindt, M.A.; DeLombaert, S.; Imwinkelried, R. *J. Am. Chem. Soc.* **1990**, *112*, 2264.
84. Schmeck, C.; Hegedus, L.S. *J. Am. Chem. Soc.* **1994**, *116*, 9927.
85. Miller, J.R.; Pulley, S.R.; Hegedus, L.S.; DeLombaert, S. *J. Am. Chem. Soc.* **1992**, *114*, 5602.
86. Debuisson, C.; Fukumoto, Y.; Hegedus, L.S. *J. Am. Chem. Soc.* **1995**, *117*, 3697.

184

87. Pulley, S.R.; Hegedus, L.S. *J. Am. Chem. Soc.* **1993**, *115*, 9037.

88. (a) Zhu, J.; Hegedus, L.S. *J. Org. Chem.* **1995**, *60*, 5831. (b) Zhu, J.; Deur, C.; Hegedus, L.S. *J. Org. Chem.* **1997**, *62*, 7704.

89. For a review see: Brookhart, M.; Studabaker, W.B. *Chem. Rev.* **1987**, *87*, 411.

90. For examples, see: Fischer, H.; Hoffmann, J. *Chem. Ber.* **1991**, *124*, 981.

91. Mattson, M.N.; Bays, J.P.; Zakutansky, J.; Stolarski, V.; Helquist, P. *J. Org. Chem.* **1989**, *54*, 2467.

92. Vargas, R.M.; Theys, R.D.; Hossain, M.M. *J. Am. Chem. Soc.* **1992**, *114*, 777.

93. Theys, R.D.; Hossain, M.M. *Tetrahedron Lett.* **1992**, *33*, 3447.

94. (a) Brookhart, M.; Liu, Y.; Goldman, E.W.; Timmers, D.A.; Williams, G.D. *J. Am. Chem. Soc.* **1991**, *113*, 927. (b) Brookhart, M.; Liu, Y. *J. Am. Chem. Soc.* **1991**, *113*, 939.

95. Baker, C.T.; Mattson, M.N.; Helquist, P. *Tetrahedron Lett.* **1995**, *36*, 7015.

96. Ishii, S.; Helquist, P. *Synlett* **1997**, 508.

97. For reviews see: (a) Adams, J.; Spero, D.M. *Tetrahedron* **1991**, *47*, 1765. (b) Doyle, M.P., "Transition Metal Carbene Complexes: Cyclopropanation" in "Comprehensive Organometallic Chemistry II", Abel, E.W.; Stone, F.G.A.; Wilkinson, G., Eds., Pergamon, Oxford, UK, 1995, Vol. 12, pp. 387-395. (c) Singh, V.K.; Dattagupta, A.; Sebar, G. *Synthesis* **1997**, 137. (d) Reissig, H-U. *Angew. Chem. Int. Ed. Engl.* **1996**, *35*, 971.

98. (a) Doyle, M.P.; Dorow, R.L.; Buhro, W.E.; Griffin, J.H.; Tamblyn, W.H.; Trudell, M.L. *Organometallics* **1984**, *3*, 44. (b) Doyle, M.P.; Griffin, J.H.; Bagheri, V.; Dorow, R.L. *Organometallics* **1984**, *3*, 53.

99. (a) Davies, H.M.L.; Clark, T.J.; Smith, H.D. *J. Org. Chem.* **1991**, *56*, 3817. (b) Cantrell, W.R., Jr.; Davies, H.M.L. *J. Org. Chem.* **1991**, *56*, 723. (c) Davies, H.M.L.; Clark, T.J.; Kinomer, G.F. *J. Org. Chem.* **1991**, *56*, 6440. (d) Davies, H.M.L.; Hu, B. *Heterocycles* **1993**, *35*, 385.

100. Taber, D.F.; Heri, R.J.; Gleave, D.M. *J. Org. Chem.* **1997**, *62*, 194.

101. Doyle, M.P.; Eismont, M.Y.; Protopopova, M.N.; Kwan, M.M.Y. *Tetrahedron* **1994**, *50*, 1665. For reviews see: (a) Doyle, M.P.; McKervey, M.A. *J. Chem. Soc., Chem. Comm.* **1997**, 983. (b) Doyle, M.P.; Forbes, D.C. *Chem. Rev.* **1998**, *98*, 911. (c) Doyle, M.P.; Protopopova, M.N. *Tetrahedron* **1998**, *54*, 7979.

102. (a) Doyle, M.P., et al. *J. Am. Chem. Soc.* **1995**, *117*, 5763. (b) Doyle, M.P.; Dyatkin, A.B.; Kalinin, A.V.; Ruppar, D.A. *J. Am. Chem. Soc.* **1995**, *117*, 11021.

103. (a) For a review see: Padwa, A.; Austin, D.J. *Angew. Chem. Int. Ed. Engl.* **1994**, *33*, 1797. (b) Rogers, D.H.; Yi, E.C.; Poulter, C.D. *J. Org. Chem.* **1995**, *60*, 941.

104. (a) Padwa, A.; Austin, D.J.; Xu, S.L. *Tetrahedron Lett.* **1991**, *32*, 4103. (b) Hoye, T.; Dinsmore, C.J. *Tetrahedron Lett.* **1991**, *32*, 3755. (c) Hoye, T.R.; Dinsmore, C.J. *J. Am. Chem. Soc.* **1991**, *113*, 4343. (d) Padwa, A.; Austin, D.J.; Xu, S.L. *J. Org. Chem.* **1992**, *57*, 1330.

105. (a) For related cases see: Evans, D.H.; Woerpel, K.A.; Hinman, M.M.; Faul, M.M. *J. Am. Chem. Soc.* **1991**, *113*, 726. (b) Evans, D.A.; Woerpel, K.A.; Scott. M.J. *Angew. Chem. Int. Ed. Engl.* **1992**, *31*, 430.

106. Pfaltz, A. *Acc. Chem. Res.* **1993**, *20*, 339.

107. Pique, C.; Fähndrich, B.; Pfaltz, A. *Synlett* **1995**, 491.

108. Nishiyama, H.; Itoh, Y.; Matsumoto, H.; Park, S-B.; Itoh, K. *J. Am. Chem. Soc.* **1994**, *116*, 2223.

109. For a review see: Doyle, M.P. "Transition Metal Carbene Complexes: Diazodecomposition Ylide and Insertion Chemistry" in "Comprehensive Organometallic Chemistry II", Abel, E.W.; Stone, F.G.A.; Wilkinson, G., Eds., Pergamon, Oxford, UK, 1995, Vol. 12, pp. 421-468.

110. For a review of ligand effects on C-H insertions see: Padwa, A. *Angew. Chem. Int. Ed. Engl.* **1994**, *33*, 1797.

111. Miah, S.; Slawin, A.M.Z.; Moody, C.J.; Sheehan, S.M.; Marino, J.P., Jr.; Semones, M.A.; Padwa, A. *Tetrahedron* **1996**, *52*, 2489.

112. Doyle, M.P.; Dyatkin, A.B.; Roos, G.H.P.; Cañas, F.; Pierson, D.A.; van Barten, A.; Müller, P.; Polleux, P. *J. Am. Chem. Soc.* **1994**, *116*, 4507.

113. Watanabe, N.; Ogawa, T.; Ohlake, Y.; Ikegami, S.; Hashimoto, S-i. *Synlett* **1996**, 85.

114. Ruediger, E.H.; Solomon, C. *J. Org. Chem.* **1991**, *56*, 3183.

115. Bhandaru, S.; Fuchs, P.L. *Tetrahedron Lett.* **1995**, *36*, 8347.

116. For reviews see: (a) Padwa, A.; Krumpe, K.E. *Tetrahedron* **1992**, *48*, 5385. (b) Padwa, A.; Hornbuckle, S.F. *Chem. Rev.* **1991**, *91*, 263. (c) Padwa, A. *Acct. Chem. Res.* **1991**, *24*, 22.

117. Kido, F.; Abiko, T.; Kato, M. *J. Chem. Soc., Perkin I* **1995**, 2989.

118. Kametani, T.; Yakawa, H.; Honda, T. *J. Chem. Soc., Chem. Comm.* **1986**, 651.

119. West, F.G.; Naidu, B.N.; Tester, R.W. *J. Org. Chem.* **1994**, *59*, 6892.

120. West, F.G.; Naidu, B.N. *J. Org. Chem.* **1994**, *59*, 6051.

121. For a review see: Padwa, A.; Weingarten, M.D. *Chem. Rev.* **1996**, *96*, 223.

122. Padwa, A.; Price, A.T. *J. Org. Chem.* **1998**, *63*, 556.

123. Padwa, A.; Kassir, J.M.; Semones, M.A.; Weingarten, M.D. *Tetrahedron Lett.* **1993**, *34*, 7853.
124. For a review see: Schuster, M.; Blechert, S. *Angew. Chem. Int. Ed. Engl.* **1997**, *36*, 2036.
125. For a review see: Schrock, R.R. *Acc. Chem. Res.* **1990**, *23*, 158.
126. Belderrain, T.R.; Grubbs, R.H. *Organometallics* **1997**, *16*, 400.
127. Wilhelm, T.E.; Belderrain, T.R.; Brown, S.N.; Grubbs, R.H. *Organometallics* **1997**, *16*, 3867.
128. Kirkland, T.A.; Grubbs, R.H. *J. Org. Chem.* **1997**, *62*, 7311.
129. For a review see: Moore, J.S., "Transition Metals in Polymer Synthesis. Ring Opening Metathesis Polymerization and Other Transition Metal Polymerization Techniques" in "Comprehensive Organometallic Chemistry II", Abel, E.W.; Stone, F.G.A.; Wilkinson, G., Eds., Pergamon, Oxford, UK, 1995, Vol. 12, pp. 1209-1233.
130. Lynn, D.M.; Kanaoka, S.; Grubbs, R.H. *J. Am. Chem. Soc.* **1996**, *118*, 784.
131. Walba, D.M.; Keller, P.; Shao, R.; Clark, N.A.; Hillmeyer, M.; Grubbs, R.H. *J. Am. Chem. Soc.* **1996**, *118*, 2740.
132. Huwe, C.M.; Kiehl, O.C.; Blechert, S. *Synlett* **1996**, 65.
133. Overkdeef, H.S.; Pandit, U.K. *Tetrahedron Lett.* **1996**, *37*, 547.
134. Miller, S.J.; Kim, S-H.; Chen, Z-R.; Grubbs, R.H. *J. Am. Chem. Soc.* **1995**, *117*, 2108.
135. Furstner, A.; Müller, T. *J. Org. Chem.* **1998**, *63*, 425.
136. Nicolaou, K.C.; Vourloumis, D.; Li, T.; Pastor, J.; Wiinsinger, N.; He, Y.; Ninkovic, S.; Sarabia, F.; Vallberg, H.; Rosehanger, F.; King, N.P.; Rays, M.; Finlay, V.; Giannakobow, P.; Verdia-Pimmard, P.; Hamel, E. *Angew. Chem. Int. Ed. Engl.* **1997**, *36*, 2097.
137. Mohr, B.; Weck, B.; Sauvage, J.P.; Grubbs, R.H. *Angew. Chem. Int. Ed. Engl.* **1997**, *36*, 1308.
138. Dietrich-Buchecher, C.; Rapenne, G.; Sauvage, J.P. *J. Chem. Soc., Chem. Comm.* **1997**, 2053.
139. McKervey, M.A.; Pitarch, M. *J. Chem. Soc., Chem. Comm.* **1996**, 1689.
140. (a) Fujimura, O.; Grubbs, R.H. *J. Org. Chem.* **1998**, *63*, 825. (b) Alexander, J.B.; La, D.S.; Cefalo, D.R.; Graf, D.D.; Hoveyda, A.H.; Schrock, R.R. *J. Am. Chem. Soc.* **1998**, *120*, 9720.
141. For a review see: Stille, J.R., "Transition Metal Carbene Complexes. Tebbe's Reagent and Related Nucleophilic Alkilidenes," in "Comprehensive Organometallic Chemistry II", Abel, E.W.; Stone, F.G.A.; Wilkinson, G., Eds., Pergamon, Oxford, UK, 1995, Vol. 12, pp. 577-600.
142. Cannizzo, L.F.; Grubbs, R.H. *J. Org. Chem.* **1985**, *50*, 2386.
143. Cannizzo, L.F.; Grubbs, R.H. *J. Org. Chem.* **1985**, *50*, 2316.
144. For a comparison of Tebbes reagent with Ph_3PCH_2, see: Pine, S.H.; Shen, G.S.; Hoang, H. *Synthesis* **1991**, 165.
145. Sun, S.; Dullaghan, C.A.; Sweigert, D.A. *J. Chem. Soc., Dalton Trans* **1996**, 4493.
146. Borrelly, S.; Paquette, L.A. *J. Am. Chem. Soc.* **1996**, *118*, 727.
147. Buchwald, S.L.; Grubbs, R.H. *J. Am. Chem. Soc.* **1983**, *105*, 5490.
148. Nicolaou, K.C.; Postema, M.H.D.; Claiborne, C.F. *J. Am. Chem. Soc.* **1996**, *118*, 1565.
149. Petasis, N.A.; Bzowej, E.I. *J. Am. Chem. Soc.* **1990**, *112*, 6392.
150. DeShong, P.; Rybczynski, P.J. *J. Org. Chem.* **1991**, *56*, 3207.
151. Petasis, N.A.; Bzowej, E.I. *J. Org. Chem.* **1992**, *57*, 1327.
152. Petasis, N.A.; Akritopoulou, I. *Synlett* **1992**, 665.
153. Petasis, N.A.; Hu, Y-H. *J. Org. Chem.* **1997**, *62*, 782.
154. Petasis, N.A.; Bzowei, E.I. *Tetrahedron Lett.* **1993**, *34*, 943.
155. Petasis, N.A.; Lu, S-P. *Tetrahedron Lett.* **1996**, *37*, 141.
156. Petasis, N.A.; Gzowei, E.I. *Tetrahedron Lett.* **1993**, *34*, 1721.
157. Reed, A.D.; Hegedus, L.S. *J. Org. Chem.* **1995**, *60*, 3787.

CHAPTER 7

Synthetic Applications of Transition Metal Alkene, Diene, and Dienyl Complexes

7.1 Introduction

Nucleophilic attack on transition metal-complexed olefin, diene, and dienyl systems is among the most useful of organometallic processes for the synthesis of complex organic molecules.[1] Not only does complexation reverse the normal reactivity of these functional groups, changing them from nucleophiles to electrophiles, but the process also results in the formation of new bonds between the nucleophile and the olefinic carbon and new metal-carbon bonds that can be further elaborated (Eq. 7.1). The fundamental features of this process were presented in Chapter 2. Here its' uses in organic synthesis will be considered.

Eq. 7.1

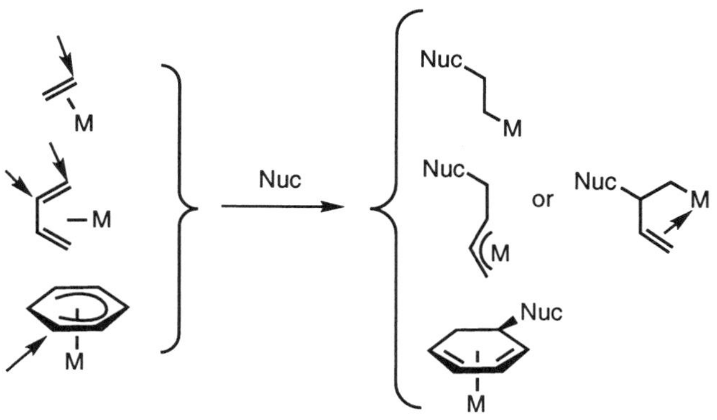

7.2 Metal Alkene Complexes[1]

a. Palladium(II) Complexes

Olefin complexes of palladium(II) were among the first used to catalyze useful organic transformations and are among the most extensively developed, at least in part because they are both easily generated and highly reactive. Palladium chloride is the most common catalyst precursor. It is a commercially available redbrown, chlorobridged oligomer, insoluble in most organic solvents (Figure 7.1). However, the oligomer is easily disrupted by treatment with alkali metal chlorides, such as LiCl or NaCl, providing the considerably more soluble, hygroscopic monomeric palladate, M_2PdCl_4. Nitriles, particularly acetonitrile and benzonitrile, also give air-stable, easily handled, soluble monomeric solids, which are the most convenient sources of palladium(II). Treatment of soluble palladium(II) salts with

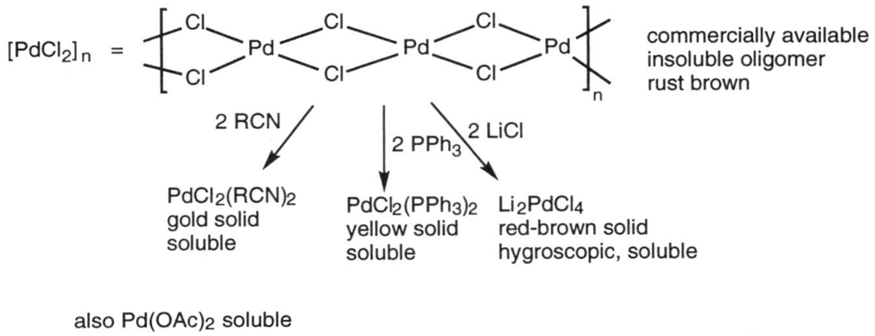

Figure 7.1 Palladium(II) Sources

alkenes results in rapid, reversible coordination of the alkene to the metal (a in Figure 7.2). Ethylene and terminal monoolefins coordinate most strongly followed by *cis* and *trans* internal olefins. Although these olefin complexes can be isolated if desired, they are normally generated *in situ* and used without isolation (Figure 7.2). Once complexed, the olefin becomes *generally* reactive towards nucleophiles ranging from Cl⁻ through Ph⁻, a range of about 10^{36} in basicity. Attack occurs from the face opposite the metal (*trans* attack) and at the more-substituted carbon, resulting in the formation of a carbon-nucleophile bond and a metal-carbon bond (b). Note that this is the opposite regiochemistry and stereochemistry from that obtained by insertion of olefins into metal-carbon σ-bonds. The regiochemistry is that corresponding to attack at the olefinic position which can *best* stabilize positive charge, and, in a sense, the highly electrophilic palladium(II) is behaving like a very expensive and selective proton towards the olefin.

The newly-formed σ-alkylpalladium(II) complex has a very rich chemistry. Warming above –20°C results in β-hydrogen elimination (c), generating an olefin which is the product of a formal nucleophilic substitution. This σ-alkylpalladium(II)

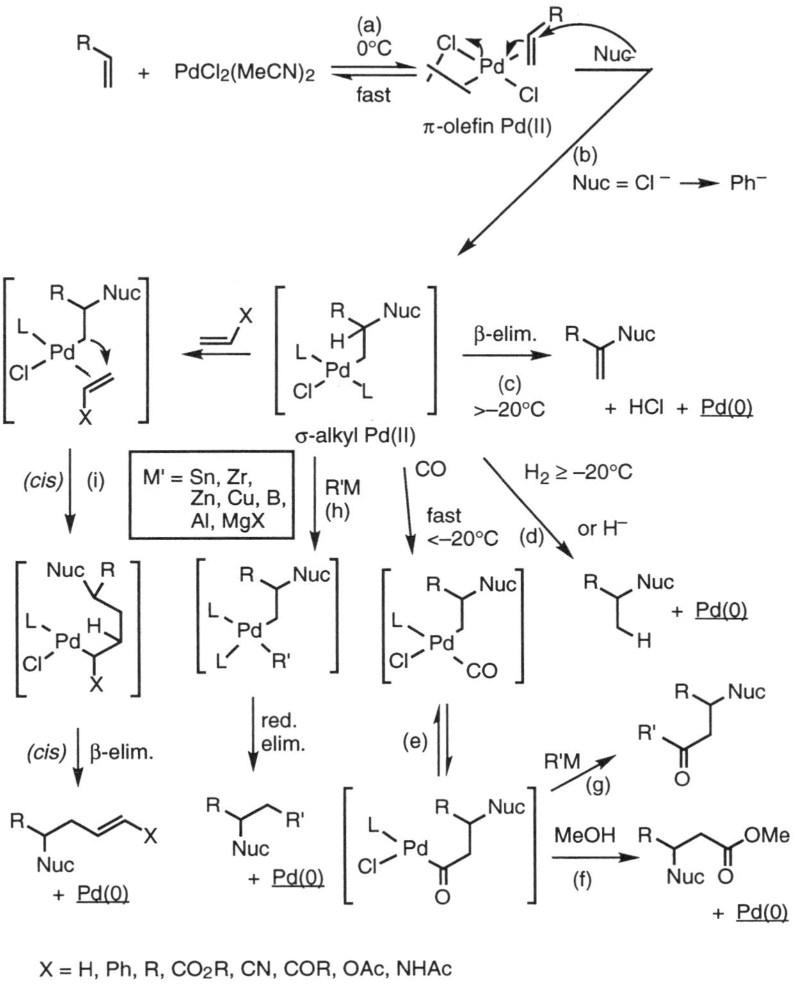

X = H, Ph, R, CO₂R, CN, COR, OAc, NHAc

Figure 7.2 Reactions of Palladium(II) Olefin Complexes

complex is easily reduced, either by hydrides or hydrogen gas at one atmosphere, resulting in a formal nucleophilic addition to the alkene (d). Carbon monoxide insertion is fast, even at low temperatures, and successfully competes with β-elimination. Thus σ-acyl species are easily produced simply by exposing the reaction solution to an atmosphere of carbon monoxide (e). Cleavage of this acyl species with methanol produces esters (f), while treatment with main group organometallics leads to ketones via a *carbonylative coupling* (g). The initially formed σ-alkylpalladium(II) complex itself will undergo transmetallation from many main group organometallics, resulting in overall *nucleophilic addition/alkylation* (h).

In principle, olefins can insert into metal-carbon σ-bonds, but in practice this usually cannot be achieved when β-hydrogens are present, since β-elimination is faster. In exceptional cases, however, olefins can insert into σ-alkylpalladium(II) complexes prepared from nucleophilic attack on olefins, resulting in an overall *nucleophilic addition/olefination* (I). In all of the above processes, palladium(II) is required to activate the alkene, but palladium(0) — from reductive elimination — is the product of the reaction. Thus, for catalysis, a procedure to oxidize Pd(0) back to Pd(II) in the presence of reactants and products is needed. Many systems have been developed ranging from the classic $O_2/CuCl_2$ redox system to more recently developed benzoquinone, or quinone/reoxidant. The choice of reoxidant is usually dictated by the stability of the products toward the oxidant.

Palladium(II) olefin complexes are particularly well-behaved in their reactions with oxygen nucleophiles such as water, alcohols, and carboxylates, and a large number of useful synthetic transformations are based on this process. One of the earliest is the Wacker process, the palladium(II)-catalyzed "oxidation" of ethylene to acetaldehyde for which the key step is nucleophilic attack of water on metal-coordinated ethylene. Although this particular (industrial) process is of little interest for the synthesis of fine chemicals, when applied to terminal alkenes it provides a very efficient synthesis of methyl ketones (Eq. 7.2).[2] The reaction is specific for terminal olefins, since they coordinate much more strongly than internal alkenes, and internal alkenes are tolerated elsewhere in the substrate (Eq. 7.3). A variety of remote functional groups are tolerated (Eq. 7.4[3] and Eq. 7.5[4]) provided they are not strong ligands for palladium(II) (e.g., amines), in which case the catalyst is poisoned. Copper(II) chloride/oxygen is used to reoxidize the palladium, since the product alcohols are stable to these reagents.

Eq. 7.2

Eq. 7.3

59%

Eq. 7.4

82%

Eq. 7.5

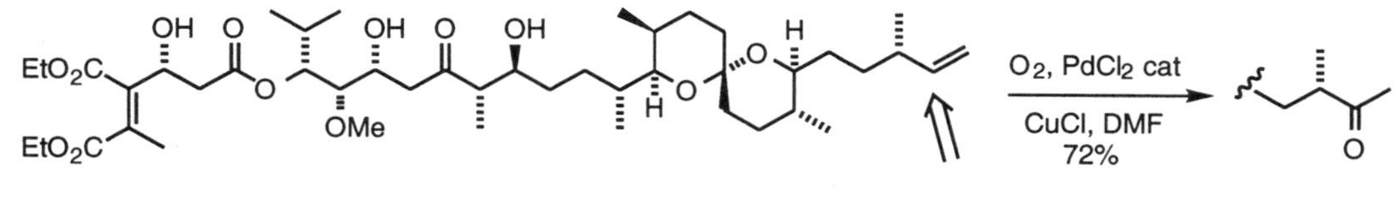

In all of these cases, nucleophilic attack occurred at the 2° position, as expected. However, if the substrate has adjacent functional groups which can coordinate to palladium to form a five or six membered palladiacyclic intermediate, the attack of the nucleophile may be directed to the terminal position (Eq. 7.6,[5] Eq. 7.7,[6] and Eq. 7.8[7]). Adjacent directing groups can even activate internal alkenes to undergo Wacker oxidation (Eq. 7.9)[7] although this is a rare occurrence.

Eq. 7.6

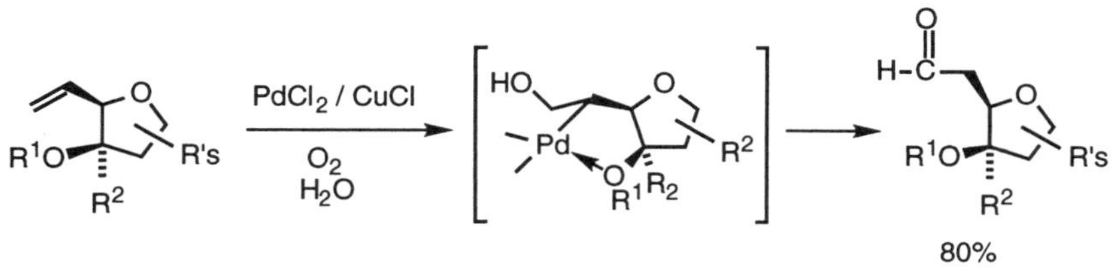

Eq. 7.7

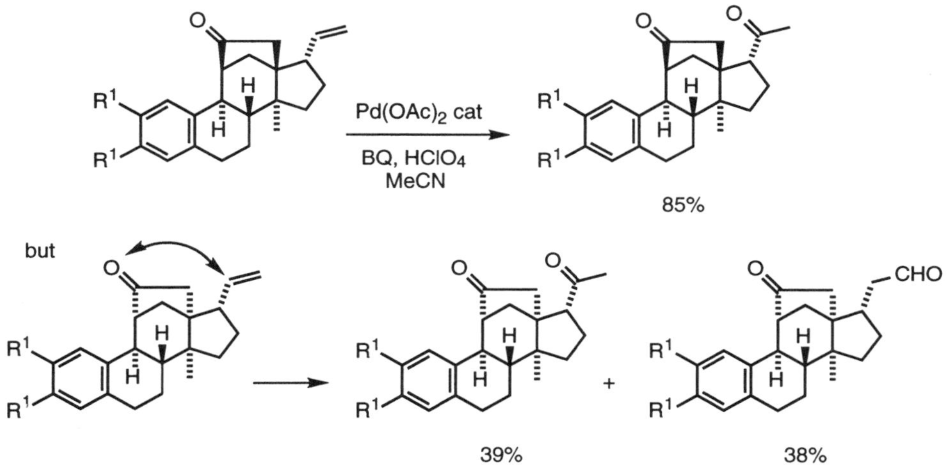

but

Eq. 7.8

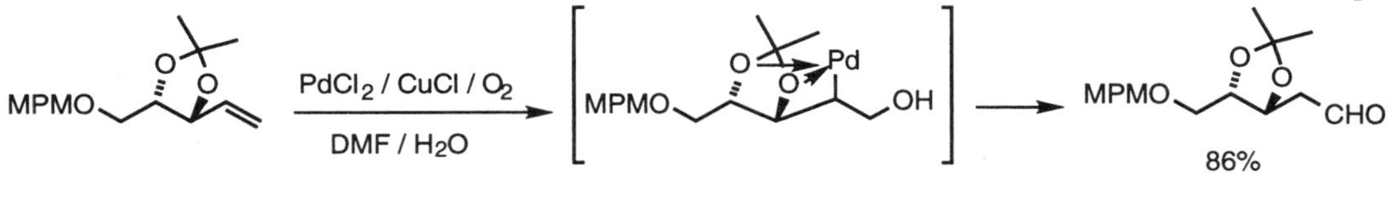

Eq. 7.9

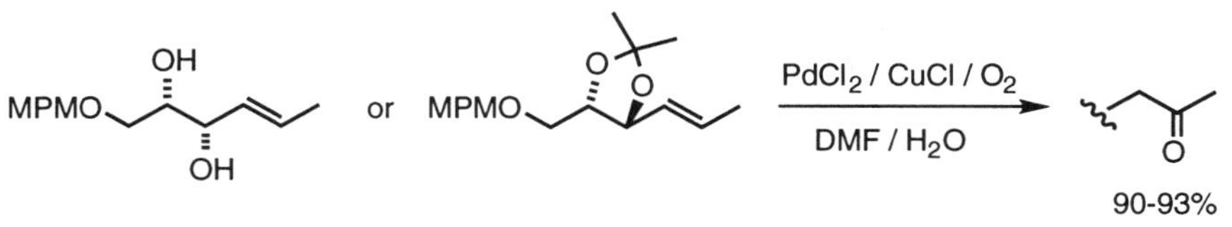

90-93%

Lacking ligand assistance, this reaction is not useful for internal olefins since they react only slowly, if at all. However, α,β-unsaturated ketones and esters are cleanly converted to β-keto compounds under somewhat specialized conditions (Eq. 7.10).[2]

Eq. 7.10

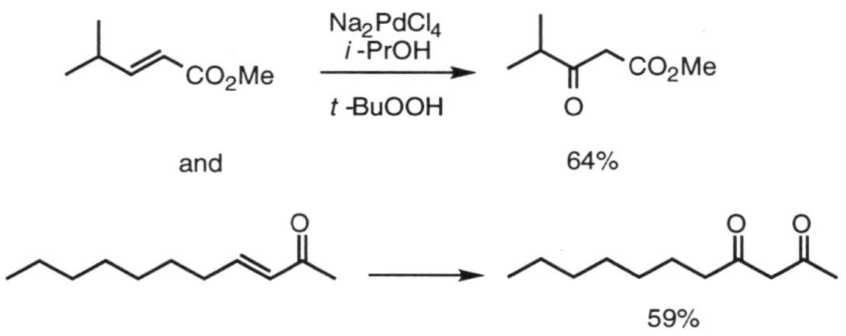

and

64%

59%

Alcohols also add cleanly to olefins in the presence of palladium(II) but the intermolecular version has found little use in synthesis. In contrast, the intramolecular version provides a very attractive route to oxygen heterocycles.[8] In intramolecular reactions, internal and even trisubstituted alkenes undergo attack and the regioselectivity is often determined more by ring size than by substitution (Eq. 7.11[9] and Eq. 7.12[10]). In the presence of chiral binap oxazoline ligands, asymmetry can be induced (Eq. 7.13[11]).

Eq. 7.11

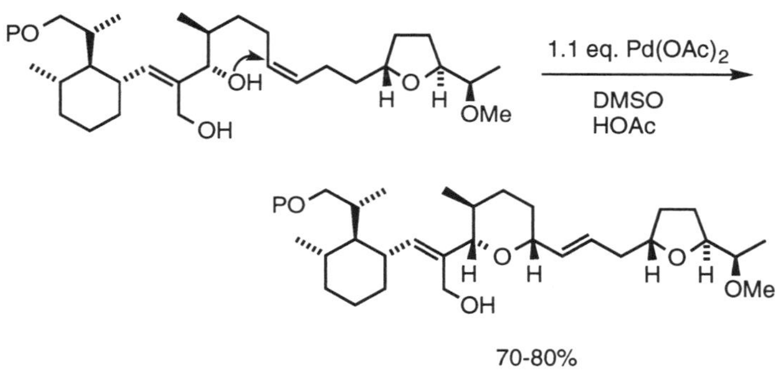

70-80%

Eq. 7.12

80%

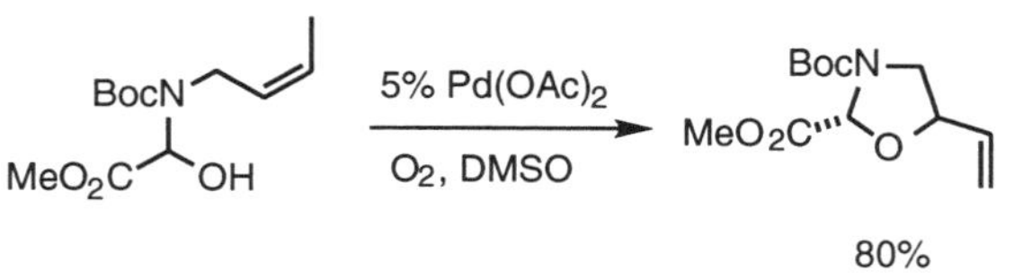

Eq. 7.13

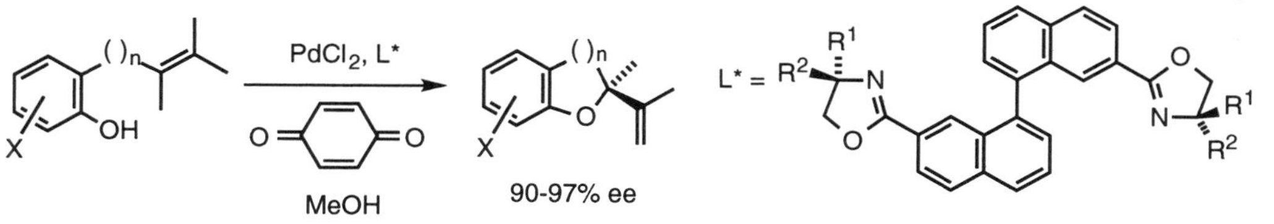

90-97% ee

Reoxidation is necessary in these palladium(II)-catalyzed processes because "HPdX" is produced in the β-elimination step (Figure 7.2) and rapidly loses (reductive elimination) HX to produce palladium(0). If some group other than hydride, such as Cl⁻ or OH⁻, can be β-eliminated, subsequent reductive elimination to form palladium(0) (and Cl_2 or HOCl) does not occur, and the resulting $PdCl_2$ or Pd(OH)(Cl) can reenter the catalytic cycle directly, obviating the need for oxidants (Eq. 7.14).[12]

Eq. 7.14

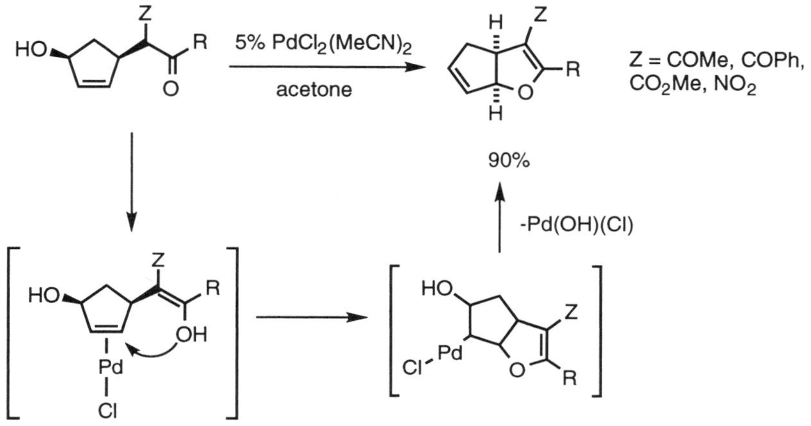

Because CO insertion into Pd–C bonds is so facile, and competes effectively with β-hydride elimination, efficient alkoxycarbonylation of alkenes has been achieved[13] (Eq. 7.15[14]).

Eq. 7.15

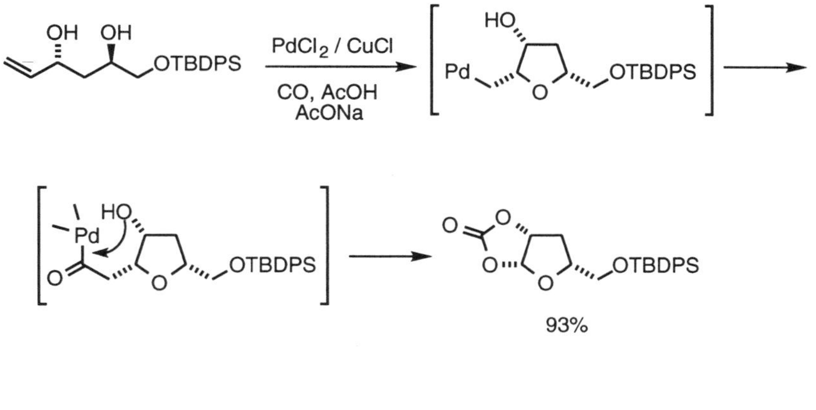

If β-elimination is blocked, even olefins can insert (Eq. 7.16[15]) although this is rare. Under exceptional circumstances, olefins can be made to insert into σ-alkylpalladium(II) complexes having β-hydrogens. Such a case is seen in Eq. 7.17.[16] The initial step is palladium(II)-assisted alcohol attack on the complexed enol ether, generating a σ-alkylpalladium(II) complex, which has a β-hydrogen potentially

Eq. 7.16

Eq. 7.17

accessible for β-elimination. However, β-elimination requires a *syn*-coplanar relationship and, if the adjacent olefin coordinates to palladium, rotation to achieve this geometry is suppressed. At the same time, coordination of the olefin permits it to insert into the σ-metal-carbon bond, which is indeed what occurs, to form the bicyclic system. The insertion is a *syn* process, and the σ-alkylpalladium(II) complex resulting from this insertion cannot achieve a *syn*-coplanar geometry with any of the β-hydrogens, since the ring system is rigid. In this special circumstance *inter*molecular olefin insertion occurs, giving the bicyclic product in excellent yield. Note that this process forms one carbon-oxygen and two carbon-carbon bonds in a one pot procedure.

Carboxylate anions are also efficient nucleophiles for palladium-assisted attack on olefins. Intramolecular versions are efficient (Eq. 7.18[17]) and, as is often the case for intramolecular reactions, even trisubstituted olefins are attacked.[18]

Amines also attack palladium-bound olefins, but catalytic intermolecular amination of olefins has not been achieved, since both the amine and the product enamine are potent ligands for palladium and act as catalyst poisons. The stoichiometric amination of simple olefins can be achieved in good yield, but three equivalents of amine are required, and the resulting σ-alkyl-palladium(II) complex must be reduced to give the free amine.[19] A careful study of intramolecular amination

Eq. 7.18

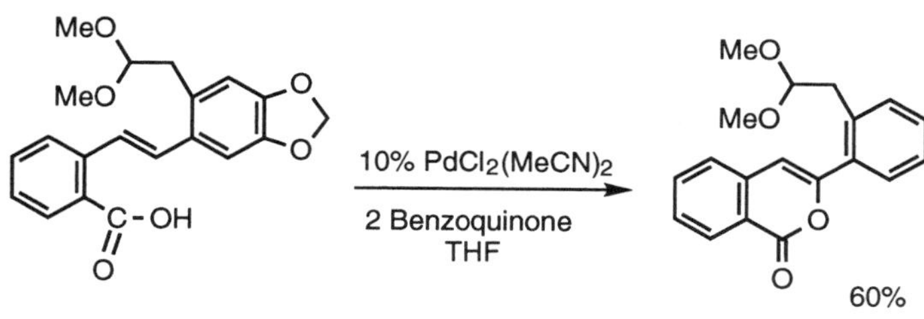

elucidated some of the difficulties (Eq. 7.19).[20] Treatment of the cyclopentyl aminoolefin with palladium(II) chloride resulted in the formation of a stable, chelated aminoolefin complex, confirming that one problem is indeed the strong

Eq. 7.19

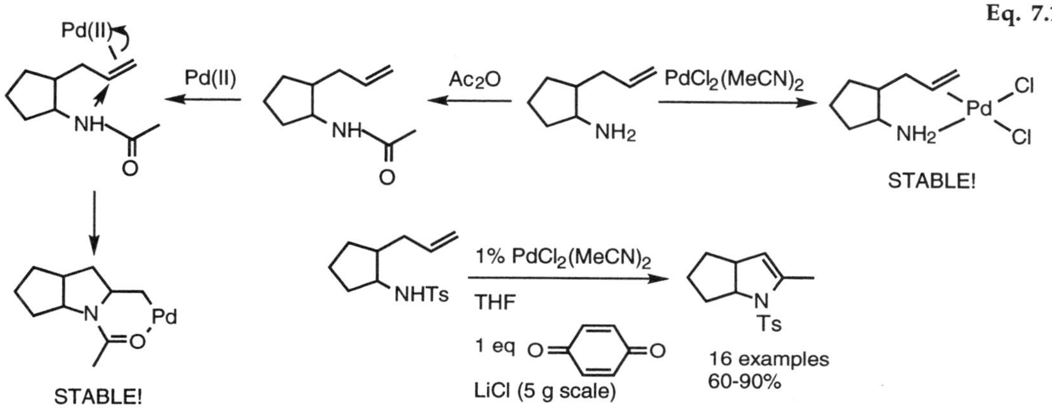

coordination of amines to Pd(II). N-Acylation of the amine dramatically decreased its basicity and coordinating ability, and this acetamide did not irreversibly bind to palladium(II). Instead intramolecular amination of the olefin ensued. However, the resulting σ-alkylpalladium(II) complex was stabilized by chelation to the amide oxygen, preventing β-hydrogen elimination, and thus catalysis. Finally, by placing the strongly electron-withdrawing sulfonyl group on nitrogen, neither the substrate nor the product coordinated strongly enough to palladium(II) to prevent catalysis, and efficient cyclization was achieved.

Aromatic amines are about 10^6 less basic than aliphatic amines, and this factor permits the efficient intramolecular amination of olefins by these amines without the need for N-acylation or sulfonation, although these N-functionalized substrates also cyclize efficiently (Eq. 7.20)[21] and (Eq. 7.21).[22] A variety of functionalized indoles can be made by this procedure.

A very mild and general catalyst system for the cyclization of both aliphatic and aromatic olefinic tosamides has recently been developed.[23] It consists of palladium acetate in DMSO and relies upon oxygen as a reoxidant. In most cases the regioselectivity parallels that of the previous systems (Eq. 7.22), but ortho-allyl-Ntosylanilines cyclize to give dihydroquinolines (attack at the less-substituted olefin terminus) rather than indoles.

Eq. 7.20

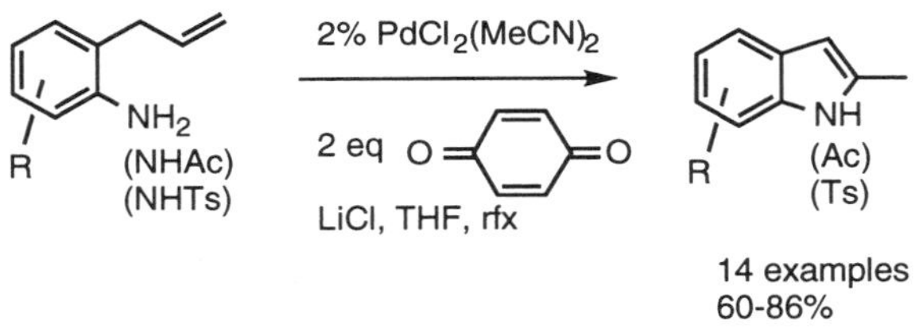

Eq. 7.21

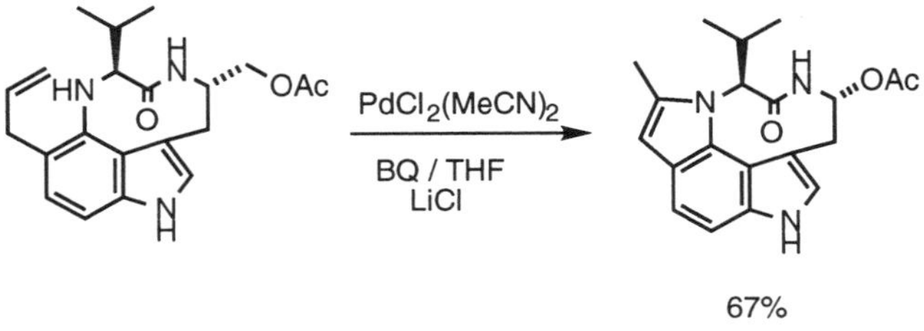

Eq. 7.22

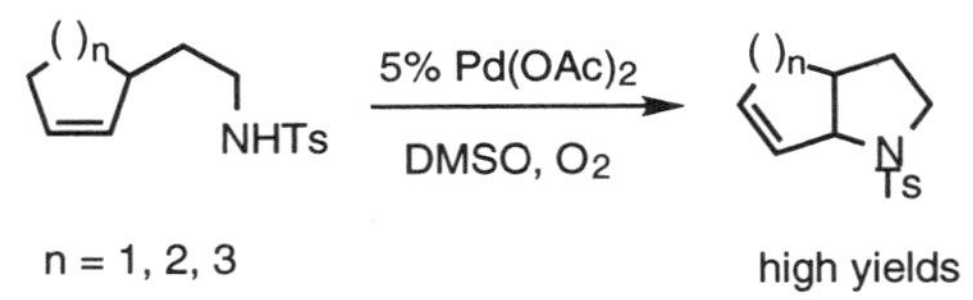

As with oxygen nucleophiles, if the olefin to be aminated has an allylic leaving group, reoxidation is not necessary and catalytic cyclizations proceed smoothly (Eq. 7.23[24]).

Eq. 7.23

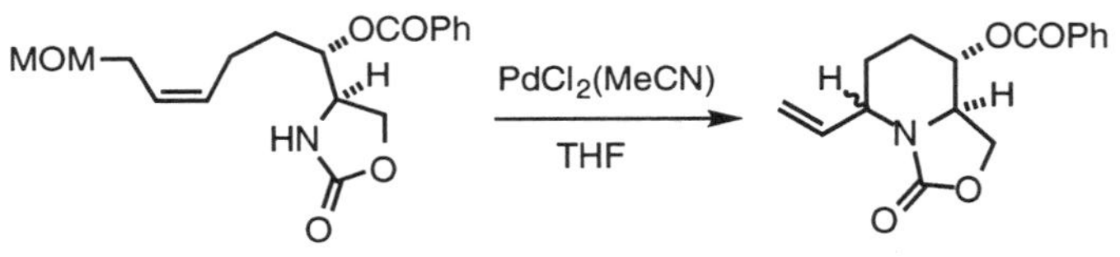

The σ-alkylpalladium(II) complex from the amination of olefins can easily be trapped by CO[25] (Eq. 7.24[26] and Eq. 7.25[27]). With careful choice of conditions even olefins will insert reasonably efficiently although β-hydride elimination competes even under the best of circumstances (Eq. 7.26).[28] Note that although there are nine sites at which the catalyst can (and almost certainly does) coordinate, only one leads to cyclization. Provided that the catalyst is not irreversibly bound to one of the other sites, catalysis ensues.

Eq. 7.24

Eq. 7.25

Eq. 7.26

The use of carbon nucleophiles results in yet another set of problems. Carbanions, even stabilized ones, are easily oxidized, and palladium(II) salts, as well as many other transition metal salts, tend to oxidatively dimerize carbanions, preventing their attack on olefins. Even if this can be prevented, catalysis is (as yet) impossible to achieve since it is necessary to find an oxidizing agent capable of reoxidizing the catalytic amount of Pd(0) present to Pd(II), but at the same time, not able to oxidize the large amount of the carbanion substrate present. Stoichiometric alkylation of olefins by stabilized carbanions can be achieved by very careful choice of conditions (Eq. 7.27).[29] Preforming the olefin complex at 0°C followed by cooling

Eq. 7.27

Applications of Transition Metal Alkene, Diene, and Dienyl Complexes • **197** to −78°C and addition of two equivalents of triethylamine generates an olefin/amine complex. The coordinated amine directs the carbanion away from the metal and to the olefin, suppressing metal-centered, electron-transfer redox chemistry, and promoting alkylation of the olefin. As usual, attack occurs at the more substituted carbon of the olefin, producing a σ-alkylpalladium(II) complex which can undergo reduction, β-elimination, or CO insertion. The Pd(0) produced can be recovered and reoxidized separately, but not concurrently with alkylation, so the process is stoichiometric in palladium, and unlikely to find extensive use in synthesis.

The exceptions to this will be cases in which this chemistry permits the efficient synthesis of products which would be considerably more difficult to prepare by classical means. Such a situation is seen in the synthesis of a relay to (+)-thienamycin (Eq. 7.28).[30] In this case, one pot alkylation/acylation of an optically active ene carbamate produced the keto diester in 70% chemical yield, and with complete control of stereochemistry at the chiral center α- to nitrogen. In this case, palladium was used to form two C–C and one O–C bond, and to generate a new chiral center. When palladium(II)-assisted alkylation of this olefin was combined with carbonylative coupling to tin reagents, even more complex structures were efficiently produced (Eq. 7.29).[31] In this case, palladium(II) promotes stereospecific alkylation of the olefin, followed by insertion of CO into the metal-carbon σ-bond, followed by transmetallation from tin to palladium, followed by reductive elimination. In this case three C–C bonds and a chiral center are efficiently generated in a one-pot reaction.

Eq. 7.28

Eq. 7.29

>95% de
77% overall

(+)-Negamycin (13% overall yield)

Palladium(II) acetate also promotes the intramolecular alkylation of olefins by silylenol ethers (Eq. 7.30[32] and Eq. 7.31[33]). It should be possible to make this process proceed catalytically since silylenol ethers themselves should not be oxidatively dimerized by Pd(II). However, some technical problem exists, since only a few turns of Pd(II) have been achieved.

Eq. 7.30

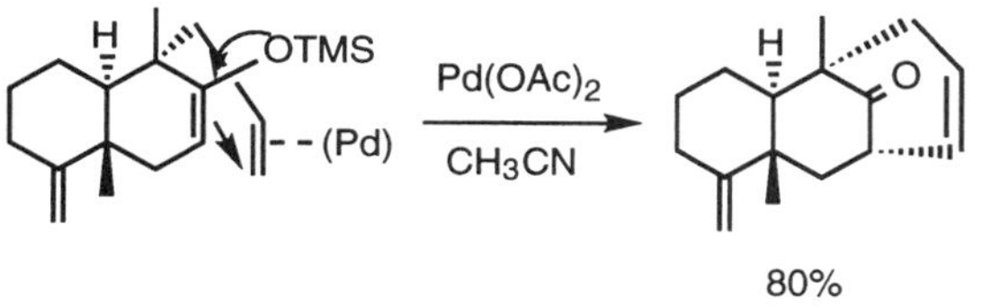

80%

Eq. 7.31

99%

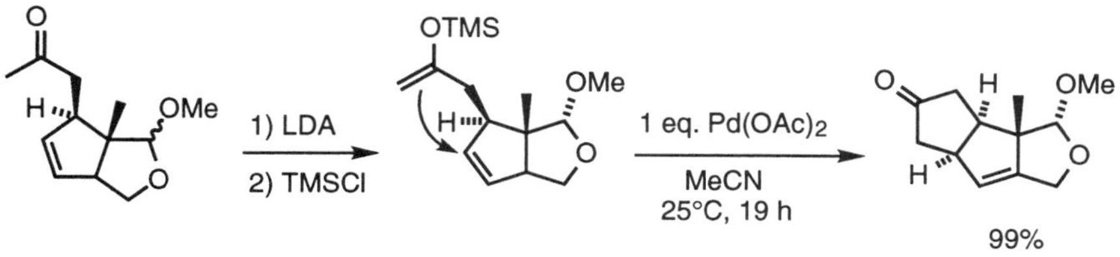

Palladium(II) complexes catalyze the rearrangement of a very broad array of allylic systems, in a process which involves intramolecular nucleophilic attack on a Pd(II)-complexed alkene as the key step (Eq. 7.32).[34] Allyl acetates rearrange under mild conditions without complication from skeletal rearrangements or formation

Eq. 7.32

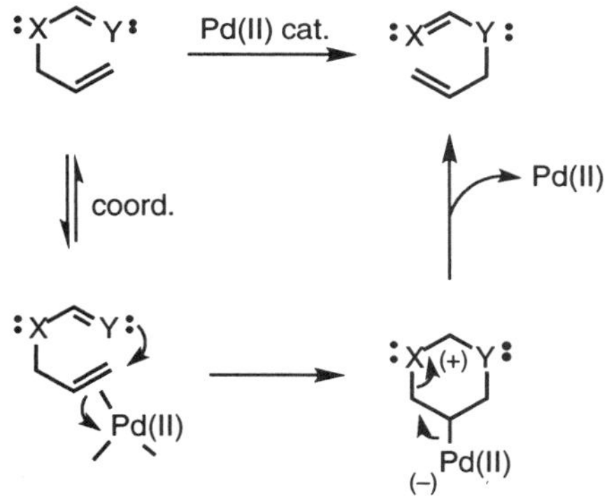

other side products. With optically active allyl acetates, rearrangement occurs with complete transfer of chirality (Eq. 7.33)[35] and (Eq. 7.34).[36] Note that the absolute stereochemistry observed at the newly-formed chiral center must result from coordination of palladium to the face of the olefin opposite the acetate, and attack of the olefin from the face opposite the metal, as expected for the mechanism shown in Eq. 7.32.

Eq. 7.33

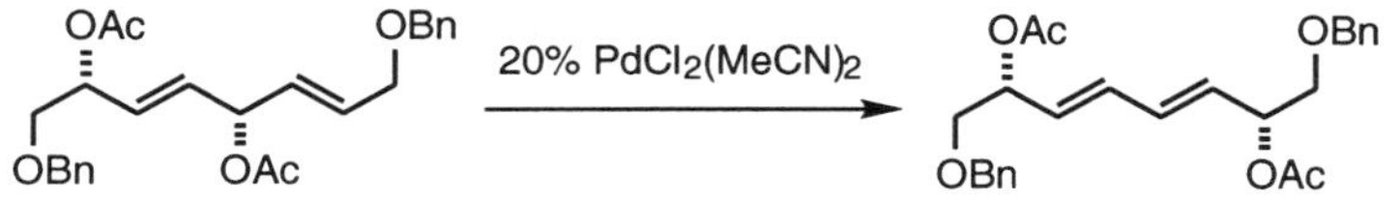

Eq. 7.34

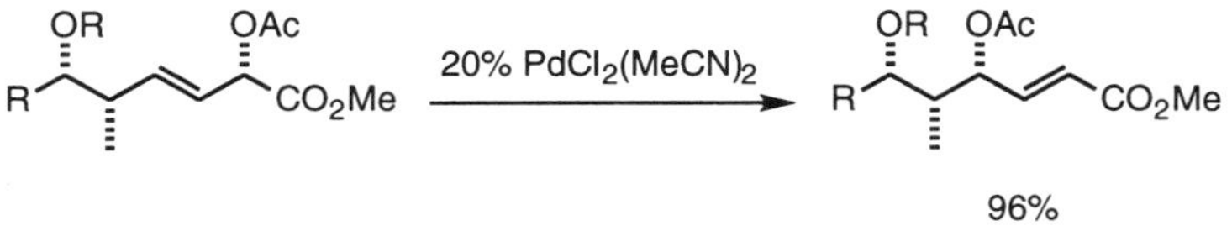

96%

Palladium(II) complexes catalyze the Cope rearrangement of 1,5-dienes under very mild conditions, leading to a roughly 10^{10} rate enhancement over the thermal process (Eq. 7.35).[34] There are some limitations to the catalyzed process, in that C-2 must bear a substituent to help stabilize the developing positive charge, and C-5 must bear a hydrogen so that efficient complexation to Pd(II) results. (Recall that trisubstituted olefins complex poorly to Pd(II), and that *gem* disubstituted alkenes undergo attack at the substituted position, making Cope rearrangements disfavored for these systems under Pd(II) catalysis.) Again, with chiral substrates, complete chirality transfer is observed.[37] Palladium(II) complexes also catalyze the oxyCope rearrangement (Eq. 7.36).[38]

Eq. 7.35

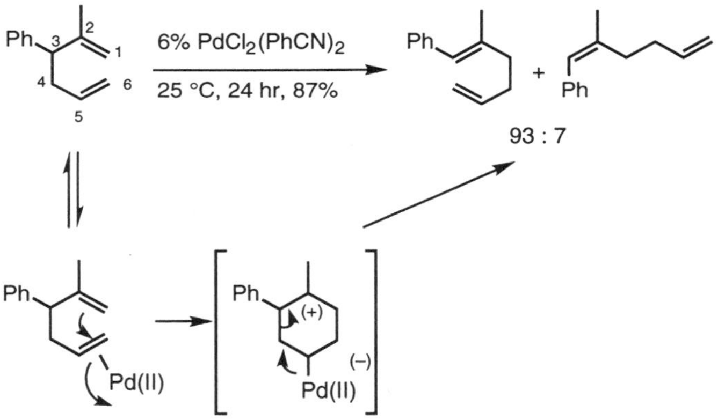

Eq. 7.36

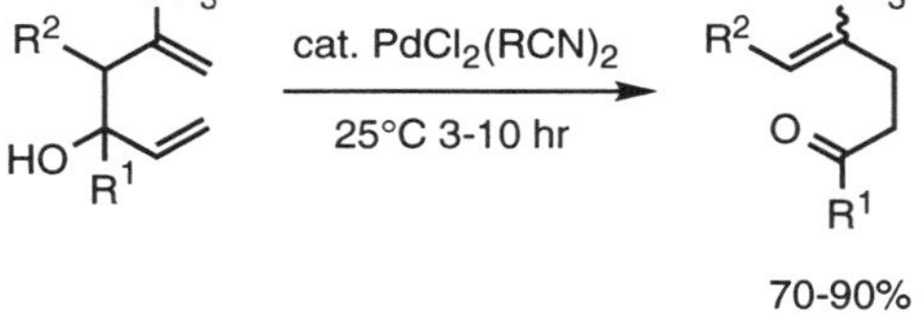

70-90%

Many other allylic systems undergo this palladium(II)-catalyzed allylic transposition. Trichloromethylimidates rearrange cleanly, and with complete transfer of chirality (Eq. 7.37).[39] O-allylimidates rearrange in a similar fashion,[40] and in the presence of chiral ligands, modest (50-60%) asymmetric induction is observed.[41] O-Allyl oximes rearrange to nitrones, which can be efficiently utilized in 1,3-dipolar cycloaddition reactions (Eq. 7.38).[42]

Eq. 7.37

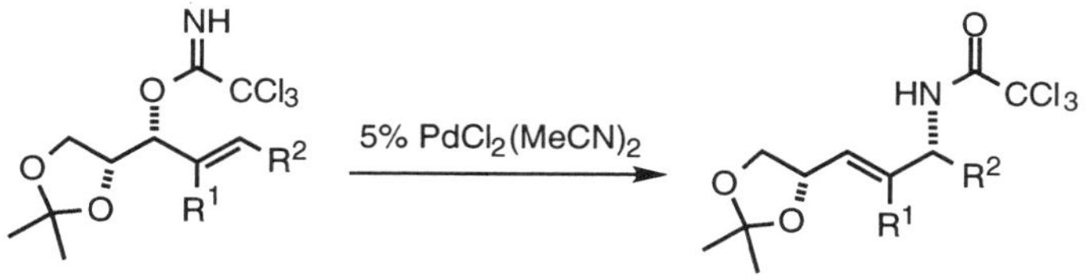

Eq. 7.38

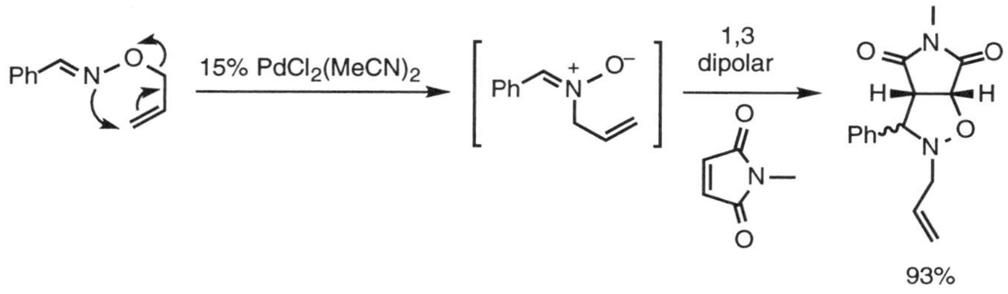

93%

b. Iron(II) Complexes

Cationic olefin complexes of iron(II) are easily made by alkene exchange with the $CpFe(CO)_2$(isobutene)$^+$ complex which is a stable gold-yellow solid, made by the reaction of Fp$^-$ (Fp = $CpFe(CO)_2$) with methallyl bromide, followed by γ-protonation with HBF$_4$ (see Chapter 2) (Eq. 7.39).[43] Once complexed, the olefin becomes reactive towards nucleophilic attack, sharing most of the reactivity, regio- and stereochemical

Eq. 7.39

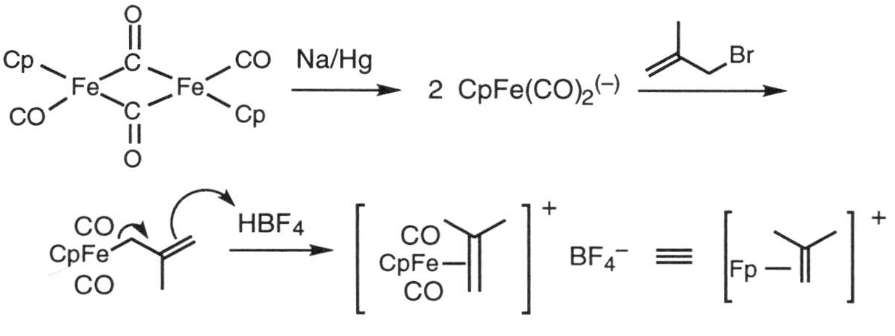

features of the neutral olefinpalladium(II) complexes discussed above. However, in contrast to the palladium analogs, the resulting σ-alkyliron(II) complex is very stable, and iron must be removed in a separate chemical step, usually oxidation. As an added complication, oxidative cleavage of Fp alkyl complexes usually results in insertion of CO into the metal-carbon σ-bond, producing carbonylated products (Eq. 7.40). For these reasons, iron-olefin complexes have been used in synthesis considerably less often than those of palladium. However, the reaction chemistry of electron-rich alkenes complexed to iron is quite unusual, and synthetically useful.

Eq. 7.40

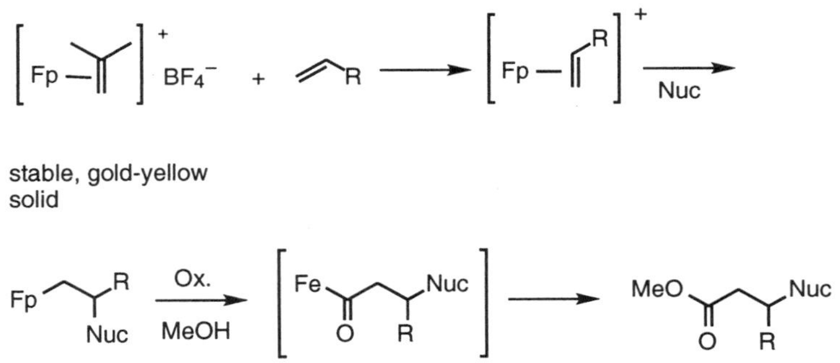

stable, gold-yellow
solid

When complexed to the Fp cation, dimethoxyethene is quite reactive towards nucleophiles, and both methoxy groups can be replaced, making this iron-olefin complex essentially a vinylidene dication ($CH^+=CH^+$) equivalent.[44] Treatment with optically active 2,3-butandiol generates the Fp^+-dioxene complex in good yield (Eq. 7.41). The optically active dioxene is easily freed by treatment with iodide,

Eq. 7.41

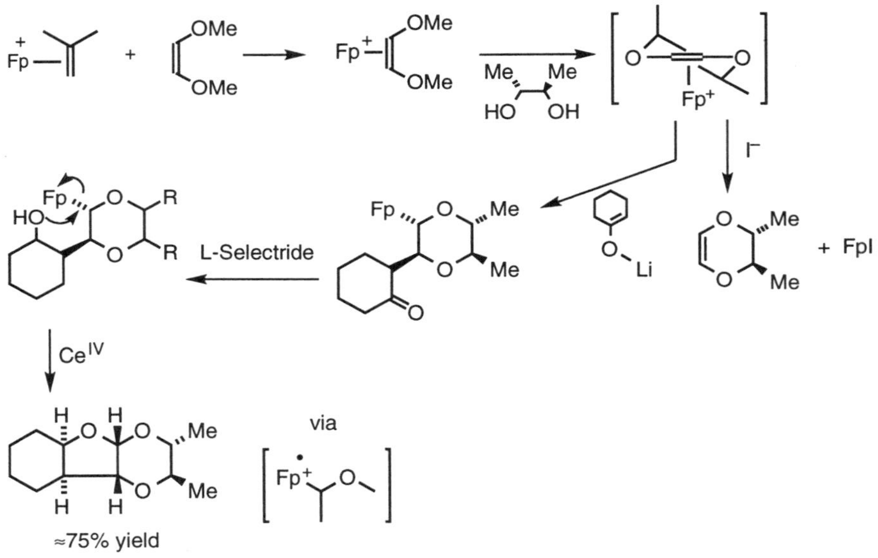

≈75% yield

providing a simple route to optically active compounds not easily prepared otherwise. More interestingly, ketone enolates attack the dioxene complex from the face opposite the metal to give the stable Fp alkyl complex with high diastereoselectivity. Stereoselective reduction of the carbonyl group with L-

Selectride, followed by oxidative removal of the metal produces the tricyclic furan in excellent yield and with high diastereoselectivity. In this case oxidation of the iron does not result in CO insertion, but rather it simply makes the iron a good leaving group (Eq. 7.41).[45]

An unusual class of reactions involving cationic iron-olefin complexes is the "3+2" cycloaddition reaction that occurs between electron-poor alkenes and η^1-allyl Fp complexes (Eq. 7.42).[46] In this case, the η^1-allyl Fp complex nucleophilically attacks the electrophilic alkene, producing a stabilized enolate and a cationic Fp olefin complex. The enolate then attacks the cationic olefin complex, closing the ring, and generating a Fp alkyl complex. Oxidative removal of the iron promotes CO insertion, and generates the highly functionalized cyclopentane ring system.

Eq. 7.42

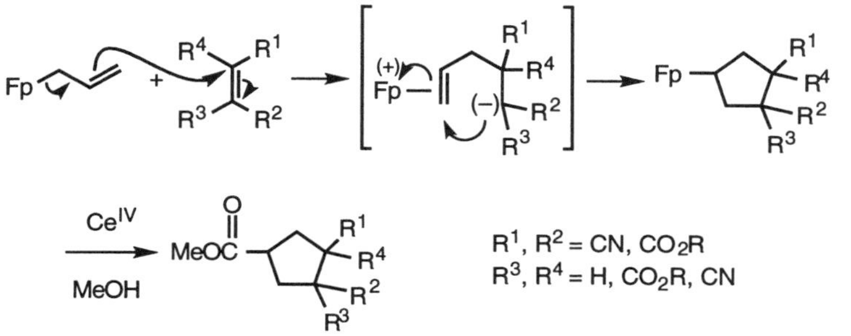

$$R^1, R^2 = CN, CO_2R$$
$$R^3, R^4 = H, CO_2R, CN$$

The same chemical features that make cationic iron-olefin complexes reactive towards nucleophiles makes them unreactive towards electrophiles and complexation of alkenes to Fp$^+$ can protect them against an array of electrophilic reactions. Complexation occurs preferentially at the less-substituted alkene, and at alkenes rather than alkynes, so selective protection is possible (Eq. 7.43, Eq. 7.44, and Eq. 7.45).[34] The olefin is freed of iron after reaction by treatment with iodide.

Eq. 7.43

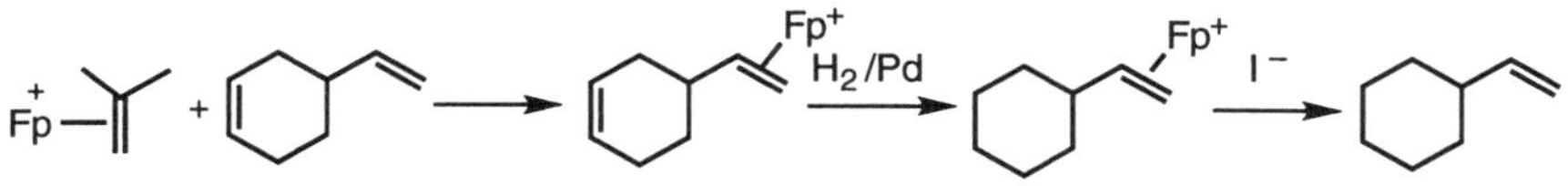

Eq. 7.44

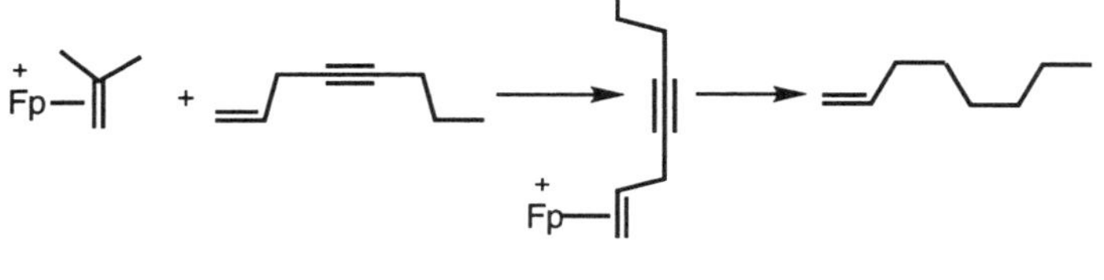

Eq. 7.45

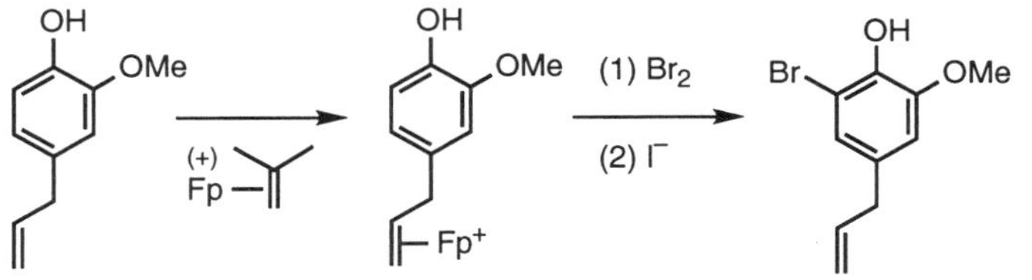

7.3   Metal Diene Complexes

a.   Fe(CO)₃ as a 1,3-Diene Protecting Group[48]

Heating $Fe_2(CO)_9$ with dienes produces iron carbonyl complexes of conjugated dienes (Eq. 7.46). Non-conjugated dienes often rearrange to give conjugate diene complexes in the presence of $Fe_2(CO)_9$ or $Fe(CO)_5$. These iron tricarbonyl complexes of dienes are very stable: they fail to undergo Diels-Alder reactions, and undergo facile Friedel-Crafts acylation without decomplexation from the metal. Although they can be made to undergo reaction with strong nucleophiles (see next section), complexation of the diene segment is most often used to protect it.

Eq. 7.46

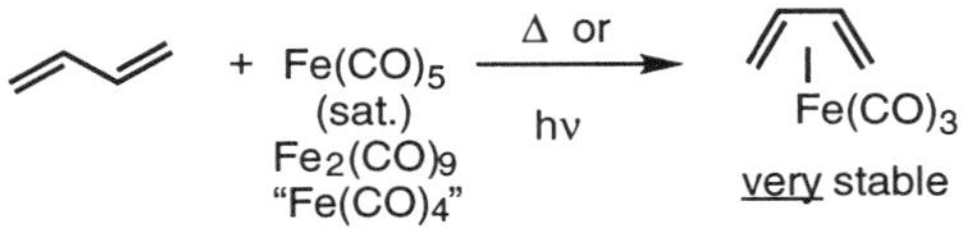

For example, dienals are highly reactive and polymerize readily, making reactions at the aldehyde difficult and inefficient. However, treatment with $Fe_2(CO)_9$ forms the stable diene complex, in which the aldehyde reacts normally. When the reactions are complete, the iron is oxidatively removed, giving the metal-free organic compound. Particularly notable is the range of reactions the iron tricarbonyl group will withstand. These include Wittig, Aldol, reduction, and even the *cis* hydroxylation of an olefin with osmium tetroxide (Eq. 7.47).[49]

Eq. 7.47

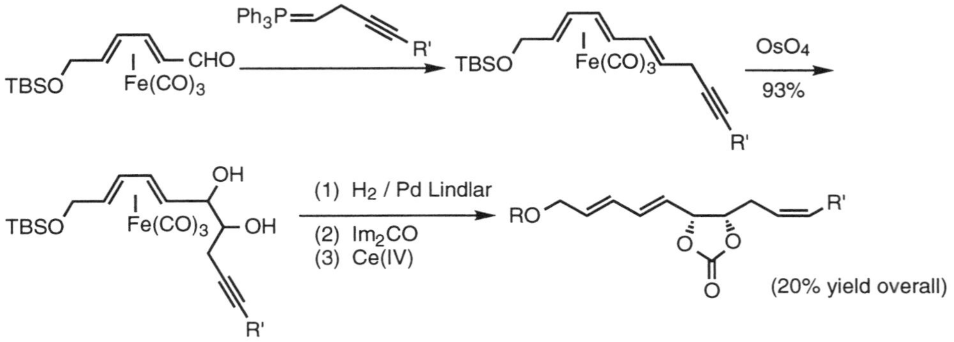

Because iron occupies a single face of these diene complexes, substituted diene complexes are intrinsically chiral, can be resolved, and the presence of the iron has a substantial stereochemical influence (Eq. 7.48,[50] Eq. 7.49,[51] and Eq. 7.50[52]).

Eq. 7.48

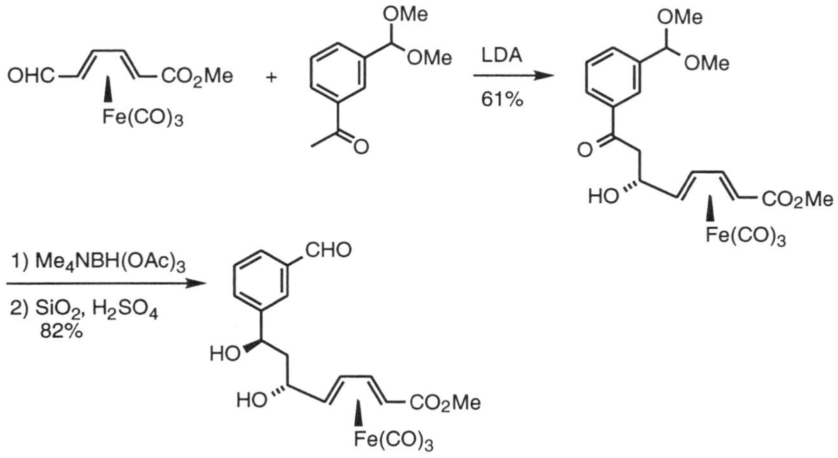

Eq. 7.49

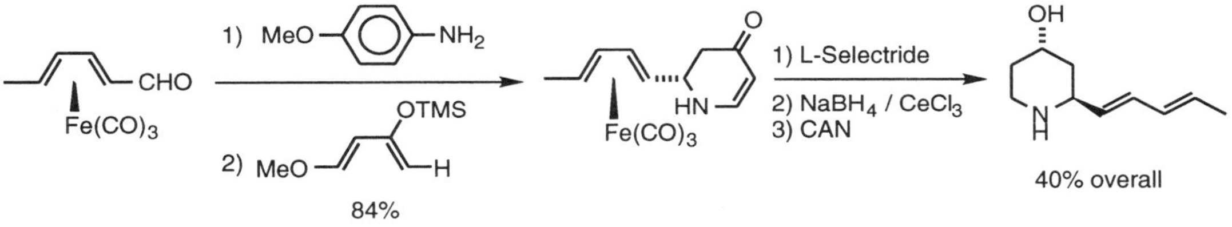

Eq. 7.50

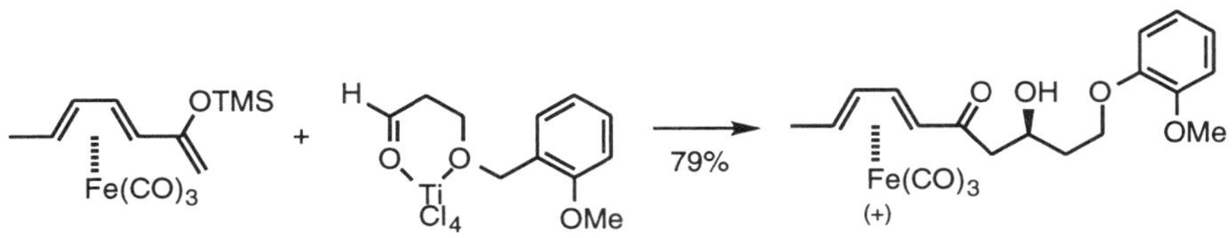

Complexation of dienes to iron tricarbonyl fragments can also be used to both activate and stabilize allylic positions, as well as to exert stereochemical control (Eq. 7.51[53] and Eq. 7.52[54]). Complexation of the cyclohepta-3,5-dienone permitted α-dialkylation with clean *cis* stereochemistry (Eq. 7.53).[55] In this case, the iron fragment prevents isomerization of the enolate and directs the alkyl groups to the face opposite the metal. Reduction of the resulting α,α'-dimethylketone also occurs stereospecifically, but, because of the flanking methyl groups, attack occurs from the same face as the metal. Similarly, two of the three conjugated double bonds of cyclohepta-2,4,6-trien-1-ol could be complexed to iron, permitting oxidation or hydroboration of the uncomplexed olefin with very high stereoselectivity (Eq. 7.54).[56]

Eq. 7.51

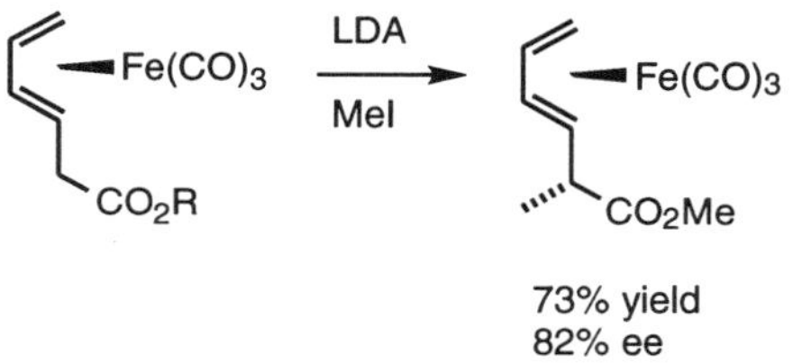

73% yield
82% ee

Eq. 7.52

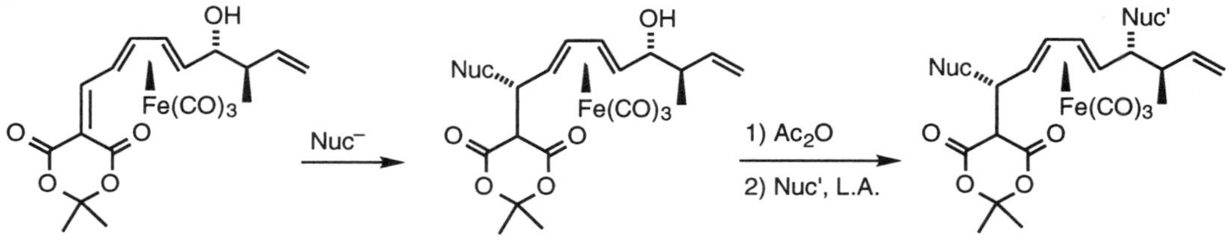

Eq. 7.53

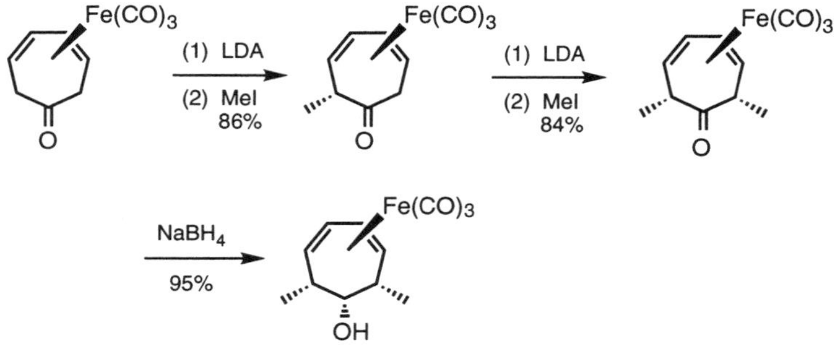

Eq. 7.54

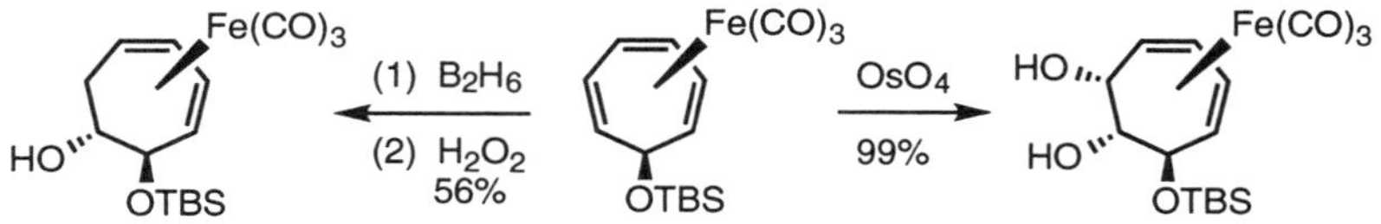

Similar activation and directing effects have been achieved using complexation of dienes to molybdenum(II) (Eq. 7.55).[57] In this case, the cationic diene

Eq. 7.55

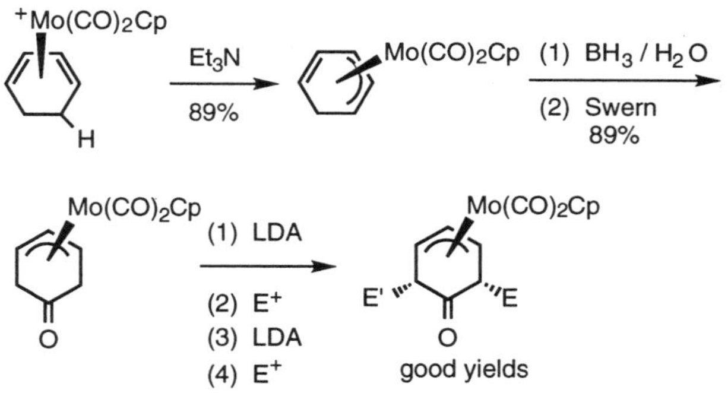

complex was converted to the neutral π-allylmolybdenum complex by allylic proton abstraction. This protected one of the double bonds of the diene, permitting functionalization of the other. Cycloheptadiene underwent a similar series of reactions.

Finally, complexation to iron can stabilize normally inaccessible tautomers of aromatic compounds permitting quite unusual synthetic transformations (Eq. 7.56)[58] and (Eq. 7.57).[59]

Eq. 7.56

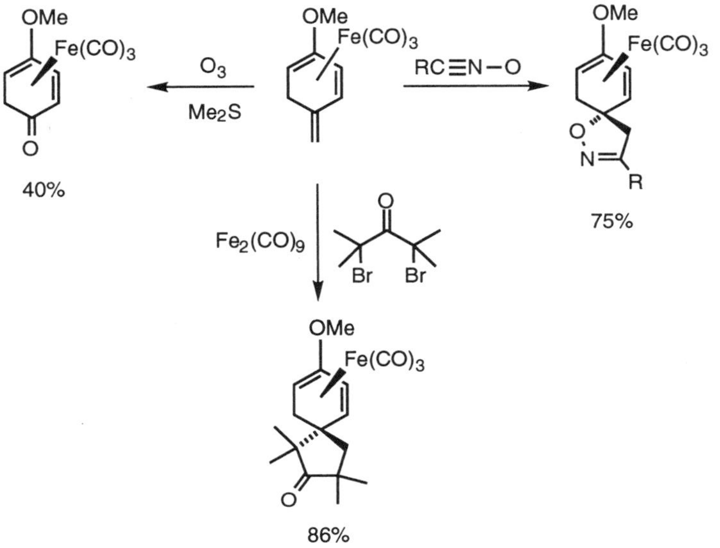

Eq. 7.57

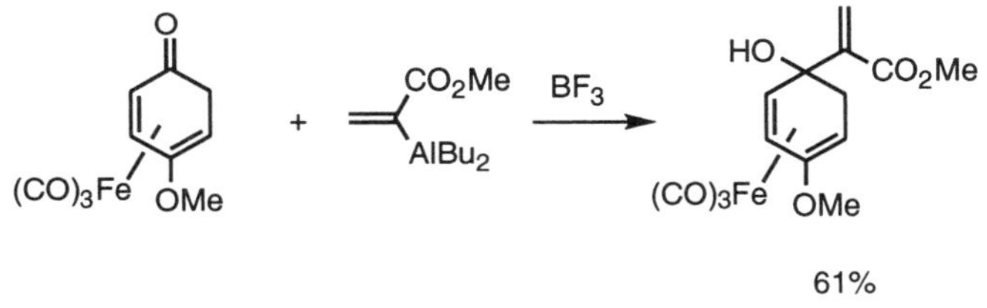

Diene complexes are also produced by the reaction of 1,4-dihalo-2-butenes with $Fe_2(CO)_9$. An especially useful example of this reaction is the preparation of (η^4-cyclobutadiene)iron tricarbonyl from dichlorocyclobutene[60] (Eq. 7.58). This reaction illustrates the stabilization of extremely unstable organic molecules by complexation to appropriate transition metals. Free cyclobutadiene itself has only

Eq. 7.58

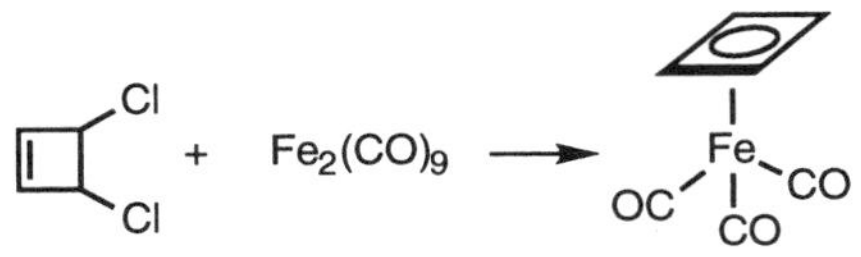

fleeting existence, since it rapidly dimerizes. However, cyclobutadieneiron tricarbonyl is a very stable complex. The complexed cyclobutadiene undergoes a variety of electrophilic substitution reactions, including Friedel Crafts acylation, formylation, chloromethylation, and aminomethylation. The keto group of the acylated cyclobutadieneiron tricarbonyl complex can be reduced by hydride reagents without decomposition.[61] On the other hand, oxidation of the complex with cerium(IV), triethylamine oxide, or pyridine N-oxide frees the cyclobutadiene ligand for use in synthesis. When combined with metathesis (Chapter 6) this chemistry provides a rapid entry into cyclooctadiene ring systems (Eq. 7.59).[62]

Eq. 7.59

b. Nucleophilic Attack on Metal-Diene Complexes[63]

Although complexation of dienes to electron-deficient metal fragments stabilizes these ligands towards electrophilic attack, it activates them towards nucleophilic attack, a reaction mode unavailable to the uncomplexed ligand. Neutral η^4-1,3-diene iron tricarbonyl complexes are reactive only towards very strong nucleophiles, and attack can occur at either the terminal or internal position of the diene. Kinetically, attack at the terminal position to form an η^3-allyliron complex is favored. However, this is a reversible process, and thermodynamically, attack at an internal position generating a η^1-alkyl-η^2-olefin iron complex is favored (Eq. 7.60).[64] As usual, attack occurs from the face opposite the metal. Protolytic cleavage of the "thermodynamic" iron complex gives the alkylation product, while treatment with carbon monoxide results in insertion into the iron-carbon σ-bond. Protolytic cleavage of this σ-acyl complex produces the *trans* aldehyde, while treatment with an electrophile in the presence of a ligand leads to acylation (Eq. 7.61).[65]

With acyclic dienes, the situation is a little more complicated. Complexed butadiene itself is converted cleanly to the corresponding cyclopentanone, by a process (Eq. 7.62) involving nucleophilic attack (a), followed by CO insertion (b), followed by olefin insertion (c). However, substitution on the diene suppresses the

Eq. 7.60

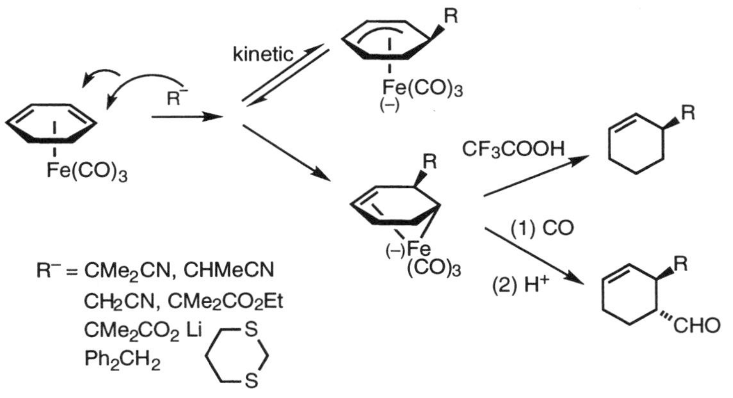
kinetic
R⁻
Fe(CO)3
(–)
CF3COOH
(1) CO
(2) H+
CHO
R⁻ = CMe2CN, CHMeCN
CH2CN, CMe2CO2Et
CMe2CO2 Li
Ph2CH2

Eq. 7.61

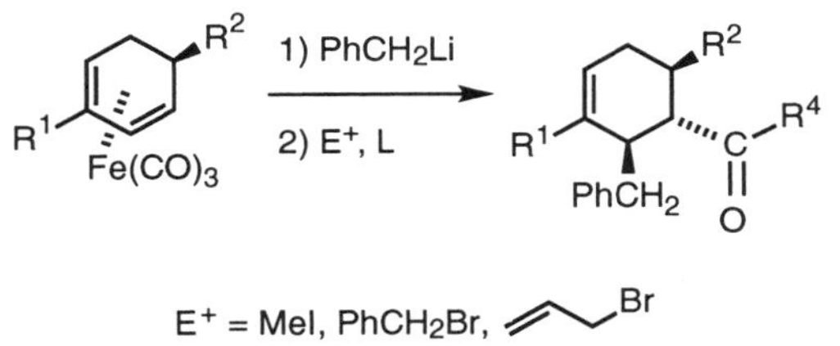
1) PhCH2Li
2) E+, L
Fe(CO)3
PhCH2
E+ = MeI, PhCH2Br,

Eq. 7.62

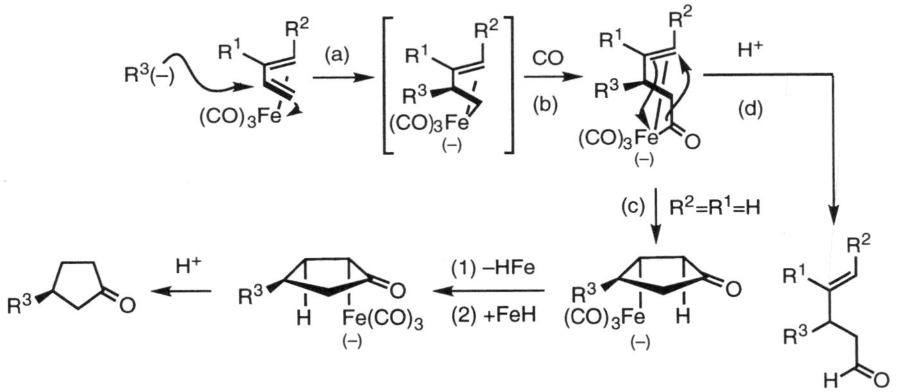
(a)
CO
(b)
H+
(d)
(c)
R2=R1=H
(1) –HFe
(2) +FeH
(CO)3Fe
Fe(CO)3

olefin insertion step, and the amount of the "normal" cleavage product (d) (eg., aldehyde) increases as substitution increases. This olefin-insertion followed by migration of iron to the position α– to the carbonyl group is precedented in the insertion of ethylene into iron complexes generated from $Fe(CO)_4^{2-}$. In the absence of CO, the σ-alkyliron complex will undergo reaction with a range of electrophiles, to result in overall 1,2-difunctionalization of the diene (Eq. 7.63).[67]

Eq. 7.63

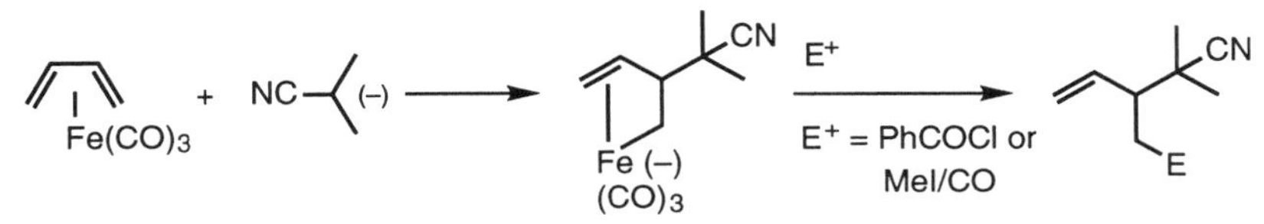
Fe(CO)3
E+
E+ = PhCOCl or
MeI/CO
Fe (–)
(CO)3

By making the metal fragment more electrophilic, the reactivity of metal diene complexes toward nucleophiles can be increased. This is often achieved by replacing a neutral metal fragment by a cationic one, such as the $CpMo(CO)_2^+$ fragment. These complexes are relatively easy to synthesize, although several steps are required (Eq. 7.64).[68] They undergo reaction with a much wider range of nucleophiles, exclusively

Eq. 7.64

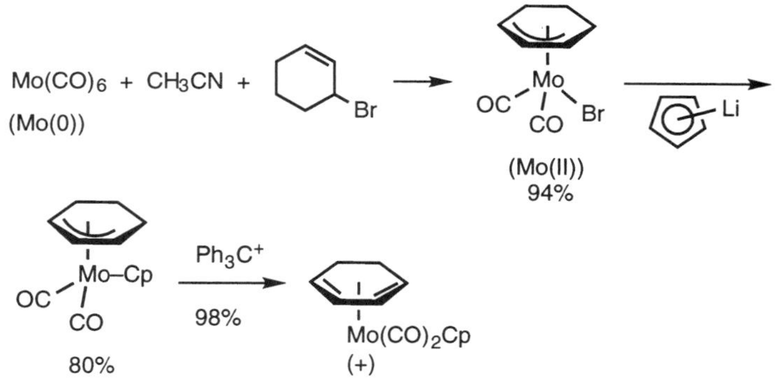

at the terminal position, to generate a neutral η^3-allylmolybdenum complex. Hydride abstraction from this complex regenerates a cationic diene complex which again undergoes nucleophilic attack from the face opposite the metal, resulting in overall *cis*-1,4 difunctionalization of the diene (Eq. 7.65).[69] Cyclohepta-1,3-diene undergoes a similar series of transformations. The major limitation, to date, is finding an

Eq. 7.65

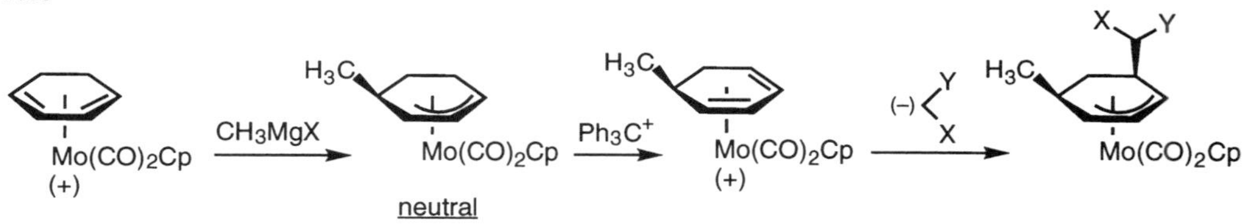

efficient way to remove the molybdenum from the final η^3-allyl complex. Modest success has been achieved by conversion of the neutral η^3-allylmolybdenum carbonyl complex to the cationic nitrosyl (exchange of neutral CO for NO$^+$) permitting nucleophilic attack on the η^3-allyl complex (Eq. 7.66)[70] or by oxidizing the π-allylmolybdenum complex with iodonium trifluoroacetate (Eq. 7.67).[71] Of particular synthetic interest is the application of this methodology to heterocyclic dienes (Eq. 7.68).[72]

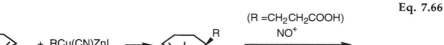

Eq. 7.66

+ RCu(CN)ZnI

Mo(CO)₂Cp
(+)

R
Mo(CO)₂Cp

$(R = CH_2CH_2COOH)$
NO^+

Cp Mo(NO)(CO)
(+)

Et₃N

35%

$R = (CH_2)_4CO_2Et, (CH_2)_3CO_2Me,$
$(CH_2)_2CO_2Me, PhCH_2, AcO(CH_2)_4$
$NC(CH_2)_3, NC(CH_2)_4$

Eq. 7.67

1) Mo(CO)₃(MeCN)₃
2) CpLi
58%

Br

CpMo(CO)₂

Ph_3C^+

Cp Mo(CO)₂
(+)

OLi

73%

CpMo(CO)₂

CF_3CO_2I
82%

Eq. 7.68

Mo(CO)₂Cp
(+)

R(−)

R   Mo(CO)₂Cp
60-90%

Ph_3C^+
92%

R   Mo(CO)₂Cp
(+)

R'(−)
92%

R   Mo(CO)₂Cp
60-90%

(1) CF₃COOH
(2) H₂/Pd
72%

R   O   R'

$R, R' = D, Me, p\,tol, RC\equiv C, CH_2CO_2t\,Bu,$

c. Nucleophilic Attack on Cationic Metal Dienyl Complexes[73]

Cationic *dienyl* complexes of iron are among the most extensively used complexes in organic synthesis. They are generally reactive towards a wide range of nucleophiles, and the presence of iron permits a high degree of both stereo- and regiocontrol. Cyclohexadienyl complexes are the most studied, at least in part because a wide array of substituted cyclohexadienes are readily available from the Birch reduction of aromatic compounds. The cyclohexadienyliron complexes are prepared by simply treating either the conjugated or nonconjugated diene with $Fe_2(CO)_9$, (Eq. 7.69). Treatment of the resulting neutral diene complex with trityl cation (hydride abstraction) produces the relatively air- and moisture-stable cationic cyclohexadienyliron complex. A very wide range of nucleophiles attack this complex at the terminus of the dienyl system from the face opposite the metal, regenerating the neutral diene complex. Treatment with trityl cation removes hydride from the unsubstituted position of the diene complex, and the resulting dienyl complex again reacts with nucleophiles at this position from the face opposite the metal, giving a cis-1,2-disubstituted cyclohexadiene. The iron is easily removed by treatment with amine oxides, giving the free organic compound (Eq. 7.69).

Eq. 7.69

Substituted complexes can be prepared and undergo reaction with high regioselectivity. The diene complex derived from 4-methyl anisole undergoes hydride abstraction α to the methoxy group, giving a single regioisomer of the dienyl complex. Nucleophilic attack occurs exclusively at the *methyl* terminus of the dienyl system, apparently because the other end is electronically deactivated by the strongly electron-donating methoxy group. Oxidative removal of the iron followed by hydrolysis gives the 4,4-disubstituted cyclohexenone (Eq. 7.70[74] and Eq. 7.71[75]).

Cationic dienyliron complexes are potent electrophiles, and even undergo electrophilic aromatic substitution reactions with electron-rich arenes (Eq. 7.72[76] and Eq. 7.73[77,78]).

Eq. 7.70

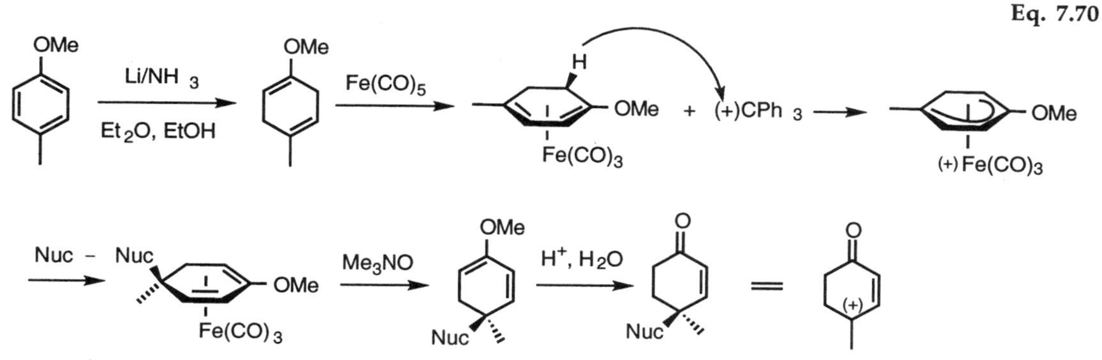

Eq. 7.71

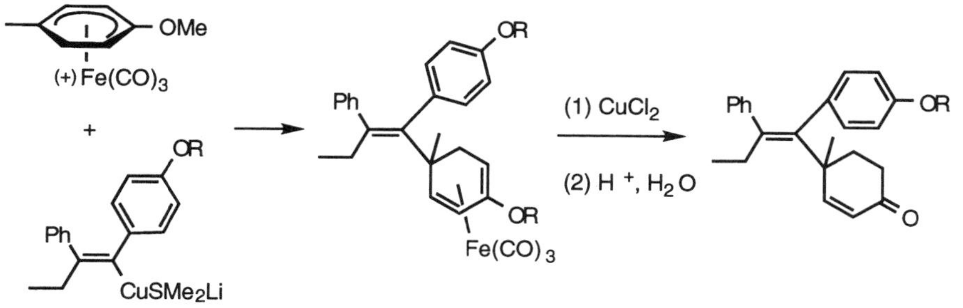

Eq. 7.72

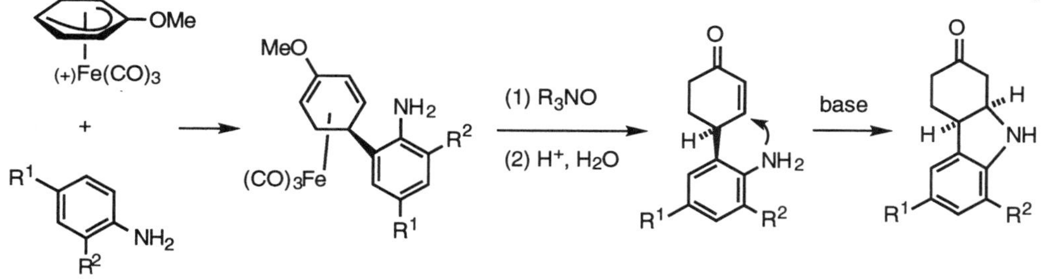

Eq. 7.73

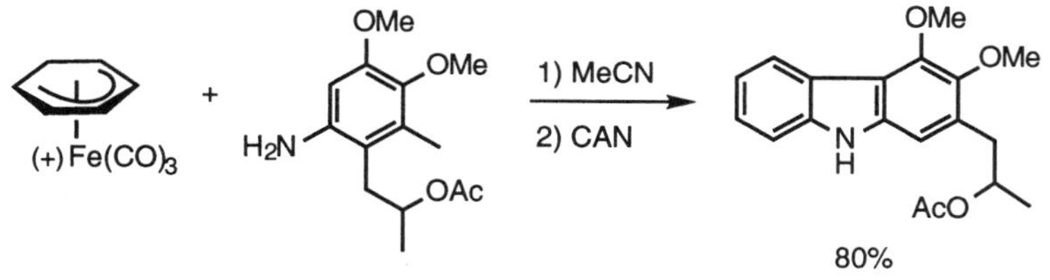

Because iron tricarbonyl occupies a single face of the η^5-dienyl complex, unsymmetrically substituted complexes are intrinsically chiral and can be resolved (Eq. 7.74).[74] Since addition of nucleophiles is stereospecific, these complexes are of use in asymmetric synthesis. Functional groups can direct complexation, as seen with the optically active diol from microbial oxidation in Eq. 7.75.[80]

Eq. 7.74

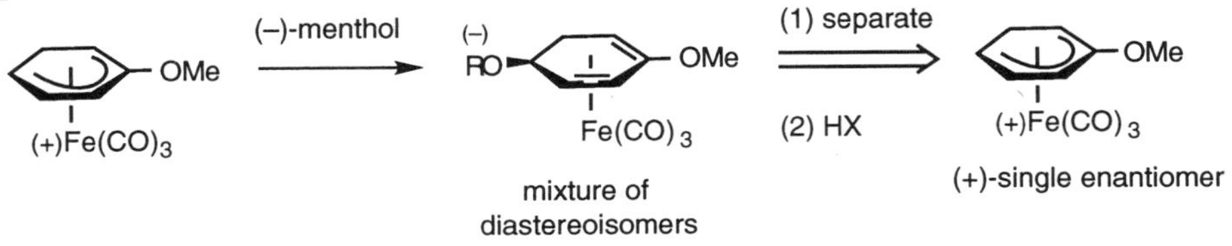

mixture of
diastereoisomers

Eq. 7.75

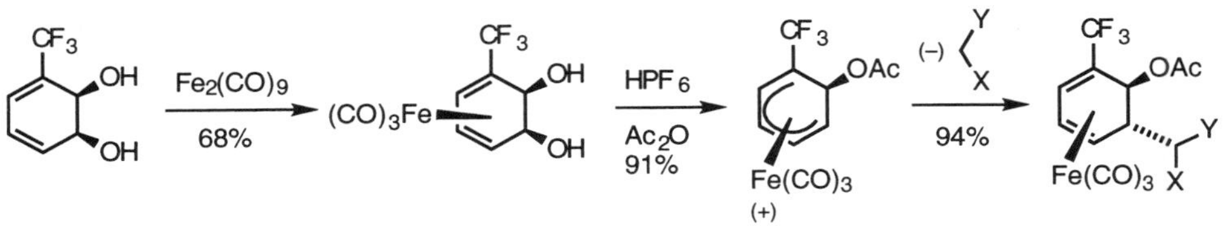

The above reaction chemistry is not restricted to cyclohexadienes. The same degree of activation, regio- and stereoselectivity is achieved with cycloheptadiene.[81] In contrast, acyclic η^5-dienyl complexes have been much less studied, and have only rarely been used in synthesis. They are more difficult to prepare, and often react with lower regioselectivity than do their cyclic analogs. However, once formed they are generally reactive towards nucleophiles (Eq. 7.76[82] and Eq. 7.77[83]).

Eq. 7.76

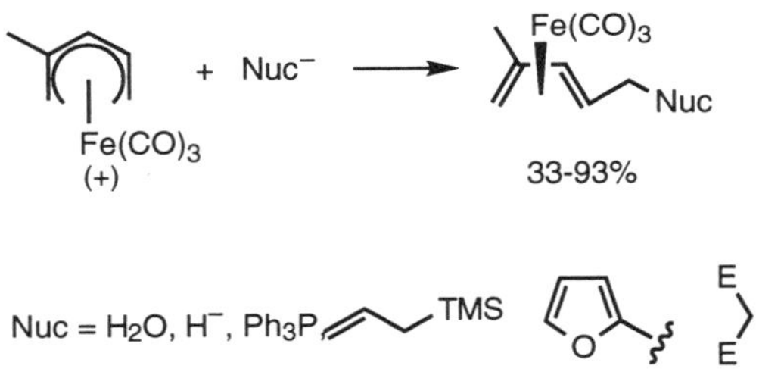

33-93%

Nuc = H$_2$O, H$^-$, Ph$_3$P, TMS

Eq. 7.77

27-64%

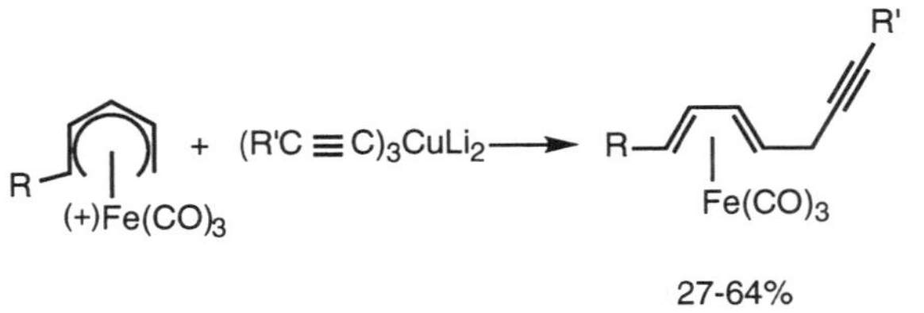

d. Metal-Catalyzed Cycloaddition Reactions[84]

Many types of cycloaddition reactions which do not proceed under normal thermal or photochemical reaction conditions, proceed readily in the presence of appropriate transition metal catalysts. These include [2+2], [3+2], homo Diels-Alder, [4+2], [4+4], [6+2], [6+4] and many others. Almost certainly none of these metal-catalyzed cycloadditions are concerted, but rather they proceed via metal-carbon σ-bonded species, and most can be viewed as "reductive coupling" reactions wherein the metal is oxidized in the initial steps of the reaction, much like the low-valent zirconium couplings discussed in Chapter 4. Few have been studied mechanistically, and the mechanisms presented below are for the most-part hypothetical. Only the synthetically interesting variants will be discussed here.

Although nickel- and ruthenium-catalyzed [2+2] cycloadditions of alkenes and alkynes to strained alkenes such as norbornene have long been known, only relatively simple compounds have been synthesized using this chemistry. However, much more complex systems are produced by palladium-catalyzed intramolecular [2+2] cycloaddition between alkenes and alkynes (Eq. 7.78).[85] The reaction almost surely proceeds through a metallacyclopentene, with palladium being in the unusual (but not unattainable) +4 oxidation state. The range of substrates is quite limited for this unusual process.

Eq. 7.78

The [3+2] cycloadditions proceeding through oxallyliron species were discussed in Chapter 5, and those proceeding through trimethylene methane complexes will be discussed in Chapter 9. An interesting example involving neither of these is shown in Eq. 7.79, which is thought to be initiated by palladacyclopentene formation as in Eq. 7.78, followed by reversion to form a vinylcarbenepalladium complex, which then undergoes a [3+2] cycloaddition ([4+2] if you count palladium) followed by reductive elimination.[86]

Eq. 7.79

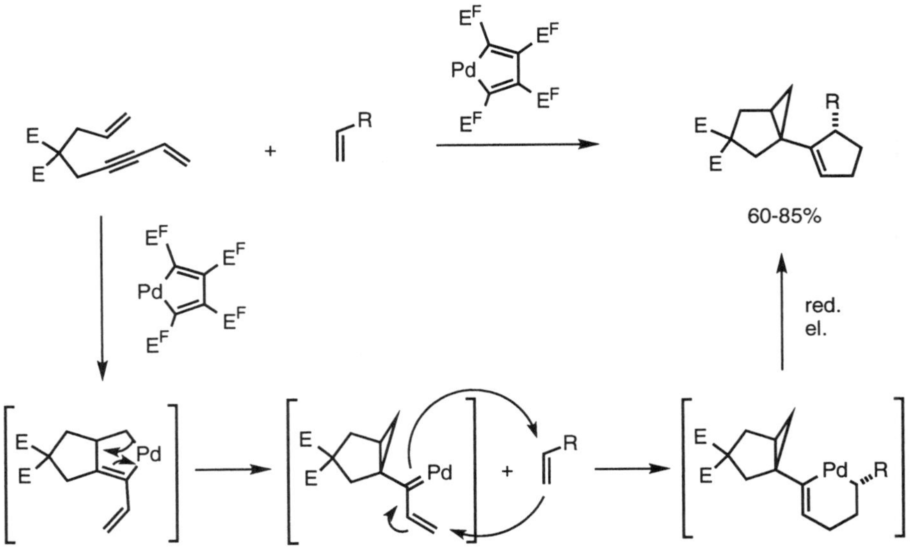

60-85%

Metal-catalyzed [2+2+2] cyclotrimerization of alkynes will be presented in the next chapter. Here, the homo-Diels Alder [2+2+2] cycloaddition will be considered (Eq. 7.80). Nickel(0)[87] and cobalt(0)[88] complexes catalyze this process, and high enantioselectivity has been observed by using chiral diphosphine ligands with the cobalt catalyst systems.[89]

Eq. 7.80

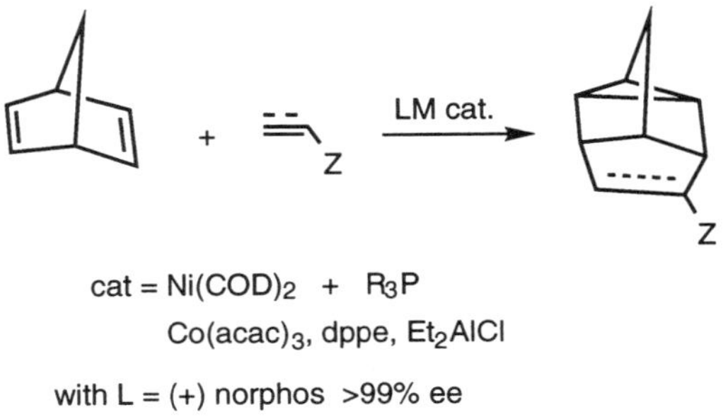

cat = Ni(COD)₂ + R₃P
Co(acac)₃, dppe, Et₂AlCl

with L = (+) norphos >99% ee

The standard Diels-Alder [4+2] cycloaddition has long been studied and is a staple in the arsenal of synthetic chemists. However this process normally is most efficient for reactions between electronically *dissimilar* dienes and dienophiles. Normally-difficult [4+2] cycloadditions between similar dienes and dienophiles can be efficiently catalyzed by a variety of low-valent metal complexes. Again, these are likely to proceed via metallacycles (Eq. 7.81).

Eq. 7.81

$$\text{\includegraphics}\;\longrightarrow\;\left[\;\underset{M}{\bigcirc}\;\right]\;\xrightarrow[\text{el.}]{\text{red.}}\;\bigcirc\;+\;M$$

Nickel(0) (Eq. 7.82)[90] and rhodium(I) (Eq. 7.83)[91] complexes are most efficient for this process and again, in the presence of chiral ligands, reasonable asymmetric induction can be achieved.[92]

Eq. 7.82

85-99%

m, n = 1
m = 2, n = 1
m = 1, n = 2

80-90%

Eq. 7.83

n = 1, 2
(11 cases)

50-75%

Nickel(0)-catalyzed [4+4] cycloadditions evolved from early studies of metal-catalyzed cyclooligomerization of butadiene but only recently has the process been applied in complex organic synthesis.[93] The mechanism again is likely to involve a "reductive" dimerization of two complexed dienes followed by reductive elimination (Eq. 7.84).[94] It has proven particularly useful in its intramolecular variant (Eq. 7.85).[95,96]

Eq. 7.84

Eq. 7.85

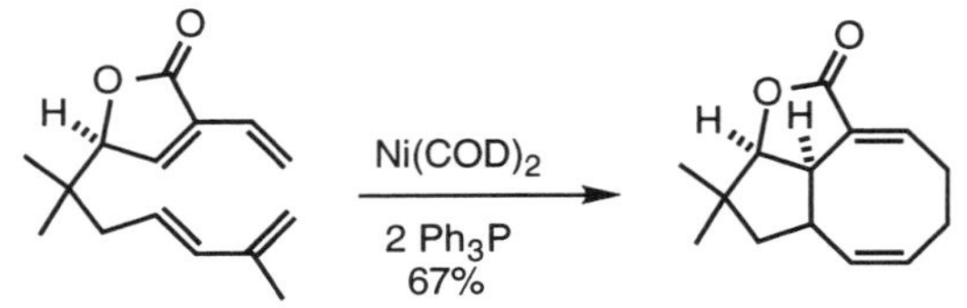

Chromium(0)-complexed trienes undergo a variety of both thermal and photochemical higher order cycloaddition reactions.[97] Both inter- and intramolecular [6+2] cycloadditions of alkenes and alkynes to chromium cycloheptatriene complexes are efficient (Eq. 7.86),[97] (Eq. 7.87),[98] (Eq. 7.88),[99] and (Eq. 7.89),[100] as are

Eq. 7.86

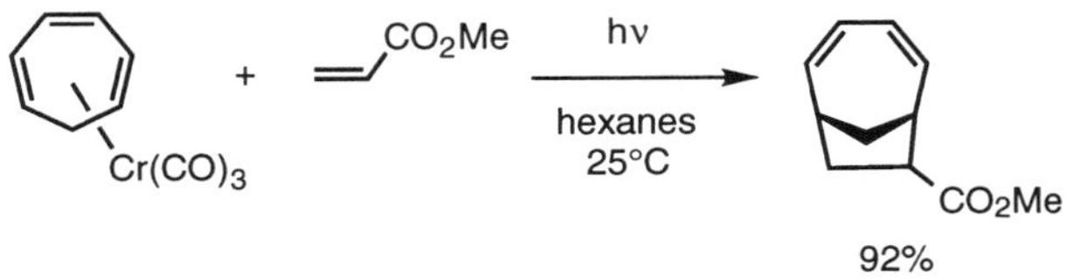

Eq. 7.87

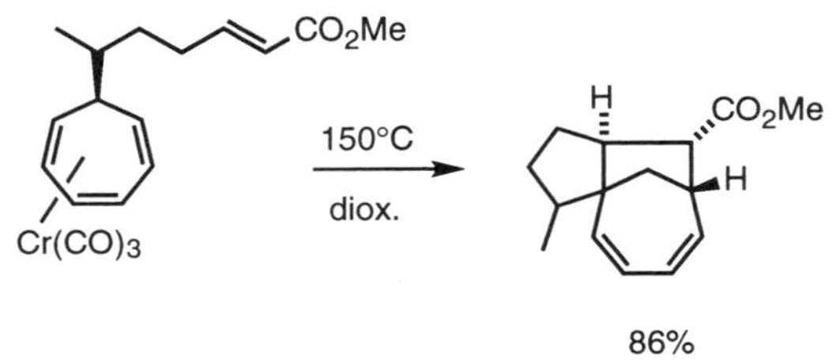

Eq. 7.88

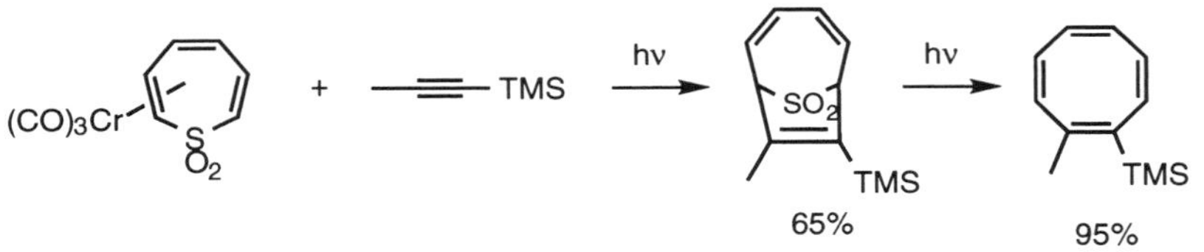

Eq. 7.89

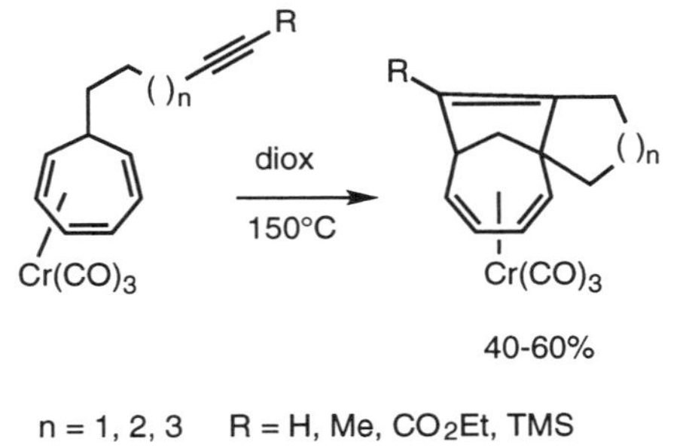

n = 1, 2, 3   R = H, Me, CO₂Et, TMS

inter (Eq. 7.90))[101] and intramolecular (Eq. 7.91)[102] [6+4] cycloadditions. Under appropriate conditions, this [6+4] cycloaddition can be catalyzed by chromium complexes (Eq. 7.92).[103] Again, these are unlikely to be concerted processes, but rather proceed via metallacyclo intermediates (Figure 7.3).

Eq. 7.90

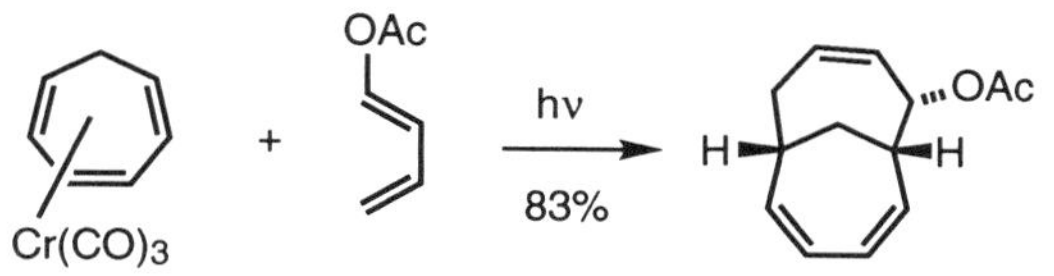

Eq. 7.91

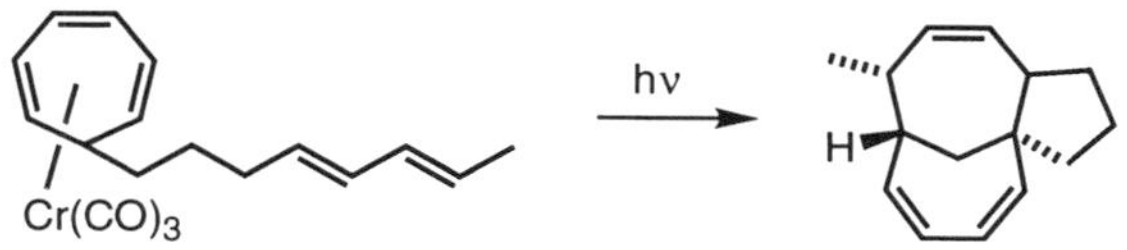

85%

Eq. 7.92

32-80%

18e

16e

+CO

σ–η⁵

with alkynes

+CO

with alkenes

[6+2]

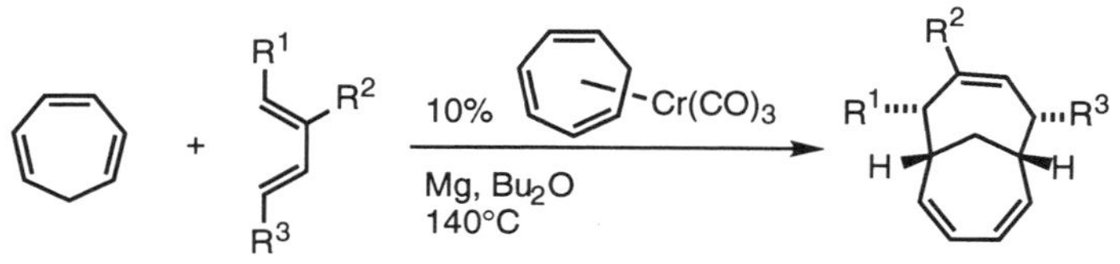

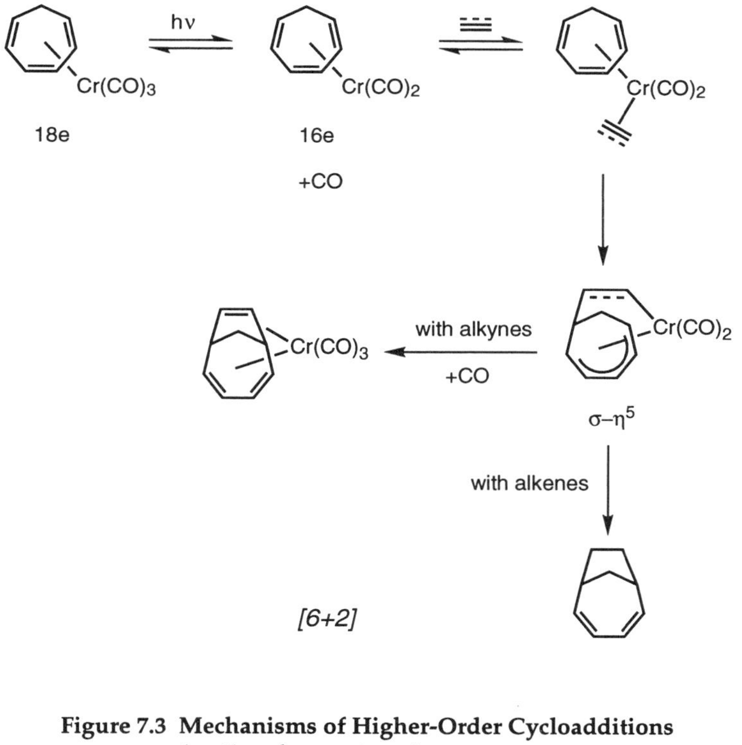

Figure 7.3 Mechanisms of Higher-Order Cycloadditions
(*continued on next page*)

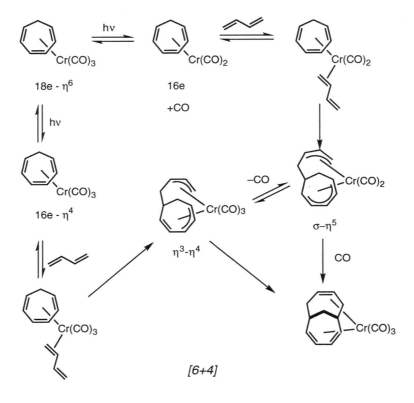

Figure 7.3 Mechanisms of Higher-Order Cycloadditions
(*continued from previous page*)

References

1. For review see: (a) McDaniel, K.F., "Transition Metal Alkene, Diene, and Dienyl Complexes: Nucleophilic Attack on Alkene Complexes," in "Comprehensive Organometallic Chemistry II", Abel, E.W.; Stone, F.G.A.; Wilkinson, G., Eds., Elsevier Science Ltd., Oxford, UK, 1995, Vol. 12, pp. 601-622; (b) Hegedus, L.S., "Palladium in Organic Synthesis" in "Organometallics in Synthesis; a Manual," 2nd Ed., Schlosser, M., Ed., Wiley Interscience, Chichester, UK, 1998.
2. Tsuji, J. *Synthesis* **1984**, 369.
3. Celimene, C.; Dhimane, H.; Saboureau, A.; Lhornmet, G. *Tetrahedron Asymmetry* **1996**, *7*, 1585.
4. Oikawa, M.; Ueno, T.; Oikawa, H.; Ichihara, A. *J. Org. Chem.* **1995**, *60*, 5048.
5. Krishnuda, K.; Krishna, P.R.; Mereyala, H.B. *Tetrahedron Lett.* **1996**, *37*, 6007.
6. Pellisier, H.; Michellys, P-Y.; Santelli, M. *Tetrahedron* **1997**, *53*, 7577.
7. Kang, S-K.; Jung, K-Y.; Chung, J-N.; Namkoong, E-Y.; Kim, T-H. *J. Org. Chem.* **1995**, *60*, 4678.
8. For review see: (a) Hosokawa, T.; Murahashi, S-i. *Acct. Chem. Res.* **1990**, *23*, 49. (b) Hosokawa, T.; Murahashi, S-i. *J. Synth. Org. Chem. Jpn.* **1995**, *53*, 1009 (in English).
9. Semmelhack, M.F.; Kim, C.R.; Dobler, W.; Meier, M. *Tetrahedron Lett.* **1989**, *30*, 4925.
10. van Bentheim, R.A.T.M.; Hiemstra, H.; Michels, J.J.; Speckamp, W.N. *J. Chem .Soc., Chem. Comm.* **1994**, 357.
11. Uozumi, Y.; Kato, K.; Hayashi, T. *J. Am. Chem. Soc.* **1997**, *119*, 5063.
12. Tenagha, A.; Kammerer, F. *Synlett* **1996**, 576. See also Saito, S.; Hara, T.; Takahashi, N.; Hirai, M.; Moriwake, T. *Synlett* **1992**, 237.
13. For a review see: Jäger, V.; Graiza, T.; Dubois, E.; Hasenöhrl, ; Hümmer, W.; Kautz, U.; Kirschbaum, B.; Liebeskind, A.; Remen, L.; Shaw, D.; Stahl, U.; Stephen, O. in "Organic Synthesis via Organometallics," Helmchen, G., Ed., F. Vieweg and Sohn Verlagsges., Braunschweig, Germany, 1997.
14. Boukowvalas, J.; Fortier, G.; Rachu, I-I. *J. Org. Chem.* **1998**, *63*, 916.

15. Semmelhack, M.F.; Epa, W.R. *Tetrahedron Lett.* **1993**, *34*, 7205.
16. Larock, R.C.; Lee, N.H. *J. Am. Chem. Soc.* **1991**, *113*, 7815.
17. Minami, T.; Nishimoto, A.; Hanaoka, M. *Tetrahedron Lett.* **1995**, *36*, 9505.
18. Korte, D.E.; Hegedus, L.S.; Wirth, R.K. *J. Org. Chem.* **1977**, *42*, 1329.
19. (a) Akermark, B.; Backvall, J-E.; Hegedus, L.S.; Zetterberg, K.; Siirala-Hansen, K.; Sjoberg, K. *J. Organomet. Chem.* **1974**, *72*, 127. (b) Akermark, B.; Akermark, G.; Hegedus, L.S.; Zetterberg, K. *J. Am. Chem. Soc.* **1981**, *103*, 3037.
20. Hegedus, L.S.; McKearin, J.M. *J. Am. Chem. Soc.* **1982**, *104*, 2444.
21. For a review see: Hegedus, L.S. *Angew. Chem. Int. Ed. Engl.* **1988**, *27*, 1113.
22. Irie, K.; Isaka, T.; Iwata, Y.; Yanai, Y.; Nakamura, Y.; Korzumi, F.; Ohigashi, H.; Wender, P.; Satomi, Y.; Nishmo, H. *J. Am. Chem. Soc.* **1996**, *118*, 10733.
23. (a) Rönn, M.; Bäckvall, J.E.; Andersson, P.G. *Tetrahedron Lett.* **1995**, *36*, 7749. (b) Larock, R.C.; Hightower, L.A.; Hasvold, L.A.; Peterson, K.P. *J. Org. Chem.* **1996**, *61*, 3584.
24. Hirai, Y.; Watanabe, J.; Nozaki, T.; Yokoyama, H.; Yamaguchi, S. *J. Org. Chem.* **1997**, *62*, 776.
25. For a review see: Tamaru, Y.; Kimura, M. *Synlett* **1997**, 749.
26. Harama, H.; Abe, A.; Sakado, T.; Kimura, M.; Fugami, K.; Tanaka, S.; Tamaru, Y. *J. Org. Chem.* **1997**, *62*, 7113.
27. Hummer, W.; Dubois, E.; Grazca, T.; Jaeger, V. *Synthesis* **1997**, 634.
28. Weider, P.R.; Hegedus, L.S.; Asada, H.; D'Andrea, S.V. *J. Org. Chem.* **1985**, *50*, 4276.
29. (a) Hegedus, L.S.; Williams, R.E.; McGuire, M.A.; Hayashi, T. *J. Am. Chem. Soc.* **1980**, *102*, 4973. (b) Hegedus, L.S.; Darlington, W.H. *J. Am. Chem. Soc.* **1980**, *102*, 4980.
30. Montgomery, J.; Wieber, G.M.; Hegedus, L.S. *J. Am. Chem. Soc.* **1990**, *112*, 6255.
31. (a) Masters, J.J.; Hegedus, L.S.; Tamariz, J. *J. Org. Chem.* **1991**, *56*, 5666. (b) Masters, J.J.; Hegedus, L.S. *J. Org. Chem.* **1993**, *58*, 4547. (c) Laidig, G.; Hegedus, L.S. *Synthesis* **1995**, 527.
32. (a) Kende, A.S.; Roth, B.; Sanfilippo, P.J. Blacklock, T.J. *J. Am. Chem. Soc.* **1982**, *104*, 5808. (b) Kende, A.S.; Roth, B.; Sanfilippo, P.J.; *J. Am. Chem. Soc.* **1982**, *104*, 1784.
33. Toyota, M.; Nishikawa, Y.; Motoki, K.; Yoshida, N.; Fukumoto, K. *Tetrahedron* **1993**, *49*, 11189.
34. For a review see: Overman, L.E. *Angew. Chem. Int. Ed. Engl.* **1984**, *23*, 579.
35. Saito, S.; Kurodo, A.; Matsunaga, H.; Ikeda, S. *Tetrahedron* **1996**, *52*, 13919. For earlier examples, see: (a) Grieco, P.A.; Takigawa, T.; Bongers, S.L.; Tanaka, H. *J. Am. Chem. Soc.* **1980**, *102*, 7587. (b) Grieco, P.A.; Tuthill, P.A.; Sham, H.L. *J. Org. Chem.* **1981**, *46*, 5005.
36. Panek, J.S.; Yang, K.M.; Solomon, J.S. *J. Org. Chem.* **1993**, *58*, 1003.
37. Overman, L.E.; Jacobsen, E.J. *J. Am. Chem. Soc.* **1982**, *104*, 7225.
38. Bluthe, N.; Malacria, M.; Gore, J. *Tetrahedron Lett.* **1983**, *24*, 1157.
39. Metz, P.; Mues, C.; Schoop, A. *Tetrahedron* **1992**, *48*, 1071.
40. Overman, L.E.; Zipp, G.G. *J. Org. Chem.* **1997**, *62*, 2288.
41. (a) Calter, M.; Hollis, T.K.; Overman, L.E.; Ziller, J.; Zipp, G.G. *J. Org. Chem.* **1997**, *62*, 1449. (b) Hollis, T.K.; Overman, L.E. *Tetrahedron Lett.* **1997**, *38*, 8837.
42. Grigg, R.; Markandu, J. *Tetrahedron Lett.* **1991**, *32*, 279.
43. Cutler, A.; Ehntholt, D.; Lennon, P.; Nicholas, K.; Marten, D.F.; Madhavarao, M.; Raghu, S.; Rosan, A.; Rosenblum, M. *J. Am. Chem. Soc.* **1975**, *97*, 3149.
44. Marsi, M.; Rosenblum, M. *J. Am. Chem. Soc.* **1984**, *106*, 7264.
45. Rosenblum, M.; Foxman, B.M.; Turnbull, M.M. *Heterocycles* **1987**, *25*, 419.
46. For a review see: Welker, M.E. *Chem. Rev.* **1992**, *92*, 97.
47. (a) Boyle, P.F.; Nicholas, K.M. *J. Org. Chem.* **1975**, *40*, 2682. (b) Nicholas, K.M. *J. Am. Chem. Soc.* **1975**, *97*, 3254.
48. For a review see: Donaldson, W.A., "Transition Metal Alkene Diene and Dienyl Complexes: Complexation of Dienes for Protection" in "Comprehensive Organometallic Chemistry II", Abel, E.W.; Stone, F.G.A.; Wilkinson, G., Eds., Elsevier Science Ltd., Oxford, UK, 1995, Vol. 12, pp. 623-637.
49. Gigou, A.; Beaucourt, J-P.; Lellouche, J-P.; Gree, R. *Tetrahedron Lett.* **1991**, *32*, 635.
50. Benvengnu, T.J.; Troupet, L.J.; Gree, R.L. *Tetrahedron* **1996**, *52*, 11811.
51. Takemoto, Y.; Ueda, S.; Takeuchi, J.; Baba, Y.; Iwata, C. *Chem. Pharm. Bull.* **1997**, *45*, 1900.
52. Frank-Neumann, M.; Bissinger, P.; Geoffroy, P. *Tetrahedron Lett.* **1997**, *38*, 4469, 4473, 4477.
53. Wasiak, J.T.; Craig, R.A.; Henry, R.; Dasgupta, B.; Li, H.; Donaldson, W.A. *Tetrahedron* **1997**, *53*, 4185.
54. Wada, C.K.; Roush, W.R. *Tetrahedron Lett.* **1994**, *35*, 7351.
55. Pearson, A.J.; Chang, K. *J. Chem. Soc., Chem. Comm.* **1991**, 394; *J. Org. Chem.* **1993**, *58*, 1228.
56. Pearson, A.J.; Srinivasan, K. *J. Chem. Soc., Chem. Comm.* **1991**, 392.
57. Pearson, A.J.; Mallik, S.; Mortezaei, R.; Perry, M.W.D.; Shively, R.J., Jr.; Youngs, W.J. *J. Am. Chem. Soc.* **1990**, *112*, 8034.

58. Ong, C.W.; Chien, T-L. *Organometallics* **1996**, *15*, 1323.
59. Hudson, R.D.A.; Osborne, S.A.; Stephenson, G.R. *Tetrahedron* **1997**, *53*, 4095.
60. Emerson, G.F.; Watts, L.; Pettit, R. *J. Am. Chem. Soc.* **1965**, *87*, 131.
61. Fitzpatrick, J.D.; Watts, L.; Emerson, G.F.; Pettit, R. *J. Am. Chem. Soc.* **1965**, *87*, 3254.
62. Snapper, M.L.; Tallarico, J.A.; Randall, M.L. *J. Am. Chem. Soc.* **1997**, *119*, 1478.
63. For a review see: Pearson, A.J., "Transition Metal Diene and Dienyl Complexes: Nucleophilic Attack on Diene and Dienyl Complexes" in "Comprehensive Organometallic Chemistry II", Abel, E.W.; Stone, F.G.A.; Wilkinson, G., Eds., Elsevier Science Ltd., Oxford, UK, 1995, Vol. 12, pp. 635-685.
64. (a) Semmelhack, M.F.; Herndon, J.W. *Organometallics* **1983**, *2*, 363. (b) Semmelhack, M.F.; Herndon, J.W.; Springer, J.P. *J. Am. Chem. Soc.* **1983**, *105*, 2497.
65. Balazs, M.; Stephenson, G.R. *J. Organomet. Chem.* **1995**, *498*, C17. For an intramolecular version, see: Yeh, M-C.P.; Chuang, L-W.; Ueng, C.H. *J. Org. Chem.* **1996**, *61*, 3874.
66. (a) Semmelhack, M.F.; Herndon, J.W.; Liu, J.K. *Organometallics* **1983**, *2*, 1885. (b) Semmelhack, M.F.; Le, H.T.M. *J. Am. Chem. Soc.* **1985**, *107*, 1455.
67. Yeh, M.C.P.; Hwu, C.C. *J. Organomet. Chem.* **1991**, *419*, 341.
68. Faller, J.W.; Murray, H.H.; White, D.L.; Chao, K.H. *Organometallics* **1983**, *2*, 400.
69. (a) Pearson, A.J.; Khan, M.N.I.; Clardy, J.C.; Ciu-Heng, H. *J. Am. Chem. Soc.* **1985**, *107*, 2748. (b) Pearson, A.J.; Khan, M.N.I. *Tetrahedron Lett.* **1984**, *25*, 3507. (c) Pearson, A.J.; Khan, M.N.I. *Tetrahedron Lett.* **1985**, *26*, 1407.
70. (a) Yeh, M.C.P.; Tsou, C-J.; Chuang, C-N.; Lin, H-W. *J. Chem. Soc., Chem. Comm* **1992**, 890. (b) Pearson, A.J.; Douglas, A.R. *Organometallics* **1998**, *17*, 1446.
71. Liebeskind, L.S.; Bombrun, A. *J. Am. Chem. Soc.* **1991**, *113*, 8736.
72. Hansson, S.; Miller, J.F.; Liebeskind, L.S. *J. Am. Chem. Soc.* **1990**, *112*, 9660.
73. For reviews see: Pearson, A.J. "Comprehensive Organic Synthesis", **1991**, *4*, 663. Also reference 63.
74. (a) Pearson, A.J. *Acc. Chem. Res.* **1980**, *13*, 463. (b) Pearson, A.J.; Richards, I.C. *Tetrahedron Lett.* **1983**, *24*, 2465.
75. Potter, G.A.; McCague, R. *J. Chem. Soc., Chem. Comm.* **1992**, 635.
76. Stephenson, G.R. *J. Organomet. Chem.* **1985**, *286*, C41.
77. (a) Knölker, H.-J.; Bauermeister, M.; Blaser, D.; Boese, R.; Pannek, J-B. *Angew. Chem. Int. Ed. Engl.* **1989**, *28*, 223. (b) Knölker, H.-J.; Bauermeister, M. *J. Chem. Soc., Chem. Comm.* **1990**, 664. (c) Knölker, H-J.; Fröhner, W. *Synlett* **1997**, 1108. (d) Knölker, H-J.; Fröhner, W. *Tetrahedron Lett.* **1998**, *39*, 2537.
78. For a review on the use of this chemistry in the synthesis of heterocycles see: Knölker, H.-J. *Synlett* **1992**, 371.
79. (a) Bandara, B.M.R.; Birch, A.J.; Kelley, L.F.; Khor, T.C. *Tetrahedron Lett.* **1983**, *24*, 2491. (b) Howell, J.A.S.; Thomas, M.J. *J. Chem. Soc. Dalton Trans* **1983**, 1401. Atton, J.G.; Evans, D.J.; Kane-Maguire, L.A.P.; Stephenson, G.R. *J. Chem. Soc. Chem. Comm.* **1984**, 1246.
80. Pearson, A.J.; Gehrmini, A.M.; Pinkerton, A.A. *Organometallics* **1992**, *11*, 936.
81. (a) Pearson, A.J.; Kole, S.L.; Yoon, J. *Organometallics* **1986**, *5*, 2075. (b) Pearson, A.J.; Ray, T. *Tetrahedron Lett.* **1986**, *27*, 3111. (c) Pearson, A.J.; Lai, Y-S.; Lu, W.; Pinkerton, A.A. *J. Org. Chem.* **1989**, *54*, 3882.
82. Donaldson, W.A. *J. Organomet. Chem.* **1990**, *395*, 187.
83. Laabassi, M.; Gree, R. *Bull Soc. Chim. Fr.* **1992**, *129*, 151.
84. For reviews see: (a) Lautens, M.; Klute, W.; Tam, W. *Chem. Rev.* **1996**, *96*, 49. (b) Frühauf, H-W. *Chem. Rev.* **1997**, *97*, 523.
85. Trost, B.M.; Yanai, M.; Hoogstein, K. *J. Am. Chem. Soc.* **1993**, *115*, 5294.
86. (a) Trost, B.M.; Hashmi, A.S.K. *Angew. Chem. Int. Ed. Engl.* **1993**, *32*, 1085. (b) Trost, B.M.; Hashmi, A.S.K. *J. Am. Chem. Soc.* **1994**, *116*, 2183.
87. Lautens, M.; Edwards, L.G.; Tam, W.; Lough, A.J. *J. Am. Chem. Soc.* **1995**, *117*, 10276 and references therein.
88. Lautens, M.; Tam, W.; Lautens, J.C.; Edwards, L.G.; Crudden, C.M.; Smith, A.C. *J. Am. Chem. Soc.* **1995**, *117*, 6863 and references therein.
89. Lautens, M.; Lautens, J.C.; Smith, A.C. *J. Am. Chem. Soc.* **1990**, *112*, 5627.
90. (a) Wender, P.B.; Smith, T.E. *J. Org. Chem.* **1996**, *61*, 824. (b) Wender, P.A.; Smith, T.E. *J. Org. Chem.* **1995**, *60*, 2962.
91. (a) Gilbertson, S.R.; Hoge, G.S. *Tetrahedron Lett.* **1998**, *39*, 2075. (b) Jolly, R.S.; Luedlke, G.; Sheehan, D.; Livinghouse, T. *J. Am. Chem. Soc.* **1990**, *112*, 4965.
92. McKinstry, L.; Livinghouse, T. *Tetrahedron Lett.* **1994**, *50*, 6145.
93. For a review see: Siebwith, S.N.; Arnard, N.T. *Tetrahedron* **1996**, *52*, 6251.
94. Jolly, P.W.; Wilke, G., "The Organic Chemistry of Nickel," Wiley, New York, 1975, Vol. 2, p. 94.

222

95. Wender, P.A.; Ihle, N.C.; Correa, C.R.D. *J. Am. Chem. Soc.* **1988**, *110*, 5904.

96. Wender, P.A.; Nuss, J.M.; Smith, D.B.; Swarez-Sobrino, A.; Vågberg, J.; DeCosta, D.; Bordner, J. *J. Org. Chem.* **1997**, *62*, 4908.

97. For a review see: Rigby, J.H. *Acc. Chem. Res.* **1993**, *26*, 579.

98. Rigby, J.H.; Kirova-Snovei, M. *Tetrahedron Lett.* **1997**, *38*, 8153.

99. Rigby, J.H.; Warshakoon, N.C. *Tetrahedron Lett.* **1997**, *38*, 2049.

100. Rigby, J.H.; Kirova, M.; Niyaz, N.; Mohanmadi, F. *Synlett* **1997**, 805.

101. Rigby, J.H.; Fales, K.R. *Tetrahedron Lett.* **1998**, *39*, 1525.

102. (a) Rigby, J.H.; Rege, S.D.; Sandanayaka, V.P.; Kirova, M. *J. Org. Chem.* **1996**, *61*, 843. (b) Rigby, J.H.; Hu, J.; Heeg, M.J. *Tetrahedron Lett.* **1998**, *39*, 2265.

103. Rigby, J.H.; Fiedler, C. *J. Org. Chem.* **1997**, *62*, 6106.

CHAPTER 8

Synthetic Applications of Transition Metal AlkyneComplexes

8.1   Introduction

Although virtually all transition metals react with alkynes, relatively few form simple, stable metal alkyne complexes analogous to the corresponding metal olefin complexes. This is because many alkyne complexes are quite reactive toward additional alkyne, and react further to produce more elaborate complexes or organic products. Those that are stable tend to be so stable that use in synthesis is precluded by their lack of reactivity. The successful development of transition-metal-alkyne complex chemistry for use in organic synthesis has involved procedures for the management of these extremes in reactivity.

8.2   Nucleophilic Attack on Metal Alkyne Complexes

In contrast to metal-olefin complexes nucleophilic attack on metal-alkyne complexes is relatively rare and has found little use in synthesis. Stable $CpFe(CO)_2^+$ alkyne complexes undergo attack by a wide range of nucleophiles, forming very stable σ-alkenyliron complexes, which can be made to insert CO by treatment with $Ag^{(+)}$ (Lewis Acid promoted insertion) (Eq. 8.1).[1] Although these complexes should have additional reactivity, little has been done with them.

Palladium(II) complexes catalyze the cyclization of homopropargyl alcohols and amines,[2] in a process that probably involves nucleophilic attack on a complexed alkyne (Eq. 8.2)[3] similar to that observed with palladium(II)-complexed alkenes (Chapter 7). However, the mechanism of this process has not been studied, and the putative protolytic cleavage of the σ-alkenylpalladium(II) complex by the protonated heterocycle is unusual. Indoles have been made by this type of amination of an alkyne (Eq. 8.3).[4]

Eq. 8.1

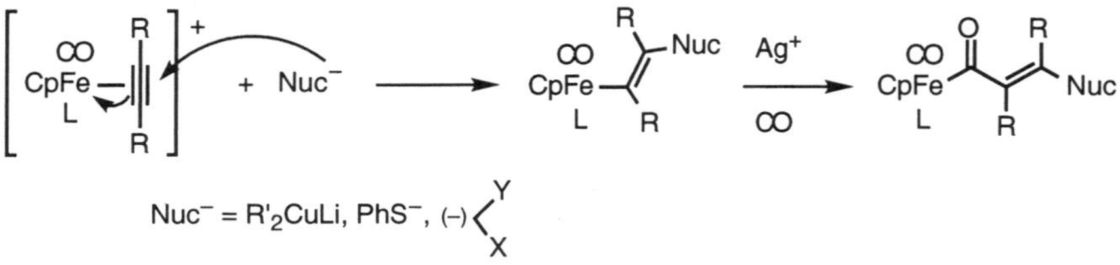

$$Nuc^- = R'_2CuLi, \ PhS^-, \ (-)\overset{Y}{\underset{X}{\diagdown}}$$

Eq. 8.2

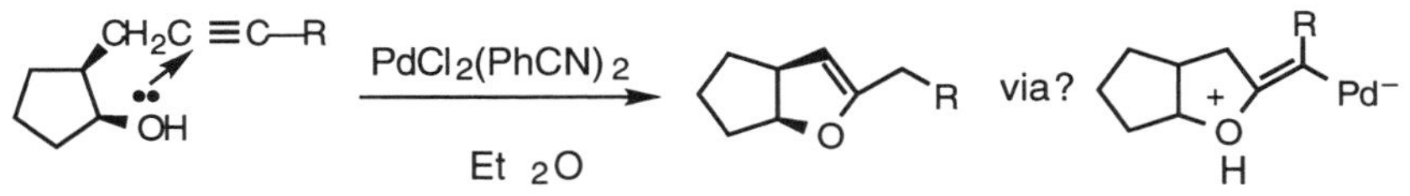

Eq. 8.3

60-80%

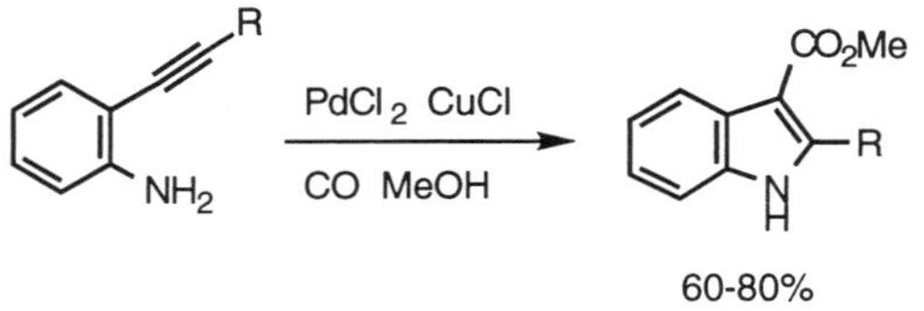

Palladium(0) complexes also catalyze a series of "alkylative" heterocyclizations of alkynes (Eq. 8.4[5] and Eq. 8.5[6]). However, these probably involve oxidative addition/insertion chemistry rather than nucleophilic attack on the alkyne, and are most closely related to the "Heck" reaction discussed in Chapter 7. Chloropalladation (palladium-catalyzed nucleophilic addition of Cl⁻ to alkynes) has been used to synthesize α-methylene lactones (Eq. 8.6).[7]

Eq. 8.4

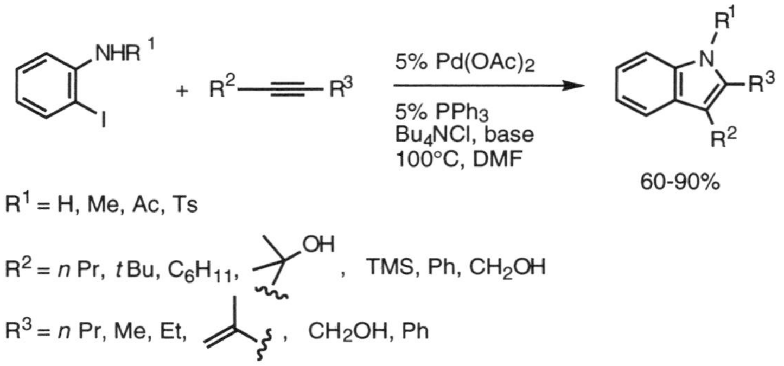

$R^1 = H, Me, Ac, Ts$

$R^2 = n \, Pr, \, t \, Bu, \, C_6H_{11}, \quad \overset{OH}{\diagup}, \quad TMS, \, Ph, \, CH_2OH$

$R^3 = n \, Pr, \, Me, \, Et, \quad \diagup\!\!\!\searrow, \quad CH_2OH, \, Ph$

60-90%

Eq. 8.5

50-80%

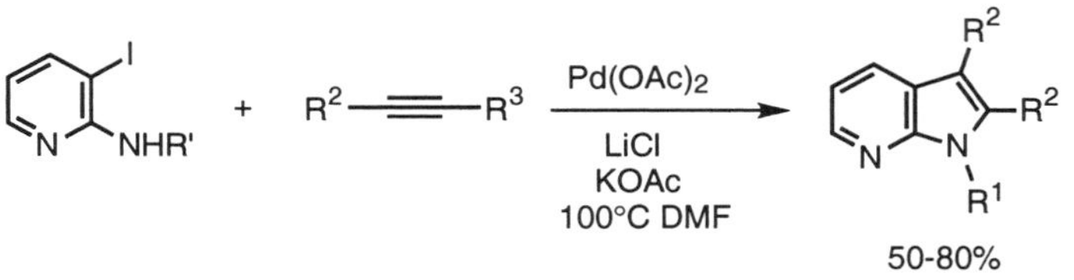

Eq. 8.6

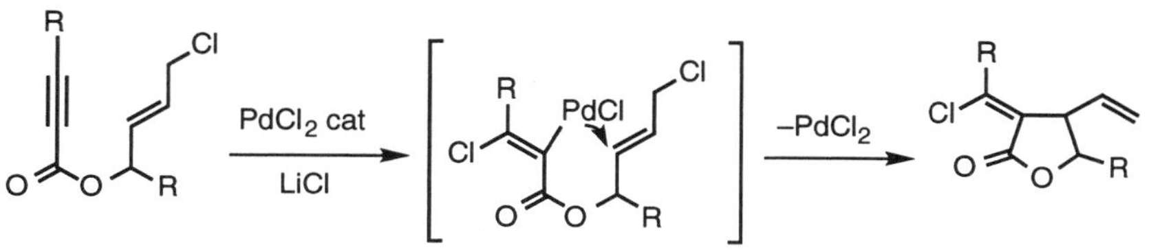

8.3 Stable Alkyne Complexes

a. As Alkyne Protecting Groups

In contrast to most other metals, dicobalt octacarbonyl forms very stable complexes with alkynes, which act as four electron donor, bridging ligands, perpendicular to the Co–Co bond (Eq. 8.7).[8] This complexation effectively reduces the reactivity of the alkyne to the extent that it can be used to protect alkynes from reduction or hydroboration (Eq. 8.8).[9] The alkyne can be regenerated by mild oxidative removal of the cobalt.

Eq. 8.7

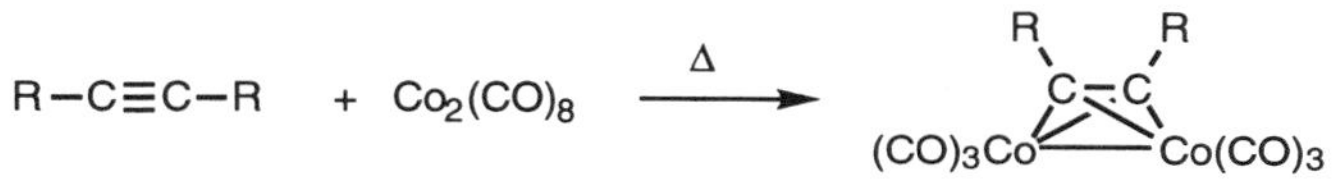

$$R-C\equiv C-R \quad + \quad Co_2(CO)_8 \quad \xrightarrow{\Delta} \quad$$

Eq. 8.8

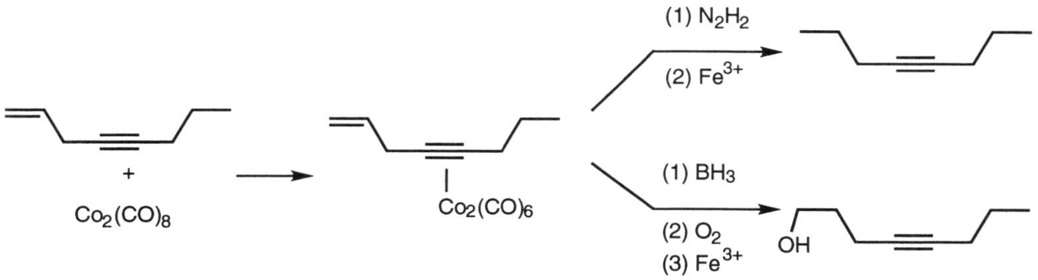

Complexation of alkynes to cobalt also stabilizes positive charge in the propargylic position, permitting synthetically useful reactions with propargyl cations (the Nicholas reaction) without the production of allenic byproducts (Eq. 8.9).[10] The cobalt-stabilized propargyl cations can be made from a variety of precursors including propargyl alcohols, ethers, epoxides, and even enynes by treatment with a Lewis acid. These undergo clean reaction with a variety of nucleophiles[11] to give the propargyl substitution product, from which the cobalt is removed by mild oxidation or by treatment with tetrabutylammonium fluoride in THF (3h, -10°C).[12]

Eq. 8.9

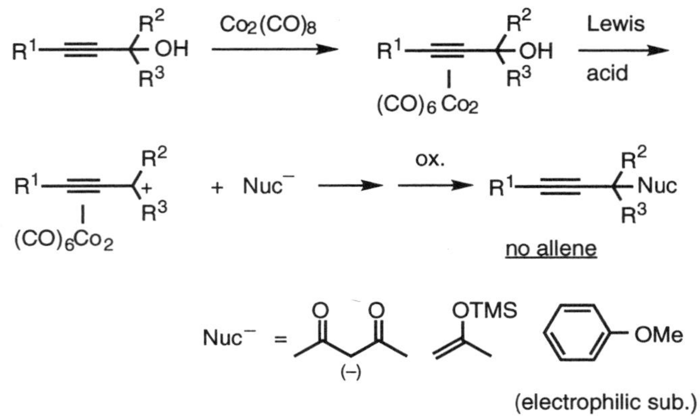

In addition, complexation of an alkyne to cobalt carbonyl also distorts its geometry towards that of an olefin. This distortion away from linearity allows the formation of relatively small cyclic alkynes via the Nicholas reactions (Eq. 8.10).[13] This process has also found extensive use in the synthesis of ene-diyne antitumor agents (Eq. 8.11),[14] and to promote more complex cyclization (Eq. 8.12).[15]

Eq. 8.10

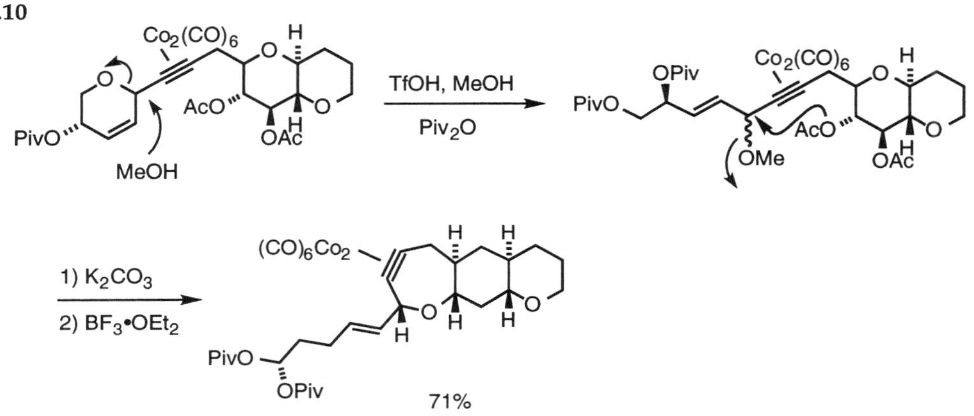

Eq. 8.11

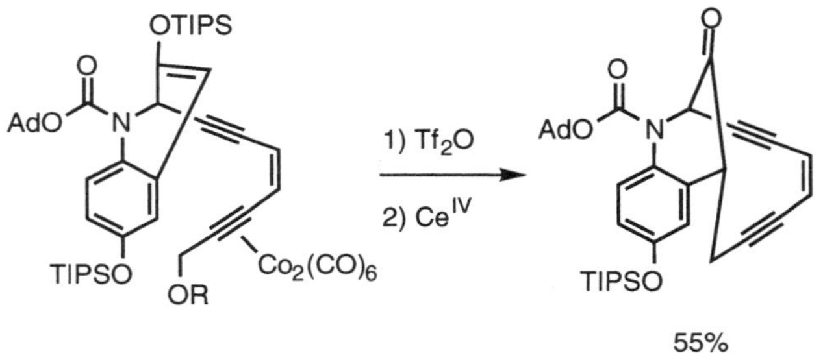

Eq. 8.12

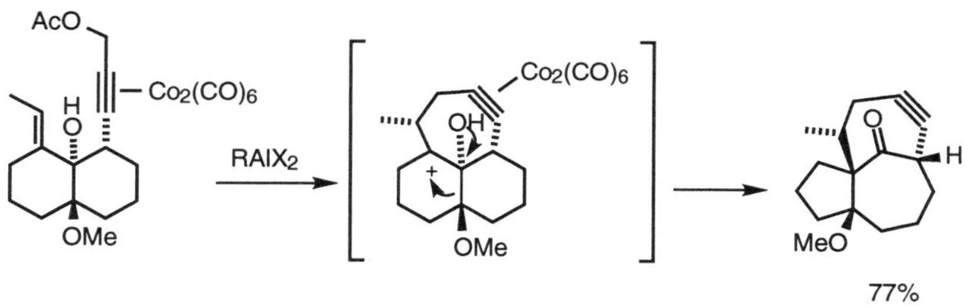

77%

With chiral boron enolates, high enantiomeric excesses can be obtained (Eq. 8.13).[13]

Eq. 8.13

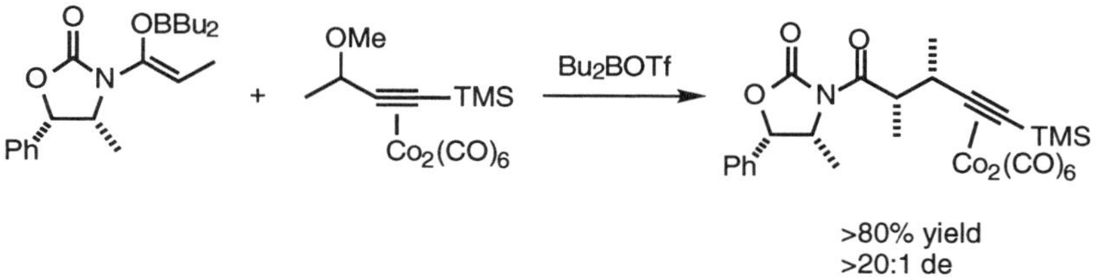

>80% yield
>20:1 de

Another use of alkynecobalt complexes is in the stereoselective aldol coupling of alkynyl aldehydes. Although free alkynyl aldehydes undergo aldol reaction with silyl enol ethers with little stereoselectivity, the complexed aldehydes react with high *syn* selectivity (Eq. 8.14).[17] With optically active allylboranes very

Eq. 8.14

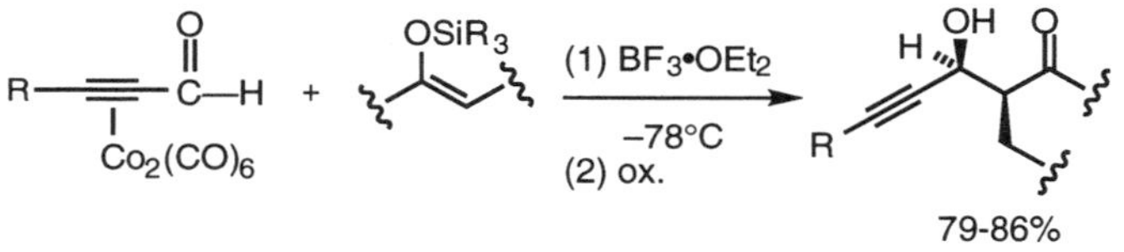

79-86%

high enantioselectivity as well as diastereoselectivity is observed and the cobalt can be readily removed by mild oxidation, giving the aldol product in good yield (Eq. 8.15).[18]

Eq. 8.15

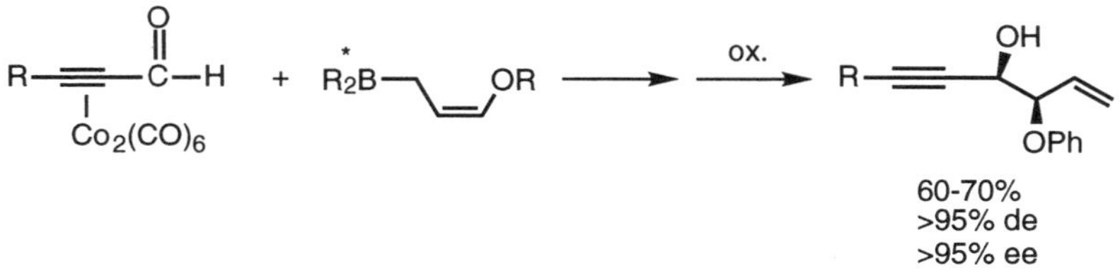

60-70%
>95% de
>95% ee

Finally, cobalt also stabilizes propargyl radicals, allowing these normally highly reactive species to be used in synthesis (Eq. 8.16).[19]

Eq. 8.16

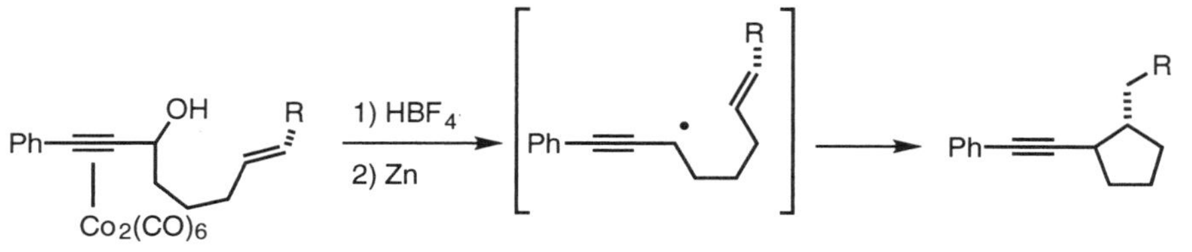

b. The Pauson-Khand Reaction[20]

Although dicobalt complexes of alkynes are quite stable to a wide variety of electrophilic and nucleophilic reagents, when they are heated in the presence of an alkene, an interesting and useful "2+2+1" cycloaddition (called the Pauson-Khand reaction after its originators), occurs, producing cyclopentenones (Eq. 8.17).[21] The reaction joins an alkyne, an alkene, and carbon monoxide in a regioselective manner,

Eq. 8.17

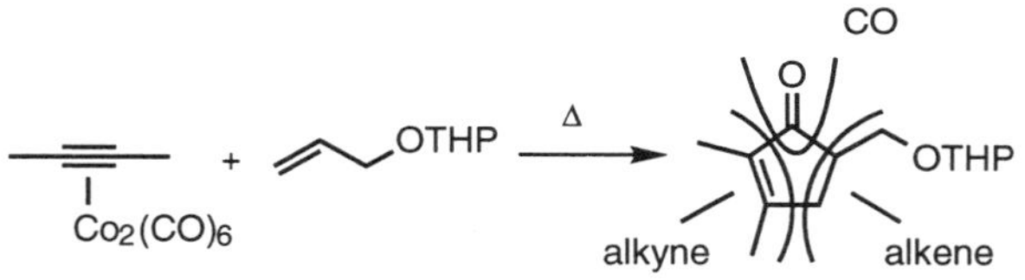

tending to place alkene substituents next to CO and large alkyne substituents adjacent to the CO, although there are many exceptions. The mechanism of this process has not been studied, but a likely one, consistent with the products formed, is shown in Eq. 8.18 and involves the (by now) expected loss of CO to generate a vacant coordination site, coordination of the alkene, insertion of the alkene into a Co–C bond, insertion of CO, and reductive elimination.

Eq. 8.18

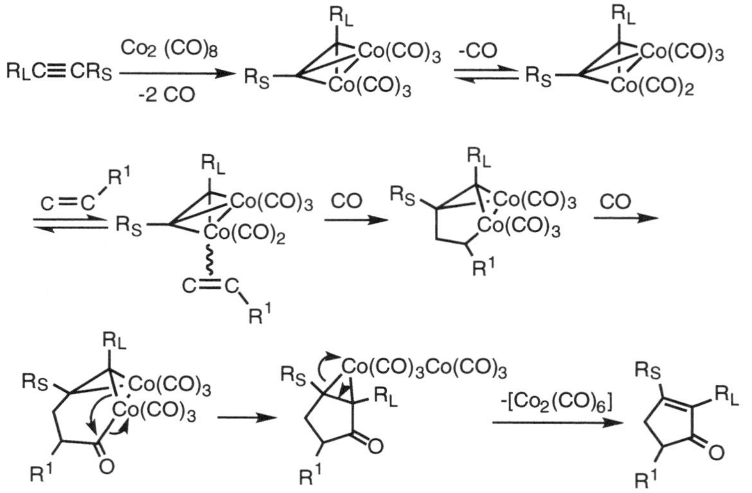

Initially, the synthetic utility of the process was compromised by the necessity of severe conditions and the resulting low yields. However, the reaction can be dramatically accelerated by addition of tertiary amine oxides,[22] or amines,[23] and the yields increased by immobilizing the system on polymers or on silica gel.[24] The reasons for this are obscure.

Although many intermolecular Pauson-Khand reactions have been reported, intramolecular versions are more synthetically interesting, and have been used to synthesize a variety of bicyclic systems (Eq. 8.19)[25] and (Eq. 8.20).[26] With chiral substrates, high diastereoselectivity can be achieved (Eq. 8.21),[27] (Eq. 8.22),[28] and (Eq. 8.23).[29] However, with a chiral auxilliary on the alkyne (chiral alkoxyalkyne or chiral propiolic ester) relatively low selectivity was observed.[30] For a very limited range of reactants (terminal alkynes, norbornene or norbornadiene) fair to excellent enantiomeric excess could be obtained utilizing a chiral phosphine ligand on the cobalt.[31]

Eq. 8.19

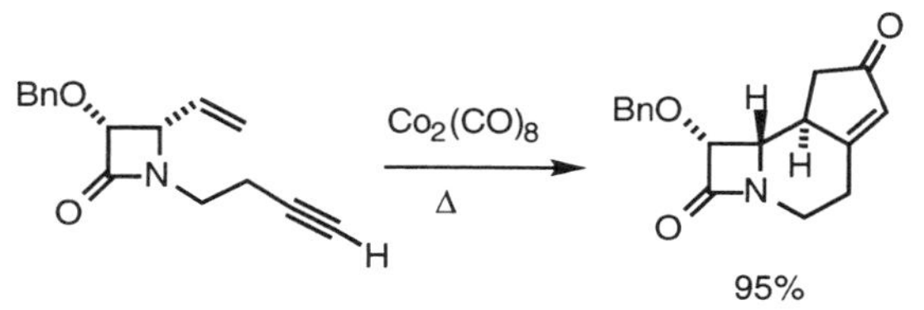

Eq. 8.20

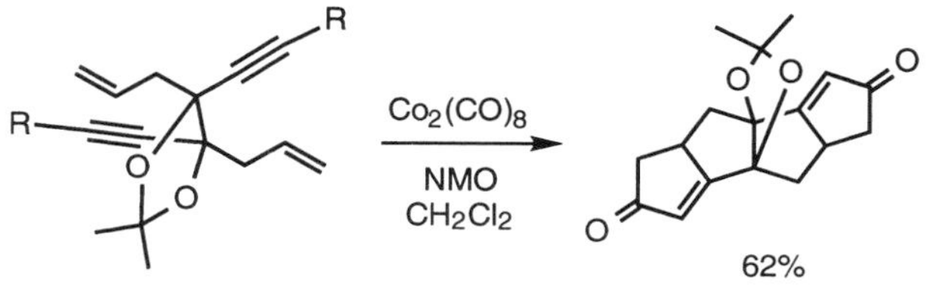

Eq. 8.21

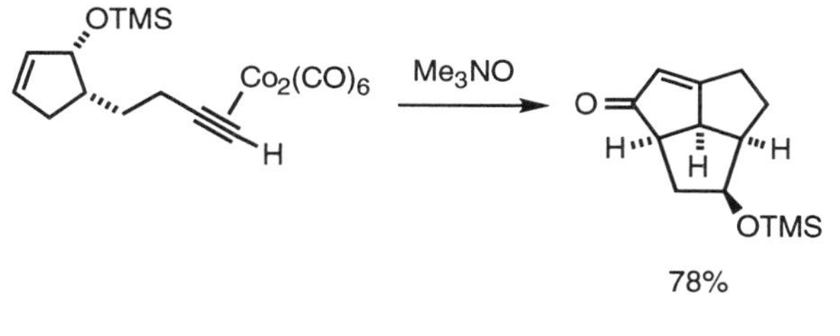

Eq. 8.22

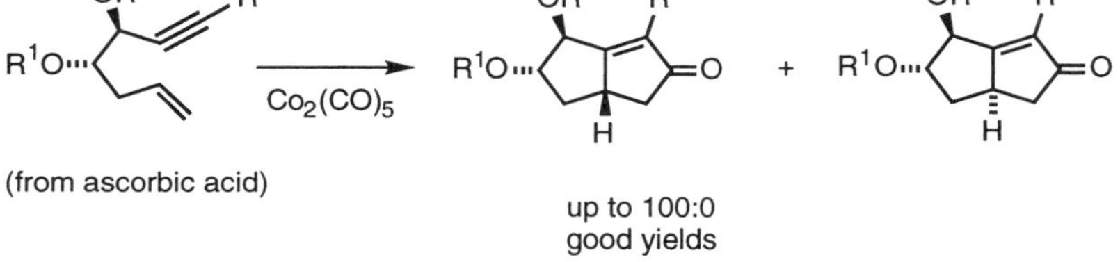

Eq. 8.23

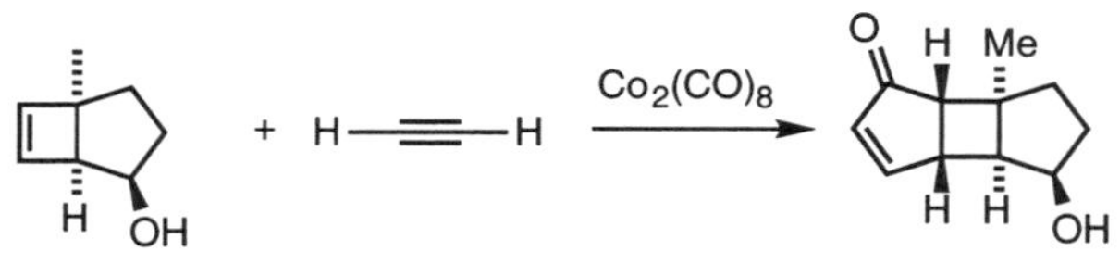

All of the above reactions rely on the use of stoichiometric amounts of dicobalt octacarbonyl. Recently a number of catalytic processes involving the use of catalytic amounts of $Co_2(CO)_8$ under high intensity visible light irradiation,[32] $Co_2(CO)_8$ in the presence of triphenyl phosphite,[33] or $Co_4(CO)_{12}$ under 10 at. of CO pressure[34] have been developed. In addition $[RhCl(CO)_2]_2$[35] and $Ru_3(CO)_{12}$[36,37] catalyze the Pauson-Khand reaction. All have been carried out on a limited range of substrates, usually that shown in Eq. 8.24 and all require carbon monoxide, usually at high pressure, but efficient catalytic systems for these very interesting transformations are indeed available.

Eq. 8.24

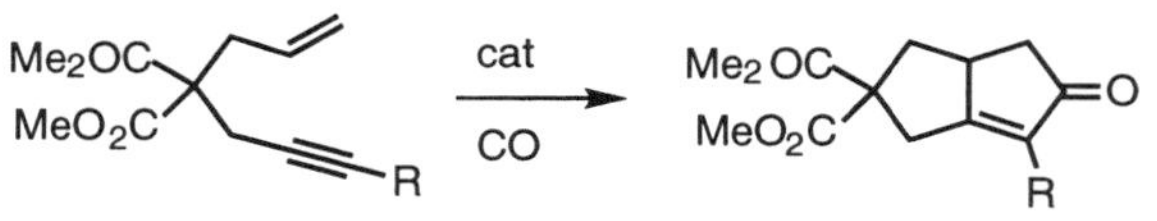

Since stable alkyne-cobalt complexes are involved in the Pauson-Khand reaction, the ability of cobalt complexation to stabilize propargyl cations can be utilized to synthesize precursors to the Pauson-Khand reaction, greatly expanding the scope of the process. An early example utilizing allylboranes as the nucleophile is shown in Eq. 8.25.[38] More recently, allylsilanes (Eq. 8.26)[39] have been used as nucleophiles. By taking advantage of the improved reaction conditions (R_3NO, SiO_2) mentioned earlier, good yields of complex products can be obtained. Note the high stereoselectivity of the allylborane reaction.

Eq. 8.25

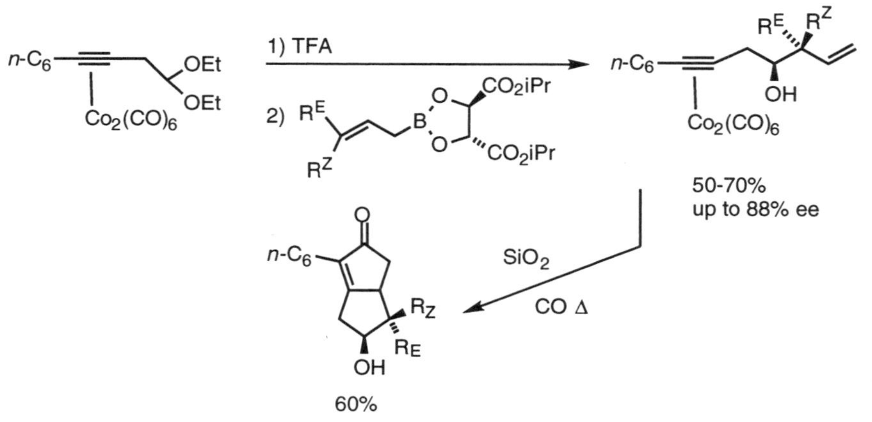

Eq. 8.26

82%

85%

Although intramolecular reactions tend to be quite regioselective, intermolecular reactions of disubstituted olefins often give mixtures of regioisomers. However, regiochemical control can often be regained by having a ligand — usually a heteroatom — in the homoallylic position of the alkene (Eq. 8.28).[40] A complete understanding of this process is not in hand, but substantial experimental data is available.[41]

Eq. 8.27

79%

homoallylic heteroatom best

single regioisomer

25:1

Eq. 8.28

40-90%

8.4 Metal-Catalyzed Cyclooligomerization of Alkynes

Although many transition metals catalyze the cyclotrimerization of alkynes to arenes, $CpCo(CO)_2$ is among the most efficient. The mechanism of the process has been studied extensively[42] and is shown in Eq. 8.29. It involves loss of CO from the catalyst (hence the requirement of relatively high temperatures to generate vacant

Eq. 8.29

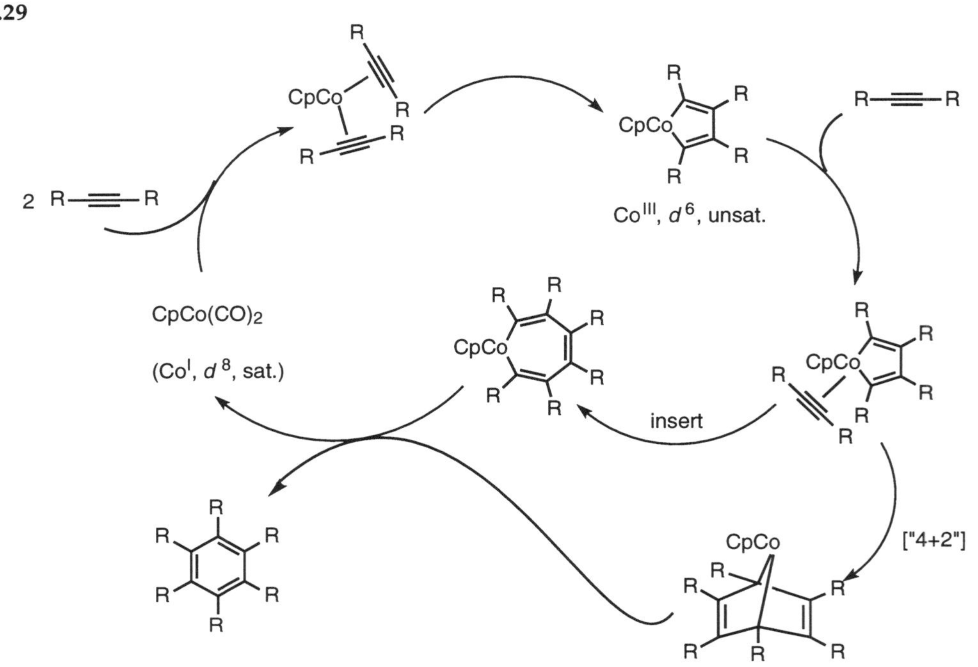

coordination sites), followed by coordination of two alkynes, and formation of the metallacyclopentadiene (a "reductive" coupling of the alkynes and concomitant "oxidation" of the metal). This unsaturated metallacyclopentadiene can then either coordinate and insert another alkyne, and undergo a reductive elimination to produce the arene and regenerate the catalytically active species, or it can undergo a 4+2 cycloaddition, achieving the same result.

Initially, the process was of little synthetic interest, since it was restricted to symmetrical, internal alkynes. Terminal alkynes gave all possible regioisomers of the trisubstituted benzenes, and "crossed" cyclotrimerization of two different alkynes gave all possible products, with no control.

An elegant solution to this latter problem and the basis of most synthetically useful applications of this methodology was to use a diyne as one component, and a huge alkyne, *bis*-trimethylsilylacetylene, as the other.[43] The diyne ensures that the metallacyclopentadiene is formed only from it, since incorporation of the second alkyne unit becomes intramolecular. The bulk of the TMS groups prevent self-trimerization of this member. These two components cleanly cocyclotrimerize to give benzocyclobutanes in excellent yield. As a bonus, these are thermally cleaved to give dienes which undergo Diels-Alder reactions to form polycyclic compounds (Eq. 8.30). When the dienophile is built into the diyne fragment, complex products are formed expeditiously as shown by the very direct synthesis of estrone (Eq. 8.31),[44] and related tricyclic compounds (Eq. 8.32)[45] by this methodology.

Eq. 8.30

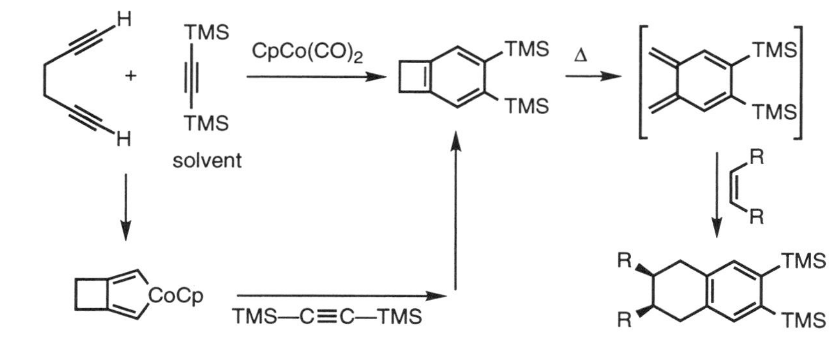

Eq. 8.31

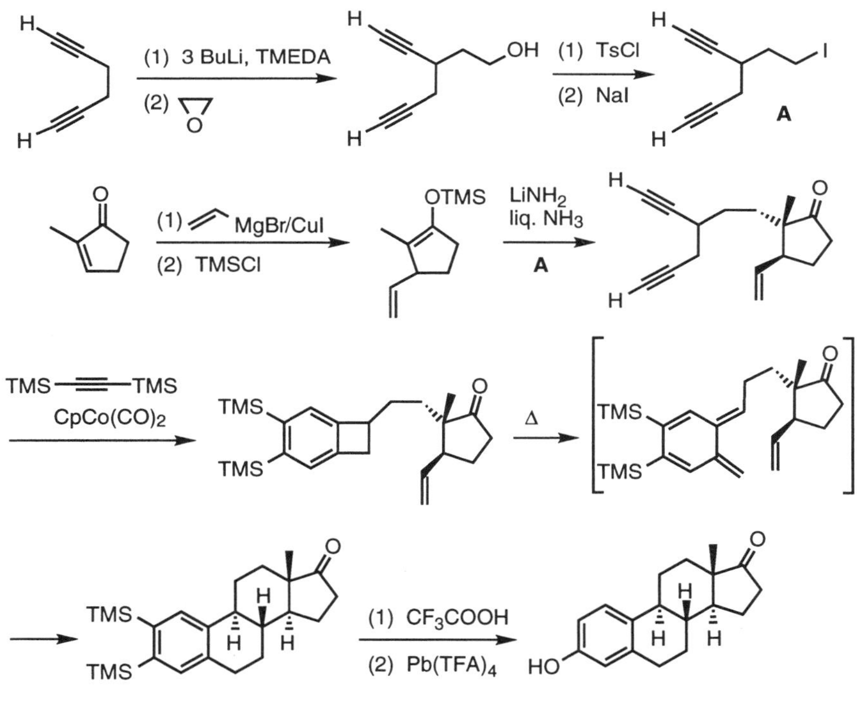

Eq. 8.32

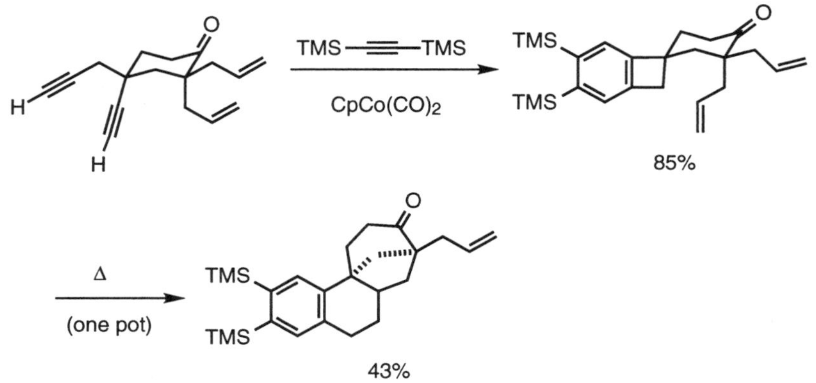

85%

43%

Although CpCo(CO)$_2$ is by far the most extensively studied alkyne cyclotrimerization system catalyst, many other metals will effect the reaction efficiently. Wilkinson's catalyst (Ph$_3$P)$_3$RhCl cocyclotrimerizes diynes with alkynes, including acetylene itself (Eq. 8.33)[46] and (Eq. 8.34),[47] as do nickel(0) complexes (Eq. 8.35).[48] Palladium(0) complexes catalyze the cocyclotrimerization of enynes with diynes (Eq. 8.36),[49] while specific hexasubstituted arenes are available by stepwise zirconium-assisted cyclotrimerization of three different alkynes (Eq. 8.37).[50]

Eq. 8.33

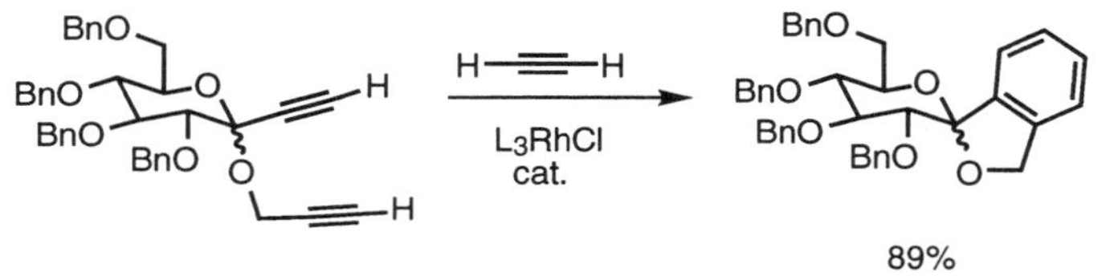

89%

Eq. 8.34

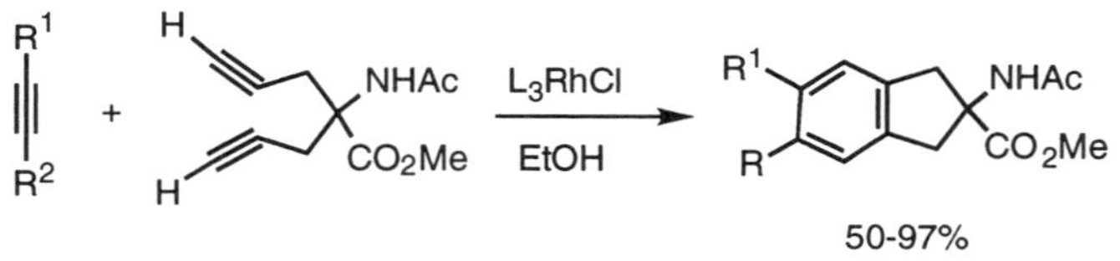

50-97%

Eq. 8.35

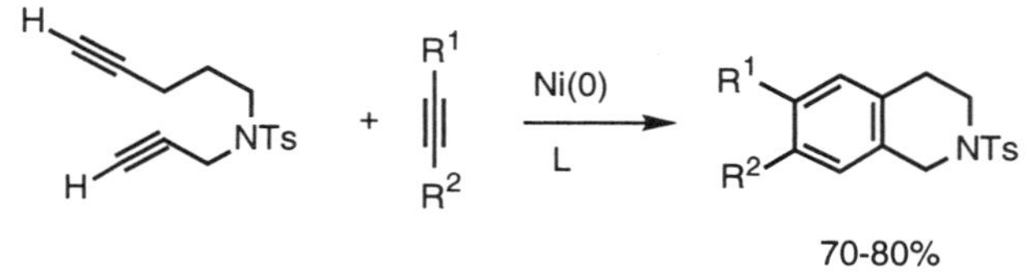

70-80%

Eq. 8.36

60-80%

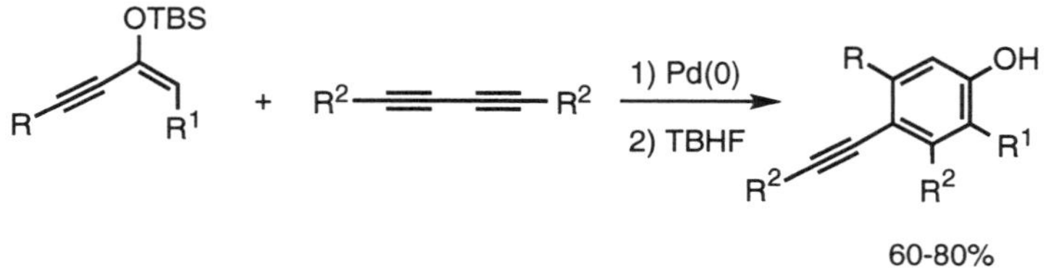

Eq. 8.37

The nitrile group is another triply-bonded species which does not self-trimerize, but can be cotrimerized with alkynes to give pyridines. As with simple alkyne cyclotrimerization, unsymmetric alkynes give all possible regioisomers. However, cocyclotrimerization with a diyne proceeds cleanly to a single pyridine (Eq. 8.38).[51] The process also works simple alkynes (Eq. 8.39)[52] and (Eq. 8.40).[53]

Eq. 8.38

Eq. 8.39

41%

Eq. 8.40

Alkenes will cocyclotrimerize with alkynes provided homotrimerization of the alkyne can be suppressed. This can be achieved by preforming the metallacyclopentadiene, and then adding the alkene (Eq. 8.41),[54] or by carrying the reaction out partially (Eq. 8.42)[55] or completely (Eq. 8.43[56] and Eq. 8.44[57]) intramolecularly.

Eq. 8.41

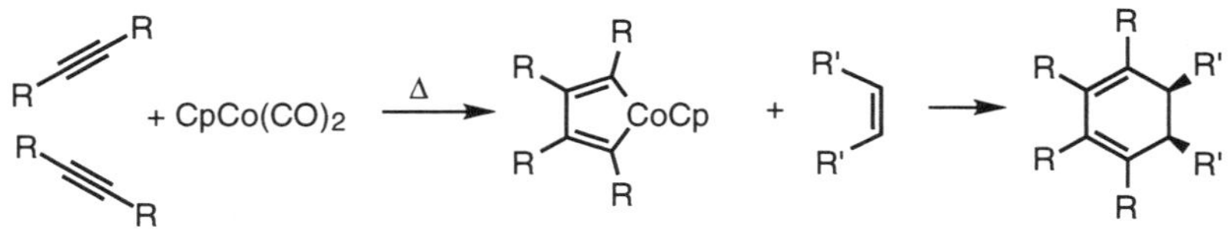

Eq. 8.42

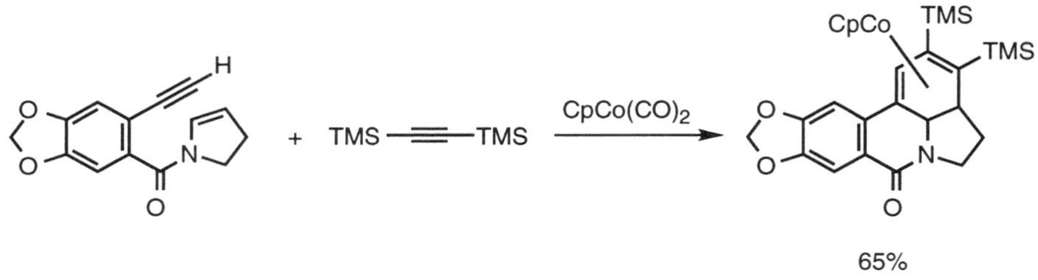

65%

Eq. 8.43

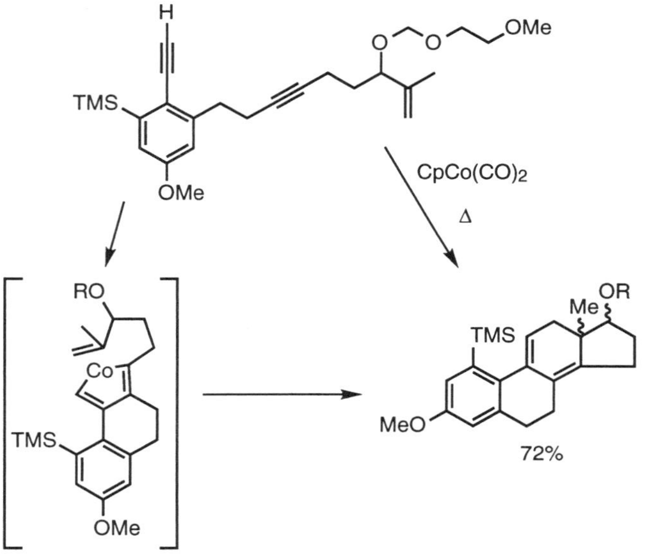

72%

Eq. 8.44

n = 1, 2

90-92%

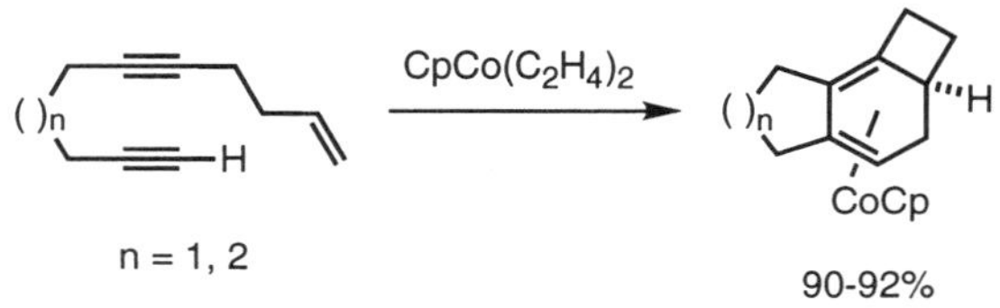

Isocyanates can fulfill the role of the double-bonded species in these cocyclotrimerizations, resulting in the production of pyridones (Eq. 8.45).[58] Nickel(0)

complexes also promote the cyclocoupling of alkynes with isocyanates (Eq. 8.46).[59] This reaction proceeds through a stable metallacycle in which one alkyne and one isocyanate have been coupled. Depending on the stoichiometry another alkyne can then insert to give the pyridone, or another isocyanate can insert to give the pyrimidine. This metallacycle can also react with CO or acids to give imides or conjugated amides (Eq. 8.47).[59]

Eq. 8.45

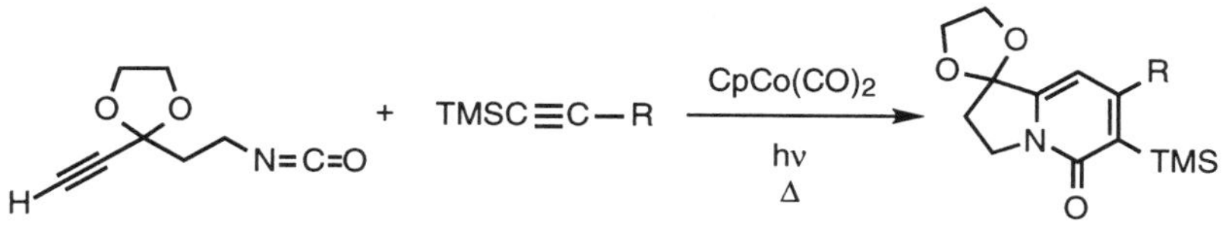

Eq. 8.46

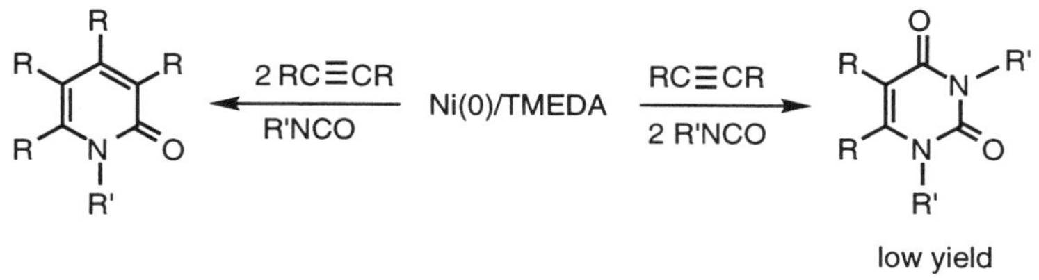

low yield

Eq. 8.47

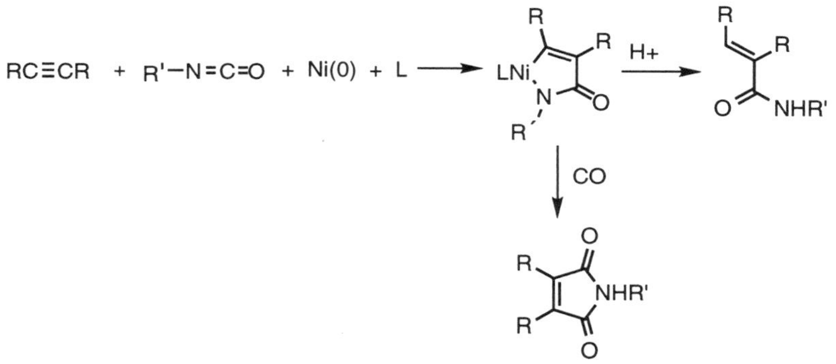

Nickel(0) complexes also "reductively couple" alkynes with carbon dioxide to give stable metallacycles. These can insert another alkyne, a molecule of carbon monoxide, or undergo protolytic cleavage (Eq. 8.48).[60] This process has not yet been used to synthesize complex molecules. However, it has been utilized to make polymers (Eq. 8.49).[61]

Alkynes react with many transition metals in the presence of carbon monoxide to produce compounds containing both carbon monoxide and alkyne-derived fragments. In addition, a profusion of organometallic complexes containing ligands constructed of alkyne-derived fragments and carbon monoxide are often obtained.[62] Reactions of this type are usually quite complex, and few have been controlled sufficiently to be of use in the synthesis of organic compounds. Quinones had long been known to be one of the products of the reaction of alkynes with

Eq. 8.48

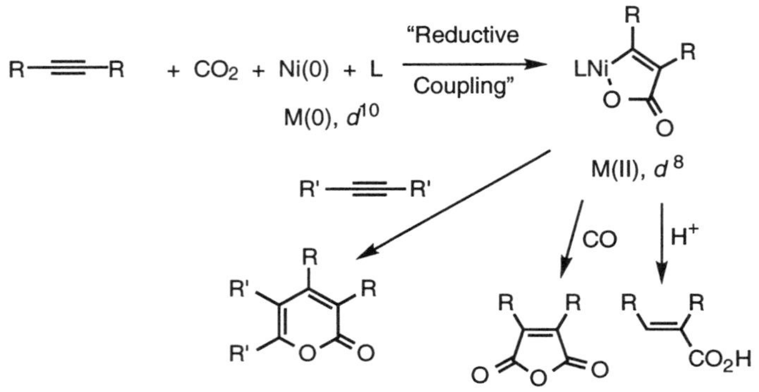

Eq. 8.49

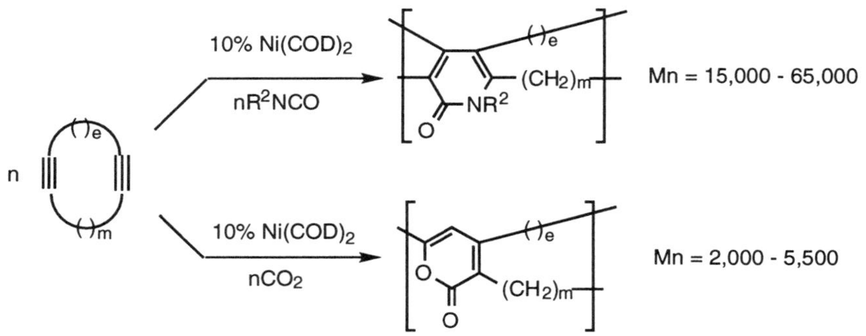

carbon monoxide, and evidence for the intermediacy of maleoyl complexes was strong (Eq. 8.50). However these processes were of little synthetic value, because a myriad of other organic and organometallic products formed as well. By devising an

Eq. 8.50

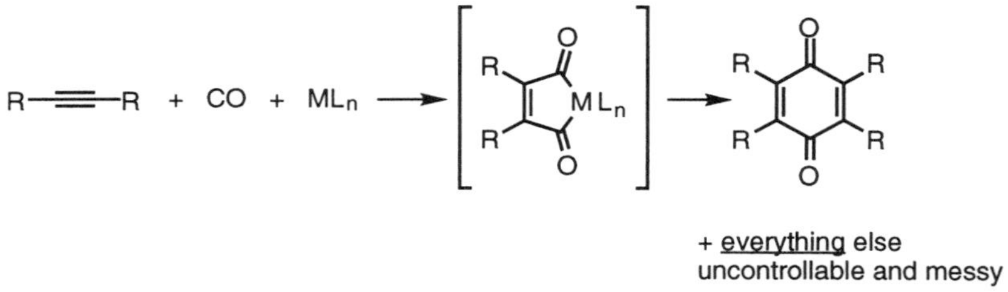

+ <u>everything</u> else
uncontrollable and messy

efficient synthesis of maleoyl and phthaloyl complexes, and permitting these discrete complexes to react with alkynes, quinones can be made efficiently and in high yield. The most efficient phthaloyl complex is that of cobalt, prepared by the reaction of benzocyclobutanediones with $CoCl(PPh_3)_3$.[63] The initially formed *bis* phosphine cobalt(III) complex is unreactive toward alkynes, since a labile site axial to the phthaloyl plane is required for alkyne complexation. However, treatment with silver(I) to produce the unsaturated cationic complex[64] or better, with dimethylglyoxime in pyridine to produce an octahedral complex with a labile *cis* pyridine ligand (x-ray) leads to active phthaloyl complexes which are efficiently converted to naphthoquinones upon reaction with a variety of alkynes (Eq. 8.51). In a similar manner, benzoquinones are efficiently synthesized from cyclobutenediones (Eq. 8.52[65] and Eq. 8.53[66]). With unsymmetric systems, mixtures of regioisomers are often obtained.

Eq. 8.51

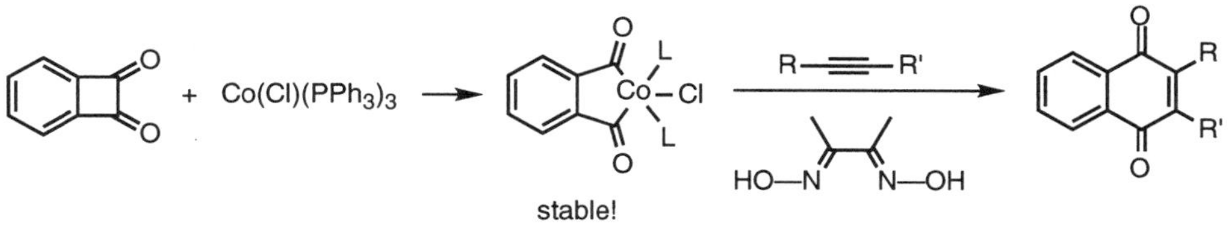

Eq. 8.52

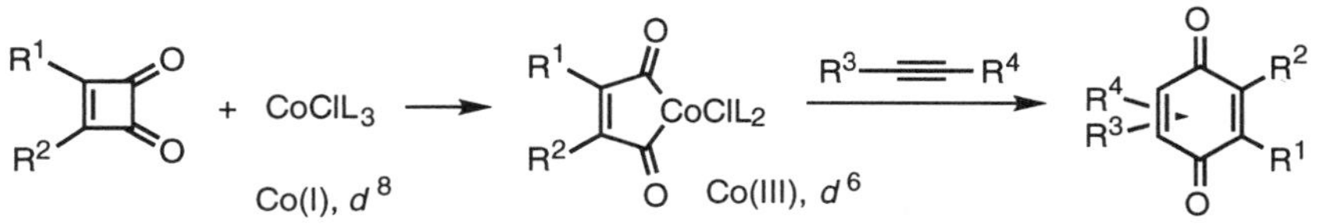

Eq. 8.53

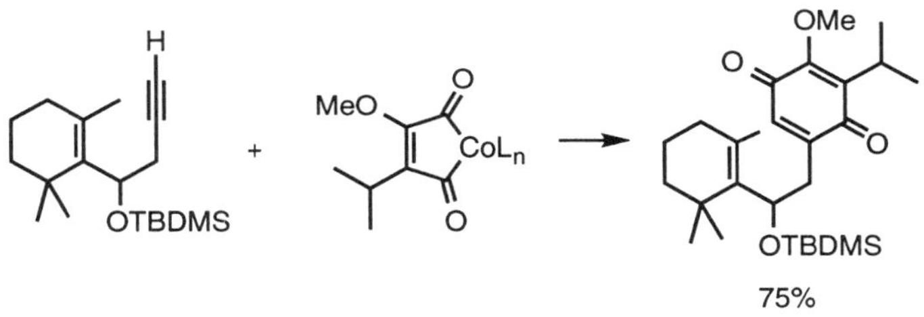

75%

8.5   Zirconobenzyne Reactions[67]

All of the alkyne chemistry discussed above involves late transition metals, with their normal manifold of reactions. However, the early transition metals also have a rich chemistry with alkynes (see for example Chapter 4) reductively dimerizing them, among other things. Zirconium forms discrete, characterizable complexes of *benzyne*, produced by treatment of aryllithium reagents with $Cp_2Zr(Me)Cl$ followed by elimination of methane (Eq. 8.54). These reactive benzyne complexes undergo a range of insertion reactions, which ultimately result in functionalization of the arene ring. When unsymmetric aryne complexes are used, these insertions are not regiospecific but insertion into the sterically less-hindered Zr–C bond is favored. Thus, isonitriles insert to give azazirconacyclopentadienes that can be cleaved to aryl ketones or iodoaryl ketones (Eq. 8.54).[68] Alkynes insert to give zirconacyclopentadienes, which are converted to benzothiophenes by treatment with sulfur dichloride.[69] Alkenes also insert to give zirconazacyclopentenes, which can be converted to a variety of substituted systems by cleavage with electrophiles (Eq. 8.55).[70,71]

Eq. 8.54

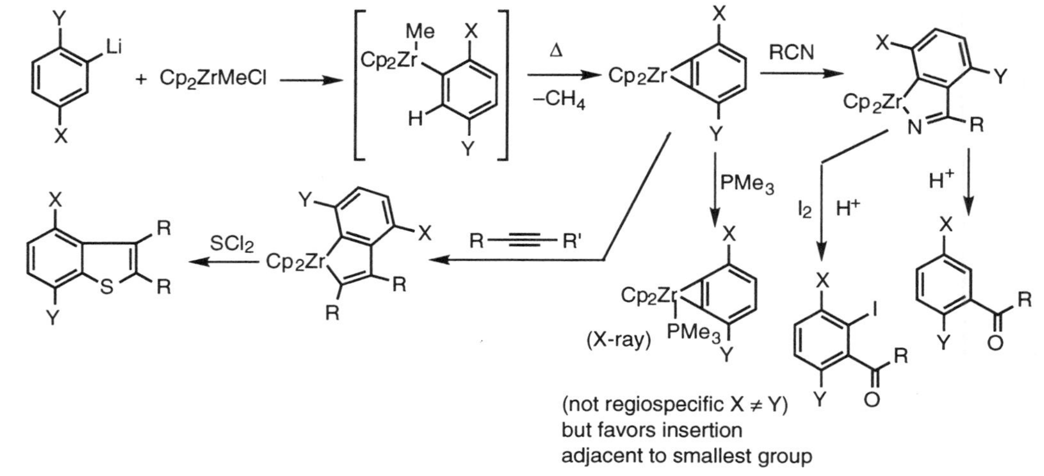

Eq. 8.55

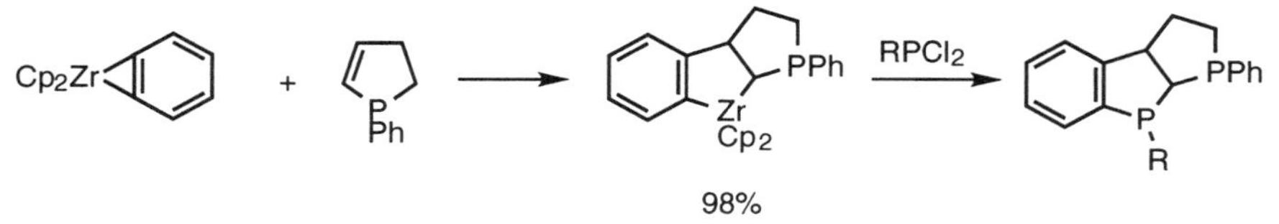

A variety of functionalized indoles can be made from zirconabenzyne intermediates (Eq. 8.56)[72] utilizing insertion chemistry similar to that in Eq. 8.54.

Eq. 8.56

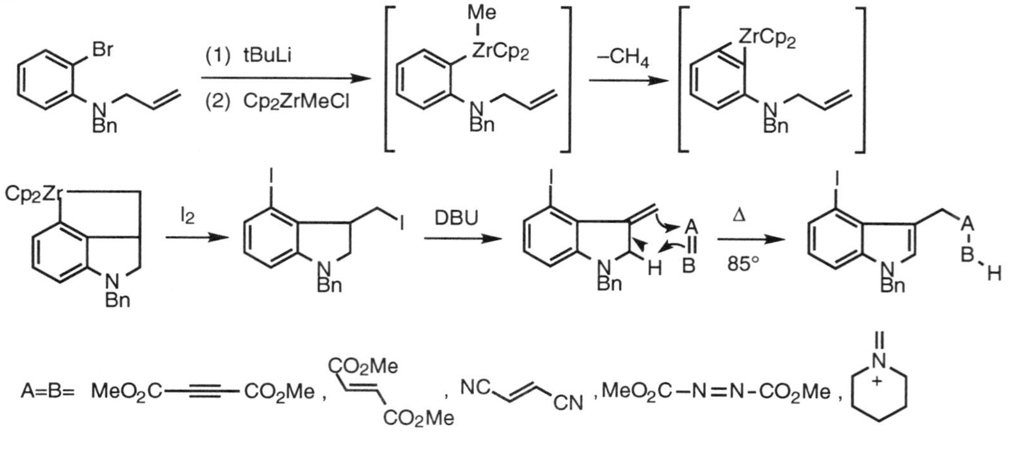

References

1. (a) Reger, D.L.; Belmore, K.A.; Mintz, E.; McElligot, P.J. *Organometallics* **1984**, *3*, 134. (b) Reger, D.L.; Mintz, E. *Organometallics* **1984**, *3*, 1759.
2. For a review see: Müller, T.E.; Beller, M. *Chem. Rev.* **1988**, *98*, 675.
3. Utimoto, K. *Pure Appl. Chem.* **1983**, *55*, 1845.
4. Kondo, Y.; Shiga, F.; Murata, N.; Sakamoto, T.; Yamanaka, H. *Tetrahedron* **1994**, *50*, 11803.
5. Larock, R.C.; Yum, E.K. *J. Am. Chem. Soc.* **1991**, *113*, 6689.
6. Park, S.S.; Chgoi, J-K.; Yum, E.K. *Tetrahedron Lett.* **1998**, *39*, 627.
7. Wang, Z.; Lu, X. *Tetrahedron Lett.* **1997**, *38*, 5213. For a review see: Lu, X.; Zhu, G.; Wang, Z.; Ma, S.; Ji, J.; Zhang, Z. *Pure Appl. Chem.* **1997**, *69*, 553.
8. Dickson, R.S.; Fraser, P.J. *Adv. Organomet. Chem.* **1974**, *12*, 323.
9. Nicholas, K.M.; Pettit, R. *Tetrahedron Lett.* **1971**, 3475.
10. For a review see: Caffyn, A.; Nicholas, K.M., "Transition Metal Alkyne Complexes. Transition Metal Stabilized Propargyl Systems" in "Comprehensive Organometallic Chemistry II", Abel, E.W.; Stone, F.G.A.; Wilkinson, G., Eds., Elsevier Science Ltd., Oxford, UK, 1995, Vol. 12, pp. 685-702.
11. For an extensive study of the reactions of cobalt-stabilized propargyl cations with a broad range of nucleophiles, see: Kuhn, O.; Raw, D.; Mayr, H. *J. Am. Chem. Soc.* **1998**, *120*, 900.
12. Jones, G.B.; Wright, J.M.; Rush, T.M.; Plourok, G.W.; Kelton, T.F.; Mathews, J.E.; Huber, R.S.; Davidson, J.P. *J. Org. Chem.* **1997**, *62*, 9379.
13. (a) Isobe, M.; Tsuboi, T.; Hosokawa, S.; Bamba, M.; Kira, K.; Shibuya, S.; Ohtani, I.I.; Nishikowa, T.; Ichikawa, Y. *Pure and Appl. Chem.* **1997**, *69*, 401. (b) Yengai, C.; Isobe, M. *Tetrahedron* **1998**, *59*, 2509.
14. For a review of these classes of compounds, including a discussion of the Nicholas reaction in this context see: (a) Nicolaou, K.C.; Dai, W.M. *Angew. Chem. Int. Ed. Engl.* **1991**, *30*, 1387. (b) Magnus, P.; Pitterna, T. *J. Chem. Soc., Chem. Comm.* **1991**, 541. (c) Magnus, P.; Miknis, G.F.; Press, N.J.; Grandjean, D.; Taylor, G.M.; Harling, J. *J. Am. Chem. Soc.* **1997**, *119*, 6739. (d) Magnus, P.; Eisenbeis, S.A.; Fourhurst, R.H.; Ikadis, T.; Magnus, T.A.; Parry, D. *J. Am. Chem. Soc.* **1997**, *119*, 5591.
15. Nakamura, T.; Matsui, T.; Tamino, K.; Kuwajima, I. *J. Org. Chem.* **1997**, *62*, 3032.
16. (a) Jacobi, P.A.; Buddu, S.C.; Fry, M D.; Rajeswari, S. *J. Org. Chem.* **1997**, *62*, 2894. (b) Jacobi, P.A.; Herradina, P. *Tetrahedron Lett.* **1997**, *38*, 6621.
17. For a review see: Mukai, C.; Hanaoka, M. *Synlett* **1996**, 11.
18. Ganesh, P.; Nicholas, K.M. *J. Org. Chem.* **1997**, *62*, 1737.
19. Salazar, K.L.; Khan, M.A.; Nicholas, K.M. *J. Am. Chem. Soc.* **1997**, *119*, 9053.
20. For reviews see: (a) Schore, N.E. *Org. React.* **1991**, *40*. 1. (b) Schore, N.E., "Transition Metal Alkyne Complexes: Pauson-Khand Reactions" in "Comprehensive Organometallic Chemistry II", Abel, E.W.; Stone, F.G.A.; Wilkinson, G., Eds., Elsevier Science Ltd., Oxford, UK, 1995, Vol. 12, pp. 703-741. (c) Gao, O.; Schmalz, H-O. *Angew. Chem. Int. Ed. Engl.* **1998**, *37*, 911.
21. Billington, D.C.; Pauson, P.L. *Organometallics* **1982**, *1*, 1560.
22. Shambayati, S.; Crowe, W.E.; Schreiber, S.L. *Tetrahedron Lett.* **1990**, *31*, 5289.
23. Rajesh, T.; Periasamy, M. *Tetrahedron Lett.* **1998**, *39*, 117.
24. (a) Smit, W.D.; Kireev, S.L.; Nefedov, O.M.; Tarasov, V.A. *Tetrahedron Lett.* **1989**, *30*, 4021. (b) Becker, D.P.; Flynn, T. *Tetrahedron Lett.* **1993**, *34*, 2087.
25. Alcoude, B.; Polanco, C.; Sierra, M.A. *Tetrahedron Lett.* **1996**, *37*, 6901.
26. (a) Van Ornum, S.G.; Cook, J.M. *Tetrahedron Lett.* **1997**, *38*, 3657. (b) Van Ornum, S.G.; Bruendel, M.M.; Cooke, J.M. *Tetrahedron Lett.* **1998**, *37*, 6649.
27. Paquette, L.A.; Borrelly, S. *J. Org. Chem.* **1995**, *60*, 6912.
28. Mukai, C.; Uchiyama, M.; Sakamoto, S.; Hanaoka, M. *Tetrahedron Lett.* **1995**, *36*, 5761.
29. Kowalczyk, B.A.; Smith, T.C.; Dauben, W.G. *J. Org. Chem.* **1998**, *63*, 1379.
30. (a) Bernardes, V.; Kam, N.; Riera, A.; Moyano, A.; Pericas, M.A.; Green, A.E. *J. Org. Chem.* **1995**, *60*, 6670. (b) Fonquerna, S.; Moyano, A.; pericas, M.A.; Riera, A. *Tetrahedron* **1995**, *51*, 4639.
31. Hay, A.M.; Kerr, W.J.; Kirk, G.G.; Middlemiss, D. *Organometallics* **1995**, *14*, 4986.
32. Pagendorf, B.L.; Livinghouse, T. *J. Am. Chem. Soc.* **1996**, *118*, 2285.
33. Jeong, N.: Hwang, S.H.; Lee, Y.; Chung, Y.K. *J. Am. Chem. Soc.* **1994**, *116*, 3159.
34. Kim, J.W.; Chung, Y.K. *Synthesis* **1998**, 142.
35. Koga, Y.; Kobayashi, T.; Narasaka, K. *Chem. Lett.* **1998**, 249.
36. Morimoto, T.; Chatani, N.; Fukumoto, Y.; Murai, S. *J. Org. Chem.* **1997**, *62*, 3762.
37. Kondo, T.; Suzuki, N.; Okada, T.; Mitsudo, T. *J. Am. Chem. Soc.* **1997**, *119*, 6187.
38. Roush, W.R.; Park, J-C. *Tetrahedron Lett.* **1991**, *32*, 6285.

39. Jameson, T.F.; Shambayati, S.; Crowe, W.E.; Schreiber, S.L. *J. Am. Chem. Soc.* **1997**, *119*, 4353.

40. Krafft, M.E.; Juliano, C.A.; Scott, I.L.; Wright, C.; McEachin, M.P. *J. Am. Chem. Soc.* **1991**, *113*, 1693.

41. (a) Krafft, M.E.; Juliano, C.A. *J. Org. Chem.* **1992**, *57*, 5106. (b) Krafft, M.E.; Scott, I.L.; Romulo, R.H. *J. Am. Chem. Soc.* **1993**, *115*, 7199.

42. (a) Wakatsuki, Y.; Kuramitsu, T.; Yamazaki, H. *Tetrahedron Lett.* **1974**, 4549. (b) McAllister, D.R.; Bercaw, J.E.; Bergman, R.G. *J. Am. Chem. Soc.* **1977**, *99*, 1666.

43. For reviews see: (a) Vollhardt, K.P.C. *Acc. Chem. Res.* **1977**, *10*, 1. (b) *Angew. Chem. Int. Ed. Engl.* **1984**, *23*, 539. (c) Grotjahn, D.B., "Transition Metal Alkyne Complexes: Transition Metal Catalyzed Cyclotrimerization", in "Comprehensive Organometallic Chemistry II", Abel, E.W.; Stone, F.G.A.; Wilkinson, G., Eds., Elsevier Science Ltd., Oxford, UK, 1995, Vol. 12, pp. 741-770, 1995. (d) Malacria, M. *Chem. Rev.* **1996**, *96*, 289. (e) Ojima, I.; Tzamarcoudaki, M.; Li, Z.; Donovan, R.J. *Chem. Rev.* **1996**, *96*, 635.

44. Funk, R.L.; Vollhardt, K.P.C. *J. Am. Chem. Soc.* **1980**, *102*, 5253.

45. Cruciani, P.; Stammler, R.; Aubert, C.; Malacria, M. *J. Org. Chem.* **1996**, *61*, 2699.

46. McDonald, F.E.; Zhu, H.Y.H.; Holmquist, C.R. *J. Am. Chem. Soc.* **1995**, *117*, 6605.

47. Kotha, S.; Brahmackary, E. *Tetrahedron Lett.* **1997**, *38*, 3561.

48. Sato, Y.; Nishimata, T.; Mori, M. *Heterocycles* **1997**, *44*, 443.

49. (a) Weibel, D.; Gevorgyn, V.; Yamamoto, Y. *J. Org. Chem.* **1998**, *63*, 1217. (b) Gevorgyn, V.; Quan, L.G.; Yamamoto, Y. *J. Org. Chem.* **1998**, *63*, 1244.

50. Takahashi, T.; Xi, Z.; Yamazuki, A.; Liu, Y.; Nakajima, K.; Kotara, M. *J. Am. Chem. Soc.* **1998**, *120*, 1672.

51. (a) Nalman, A.; Vollhardt, K.P.C. *Angew. Chem. Int. Ed. Engl.* **1977**, *16*, 708. (b) For a review see: Bonnemann, H.; Brijoux, W. *Adv. Het. Chem.* **1990**, *48*, 177.

52. (a) Boese, R.; Harvey, D.F.; Malaska, M.J.; Vollhardt, K.P.C. *J. Am. Chem. Soc.* **1994**, *116*, 11153. (b) Boese, R.; van Sickle, A.P.; Vollhardt, K.P.C. *Synthesis* **1994**, 1375.

53. Cheluri, G. *Tetrahedron Asymmetry* **1995**, *6*, 811.

54. Wakatsuki, Y.; Yamazaki, H. *J. Organomet. Chem.* **1977**, *139*, 169.

55. Grotjahn, D.B.; Vollhardt, K.P.C. *Synthesis* **1993**, 579.

56. Butenschon, H.; Winkler, M.; Vollhardt, K.P.C. *Chem. Comm.* **1986**, 388.

57. Cammack, J.K.; Jalisatgi, S.; Matzger, A.J.; Negron, A.; Vollhardt, K.P.C. *J. Org. Chem.* **1996**, *61*, 2699.

58. (a) Earl, R.A.; Vollhardt, K.P.C. *J. Org. Chem.* **1984**, *49*, 4786. (b) Earl, R.A.; Vollhardt, K.P.C. *J. Am. Chem. Soc.* **1983**, *105*, 6991.

59. (a) Hoberg, H.; Oster, B.W. *J. Organomet. Chem.* **1982**, *234*, C35. (b) Hoberg, H.; Oster, B.W. *J. Organomet. Chem.* **1983**, *252*, 359.

60. (a) Hoberg, H.; Schaefer, P.; Burkhart, G.; Kruger, C.; Ramao, M.J. *J. Organomet. Chem.* **1984**, *266*, 203. (b) Hoberg, H.; Apotecher, B. *J. Organomet. Chem.* **1984**, *270*, C15.

61. Tsuda

62. Pino, P.; Braca, G. "Carbon Monoxide Addition to Acetylenic Substrates," in *Organic Synthesis via Metal Carbonyls*; Wender, I.; Pino, P., Eds.; Wiley: New York, 1977; Vol. 2, pp 420-516; Hubel, W. "Organometallic Derivatives from Metal Carbonyls and Acetylene Compounds," in *Organic Synthesis via Metal Carbonyls*; Wender, I.; Pino, P., Eds.; Wiley: 1968; Vol. 1, pp 273-340.

63. Liebeskind, L.S.; Baysdon, S.L.; South, M.S.; Blount, J.F. *J. Organomet. Chem.* **1980**, *202*, C73.

64. Liebeskind, L.S.; Baysdon, S.L.; South, M.S.; Iyer, S.; Leeds, J.P. *Tetrahedron* **1985**, *41*, 5839.

65. Liebeskind, L.S.; Jewell, C.F. *J. Organomet. Chem.* **1985**, *285*, 305.

66. Liebeskind, L.S.; Chidambaram, R.; Nimkar, S.; Liotta, D. *Tetrahedron Lett.* **1990**, *31*, 3723.

67. For reviews see: (a) Buchwald, S.L.; Nielsen, R.B. *Chem. Rev.* **1988**, *88*, 1047. (b) Buchwald, S.L.; Broene, R.D., "Transition Metal Alkyne Complexes - Zirconium-Benzyne Complexes" in "Comprehensive Organometallic Chemistry II", Abel, E.W.; Stone, F.G.A.; Wilkinson, G., Eds., Elsevier Science Ltd., Oxford, UK, 1995, Vol. 12, pp. 771-784.

68. Buchwald, S.L.; King, S.M. *J. Am. Chem. Soc.* **1991**, *113*, 258.

69. Buchwald, S.L.; Fang, Q. *J. Org. Chem.* **1989**, *54*, 2793.

70. Cuny, G.D.; Gutierrez, A.; Buchwald, S.L. *Organometallics* **1991**, *10*, 537.

71. Sun, S.; Dullaghan, C.A.; Sweigart, D.A. *J. Chem. Soc., Dalton Trans* **1996**, 4493.

72. (a) Tidwell, J.H.; Senn, D.R.; Buchwald, S.L. *J. Am. Chem. Soc.* **1991**, *113*, 4685. (b) Tidwell, J.H.; Buchwald, S.L. *J. Am. Chem. Soc.* **1994**, *116*, 11797. (c) Tidwell, J.H.; Peat, A.J.; Buchwald, S.L. *J. Org. Chem.* **1994**, *59*, 7164.

CHAPTER 9

Synthetic Applications of η³-Allyl Transition Metal Complexes

9.1   Introduction

Although η³-allyl complexes are known for virtually all of the transition metals, relatively few have found use in organic synthesis. However, those that have, mainly those of Pd, have broad utility. η³-Allyl metal complexes can be made from a wide range of organic substrates in a variety of ways (Figure 9.1). These include (1)

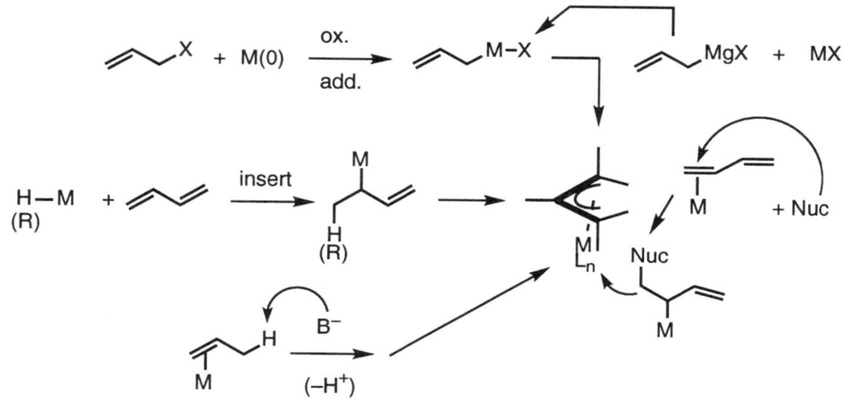

Figure 9.1   Preparation of η³-allylmetal Complexes

oxidative addition of allylic substrates to metal(0) complexes, (2) reaction of main-group allyl metal complexes with transition metals, (3) nucleophilic attack on a 1,3-diene metal complex, (4) insertion of 1,3-dienes into a metal hydride (or metal alkyl), (5) acidic cleavage of complexed allylic ethers, and (6) allylic proton abstraction from a π-olefin complex. Although many η³-allyl complexes are stable and isolable, they also are quite reactive, under appropriate conditions. The scope of reactions in which these complexes participate is presented in the following sections.

9.2 Transition-Metal Catalyzed Telomerization of 1,3-Dienes[1]

Treatment of 1,3-dienes with nucleophiles in the presence of palladium(0) catalysts results in the production of functionalized octadienes, made by the joining of two 1,3-dienes with incorporation of the nucleophile (Eq. 9.1). The mechanism of this process has not been closely studied but is thought to involve complexation of

Eq. 9.1

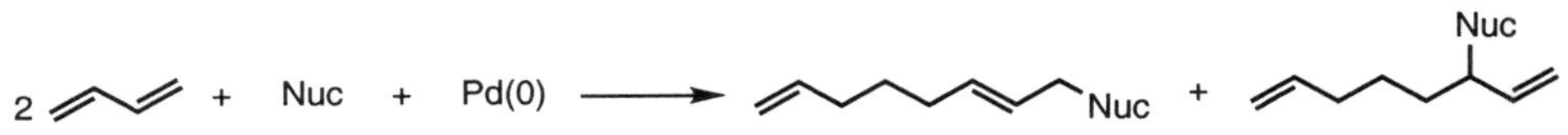

two dienes to the Pd(0) followed by the (by now) familiar "reductive dimerization" of the diene units (actually an oxidative addition of two dienes to the metal) joining them and generating a *bis*-η^3-allylpalladium species. As we shall soon see, η^3-allylpalladium complexes are generally subject to nucleophilic attack, usually at the less-substituted terminus. Nucleophilic attack produces an anionic π-olefin-η^3-η^1 allyl complex which undergoes proton transfer to the metal followed by reductive elimination to produce the diene "telomer" (an imprecise term used here to denote the combination of two diene units with a nucleophile) (Eq. 9.2). This procedure offers a convenient way to assemble functionalized chains (Eq. 9.3),[2] but the intermolecular version has found only modest use in complex molecule synthesis.

Eq. 9.2

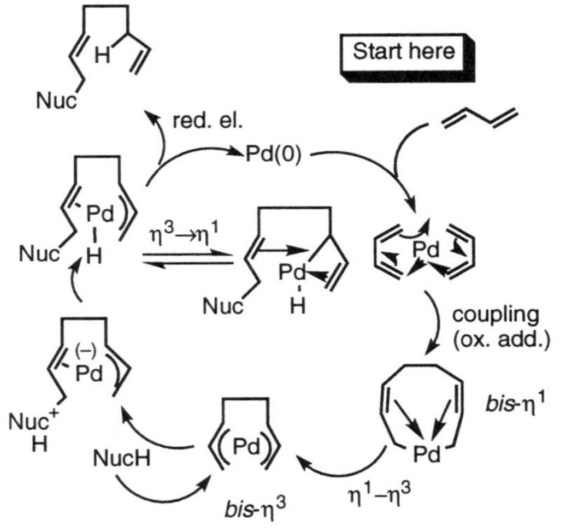

NucH = AcOH, H_2O, ROH, RNH_2, $CH_2\begin{smallmatrix}Y\\X\end{smallmatrix}$ where X, Y = CO_2Et, CN, NO_2, COR

Eq. 9.3

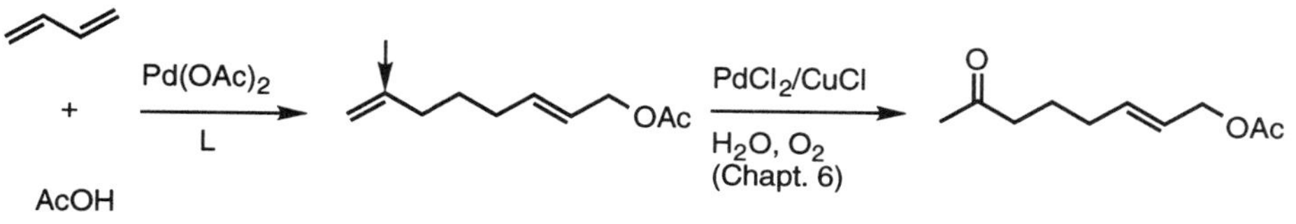

The synthetic utility of the above process was dramatically enhanced when it was applied in an intramolecular fashion (Eq. 9.4),[3] assembling five-membered rings with appended functionality quickly and efficiently. Intramolecular trapping is also efficient (Eq. 9.5).[4]

Eq. 9.4

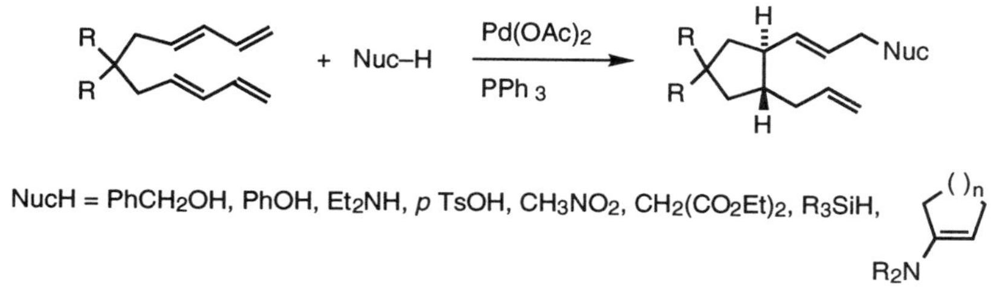

NucH = PhCH$_2$OH, PhOH, Et$_2$NH, p TsOH, CH$_3$NO$_2$, CH$_2$(CO$_2$Et)$_2$, R$_3$SiH,

Eq. 9.5

73%

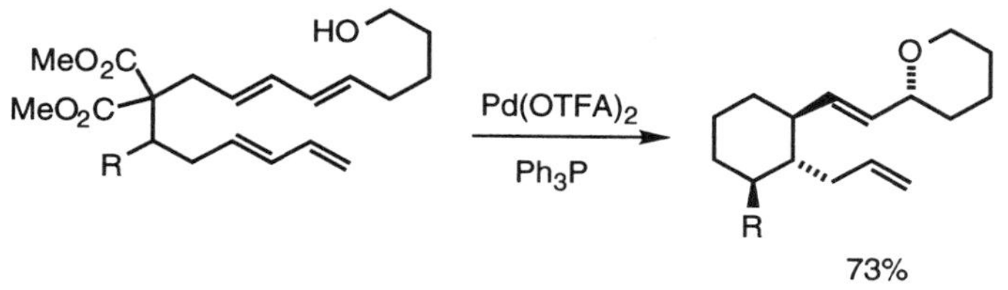

Provided the substrate has appropriately-situated β-hydrogens, this cyclization can take place without nucleophilic attack, producing trienes instead of dienes (Eq. 9.6).[5] This process is closely related to the eneyne cyclizations proceeding

Eq. 9.6

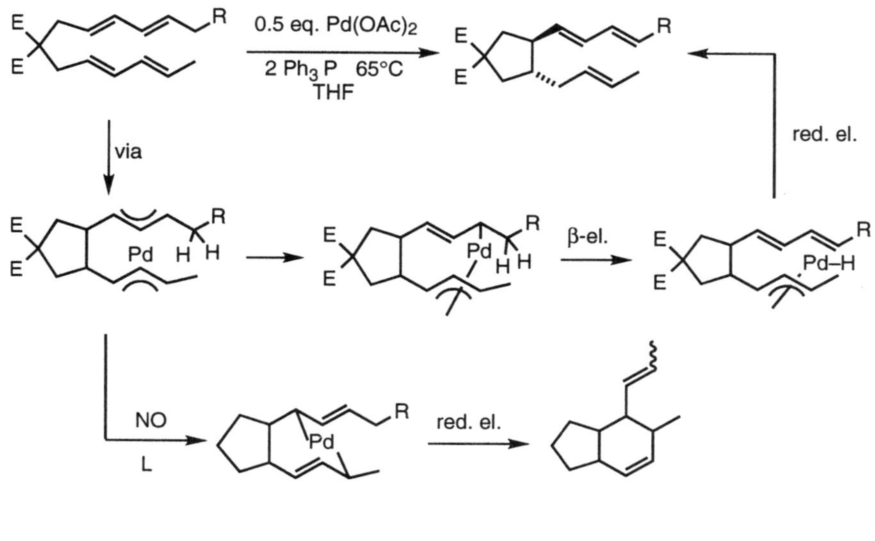

Synthetic Applications of η^3-Allyl Transition Metal Complexes • 247

by "hydrometallation" (Chapter 3), as well as to a number of other cyclization reactions catalyzed by low-valent transition metals. Interestingly, if this reaction is run with only 5% catalysts in the absence of phosphine, a second cyclization takes place.[6]

Most closely related are the iron(0) coupling reactions of trienes as exemplified by Eq. 9.7,[7] Eq. 9.8,[8] and Eq. 9.9.[8] Again, the mechanism has not been studied but is likely to be similar to that in Eq. 9.6.

Eq. 9.7

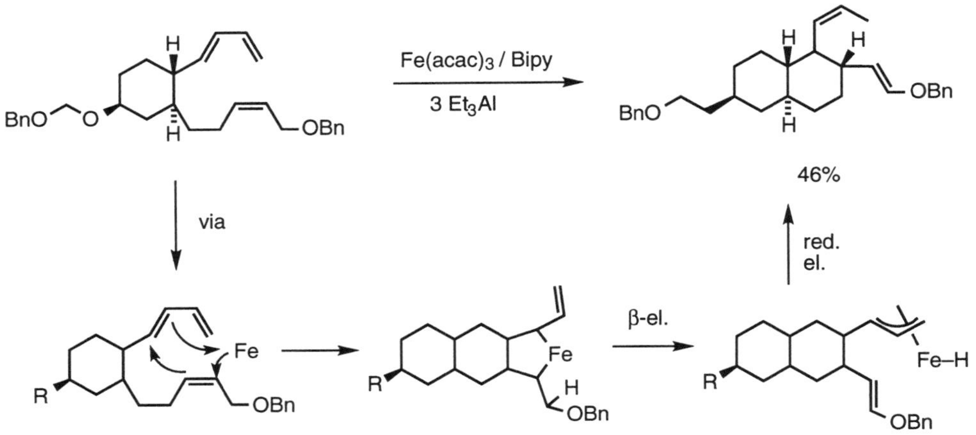

Eq. 9.8

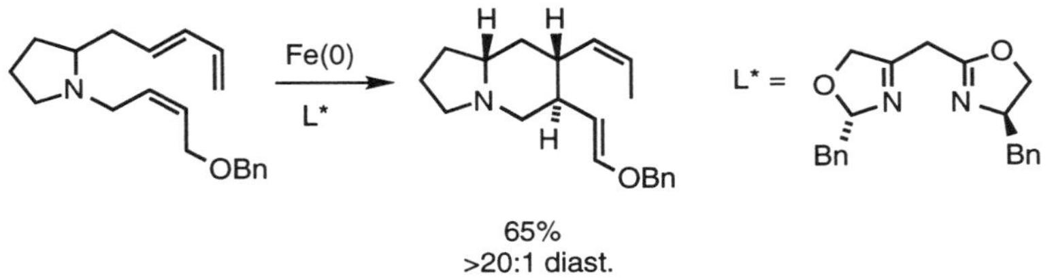

Eq. 9.9

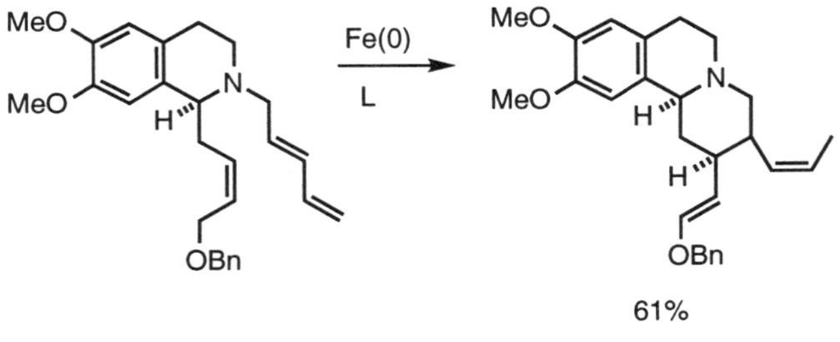

Perhaps the earliest example of this type of process is the commercially important nickel(0)-catalyzed cyclooligomerization of 1,3-dienes to produce a wide variety of compounds depending on conditions (Figure 9.2).[9] Whatever the ultimate product, they all derive from an initial dimerization of the diene to form the *bis*-η[3]-

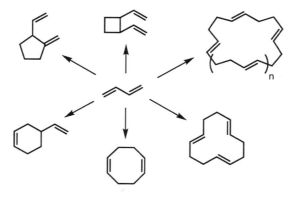

Figure 9.2   Ni(0) Catalyzed Cyclooligomerization of Butadiene

allyl nickel complex (much as in Eq. 9.2 with Pd). Insertion of an additional diene leads to the cyclic trimers, while reductive elimination from one of the several *bis*-η^1-allyl species produces cyclooctadienes, vinylcyclohexanes or divinylcyclobutanes (Eq. 9.10). Although this chemistry is very well developed with simple substrates, its

Eq. 9.10

real synthetic utility for complex systems resides in intramolecular reactions directed toward forming eight-membered rings, such as in the synthesis of (+)-astericanolide (Eq. 9.11).[10] Alkynes are also efficient partners in a similar process for forming bicyclic compounds (Eq. 9.12).[11]

Eq. 9.11

Eq. 9.12

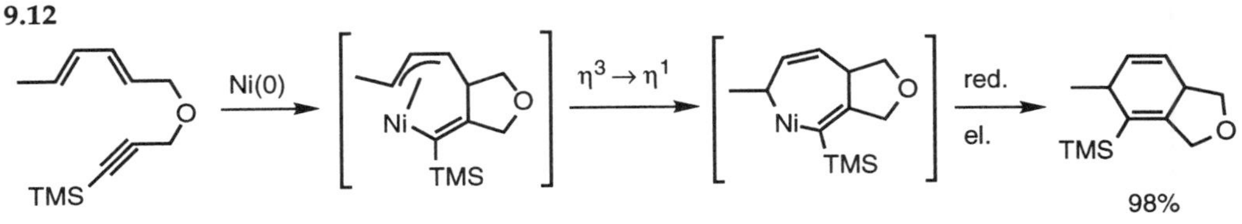

9.3 Palladium-Catalyzed Reactions of Allylic Substrates

a. General[12]

Palladium(0) complexes catalyze a wide variety of synthetically useful reactions of allylic substrates which proceed through η^3-allyl intermediates (Figure 9.3). Whatever the ultimate product, they all involve the same manifold of familiar steps, and most involve nucleophilic attack on the η^3-allylpalladium complex. The reaction begins by the oxidative addition of the allylic substrate to the palladium(0) catalyst (a), a process that goes with clean inversion to initially give the η^1-allyl

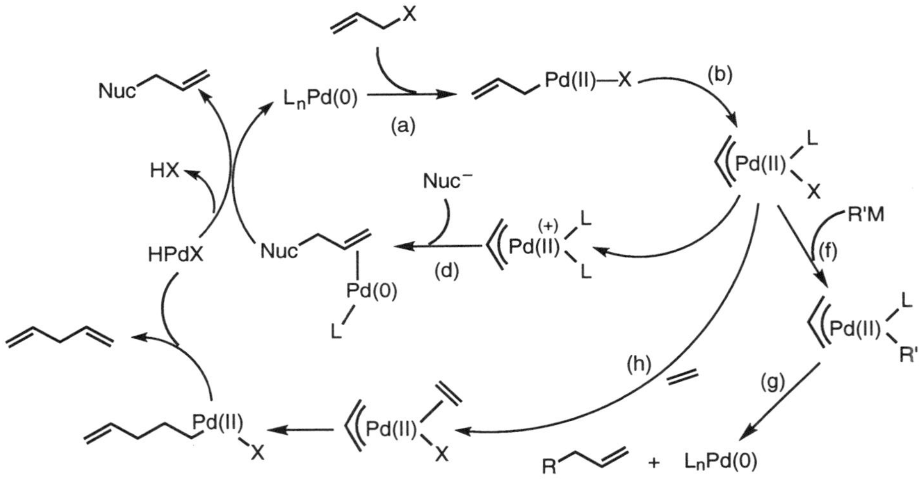

Figure 9.3 Palladium Catalyzed Reactions of Allylic Substrates

complex which is in equilibrium with the η^3-allyl complex (b) and is rarely detected. In contrast, the η^3-allyl complexes are quite stable, yellow solids which can easily be isolated if desired. However in the presence of excess ligand, low equilibrium concentrations of a cationic η^3-allyl complex (c) are generated. This is very reactive towards a wide range of nucleophiles, and undergoes nucleophilic attack from the face opposite the metal (d) (inversion) to produce the allylated nucleophile, and to regenerate palladium(0) (e) to reenter the catalytic cycle. The net stereochemistry of this allylic substitution is retention, which is the result of two inversions. The regiochemistry with unsymmetric allyl substrates, in general, favors attack at the less substituted allyl terminus, but is somewhat dependent on the nucleophile, the metal (see below), the cation,[13] as well as the ligands.[14]

η^3-Allyl complexes undergo transmetallation with main-group organometallics generating an η^3-allyl-η^1-alkyl palladium(II) complex (f), which undergoes reductive elimination (g) to produce the alkylated allyl compound. Since, in this case, the nucleophile (R' of R'M) is delivered first to the metal and then to the η^3-allyl group from the *same* face as the metal, this step occurs with retention, and the overall alkylation goes with net inversion. η^3-Allyl complexes can also insert alkenes (h) (probably via their η^1-isomers), producing 1,5-dienes after β-hydrogen elimination. All of these processes have been applied to the synthesis of complex molecules and are discussed in detail below.

A number of palladium(0) complexes and palladium(II) precatalysts can be used, and the choice of catalyst is often not important. The most frequently utilized catalyst is (Ph₃P)₄Pd, which is commercially available. However, commercial material is often of widely varying activity and it is best to prepare one's own. η^3-Allylpalladium(II) complexes are themselves catalysts, and because of their ease in preparation and handling, they are often used. A very convenient catalyst results from the treatment of the air-stable, easily prepared and handled Pd(dba)₂ (dba = PhCH=CH–CO–CH=CHPh, dibenzylidene acetone) with varying amounts of triphenylphosphine, generating PdL$_n$ *in situ*. Finally, a variety of palladium(II) salts are readily reduced in the presence of substrate and ligand, and are often efficient precatalysts for the process.

b. Allylic Alkylation[15]

Alkylation of allylic substrates by stabilized carbanions is one of the synthetically most useful reactions catalyzed by palladium(0) complexes (Eq. 9.13). A wide range of allylic substrates undergo this reaction with a reasonable range of carbanions, making this an important process for the formation of carbon-carbon bonds. The reaction is very stereoselective, and proceeds with overall retention of configuration, the result of two inversions.[16] The reaction is also quite regioselective, with attack at the less-substituted terminus of the η^3-allyl intermediate favored, regardless of the initial position of the allylic leaving group.

Eq. 9.13

$$Z = Br, Cl, OAc, -OCOR, OP(OEt)_2, O–S–R, OPh, OH, R_3N^+, NO_2, SO_2Ph, CN$$

X, Y = CO₂R, COR, SO₂Ph, CN, NO₂

This process has found extensive use in synthesis over the years, and the literature abounds with applications to highly complex systems, so the methodology is well-established and reliable. Examples are seen in Eqs. 9.14,[17] 9.15,[18] and 9.16.[19]

Eq. 9.14

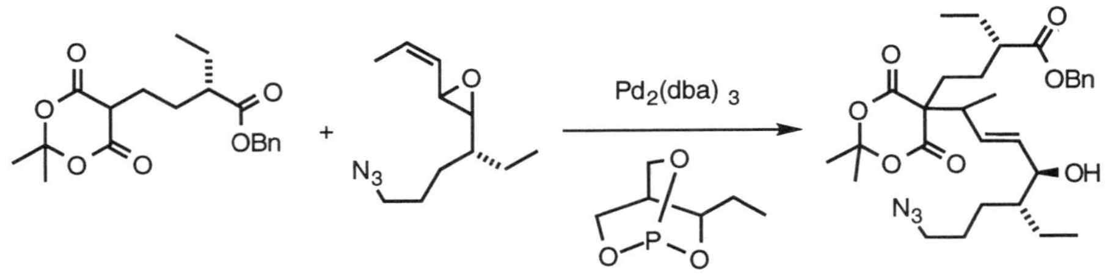

Eq. 9.15

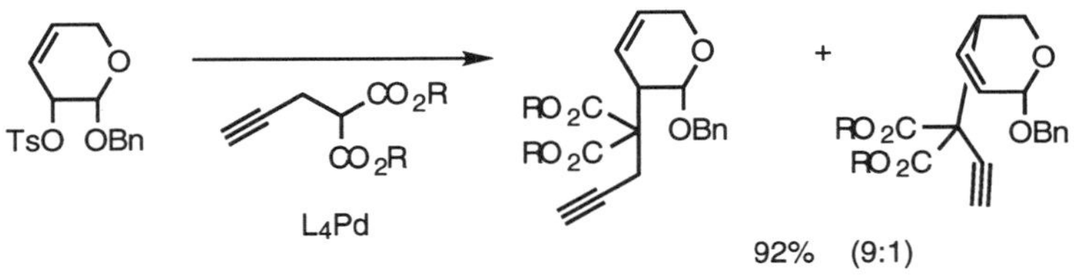

Eq. 9.16

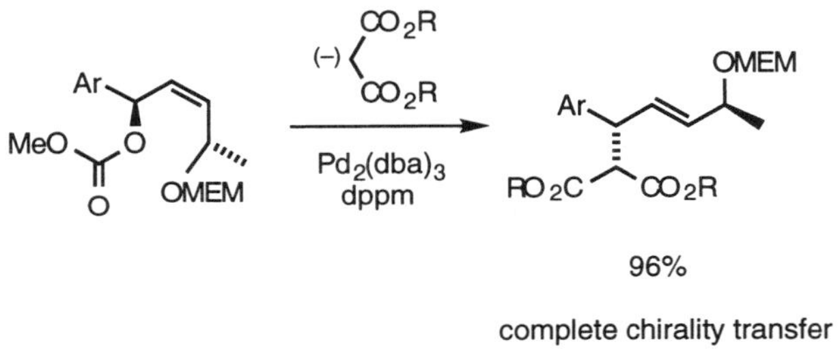

complete chirality transfer

Intramolecular versions of this process are efficient,[20] and have been used to make rings from three to eleven members, as well as macrocycles. Examples are shown in Eqs. 9.17,[21] 9.18,[22] and 9.19.[23]

Eq. 9.17

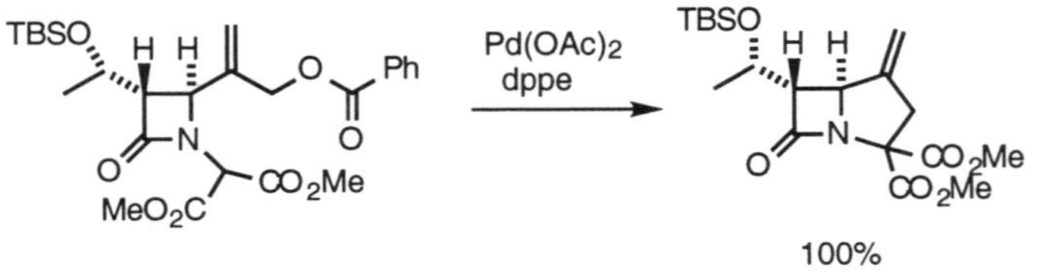

Eq. 9.18

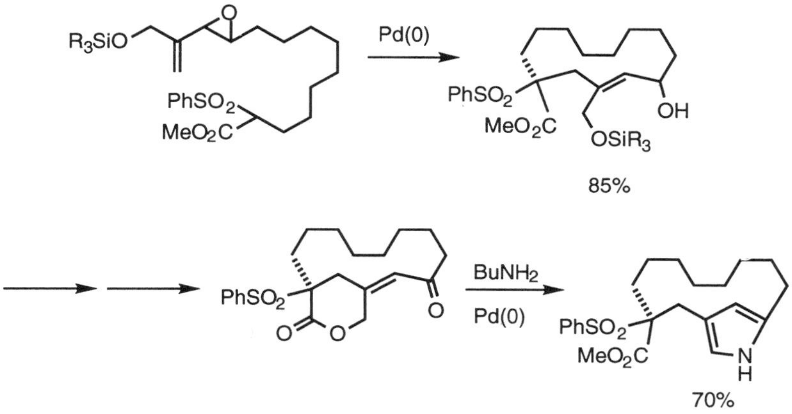

85%

70%

Eq. 9.19

good yields

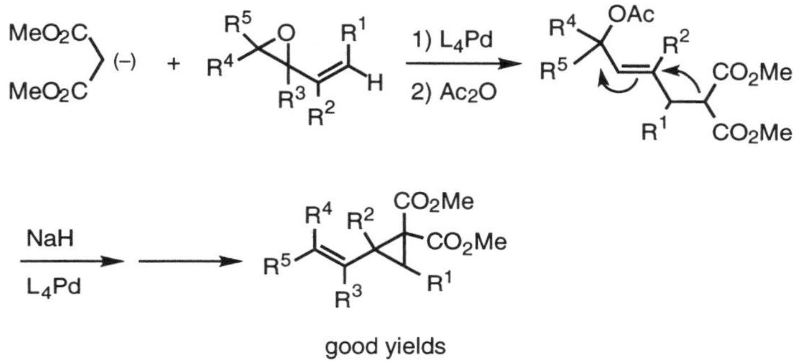

Because of the difference in reactivity of various allylic functional groups (e.g., Cl > OCO$_2$R > OAc >> OH) high regioselectivity can be achieved with *bis*-allyl substrates, in addition to high stereoselectivity (Eq. 9.20).[24]

Eq. 9.20

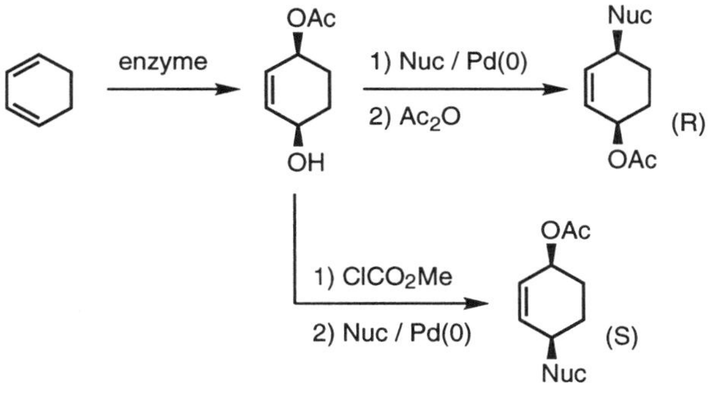

The development of procedures to induce asymmetry into palladium-catalyzed allylic reactions has dramatically enhanced the synthetic utility of this process.[25] The situation for asymmetric induction in these systems is complex, and the requirements differ with different classes of substrates. The most difficult substrates are chiral, racemic, unsymmetrically 1,3-disubstituted allylic compounds

(Eq. 9.21). Since oxidative addition goes with clean inversion, the reaction of optically active palladium(0) complexes with chiral, racemic allylic acetates will lead to *two* diastereoisomeric π-allylpalladium complexes. Since nucleophilic attack also occurs with clean inversion, to achieve high asymmetric induction, one of the two diastereoisomers must react substantially faster than the other, *and the less reactive diastereoisomer must be able to isomerize to the more reactive one at a rate that exceeds nucleophilic attack.* (Note, this is the same situation observed for asymmetric hydrogenation in Chapter 4.) With acyclic systems this isomerization is readily accomplished via a π → σ rearrangement, followed by rotation about the σ-alkylpalladium bond, followed by σ → π rearrangement, which effectively results in enantioface exchange (Eq. 9.21 and Eq. 9.22).[26]

Eq. 9.21

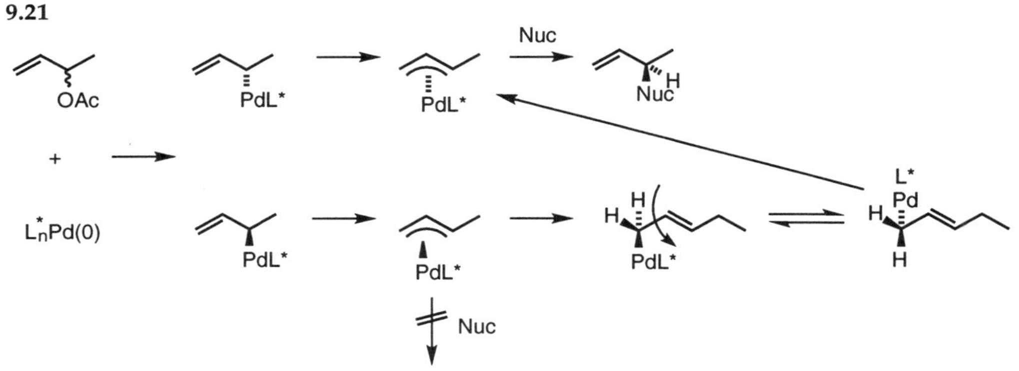

Eq. 9.22

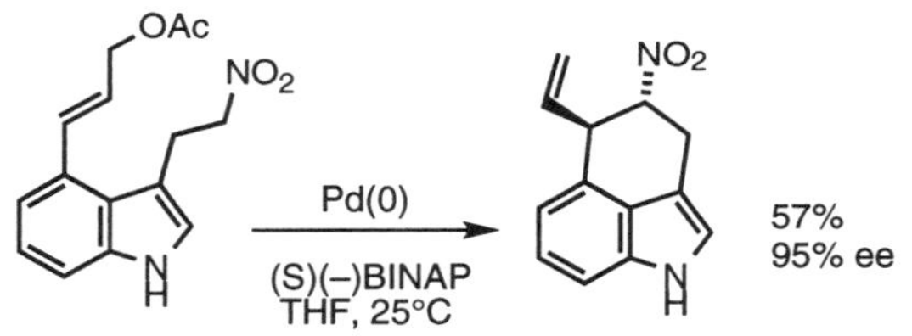

This π → σ → π-isomerization route is *not* available to cyclic systems, since the ring prevents rotation about the palladium-carbon σ-bond. However, asymmetric induction in the allylic alkylation of racemic chiral cyclic substrates has been achieved. In this case, diastereoface exchange is likely to occur by attack of the palladium *catalyst* on an η³-allylpalladium complex (Eq. 9.23). This also provides a mechanism for *erosion* of selectivity when optically active substrates are used (e.g., "racemization") as has sometimes been observed when relatively high concentrations of catalyst are used.[27]

Eq. 9.23

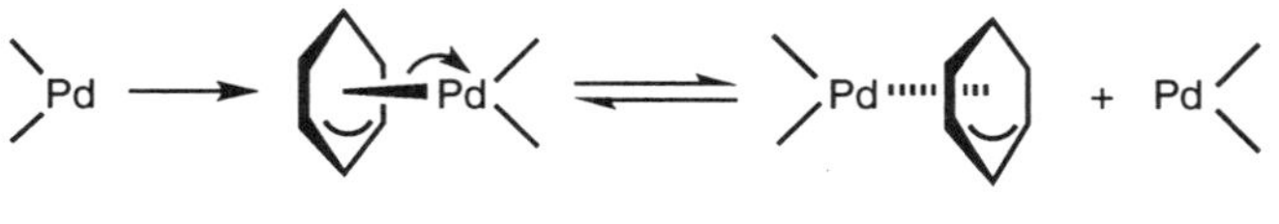

Considerably less challenging (and thus much more extensively studied) are symmetrical allyl substrates in which only discrimination between the two enantiotopic terminii of the η^3-allyl complex is required.[28] A very large number of ligands have been (and continue to be) developed for this class of substrates and most of them result in high asymmetric induction (Eq. 9.24).[29]

Eq. 9.24

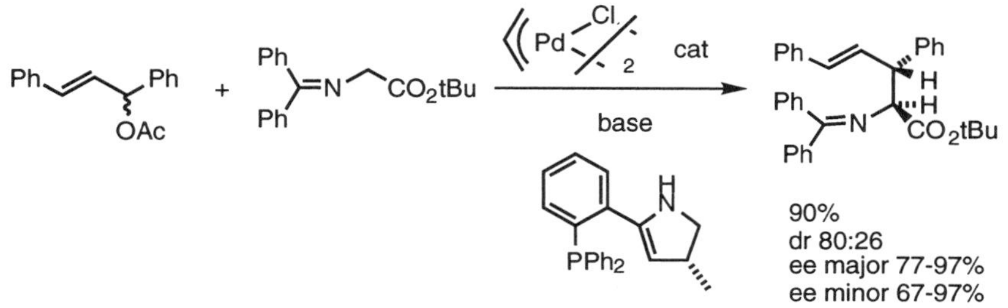

Perhaps the most useful class of asymmetric π-allylpalladium reactions involves catalytic desymmetrizations of meso substrates (Eq. 9.25).[25a] In these cases the catalyst discriminates between the two allylic leaving groups, resulting again in

Eq. 9.25

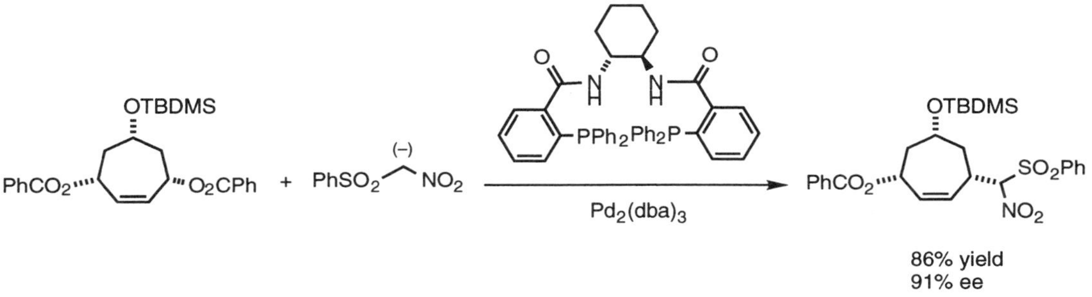

high asymmetric induction. In many synthetic applications, the second leaving group is displaced in a subsequent step, resulting in rapid construction of cyclic systems (Eq. 9.26).[25a,30] A particularly impressive example utilized two palladium-catalyzed allylic displacements followed by a Heck olefin arylation (Chapter 4) (Eq. 9.27).[31]

Eq. 9.26

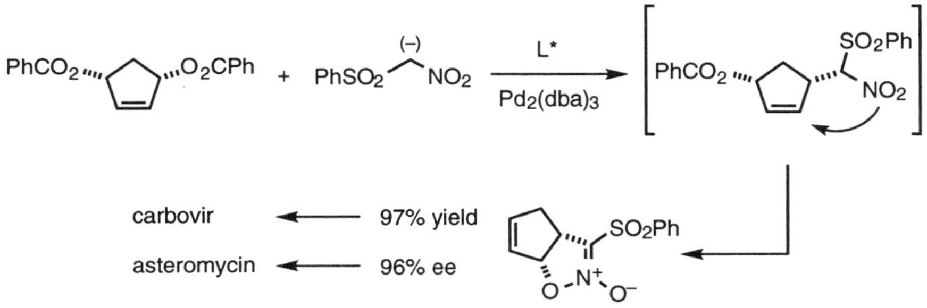

Synthetic Applications of η^3-Allyl Transition Metal Complexes • 255

Eq. 9.27

c. Allylic Alkylation by Transmetallation

Transmetallation has been used much less frequently in allylic systems than with aryl or vinyl systems, for several practical reasons. Allyl acetates are the most attractive class of allylic substrates, because of their ready availability from the corresponding alcohols, but they are substantially less reactive towards oxidative addition to Pd(0) than are allylic halides, and phosphine ligands are required to promote this process. Acetate coordinates quite strongly to palladium and this, along with the presence of excess phosphine, slows down the crucial transmetallation step, already the rate limiting step. Finally, in contrast to dialkyl palladium(II) complexes, σ-alkyl-π-allylpalladium(II) complexes undergo reductive elimination only slowly, compromising this final step in the catalytic cycle. It is thus not surprising that alkylation of allylic acetates via transmetallation has been slow to develop. However under appropriate reaction conditions (polar solvents such as DMF, "ligandless" (no phosphine) catalysts such as Pd(dba)$_2$ or PdCl$_2$(MeCN)$_2$, excess LiCl to facilitate transmetallation) allyl acetates can be alkylated by a variety of aryl- and alkenyltin reagents (Eq. 9.28). Coupling occurs at the less-substituted terminus of the allyl system with clean inversion (inversion in the oxidative addition step, retention in the transmetallation/reductive elimination step), and the geometry of the alkene in both the allylic substrate and alkenyltin partner is maintained (Eq. 9.29).[31]

Eq. 9.28

Eq. 9.29

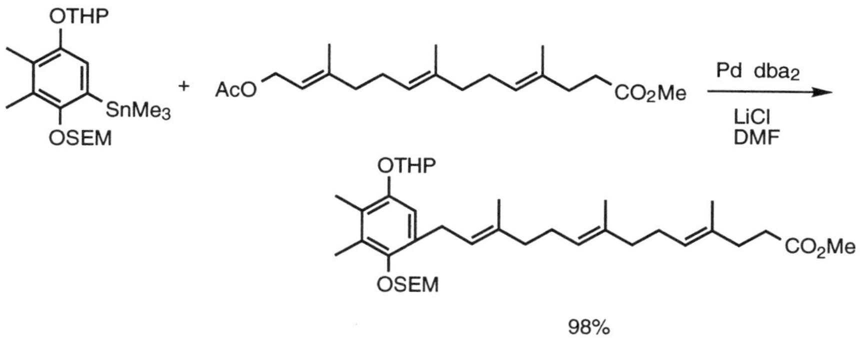

Allyl epoxides and carbonates are better substrates for this process, and are alkylated by a variety of aryl- and alkenyltin reagents (Eq. 9.30[32] and Eq. 9.31[33]). Acetylenic, benzylic or allylic tin compounds fail to couple. Allylic chlorides are the most reactive of substrates, undergoing this coupling reaction under "normal"

Eq. 9.30

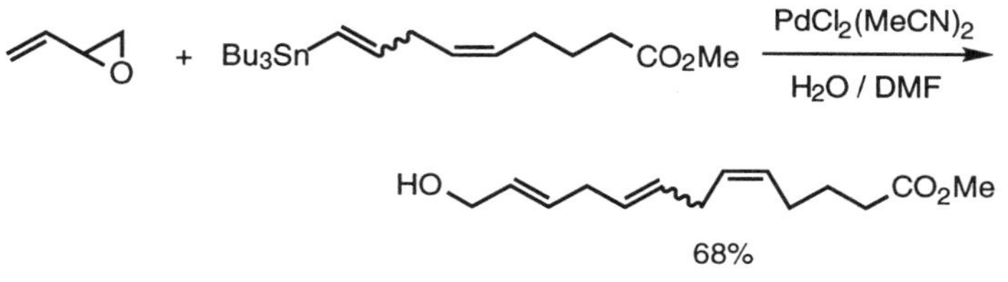

Eq. 9.31

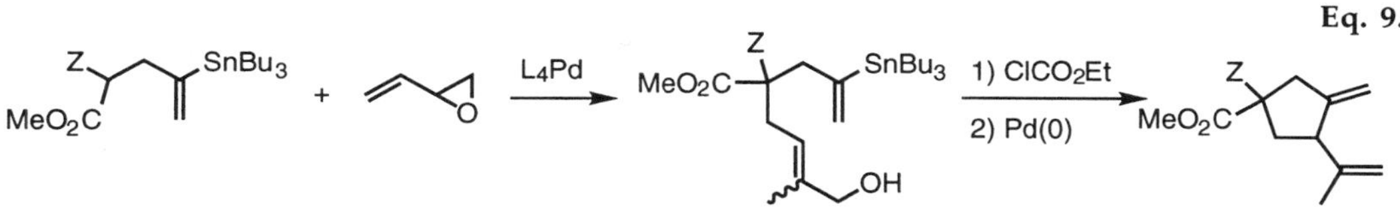

reaction conditions (phosphine present) in excellent yield with highly functionalized substrates (Eq. 9.32).[34] The process is not restricted to allyl chlorides nor to organostannanes (Eq. 9.33).[35]

Eq. 9.32

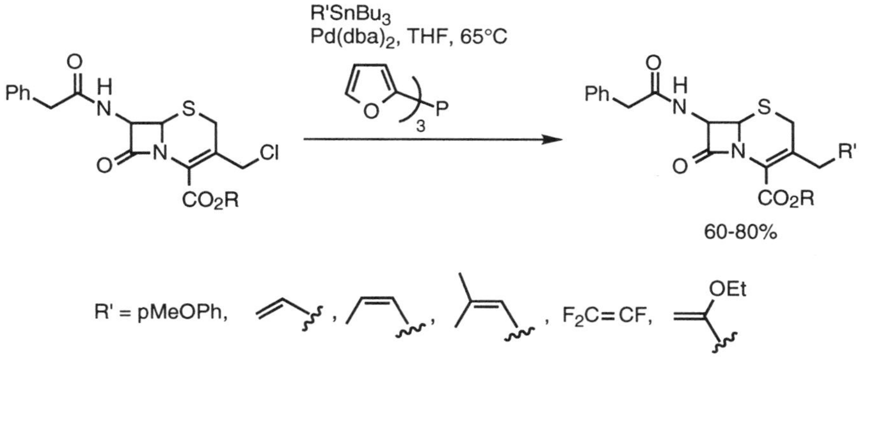

Eq. 9.33

$$\text{TMS}\diagdown\diagup\overset{R}{\underset{\underset{\underset{O}{\overset{\|}{\text{O}}}}{\overset{|}{\text{OP(OEt)}_2}}}{|} \xrightarrow[\text{PdCl}_2\text{dppf}]{\text{MeMgBr}} \text{TMS}\diagdown\diagup\overset{R}{\underset{\text{Me}}{|}}$$

60-80%

d. Allylic Amination,[36b] Alkoxylation, Reduction, and Deprotonation

The reactions of nitrogen nucleophiles with allylic substrates are among the most useful for organic synthesis, and both 1° and 2° amines (but *not* NH_3) efficiently aminate allylic substrates, as do sulfonamide anions. The stereo- and regioselectivity of allylic amination parallels that of allylic alkylation, and the range of reactive substrates is comparable. The nitrogen almost invariably ends up at the less hindered terminus of the allyl system, although that may not be the initial site of attack, since Pd(0) complexes catalyze the allylic transposition of allyl amines. The reaction proceeds efficiently with polymer-bound allylic substrates and is finding increasing applications in combinatorial chemistry.[36b]

Intermolecular aminations are efficient and a wide range of functionality is tolerated. The process is used extensively in the synthesis of carbocyclic nucleoside analogs (Eq. 9.34[37] and Eq. 9.35).[38] By using appropriate reaction sequences, diols can be aminated at either allylic position (Eq. 9.36).[39]

Eq. 9.34

53-69%

Eq. 9.35

Asymmetric induction in the palladium-catalyzed allylic amination process is also efficient, with very much the same classes of substrates as utilized in asymmetric allylic alkylation, particularly symmetrically 1,3-disubstituted substrates (Eq. 9.37).[40] It is most extensively used for the desymmetrization of meso-allylic substrates, and is efficient for the synthesis of carbocyclic nucleoside analogs (Eq. 9.38)[41] and highly functionalized cyclohexenes (Eq. 9.39)[42] and (Eq. 9.40).[43]

Eq. 9.36

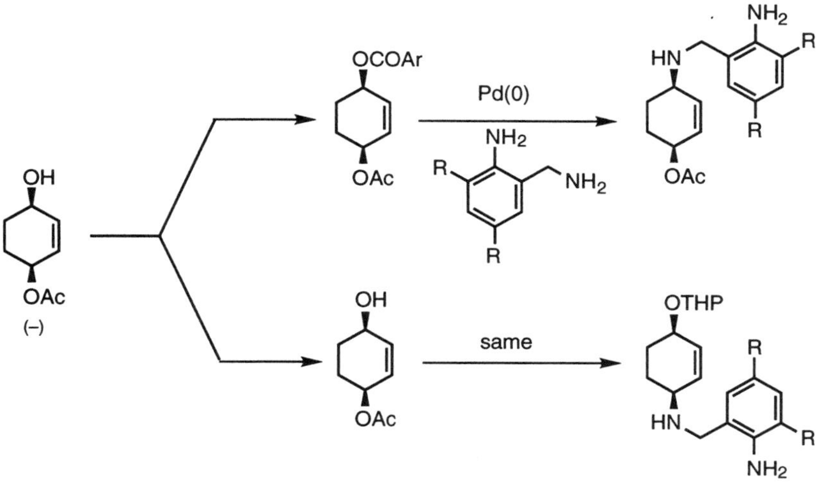

Eq. 9.37

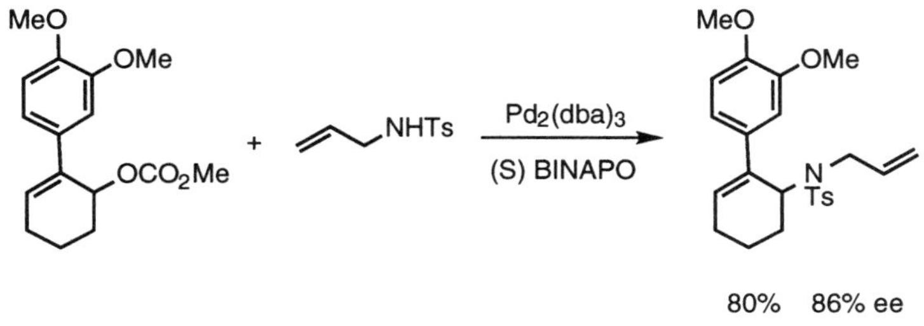

80%   86% ee

Eq. 9.38

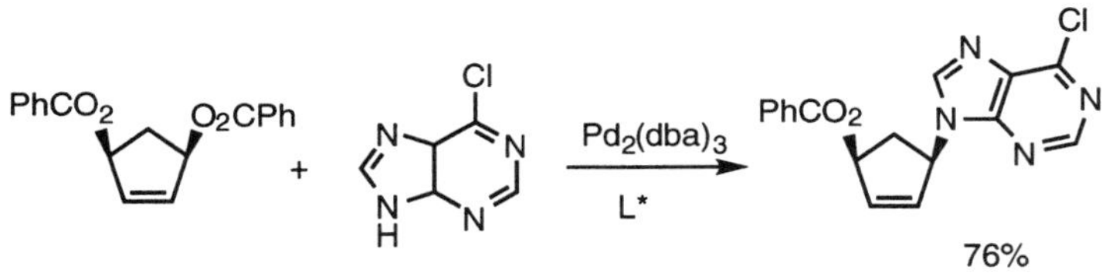

76%

Eq. 9.39

(+)-Pancratistatin

L*
TMSN$_3$
98% ee
83% yield

Conduramine A'

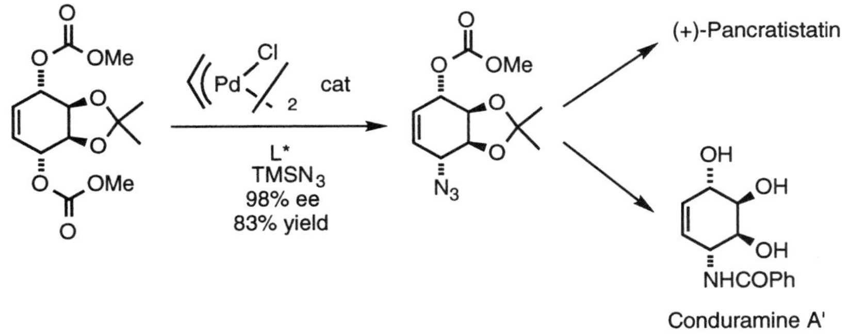

Eq. 9.40

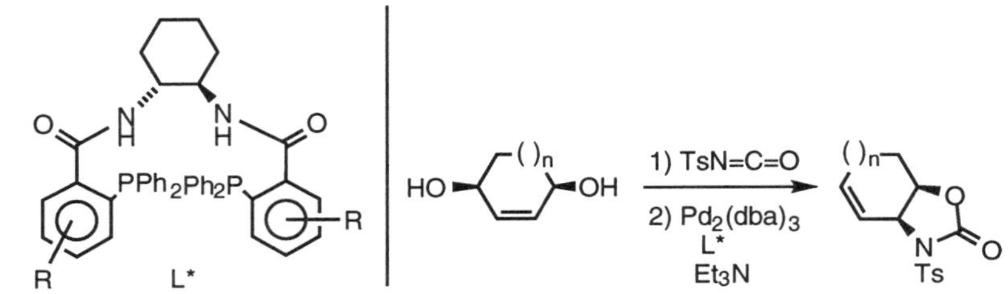

Intramolecular aminations are particularly effective, and a wide range of fused (Eq. 9.41),[44] spiro (Eq. 9.42),[45] and macrocyclic systems (Eq. 9.43)[46] can be made efficiently.

Eq. 9.41

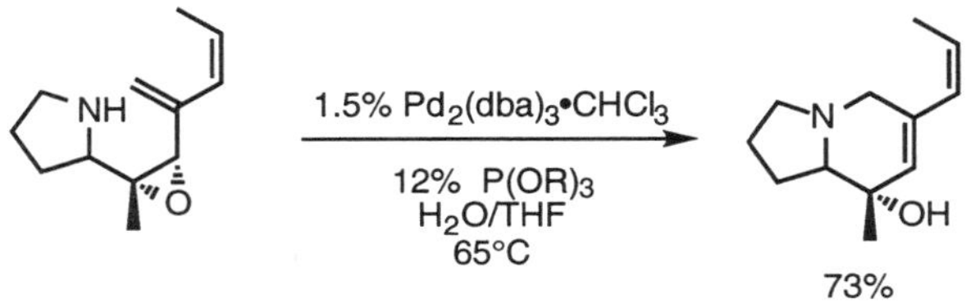

Eq. 9.42

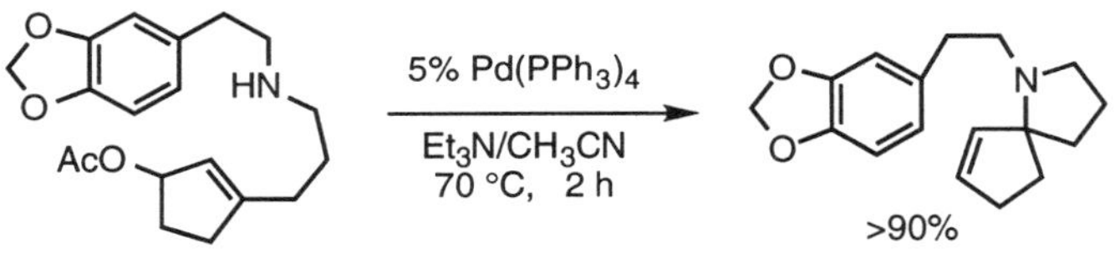

Eq. 9.43

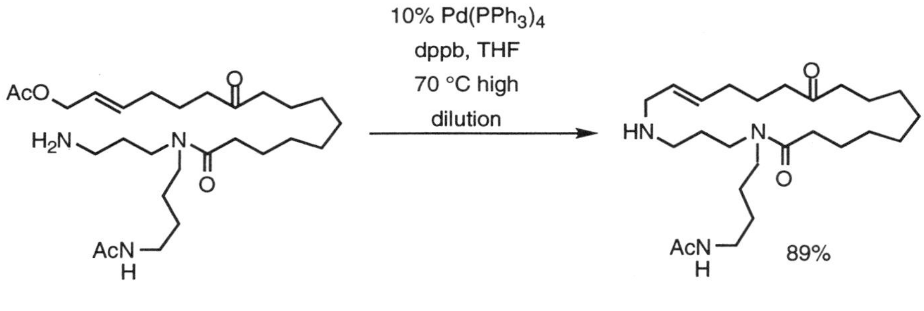

Although carbon and nitrogen nucleophiles have been most extensively used in palladium-catalyzed allylic substitution processes, a range of oxygen nucleophiles can also participate. Phenols attack allyl epoxides[47] and allyl carbonates[48] in the presence of palladium(0) catalysts, and in the presence of chiral ligands, high asymmetric induction is observed (Eq. 9.44).[49] Glycal acetates were coupled to the anomeric OH group of another carbohydrate (Eq. 9.45)[50] and intramolecular alkoxylations were efficient (Eq. 9.46).[51,52] Although hydroxide (or water) has not been used in this process, triphenylsilanol is an efficient surrogate (Eq. 9.47).[53,54]

Palladium(0) complexes even catalyze the O-alkylation of enolates (Eq. 9.48 and 9.49).[55]

Eq. 9.44

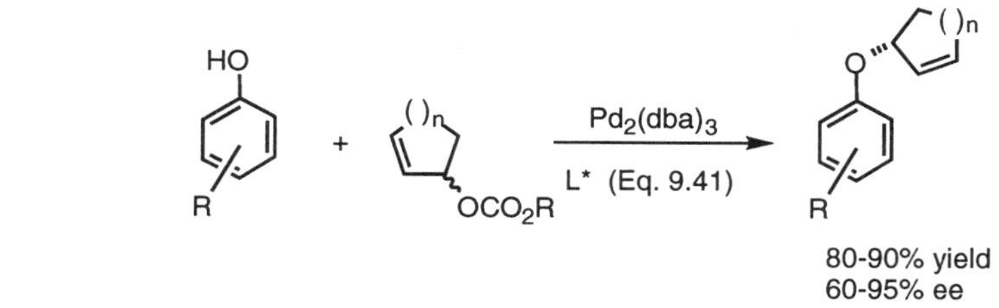

80-90% yield
60-95% ee

Eq. 9.45

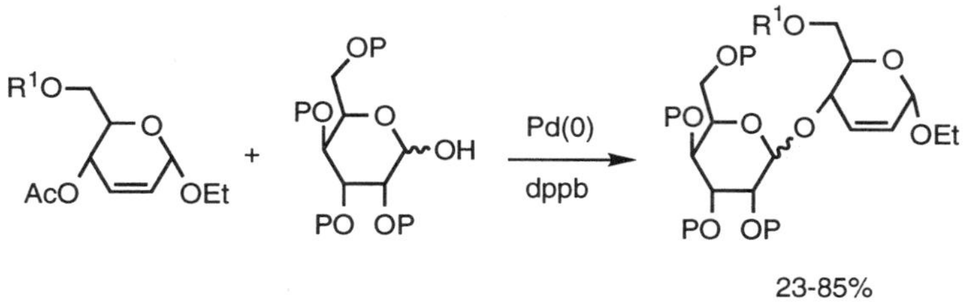

23-85%

Eq. 9.46

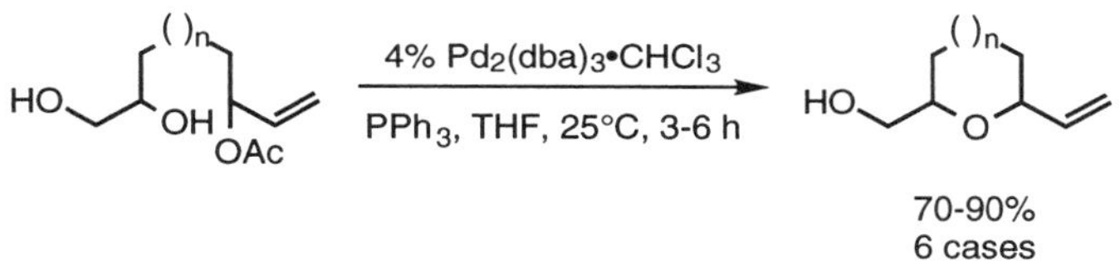

70-90%
6 cases

Eq. 9.47

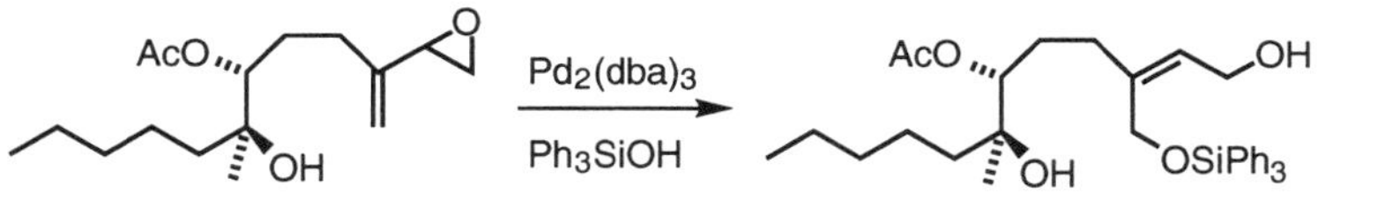

Eq. 9.48

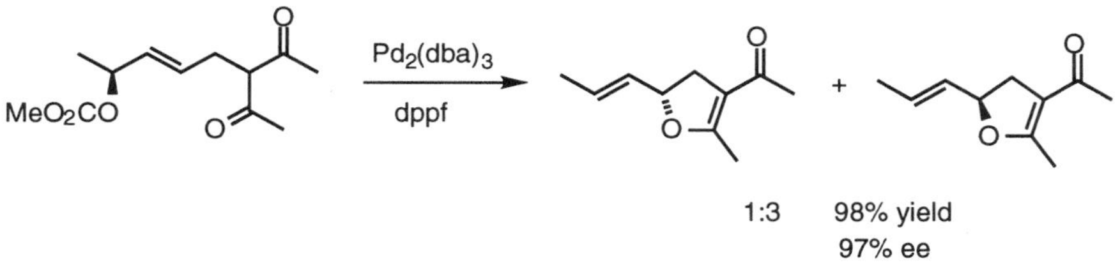

1:3   98% yield
97% ee

Eq. 9.49

n = 1, 2
Z = COMe, COPh, CO2Me, NO2

90%

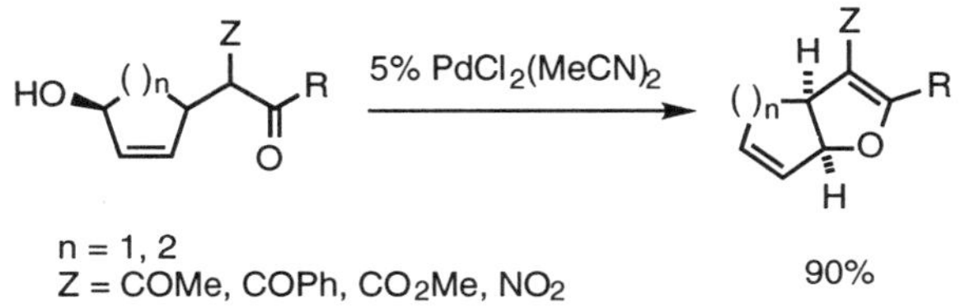

Acetate is rarely used as a nucleophile since it is most often the leaving group in palladium-catalyzed allylic substitution reactions. However, carboxylates are capable of attacking π-allylpalladium complexes and, in contrast to most nucleophiles, can either directly attack the π-allyl ligand from the face opposite the metal, resulting in inversion in this step, or it can attack the metal, resulting in retention in this step. Attack by acetate plays a major role in a useful variant involving palladium(II)-catalyzed bis-acetoxylation or chloroacetoxylation of dienes (Eq. 9.50). The process begins with a palladium(II)-assisted acetoxylation of one of the two double bonds of the complexed diene, generating a π-allylpalladium(II) complex. In the absence of added chlorides, the second acetate is delivered from the metal leading to the trans diacetate. Added chloride blocks this coordination site on the metal and leads to nucleophilic attack by uncomplexed acetate, from the face opposite the metal, giving the corresponding cis-diacetate and palladium(0).[56] The reoxidation of palladium(0) to palladium(II) can be carried by benzoquinone-MnO_2 or more efficiently by O_2-catalytic hydroquinone-catalytic Co(salophen).[57] Because the products of this reaction are allyl acetates, and because of the wide variety of palladium(0)-catalyzed reactions of allyl acetates this is a useful process that has been extensively studied.

Eq. 9.50

This chemistry has been used to develop quite efficient and stereoselective approaches to fused tetrahydrofurans and tetrahydropyrans (Eq. 9.51).[58] In this case, the tethered alcohol group is the nucleophile which initially attacks the π-olefin complex, giving the *trans* η³-allyl complex. Inter- or intramolecular attack of acetate, chloride, or alkoxide on this η³-allyl complex leads to the observed products. Other applications of this very useful process will be presented later in this chapter.

Eq. 9.51

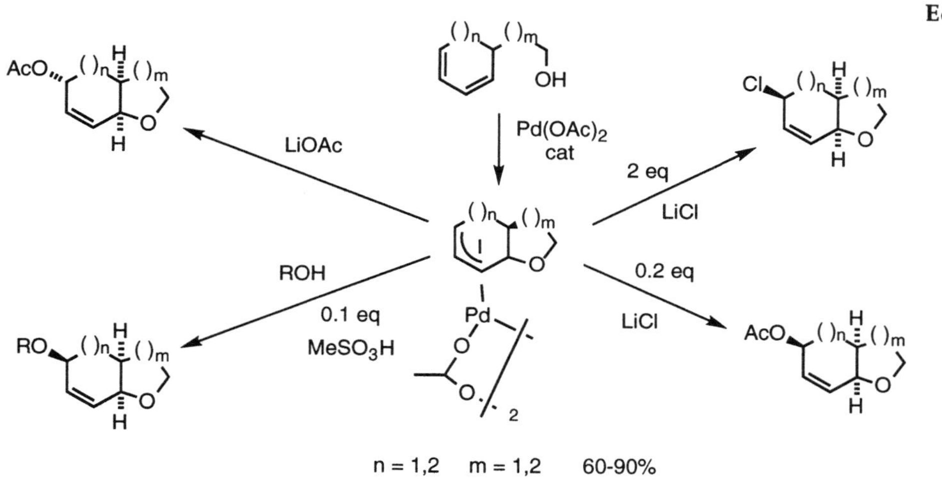

n = 1,2   m = 1,2   60-90%

Palladium(0) complexes catalyze a number of other useful reactions of allylic substrates. Allyl acetates are easily reduced to alkenes, usually with allylic transposition and clean inversion, in the presence of a hydride sources which attack the η^3-allylpalladium intermediate (Eq. 9.52).[59] Ammonium formates are particularly effective in complex systems (Eq. 9.53)[60] and (Eq. 9.54).[61] With chiral ligands, asymmetry can be induced (Eq. 9.55).[62]

Eq. 9.52

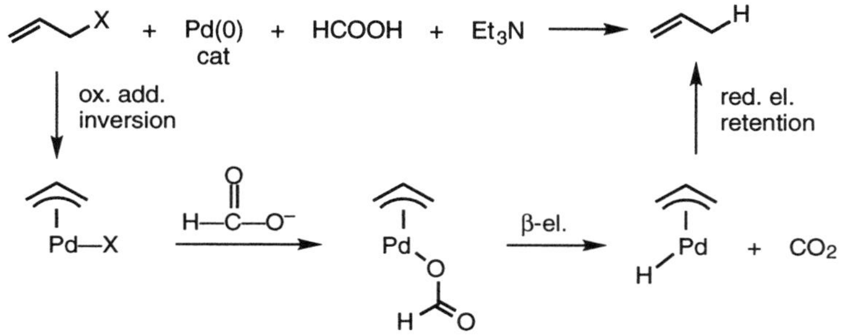

Eq. 9.53

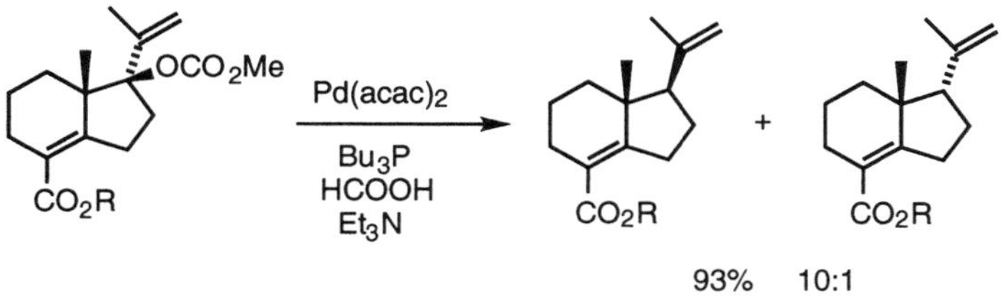

93%   10:1

Eq. 9.54

96%

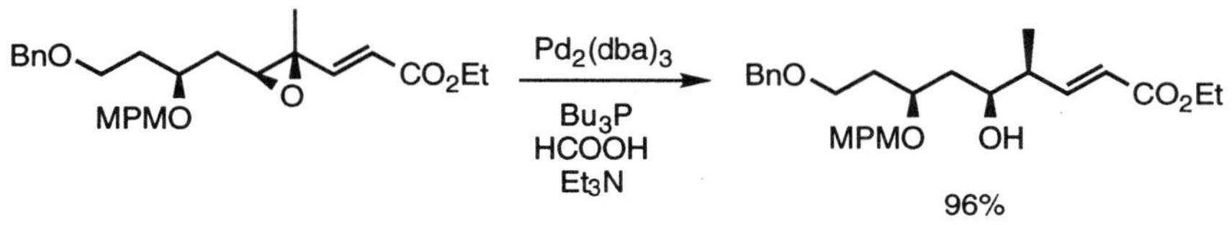

Eq. 9.55

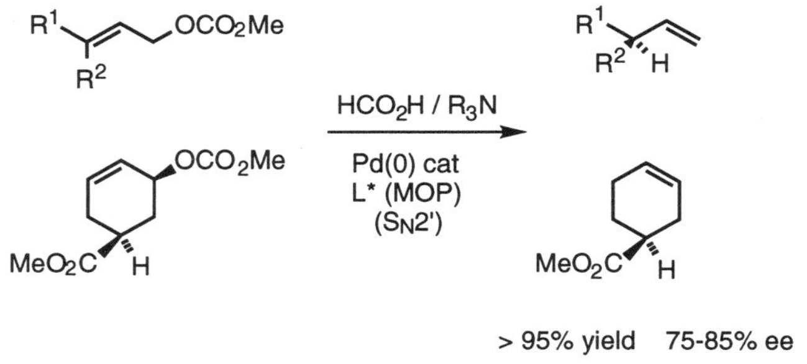

> 95% yield   75-85% ee

Allyl formates undergo intramolecular reduction, with formate acting as the hydride source (Eq. 9.56).[60] In these cases the hydride is delivered from the same face as the palladium.

Eq. 9.56

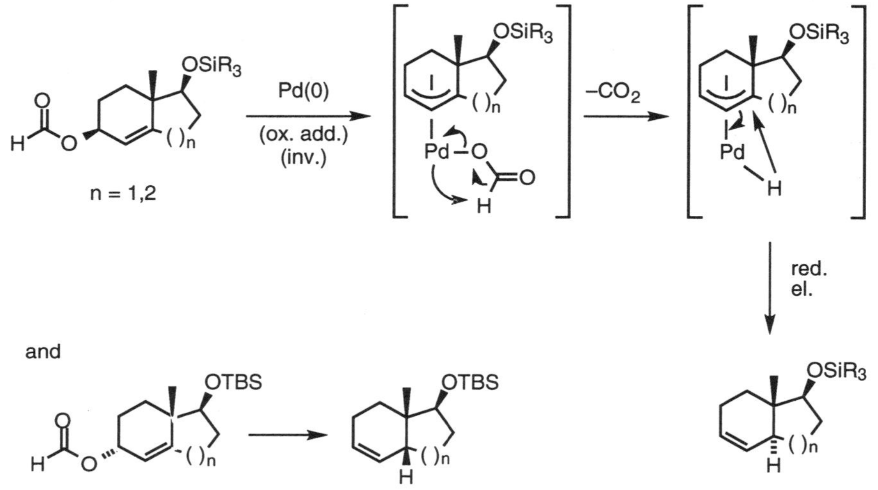

$n = 1,2$

and

In the absence of nucleophiles, palladium(0) complexes, particularly the one derived from treating palladium acetate with one equivalent of tributylphosphine,[64] efficiently eliminates allyl carbonates to dienes. The process occurs by oxidative addition with inversion, followed by *syn* elimination. In sterically-biased systems, high regioselectivity is observed (Eq. 9.57).[65] Decent asymmetric induction has been observed in appropriate symmetrical systems (Eq. 9.58).[66] (Note that in the presence of base an *anti* elimination is possible.[67])

Eq. 9.57

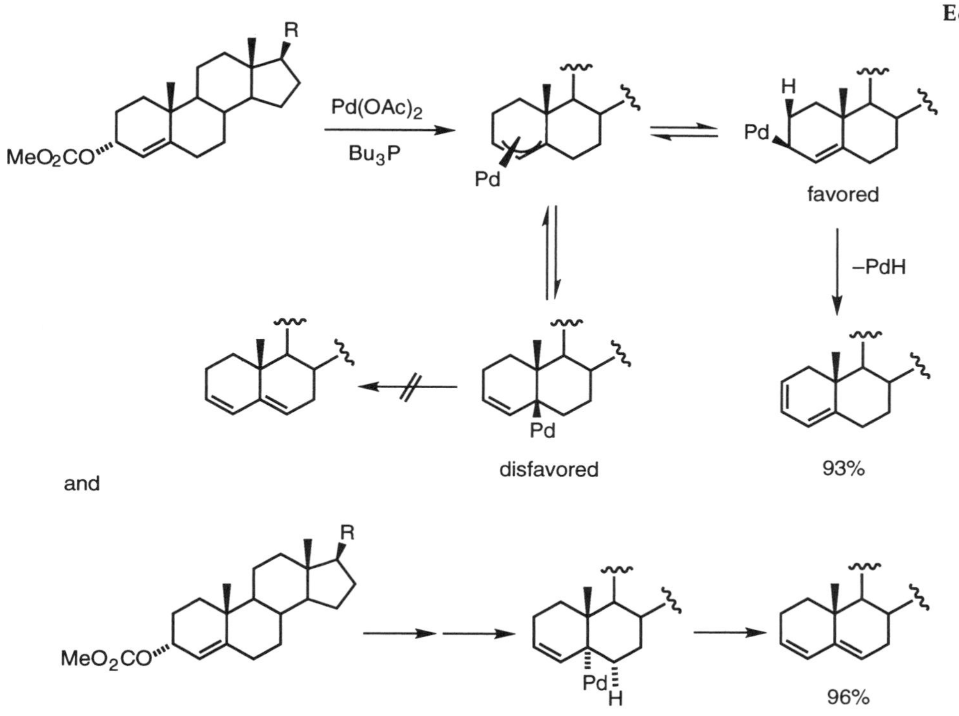

Eq. 9.58

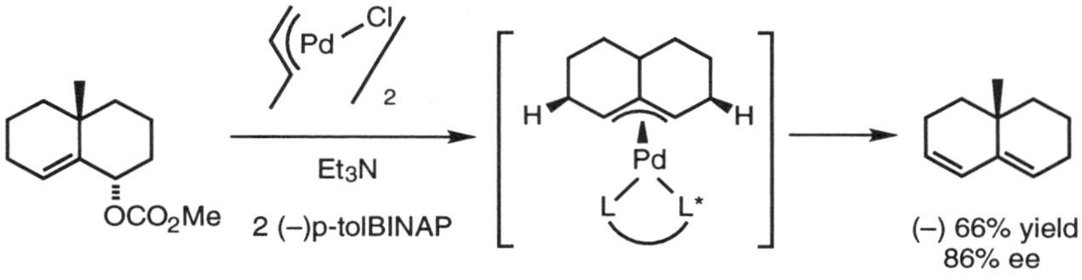

Finally, allylic carbonates and esters make excellent protecting groups[68] for carboxylic acids, alcohols and amines, because they are stable to a wide range of reaction conditions, and can be exclusively removed in the presence of other labile protecting groups. In these cases, a nucleophile must be present to regenerate the catalyst from the initially formed η^3-allyl complex (Eq. 9.59). This system can be used with quite complex substrates (Eq. 9.60).[69] The most spectacular example of the utility of this protecting group strategy was the removal of 104 allylic protecting groups from NH$_2$ and phosphate moieties in a 60 mer oligonucleotide in almost 100% overall yield in a *single* palladium-catalyzed reaction.[70]

Eq. 9.59

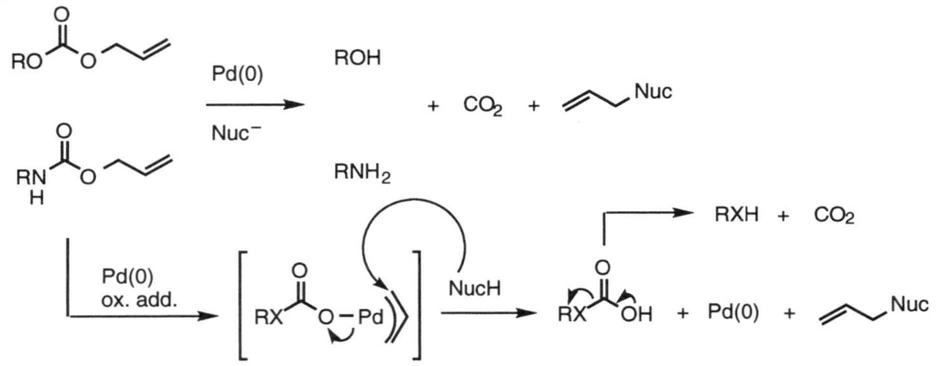

Eq. 9.60

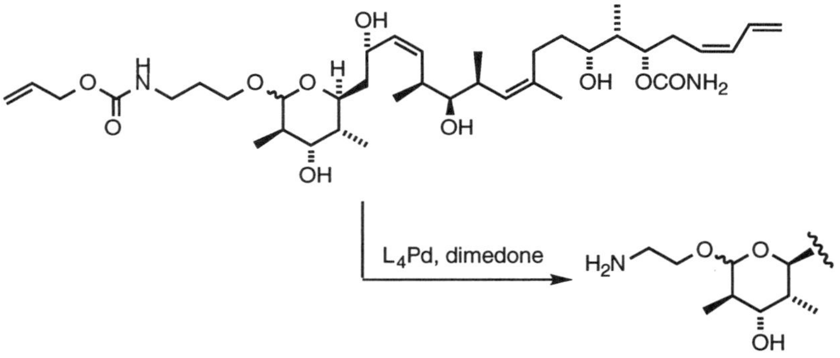

e. Insertion Processes Involving η^3-Allylpalladium Complexes

Alkenes, alkynes, and carbon monoxide insert into η^3-allylpalladium complexes (perhaps via the η^1 isomer), generating σ-alkylpalladium complexes which can undergo the very rich chemistry described in Chapter 4. When carried out in an intramolecular manner efficient cyclization processes can be developed (Eq. 9.61).[71] These reactions are synthetically very powerful, and demonstrate the real potential for the use of transition metals in organic synthesis, since often even the substrates are synthesized using transition metals. For example, in Eq. 9.62, the 1-

Eq. 9.61

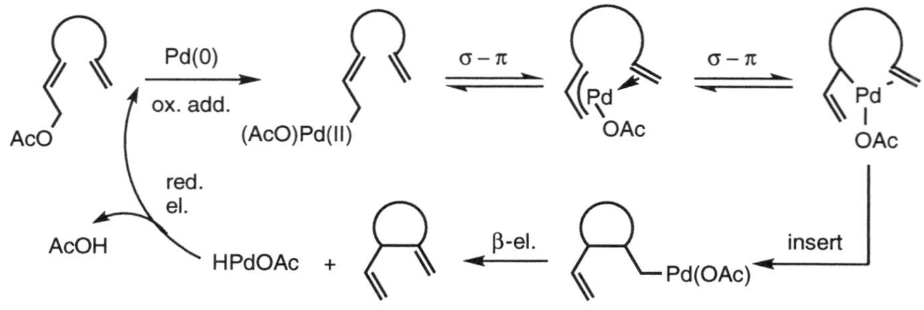

acetoxy-4-chlorobut-2-ene was synthesized by the Pd(OAc)$_2$-catalyzed chloroacetoxylation of dienes and the diallyl *bis*-sulfone prepared by Pd-catalyzed allylic alkylation of allyl halides. Treatment of this compound with Pd(dba)$_2$ in acetic acid results in efficient cyclization.[72] The reaction is quite general, and tolerates a variety of functional groups (Eq. 9.63).[73]

Eq. 9.62

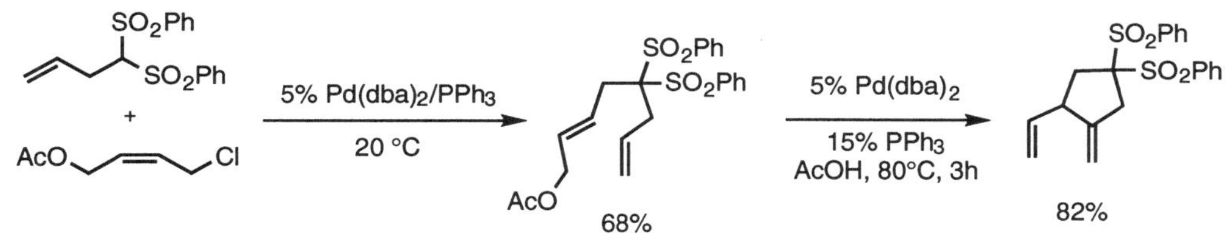

68%   82%

Eq. 9.63

90%   75%

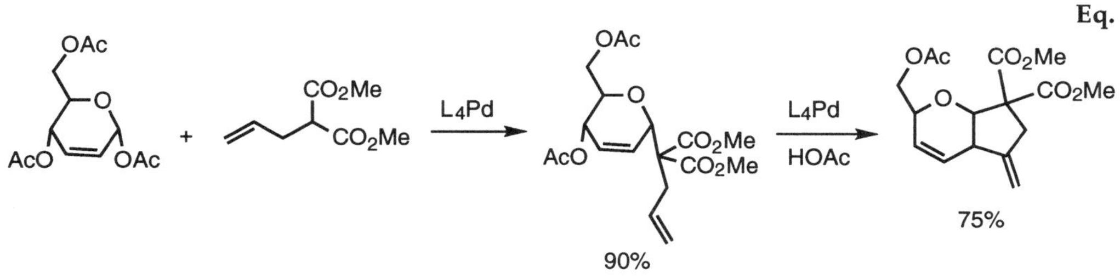

When additional unsaturation is appropriately situated, additional insertions can occur, resulting in the formation of several rings in a single efficient process (Eq. 9.64[74] and Eq. 9.65[75]).

Eq. 9.64

70%

Eq. 9.65

80%

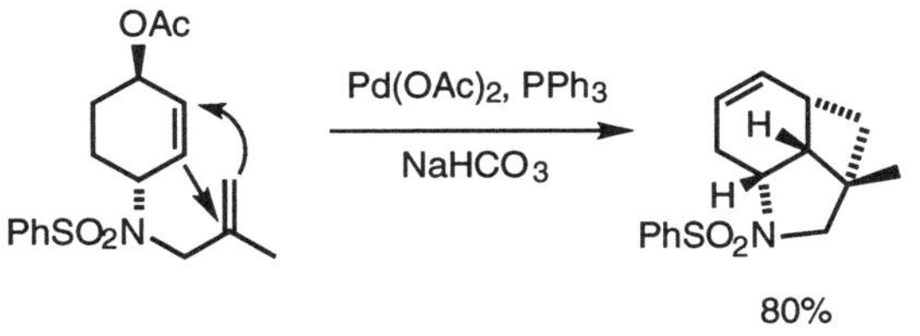

Dienes also insert into η^3-allylpalladium complexes, producing another η^3-allyl complex, which can undergo nucleophilic attack by acetate to form diene acetates (Eq. 9.66).[76] Again, palladium catalysis is also used to synthesize the starting materials. In a related process, asymmemtric induction using BINAP ligands was achieved, and the η^3-allyl intermediate was trapped by a stabilized carbanion (Eq. 9.67).[77]

Eq. 9.66

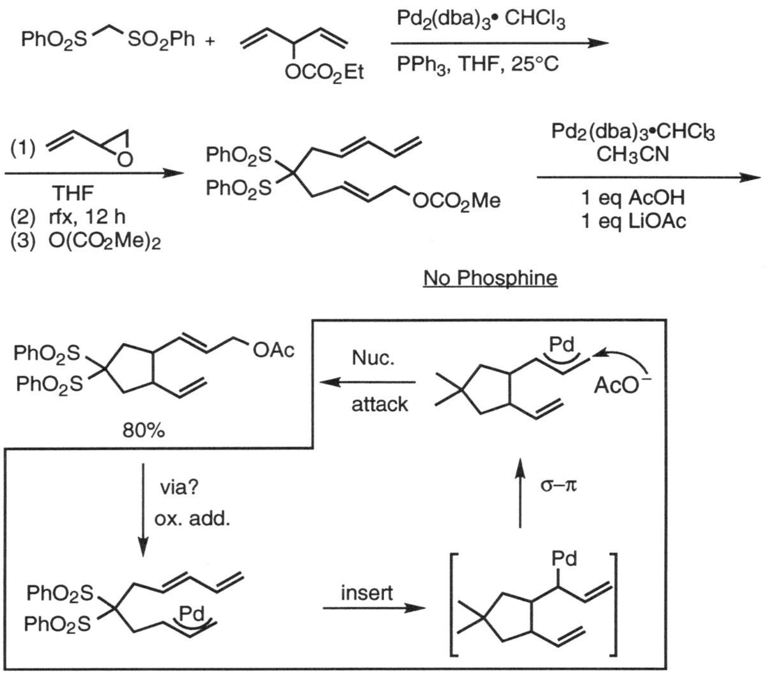

No Phosphine

Eq. 9.67

75-91%
66-80% ee

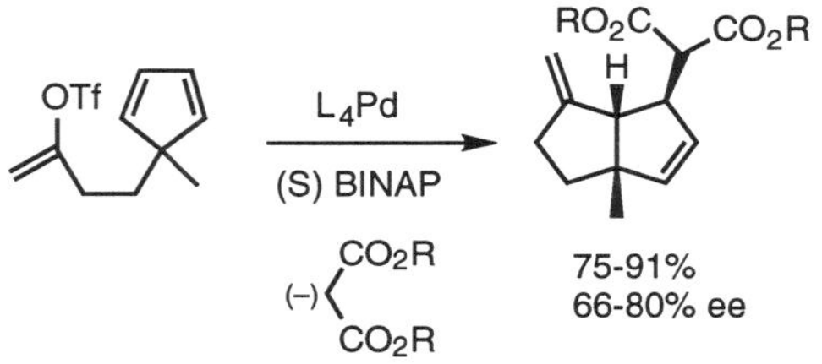

Insertion of an alkene into an η^3-allylpalladium complex generates an η^1-alkylpalladium complex, which itself is very reactive towards insertion of carbon monoxide (Chapter 4). By carrying out these cyclizations under an atmosphere of carbon monoxide the η^1-alkylpalladium intermediate can be intercepted making an η^1-acylpalladium complex which can undergo further insertion. With appropriate adjustment of substrate structure and reaction conditions, impressive polycyclizations can be achieved (Eq. 9.68[78] and Eq. 9.69[79]).

Eq. 9.68

Eq. 9.69

The process can also be truncated by transmetallation (Eq. 9.70).[80]

Eq. 9.70

Synthetic Applications of η^3-Allyl Transition Metal Complexes • **269**

f. Palladium(0)-Catalyzed Cycloaddition Reactions via Trimethylenemethane Intermediates

Although stable trimethylenemethane complexes of transition metals have been known for a long time, they have found little use in synthesis because of their lack of reactivity. However, 1,1'-bifunctional allyl compounds having an allylic acetate and an allyl silane undergo reaction with palladium(0) complexes to produce unstable, uncharacterized intermediates that react as if they were zwitterionic trimethylenemethane complexes, undergoing [3+2] cycloadditions to a range of electron-deficient alkenes (Eq. 9.71).[81] The reaction is thought to involve oxidative

Eq. 9.71

addition of the allyl acetate to the Pd(0) complex, generating a cationic η^3-allyl palladium complex. Displacement of the remaining allyl silane group, perhaps by acetate, produces a zwitterionic trimethylene methane complex, the anionic end of which attacks the β-position of a conjugated enone, generating an α-enolate which in turn attacks a terminus of the electrophilic cationic η^3-allylpalladium complex.

It is not known if this cycloaddition is concerted; however, since the stereochemistry of the alkene partners is maintained in the product, ring closure must occur faster than rotation about an incipient single bond. In addition, the process is quite diastereoselective (Eq. 9.72)[82] and (Eq. 9.73).[83] With unsymmetrical bifunctional allyl acetates, regioselective coupling occurs, with coupling at the more substituted terminus regardless of the electronic nature of the functional group and the initial position of the acetate and silyl leaving groups (Eq. 9.74 and Eq. 9.75).[84] This indicates that equilibration of the three termini of the trimethylmethane fragment must occur more rapidly than does coupling. With more extended acceptors such as pyrones or tropone, [3+2], [3+4], or [3+6] cycloadditions may occur, depending on the substituents on the acceptor (Eq. 9.76 and Eq. 9.77),[85] while β-ketoesters can participate in [3+3] cycloaddition (Eq. 9.78).[86] Intramolecular versions of this reaction are quite efficient for the production of highly functionalized polycyclic compounds (Eq. 9.79[87] and Eq. 9.80).[88]

Eq. 9.72

87%
4:1 diastereoselectivity

69%
>99:1 diastereoselectivity

Eq. 9.73

75%   >99% de

Eq. 9.74

55%

Eq. 9.75

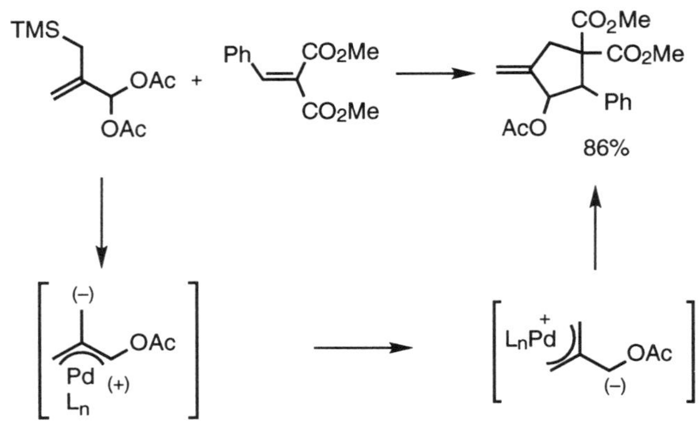

Eq. 9.76

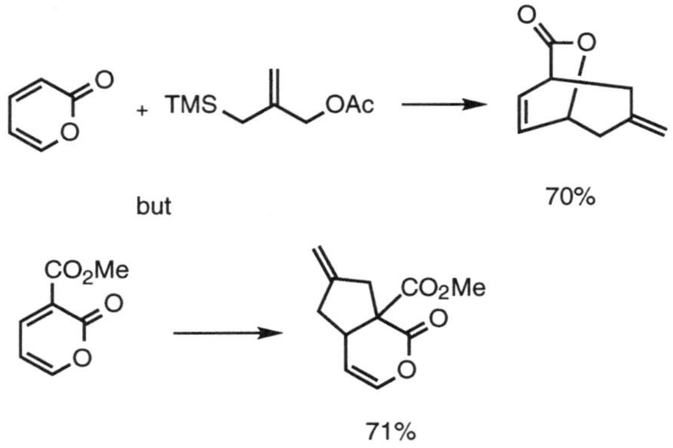

but

70%

71%

Eq. 9.77

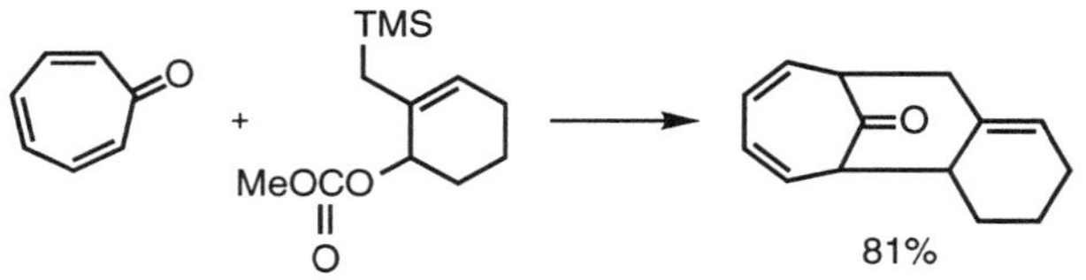

81%

Eq. 9.78

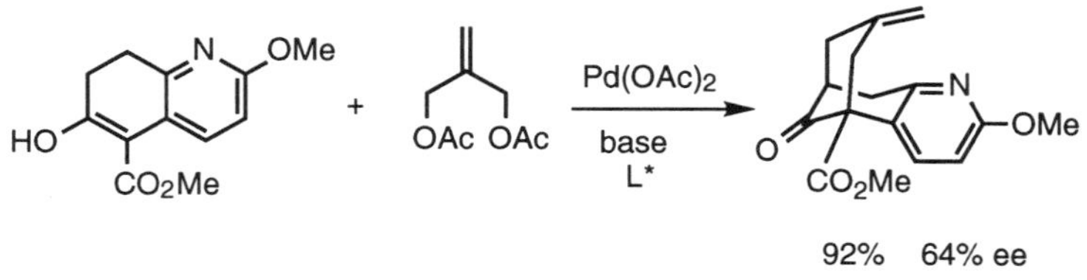

92%   64% ee

Eq. 9.79

100%

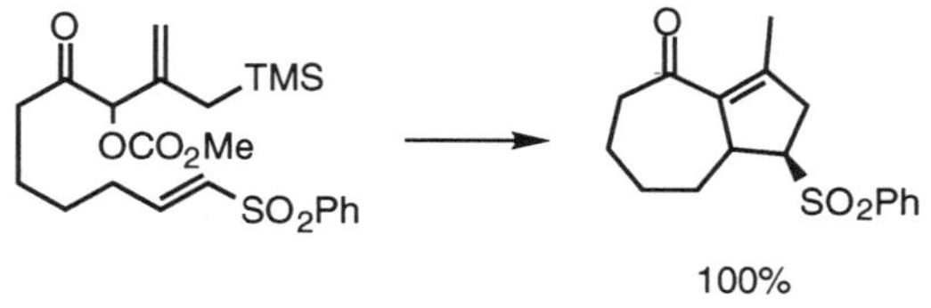

Eq. 9.80

50-70%

Aldehydes can also participate in [3+2] cycloaddition reactions with bifunctional allyl acetates in the presence of palladium(0) catalyst, forming five-membered oxygen heterocycles (Eq. 9.81).[89] With unsymmetrical trimethylene

Eq. 9.81

R = Me, Ph, CH=CH2, CN, $\overset{O}{\overset{\|}{C}}$Et, OAc

methane precursors, mixtures of regioisomers are obtained, in contrast to the observations with conjugated enones. This is rationalized by asserting that cyclization competes effectively with rearrangement of the initially formed trimethylenemethane complexes. Tin cocatalysts dramatically improve the yields and regioselectivity of the process, perhaps by acting as a Lewis acid to increase the reactivity of the aldehyde. A more profound effect is noted when an In(III) cocatalyst is used with conjugated enones as acceptors. This additive completely changes the regioselectivity of this reaction, from 1,4 (addition to the olefin) to 1,2 (addition to the carbonyl group) (Eq. 9.82).[90] This change is thought to be the result of coordination of the highly electropositive In(III) to the carbonyl oxygen, enhancing reaction at this site over attack of an olefinic position.

Eq. 9.82

Methylenecyclopropanes also undergo palladium(0)-catalyzed [3+2] cycloaddition reactions, again, most likely through trimethylenemethane intermediates (Figure 9.3)[91] formed by "oxidative addition" into the activated

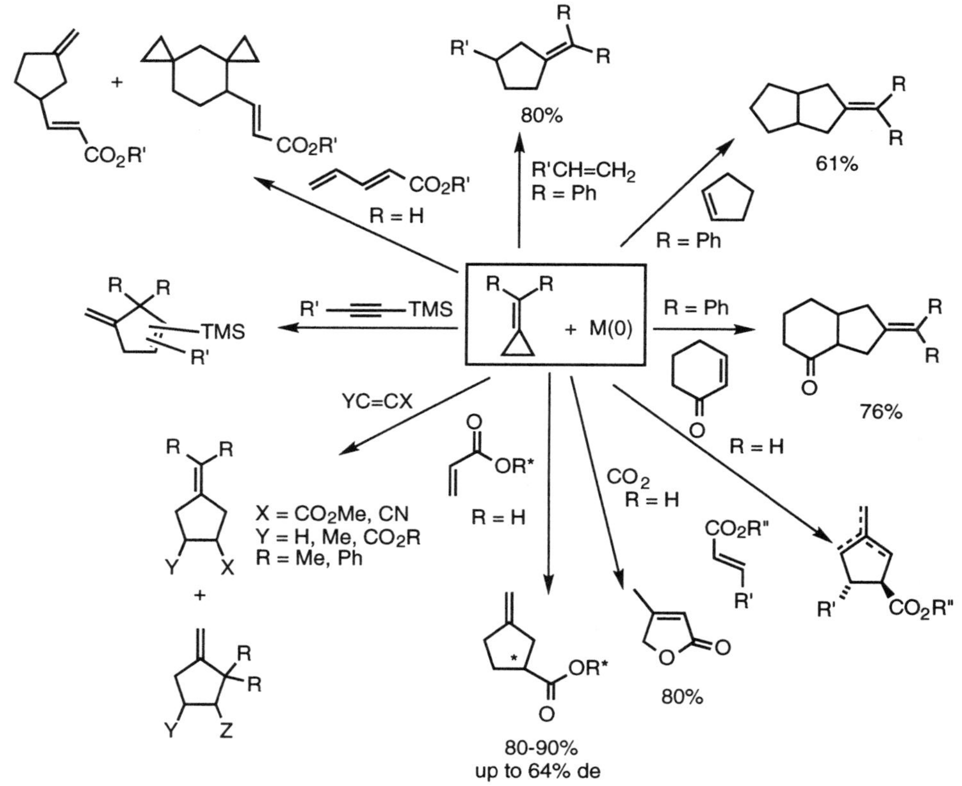

Figure 9.3 "Trimethylenemethane" Reactions of Methylene Cyclopropenes

cyclopropane carbon–carbon bond. Nickel(0) complexes catalyze similar processes. Although most of the studies of this system have been carried out on relatively simple substrates, a few, more complex, systems have been examined (Eq. 9.83).[92]

Eq. 9.83

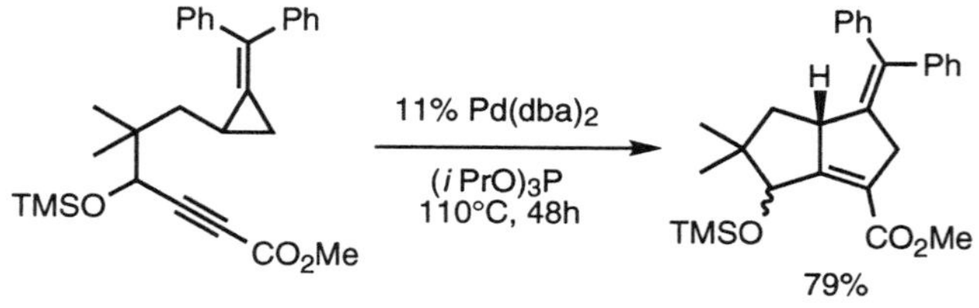

With diastereoisomerically pure methylenecyclopropanes the reaction is stereospecific and proceeds with retention of configuration at the chiral carbon center at which reaction occurred (Eq. 9.84).[93] With alkene rather than alkyne acceptors, the reaction is stereospecific (retention) with respect to the preexisting cyclopropane stereocenter, and the configuration at the α-position determines the stereochemistry of the newly-formed ring junction (Eq. 9.85).[94]

Eq. 9.84

R^4O_2C

R^3

R^3

H R^1 R^2

O

2% $Pd_2(dba)_3$

$P(OiPr)_3$
$PhCH_3$
rfx

CO_2R^4

R^2

R^1

H

R^3

R^3

retention

58-85%

$R^1 = nC_7H_{15}$, H, cyclohex
$R^2 = H$, nC_7H_{15}, cyclohex $(CH_2)_5$
$R^3 = H$, nPr

Eq. 9.85

CO_2R

H

C_7H_{15}

10% L_4Pd

$PhCH_3$ rfx

RO_2C H

cis

O

H

C_7H_{15}

retention

CO_2R

H

C_7H_{15}

RO_2C H

trans

O

H

C_7H_{15}

87-91%

retention

9.4 η^3-Allyl Complexes of Metals Other than Palladium

a. Molybdenum, Tungsten, Rhodium, and Iridium

Molybdenum and tungsten hexacarbonyl both catalyze the alkylation of allylic acetates by stabilized carbanions. However, the regioselectivity is quite different from that observed with palladium(0) (Figure 9.4).[95] Whereas palladium directs the nucleophile to the less-substituted terminus of the η^3-allyl system, molybdenum leads to reaction at the more-substituted end when malonate is the attacking anion, and the less-substituted end when more sterically demanding nucleophiles are used. Tungsten catalysts result in attack at the more-substituted terminus, regardless of the steric bulk of the nucleophile. The complex $Mo(RNC)_4(CO)_2$ is a more efficient catalyst than $Mo(CO)_6$, but, in this case, attack at the less substituted allyl terminus predominates.[96] With sterically bulky nucleophiles in the presence of chiral diamine ligands, $Mo(CO)_3(EtCN)_3$ catalyzes allylic alkylation at the more substituted terminus with high ee.[97] Both Rh(I)[98] and Ir(I)[99] catalyze the allylic alkylation of allyl carbonates at the *more* substituted position, and with high enantiomeric excess in the presence of chiral oxazolidine-phosphine ligands.[100] Clearly, the issue of regioselectivity in these processes is not yet resolved.

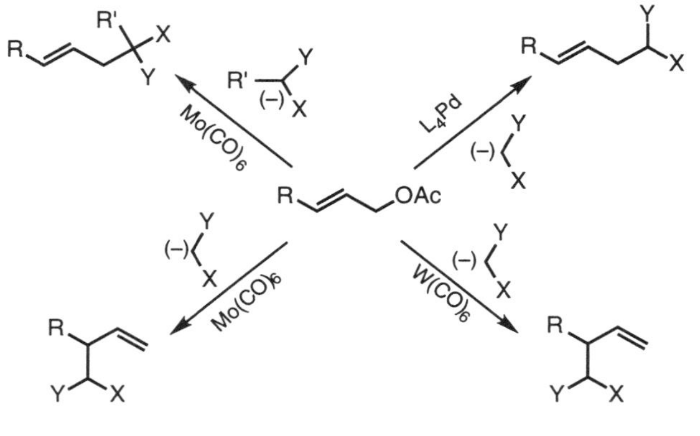

Figure 9.4   Group 6 Metal-Catalyzed Reactions of Allyl Acetates

Preformed η^3-allylmolybdenum complexes can be used both to activate adjacent functional groups and to control stereochemistry. For example, the acetyl complex in Eq. 9.86 undergoes facile aldol condensation with benzaldehyde followed by reduction by $NaBH_4$ from the face opposite the metal. Replacement of the relatively inert CO ligands by nitrosyl (formally NO+) and chloride produces a more labile complex which slowly condenses with benzaldehyde to give the 1,3-diol in modest yield after hydrolysis (Eq. 9.86).[101]

Eq. 9.86

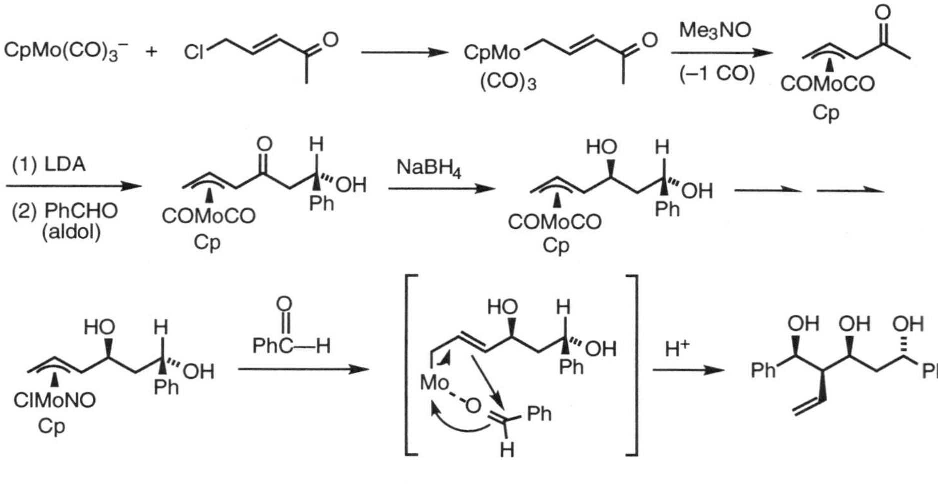

b. η³-Allyliron Complexes

Cationic η³-allyliron tricarbonyl complexes undergo nucleophilic attack by stabilized carbanions in much the same way as η³-allylpalladium complexes, and NaFe(CO)₃(NO) will catalyze the allylic alkylation of allyl chlorides and acetates.[102] However, perhaps because of the efficiency of the Pd-catalyzed processes, this related iron chemistry has found little use in organic synthesis.

In contrast to catalytic systems, η³-allyliron complexes preformed from γ-alkoxy or acetoxy enones are becoming quite useful in organic synthesis.[103] This utility relies on the ability of iron to stereoselectively complex a single prochiral face of the alkene, and to direct nucleophilic attack to the face opposite the metal, thereby leading to highly stereoselective processes. Early studies centered on unsaturated lactams and indicated that the γ-alkoxy group directed complexation primarily to the same face it occupied (Eq. 9.87).[104] The diastereoisomers could be separated, treated with allylsilane and BF₃•OEt₂, and decomplexed to give a single diastereoisomer of allylated product resulting from exclusive attack from the face opposite the metal. Acyclic γ-alkoxyenones show similar selectivity and reactivity, and have been much more widely utilized in synthesis (Eq. 9.88).[103a]

Eq. 9.87

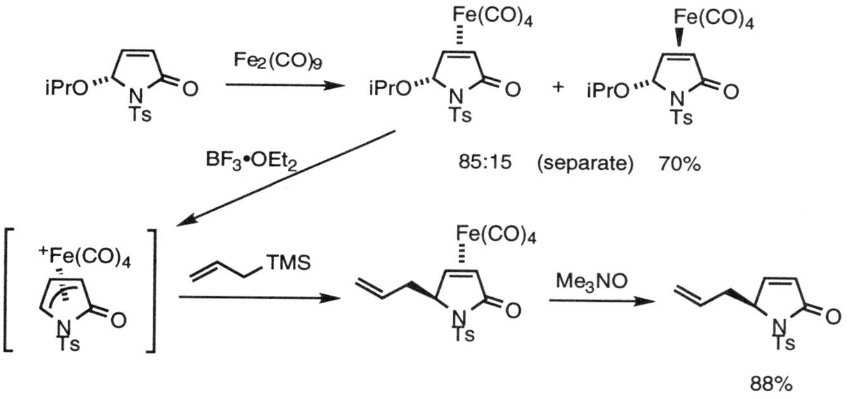

Eq. 9.88

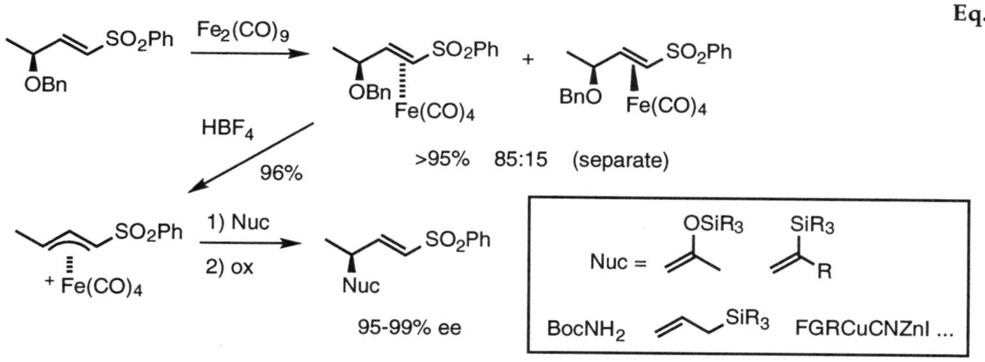

c. η³-Allylcobalt Complexes

The cobalt carbonyl anion $[Co(CO)_4]^-$ is a weak base but modest nucleophile which reacts with allylic halides to produce η³-allylcobalt carbonyl complexes (Eq. 9.89).[105] These are relatively unstable, deep red oils, and little synthetic use for them has been found. When treated with methyl iodide, the $[Co(CO)_4]^-$ anion reacts to form the η¹-methyl complex, which readily inserts CO to form the η¹-acyl complex. Treatment of this complex with butadiene results in another insertion, producing the β-acyl-η³-allyl cobalt complex (Eq. 9.90). Treatment with a strong base abstracts the quite acidic α-proton, producing the acyl diene, and regenerating the cobalt carbonyl anion.[106] In contrast, treatment with a stabilized carbanion results in nucleophilic attack at the unsubstituted end of the η³-allyl group, resulting in an overall alkylation-acylation of the 1,3-diene.[107] Neither of these processes have been utilized with complex substrates.

Eq. 9.89

$$Co_2(CO)_8 \; + \; Na/Hg \longrightarrow 2 \; Na \; Co(CO)_4 \xrightarrow{\quad} \langle\langle Co(CO)_3$$

$$M^0 \, d^9 \qquad\qquad M^{-1} \, d^{10}$$

Eq. 9.90

$$NaCo(CO)_4 \; + \; CH_3I \longrightarrow CH_3Co(CO)_4 \xrightarrow{CO} CH_3\overset{O}{\overset{\|}{C}}-Co(CO)_4 \xrightarrow{\quad}$$

$$M^{-1} \, d^{10} \qquad\qquad M^{+1} \, d^8 \qquad\qquad M^{+1} \, d^8$$

d. η³-Allylnickel Complexes[108]

η³-Allylnickel halides are generated in high yield by the reaction of allylic halides with nickel(0) complexes, usually nickel carbonyl or *bis*(cyclooctadiene)nickel in nonpolar solvents such as benzene (Eq. 9.91). This reaction tolerates a range of functional groups in the allyl chain, and allows the preparation of a number of synthetically useful functionalized complexes. η³-Allylnickel halides are not directly accessible from olefins, in contrast to the corresponding palladium complexes, at least in part because nickel(II)-olefin complexes are virtually unknown, allowing no pathway for activation of the allylic position for proton removal. η³-Allylnickel halide complexes are deep red to red-brown crystalline solids, which are quite air-sensitive in solution, but are stable in the absence of air.

Eq. 9.91

$$\text{allyl–X} + \begin{array}{l}\text{Ni(CO)}_4\\ \text{Ni(COD)}_2\end{array} \longrightarrow \left(\!\!\left(\text{Ni}\underset{X}{\overset{X}{<}}\text{Ni}\right)\!\!\right)$$

red, air-sensitive
solid

Although η^3-allylpalladium halides are subject to nucleophilic attack, η^3-allylnickel halides are, in most cases, not. Instead they behave, at least superficially, as if they were nucleophiles themselves, reacting with organic halides, aldehydes, and ketones, to transfer the allyl group. However, these reactions are radical-chain processes rather than nucleophilic reactions, and the chemistry of η^3-allylnickel halides is drastically different from that of the corresponding palladium complexes.

The best-established and most widely used reaction of η^3-allylnickel halide complexes is their reaction with organic halides to replace the halogen with the allyl group (Eq. 9.92). This reaction proceeds only in polar coordinating solvents, such as DMF, HMPA, or N-methylpyrrolidone. Aryl, alkenyl, primary, and secondary alkyl

Eq. 9.92

$$\left(\!\!\left(\text{Ni}\underset{}{\overset{Br}{<}}\right)\!\!\right)_2 + \text{RX} \xrightarrow{\text{DMF}} \text{R}\diagup\!\!\diagdown + \text{NiBrX}$$

R = alkyl, aryl, alkenyl, benzyl

tolerates OH, NH$_2$, CO$_2$R, CO$_2$H, CHO, COR, CN;
I > Br > Cl

bromides and iodides react in high yields, with aryl and alkenyl halides being considerably more reactive than the alkyl halides. Chlorides react much more slowly than bromides or iodides. This reaction tolerates a wide variety of functional groups including hydroxyl, ester, amide, and nitrile. These complexes will react with bromides in preference to chlorides in the same molecule, and will tolerate ketones and aldehydes in some instances. With unsymmetric η^3-allyl groups, coupling occurs exclusively at the less-substituted terminus, in contrast to most main group allyl organometallics. This property was used to an advantage in the synthesis of (+)-cerulenine (Eq. 9.93)[109] and an indole alkaloid (Eq. 9.94).[110] Although allylic halides

Eq. 9.93

$$\text{(furanone-CH}_2\text{CH=CHCH}_2\text{I)} + \left(\!\!\left(\text{Ni}\overset{Br}{<}\right)\!\!\right)_2 \xrightarrow{\text{DMF}} \text{(product)}$$

87%

are among the most reactive toward these complexes, coupling normally results to give all possible products because of rapid exchange of η^3-allyl ligand with the allyl halide. If the two allyl groups are somewhat different electronically, selective cross-coupling is sometimes observed in reasonable yield.[111] The stereochemistry of the olefin in the η^3-allyl system is lost in these allyl transfer reactions, but the stereochemistry of the olefin in alkenyl halide substrates is normally maintained.

Eq. 9.94

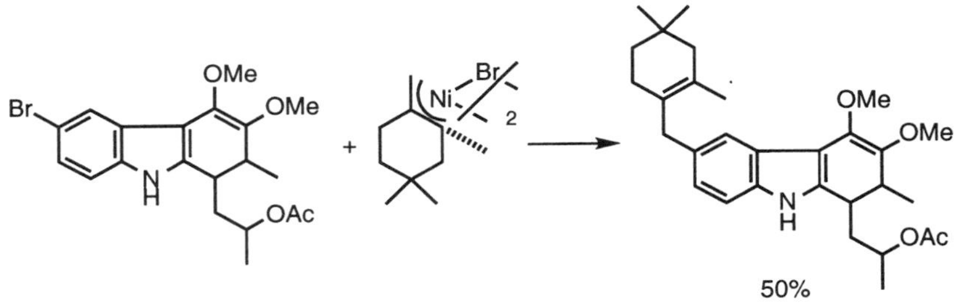

The ability of η^3-allylnickel halides to react with aryl halides under very mild conditions and to tolerate a wide range of functional groups permits the introduction of allyl groups into a very broad array of substrates (Eq. 9.95[111a] and Eq. 9.96[112]).

Eq. 9.95

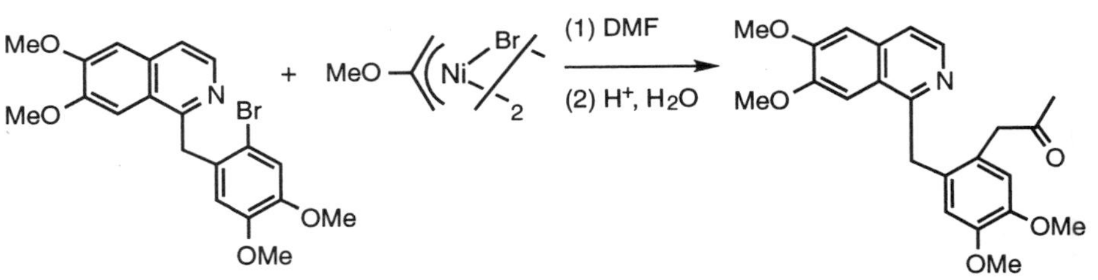

Eq. 9.96

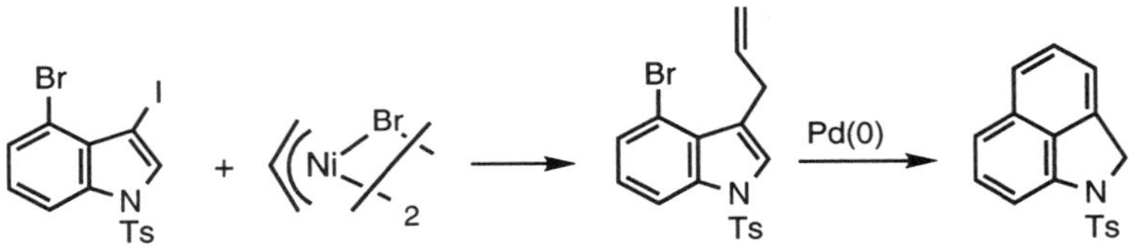

Conjugated enones react with nickel(0) complexes in the presence of trimethylsilyl chloride to produce 1-trimethylsilyloxy-η^3-allylnickel complexes. These undergo typical coupling reactions with organic halides to produce silylenol ethers (Eq. 9.97).[113] This product is one that would arise by nucleophilic (R-) addition to the β-position of the starting enone. However, by complexation to nickel, the normal reactivity patterns are reversed, and the β-alkyl group is introduced as an electrophile (RX = "R+"). By using η^3-allylnickel complexes as intermediates, the conjugate addition of alkyltin reagents to conjugated enones can be catalyzed. Although these two processes are related, they differ somewhat mechanistically (Eq. 9.98).[114]

Eq. 9.97

R[1] = H, Me, Ph
R[2] = H, Me

80-97%

60-80%

R = n Pr, n Bu,

, Ph, i Pr, p CHOPr

Eq. 9.98

48-79%

R =   , Ph   ,   , CH_3-C-

Although the products of the reaction between η^3-allylnickel halides and organic halides are simple, the process by which they form is complex. The process appears to be a radical-chain reaction, initiated by light or added reducing agents and completely inhibited by less than one mole percent of *m*-dinitrobenzene, an efficient radical anion scavenger. Chiral secondary halides racemize upon allylation, implying the intermediacy of free carbon-centered radicals, but alkenyl halides maintain their streochemistry, implying an absence of free carbon-centered radicals.[115]

Although η^3-allylnickel halides react with organic halides in preference to aldehydes or ketones, under slightly more vigorous conditions (50°C versus 20°C) carbonyl compounds do react to produce homoallylic alcohols.[116] α-Diketones are the most reactive substrates, producing α-ketohomoallylic alcohols. Aldehydes and cyclic ketones, including cholestanone, progesterone, and 5-α-androstane-3,17-dione, react well (the steroids at the most reactive carbonyl group), but simple aliphatic and α,β-unsaturated ketones react only sluggishly. Again, reaction occurs at the less-substituted end of the allyl group, in contrast to main group allyl organometallics.

The reaction of η^3-(2-carboethoxyallyl)nickel bromide with aldehydes and ketones produces α-methylene-γ-butyrolactones (Eq. 9.99).

Eq. 9.99

In a process that must involve η^3-allylnickel complex intermediates and oxidative addition/transmetallation/reductive elimination cycles, nickel(II) phosphine complexes catalyzed the alkylation of allylic acetates by aryl[117] and alkenyl boronates (Eq. 9.100).[118] These processes bear a striking resemblence to π-allylpalladium chemistry discussed above, and are likely to increase the scrutiny of nickel complexes to catalyze other processes thought to be the exclusive domain of palladium chemistry, such as allylic amination (Eq. 9.101).[119]

Eq. 9.100

Eq. 9.101

R = Ac, PhCO, Ph

References

1. (a) Tsuji, J. *Acct. Chem. Res.* **1973**, *6*, 8. (b) Tsuji, J. *Pure Appl. Chem.* **1981**, *53*, 2371. (c) Tsuji, J. *Pure Appl. Chem.* **1982**, *54*, 197. (d) Takacs, J.M. "Transition Metal Allyl Complexes: Telomerization of Dienes" in *Comprehensive Organometallic Chemistry*, Wilkinson, G.; Stone, F.G.A.; Abel, E.W. eds., Pergamon, New York, 1982, vol. 6, pp. 785-797.
2. Takahashi, T.; Minami, I.; Tsuji, J. *Tetrahedron Lett.* **1981**, *22*, 2651.
3. (a) Takacs, J.M.; Zhu, J. *J. Org. Chem.* **1989**, *54*, 5193. (b) Takacs, J.M.; Chandramouli, S. *Organometallics* **1990**, *9*, 2877. (c) Takacs, J.M.; Zhu, J. *Tetrahedron Lett.* **1990**, *31*, 1117.
4. Takacs, J.M.; Chandramouli, S.V. *J. Org. Chem.* **1993**, *58*, 7315.
5. Takacs, J.M.; Zhu, J.; Chandramouli, S. *J. Am. Chem. Soc.* **1992**, *114*, 773.
6. Takacs, J.M.; Clement, F.; Zhu, J.; Chandramouli, S.V.; Gong, X. *J. Am. Chem. Soc.* **1997**, *119*, 5804.
7. Takacs, J.M.; Weidner, J.J.; Newsome, P.W.; Takacs, B.E.; Chidambaraw, R.; Shoemaker, R. *J. Org. Chem.* **1995**, *60*, 3473.
8. Takacs, J.M.; Boito, S.C. *Tetrahedron Lett.* **1995**, *36*, 2941.

9. Jolly, P.W. "Nickel-Catalyzed Oligomerization of Alkenes and Related Reactions" in *Comprehensive Organometallic Chemistry*, Wilkinson, G.; Stone, F.G.A.; Abel, E.W. eds., Pergamon, New York, 1982, vol. 6, pp. 615-48.

10. (a) Wender, P.A.; Ihle, N.C.; Corriea, C.R.D. *J. Am. Chem. Soc.* **1988**, *110*, 5904. (b) For related cycloadditions see: Wender, P.M.; Nuss, J.M.; Smith, D.B.; Swarez-Sobrino, A.; Vagberg, J.; DeCosta D.; Bordner, J. *J. Org. Chem.* **1997**, *62*, 4908.

11. Wender, P.A.; Jenkins, T.E. *J. Am. Chem. Soc.* **1989**, *118*, 6432.

12. For a review on selectivity in these reactions see: Frost, C.G.; Howarth, J.; Williams, J.M.J. *Tetrahedron Asymm.* **1992**, *3*, 1089. For a general review see: Harrington, P.M., "Transition Metal Allyl Complexes: Pd, W, Mo-Assisted Nucleophilic Attack" in "Comprehensive Organometallic Chemistry II", Abel, E.W.; Stone, F.G.A.; Wilkinson, G., Eds., Elsevier Science Ltd., Oxford, UK, 1995, Vol. 12, pp. 797-904.

13. (a) Trost, B.M.; Bant, R.C. *J. Am. Chem. Soc.* **1998**, *120*, 70. (b) Kawatsura, M.; Uozumi, Y.; Hayashi, T. *J. Chem. Soc., Chem. Comm.* **1998**, 217.

14. Hayashi, T.; Kawatsura, M.; Uozumi, Y. *J. Am. Chem. Soc.* **1998**, *120*, 1681.

15. For old reviews see: (a) Trost, B.M.; Verhoeven, T.R. *J. Am. Chem. Soc.* **1980**, *102*, 4730. (b) Yamamoto, T.; Saito, O.; Yamamoto, A. *J. Am. Chem. Soc.* **1981**, *103*, 5600. (c) Trost, B.M. *Acc. Chem. Res.* **1980**, *13*, 385.

16. (a) Hayashi, T.; Konishi, M.; Kumada, M. *J. Chem. Soc., Chem. Comm.* **1984**, 107. (b) Hayashi, T.; Hagihara, T.; Konishi, M.; Kumada, M. *J. Am. Chem. Soc.* **1983**, *105*, 7767. (c) Leutenegger, V.; Umbricht, G.; Fahrni, C. von Matt, P.; Pfaltz, A. *Tetrahedron* **1992**, *48*, 2143.

17. Trost, B.M.; Ceschi, M.A.; König, B. *Angew. Chem. Int. Ed. Engl.* **1997**, *36*, 1486.

18. Naz, N.; Al-Tey, T.H.; Al-Abed, Y.; Voelter, W.; Fikes, R.; Hiller, W. *J. Org. Chem.* **1996**, *61*, 3230.

19. Braun, M.; Onuma, H.; Arinaga, Y. *Chem. Lett.* **1995**, 1099.

20. For a review see: Heumann, A.; Regher, M. *Tetrahedron* **1995**, *51*, 975.

21. Roland, S.; Durand, J.O.; Savignac, M.; Genet, J.P. *Tetrahedron Lett.* **1995**, *36*, 3007.

22. Fürstner, A.; Weintritt, H. *J. Am. Chem. Soc.* **1998**, *120*, 2817.

23. (a) Boeckman, R.K.; Shair, M.D.; Vargas, J.R.; Stoltz, L.A. *J. Org. Chem.* **1993**, *58*, 1295. (b) Michelet, V.; Besner, I.; Genet, J.P. *Synlett* **1996**, 215.

24. (a) Schink, H.E.; Backvall, J-E. *J. Org. Chem.* **1992**, *57*, 1588. (b) Bäckvall, J-E.; Gatti, R.; Shink, H.E. *Synthesis* **1993**, 343.

25. For reviews see: (a) Trost, B.M.; Van Vranken, D.L. *Chem. Rev.* **1996**, *96*, 395. (b) Trost, B.M. *Acc. Chem. Res.* **1996**, *29*, 355. (c) Williams, J.M.J. *Synlett* **1996**, 705. (d) Helmchen, G.; Kudis, S.; Sennhenn, P.; Steinhaugher, H. *Pure Appl. Chem.* **1997**, *69*, 513.

26. Kardos, N.; Genet, J.P. *Tetrahedron Asymmetry* **1994**, *5*, 1525. See also Pretot, R.; Pfaltz, A. *Angew. Chem. Int. Ed. Engl.* **1998**, *37*, 323.

27. Granberg, K.L.; Bäckvall, J-E. *J. Am. Chem. Soc.* **1992**, *114*, 6858.

28. For a mechanistic study see: Seebach, D.; Devaquat, E.; Ernst, A.; Hayakawa, M.; Kühnle, F.N.M.; Schweizer, W.B.; Weber, B. *Helv. Chim. Acta* **1995**, *78*, 1636.

29. Baldwin, J.C.; Williams, J.M.J.; Beckett, R.P. *Tetrahedron Asymmetry* **1995**, *6*, 1515.

30. For the cyclohexenyl version of this process see: Trost, B.M.; Chipak, L.S.; Lübbers, T. *J. Am. Chem. Soc.* **1998**, *120*, 1732.

31. Yoshigaki, H.; Satoh, H.; Sato, Y.; Nukui, S.; Shibasaki, M.; Mori, M. *J. Org. Chem.* **1995**, *60*, 2016.

31. (a) Takanashi, S-i.; Mori, K. *Liebigs Ann Chem./Rec.* **1997**, 825. (b) For the original work in this area see: Del Valle, J.K.; Hegedus, L.S. *J. Org. Chem.* **1990**, *55*, 3019.

32. (a) White, J.D.; Jensen, M.S. *Synlett* **1996**, 31. (b) Echavarren, A.M.; Tueting, D.R.; Stille, J.K. *J. Am. Chem. Soc.* **1988**, *110*, 4039. (c) Teuting, D.R.; Echavarren, A.M.; Stille, J.K. *Tetrahedron* **1989**, *45*, 979.

33. Castaño, A.M.; Ruano, M.; Echavarren, A.M. *Tetrahedron Lett.* **1996**, *37*, 6591.

34. Farina, V.; Baker, S.R.; Benigni, D.A.; Sapino, C., Jr. *Tetrahedron Lett.* **1988**, *29*, 5739.

35. Urabe, H.; Inami, H.; Sato, F. *J. Chem. Soc., Chem. Comm.* **1993**, 1595.

36. (a) For a review on allylic amination see: Johannsen, M.; Jørgensen, *Chem. Rev.* **1998**, *98*, 1689. (b) Flegelova, Z.; Patek, M. *J. Org. Chem.* **1996**, *61*, 6735.

37. Ohno, H.F.; Yu, J. *J. Chem. Soc., Perkin Trans I* **1998**, 391.

38. Kapeller, H.; Marschener, C.; Weissenbacher, M.; Griengl, H. *Tetrahedron* **1998**, *54*, 1439.

39. Larsson, A.L.E.; Gatti, R.G.P.; Backväll, J-E. *J. Chem. Soc., Perkin Trans I* **1997**, 2873.

40. Mori, M.; Kuroda, S.; Zhang, C-S.; Sato, Y. *J. Org. Chem.* **1997**, *62*, 3263.

41. Aggarwal, V.K.; Montiero, N. *J. Chem. Soc., Perkin Trans I* **1997**, 2349.

42. (a) Trost, B.M.; Pulley, S.R. *J. Am. Chem. Soc.* **1995**, *117*, 10143. (b) Trost, B.M.; Pulley, S.R. *Tetrahedron Lett.* **1995**, *36*, 8737.

43. Trost, B.M.; Patterson, D.E. *J. Org. Chem.* **1998**, *63*, 1339.

44. Trost, B.M.; Scanlan, T.S. *J. Am. Chem. Soc.* **1989**, *111*, 4988.
45. (a) Godleski, S.A.; Meinhart, J.D.; Miller, D.J.; Van Wallendae, S. *Tetrahedron Lett.* **1981**, *22*, 2247. (b) Tietze, L.F.; Schrock, H. *Angew. Chem. Int. Ed. Engl.* **1997**, *36*, 1124.
46. Trost, B.M.; Cossy, J. *J. Am. Chem. Soc.* **1982**, *104*, 6881.
47. Bradette, T.; Esher, J.L.; Johnson, C.R. *Tetrahedron Asymmetry* **1995**, *36*, 6251.
48. Goux, C.; Massacret, M.; Lhoste, P.; Sinow, D. *Organometallics* **1995**, *14*, 4585.
49. Trost, B.M.; Toste, F.D. *J. Am. Chem. Soc.* **1998**, *120*, 815.
50. Sinou, D.; Frappa, I.; Lhoste, P.; Porwanski, S.; Krycza, B. *Tetrahedron Lett.* **1995**, *36*, 6251.
51. Trost, B.M.; Tenaglia, A. *Tetrahedron Lett.* **1988**, *29*, 2974.
52. Fournier-Nguefack, C.; Lhoste, P.; Sinow, D. *J. Chem. Res. (S)* **1998**, 105.
53. Trost, B.M.; Greenspan, P.D.; Geissler, H.; Kim, J.H.; Greeves, N. *Angew. Chem. Int. Ed. Engl.* **1994**, *33*, 2182.
54. Shimizu, I.; Omura, T. *Chem. Lett.* **1993**, 1759.
55. (a) Hayashi, T.; Yamane, M.; Ohno, A. *J. Org. Chem.* **1997**, *62*, 204. (b) Tenaglia, A.; Kammerer, F. *Synlett* **1996**, 576.
56. (a) Bäckvall, J.E. in *Advances in Metal-Organic Chemistry*, Liebeskind, L.S.; ed. JAI Press, London, 1989, vol. 1, p. 135. (b) Bäckvall, J.E.; Bystrom, S.E.; Nordberg, R.E. *J. Org. Chem.* **1984**, *49*, 4619. (c) Nystrom, J-E.; Rein, T.; Bäckvall, J-E. *Org. Syn.* **1989**, *67*, 105.
57. (a) Grennberg, H.; Gogall, A.; Bäckvall, J-E. *J. Org. Chem.* **1991**, *56*, 5808. (b) Bergstad, K.; Grennberg, H.; Bäckvall, J.E. *Organometallics* **1998**, *17*, 45.
58. (a) Bäckvall, J.E.; Andersson, P.G. *J. Am. Chem. Soc.* **1992**, *114*, 6374. (b) Itami, K.; Palmgren, A.; Bäckvall, J.-E. *Tetrahedron Lett.* **1998**, *39*, 1223.
59. For a review see: Tsuji, J.; Mandai, T. *Synthesis* **1996**, 1.
60. (a) Mandai, T.; Kaihara, Y.; Tsuji, J. *J. Org. Chem.* **1994**, *59*, 5847. 9b) Lautens, M.; Delanghi, P.H.M. *Angew. Chem. Int. Ed. Engl.* **1994**, *33*, 2448.
61. Nagasawa, K.; Shimizu, I.; Nakata, T. *Tetrahedron Lett.* **1996**, *37*, 6881.
62. (a) Hayashi, T.; Iwamura, I.; Naito, M.; Matsumoto, Y.; Nozumi, Y.; Miki, M.; Yanagai, K. *J. Am. Chem. Soc.* **1994**, *116*, 775. (b) Hayashi, T.; Kawatsura, M.; Iwamura, H.; Yamura, Y.; Uozumi, Y. *J. Chem. Soc., Chem. Comm.* **1996**, 1767.
63. (a) Mandai, T.; Matsumoto, T.; Kawada, M.; Tsuji, J. *J. Org. Chem.* **1992**, *57*, 1326. (b) Mandai, T.; Suzuki, S.; Murakami, T.; Fujita, M.; Kawada, M.; Tsuji, J. *Tetrahedron Lett.* **1992**, *33*, 2987.
64. Mandai, T.; Matsumoto, T.; Tsuji, J. *Tetrahedron Lett.* **1993**, *34*, 2513.
65. Mandai, T.; Matsumoto, M.; nakao, Y.; Teramoto, A.; Kawada, M.; Tsuji, J. *Tetrahedron Lett.* **1992**, *33*, 2549.
66. Shimizu, I.; Matsumoto, Y.; Ono, T.; Satake, A.; Yamamoto, A. *Tetrahedron Lett.* **1996**, *37*, 7115.
67. Andersson, P.G.; Schab, S. *Organometallics* **1995**, *14*, 1.
68. For a review see: Guibe, F. *Tetrahedron* **1998**, *54*, 2967.
69. Hang, D.T.; Nerenberg, J.B.; Schreiber, S.L. *J. Am. Chem. Soc.* **1996**, *118*, 11054.
70. Hayakawa, Y.; Wakabayashi, S.; Kato, H.; Noyori, R. *J. Am. Chem. Soc.* **1990**, *112*, 1691.
71. For reviews see: (a) Oppolzer, W. in "Comprehensive Organometallic Chemistry II", Abel, E.W.; Stone, F.G.A.; Wilkinson, G., Eds., Pergamon, Oxford, UK, 1995, Vol. 12, pp. 905-921. (b) Oppolzer, W. *Angew. Chem. Int. Ed. Engl.* **1989**, *28*, 38. (c) *Pure Appl. Chem.* **1990**, *62*, 1941.
72. Oppolzer, W.; Xu, J.-Z.; Stone, C. *Helv. Chim. Acta* **1991**, *74*, 465.
73. Holzapfel, C.W.; Marais, L. *J. Chem. Res. (S)* **1998**, 60.
74. Oppolzer, W.; DeVita, R.J. *J. Org. Chem.* **1991**, *56*, 6256.
75. Grigg, R.; Sridharan, V.; Surkirthalingam, S. *Tetrahedron Lett.* **1991**,*32*, 3855.
76. Trost, B.M.; Luengo, J.I. *J. Am. Chem. Soc.* **1988**, *110*, 8239.
77. Ohshima, T.; Kagechika, K.; Adachi, M.; Sodeoka, M.; Shibasaki, M. *J. Am. Chem. Soc.* **1996**, *118*, 7108.
78. (a) Oppolzer, W.; Bienayme, H.; Genevois-Borella, A. *J. Am. Chem. Soc.* **1991**, *113*, 9660. (b) Oppolzer, W.; Xu, J-Z.; Stone, C. *Helv. Chim. Acta* **1991**, *74*, 465.
79. Keese, R.; Guidetti-Grept, R.; Herzog, B. *Tetrahedron Lett.* **1992**, *33*, 1207.
80. Holzapfel, C.W.; Marais, L. *Tetrahedron Lett.* **1998**, *39*, 2179.
81. For reviews see: (a) Trost, B.M. *Angew. Chem. Int. Ed. Engl.* **1986**, *25*. 1. (b) Trost, B.M. *Pure Appl. Chem.* **1988**, *60*, 1615. (c) Harrington, P.J. "Transition Memtal Allyl Complexes: Trimethylene Methane Complexes" in "Comprehensive Organometallic Chemistry II", Abel, E.W.; Stone, F.G.A.; Wilkinson, G.; Eds., Pergamon, Oxford, UK, 1995, Vol. 12, pp. 923-945. (d) Lautens, M.; Klute, W.; Tam, W. *Chem. Rev.* **1996**, *96*, 49.
82. (a) Trost, B.M.; Lynch, J.; Renaut, P.; Steinman, D.H. *J. Am. Chem. Soc.* **1986**, *108*, 284. (b) Trost, B.M.; Mignani, S.M. *Tetrahedron Lett.* **1986**, *27*, 4137.
83. Holzapfel, C.W.; van der Morwe, T.L. *Tetrahedron Lett.* **1996**, *37*, 2303.

84. Trost, B.M.; Nanninga, T.N.; Satoh, T. *J. Am. Chem. Soc.* **1985**, *107*, 721.
85. Trost, B.M.; Seoane, P.R. *J. Am. Chem. Soc.* **1987**, *109*, 615.
86. Kaneko, S.; Yoshino, T.; Katoh, T.; Terashima, S. *Tetrahedron* **1998**, *54*, 5471. For the first report of this transformation see: Campiani, G.; Sun, L-Q.; Kozikowski, A.P.; Aagaard, P.; McKinney, M. *J. Org. Chem.* **1993**, *58*, 7660.
87. Trost, B.M.; Grese, T.A. *J. Org. Chem.* **1992**, *57*, 686.
88. Trost, B.M.; Higuchi, R.L. *J. Am. Chem. Soc.* **1996**, *118*, 10094 and references therein.
89. Trost, B.M.; King, S.A. *J. Am. Chem. Soc.* **1990**, *112*, 408.
90. (a) Trost, B.M.; Sharma, S.; Schmidt, T. *J. Am. Chem. Soc.* **1992**, *114*, 7903. (b) Trost, B.M.; Sharma, S.; Schmidt, T. *Tetrahedron Lett.* **1993**, *34*, 7183.
91. Binger, P.; Büch, H.M. *Top. Curr. Chem.* **1987**, *135*, 77. For stereochemical and mechanistic studies see: Corley, H.; Motherwell, W.B.; Pennell, A.M.K.; Shipman, M.; Slawin, A.M.Z.; Williams, D.J.; Bingh, P.; Stepp, M. *Tetrahedron* **1996**, *52*, 4883.
92. Lewis, R.T.; Motherwell, W.B.; Shipman, M.; Slawin, A.M.Z.; Williams, D.J. *Tetrahedron* **1995**, *51*, 3285.
93. Lautens, M.; Ren, Y. *J. Am. Chem. Soc.* **1996**, *118*, 9597.
94. Lautens, M.; Ren, Y. *J. Am. Chem. Soc.* **1996**, *118*, 10668.
95. Trost, B.M.; Hung, M-H. *J. Am. Chem. Soc.* **1984**, *106*, 6837.
96. Trost, B.M.; Merlic, C.A. *J. Am. Chem. Soc.* **1990**, *112*, 9590.
97. Trost, B.M.; Hachiya, I. *J. Am. Chem. Soc.* **1998**, *120*, 1104.
98. Evans, P.A.; Nelson, J.D. *Tetrahedron Lett.* **1998**, *39*, 1729.
99. (a) Takeuchi, R.; Kashio, M. *Angew. Chem. Int. Ed. Engl.* **1997**, *36*, 263. (b) Takeuchi, R.; Kashio, M. *J. Am. Chem. Soc.* **1998**, *120*, 8647.
100. Jansson, J.P.; Helmchen, G. *Tetrahedron Lett.* **1997**, *38*, 8025.
101. Vong, W-J.; Peng, S-M.; Lin, S-H.; Lin, W-J.; Liu, R-S. *J. Am. Chem. Soc.* **1991**, *113*, 573.
102. Roustan, J.L.; Merour, J.Y.; Houlihan, F. *Tetrahedron Lett.* **1979**, 3721.
103. For reviews see: (a) Enders, D.; Jandeleit, B.; von Berg, S. *Synlett* **1997**, 421. (b) Speckamp, W.N. *Pure Appl. Chem.* **1996**, *68*, 695. (c) For early reports see: Green, J.; Carroll, M.K. *Tetrahedron Lett.* **1991**, *32*, 1141.
104. Hopman, J.C.P.; Hiemstra, H.; Speckamp, W.N. *J. Chem. Soc., Chem. Comm.* **1995**, 617.
105. Heck, R.F.; Breslow, D.S. *J. Am. Chem. Soc.* **1963**, *85*, 2779.
106. Heck, R.F. in *Organic Synthesis via Metal Carbonyls*, Wender, P. and Pino, P., eds. Wiley, New York, 1968, vol. 1, pp. 379-384.
107. (a) Hegedus, L.S.; Inoue, Y. *J. Am. Chem. Soc.* **1982**, *104*, 4917. (b) Hegedus, L.S.; Perry, R.J. *J. Org. Chem.* **1984**, *49*, 2570.
108. For reviews, see: (a) Semmelhack, M.F. *Org. React.* **1972**, *19*, 115. (b) Billington, D.C. *Chem. Soc. Rev.* **1985**, *14*, 93. (c) Krysan, D.J. "Transition Metal Allyl Complexes π-Allylnickel halides and Other π-Allyl Complexes Excluding Palladium" in "Comprehensive Organometallic Chemistry II", Abel, E.W.; Stone, F.G.A.; Wilkinson, G., Eds., Elsevier Science Ltd., Oxford, UK, 1995, Vol. 12, pp. 959-978.
109. Kedar, T.E.; Miller, M.W.; Hegedus, L.S. *J. Org. Chem.* **1996**, *61*, 6121.
110. Knölker, H-J.; Fröhner, W. *Tetrahedron Lett.* **1998**, *39*, 2537.
111. (a) Hegedus, L.S.; Stiverson, R. *J. Am. Chem. Soc.* **1974**, *96*, 3250. (b) Guerrieri, F.; Chinsoli, G.P.; Merzoni, S. *Gazz. Chim. Ital.* **1974**, *104*, 557.
112. Hegedus, L.S.; Sestrick, M.R.; Michaelson, E.T.; Harrington, P.J. *J. Org. Chem.* **1989**, *54*, 4141.
113. Johnson, J.R.; Tully, P.S.; Mackenzie, P.B.; Sabat, M. *J. Am. Chem. Soc.* **1991**, *113*, 6172.
114. Grisso, B.A.; Johnson, J.R.; Mackenzie, P.B. *J. Am. Chem. Soc.* **1992**, *114*, 5160.
115. Hegedus, L.S.; Thompson, D.H.P. *J. Am. Chem. Soc.* **1985**, *107*, 5663.
116. Hegedus, L.S.; Wagner, S.D.; Waterman, E.L.; Siirala-Hansen, K. *J. Org. Chem.* **1975**, *40*, 593.
117. Kobayashi, Y.; Takahisa, E.; Usimani, S.B. *Tetrahedron Lett.* **1998**, *39*, 597.
118. Kobayashi, Y.; Takahisa, E.; Usmani, S.B. *Tetrahedron Lett.* **1998**, *39*, 601.
119. Bricot, H.; Carpentier, J-F.; Mortreux, A. *Tetrahedron* **1998**, *54*, 1073.

CHAPTER 10

Synthetic Applications of Transition Metal Arene Complexes

10.1   Introduction

Arenes form stable, isolable complexes with a range of transition metals, including Cr, Mo, W, Fe, Ru, Os, and Mn. By far, the most common type of arene-transition metal complex is the η^6-coordinated type (Figure 10.1) in which the entire π-system of the arene is complexed. These η^6-arene complexes have a very rich reaction chemistry which constitutes the bulk of this chapter. η^2-Arene complexes are much less common, and only those of Os have found use in organic synthesis. With this type of arene complex, only two carbons of the π-arene system are complexed, in essence deconjugating the arene. The uncomplexed portion of the π-system then displays reactivity common to the residual degree of unsaturation (e.g. 1,3-dienes for Os-complexed benzenes). Both types of arene complexation – η^6 and η^2 – result in dramatic alteration in the reactivity of the complexed arene; and therein lies their utility for organic synthesis.

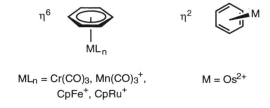

$$\eta^6$$

$$ML_n = Cr(CO)_3, Mn(CO)_3^+,$$
$$CpFe^+, CpRu^+$$

$$\eta^2$$

$$M = Os^{2+}$$

Figure 10.1   Modes of Arene Complexation

10.2 η^6-Arene Complexes[1]

a. Preparation

Arenechromium tricarbonyl complexes are normally prepared by heating $Cr(CO)_6$ in the arene as solvent, although the procedure is complicated by the sublimation of $Cr(CO)_6$.[2] Alternatively, one can start with preformed amine complexes such as $Cr(CO)_3(NH_3)_3$ or $Cr(CO)_5$(2-picoline) to avoid this problem (Eq. 10.1).[3] Perhaps the most convenient method is to carry out an arene exchange with the (naphthalene) $Cr(CO)_3$ complex (Eq. 10.2),[4] a procedure which goes under mild conditions and does not require large excesses of arene.

Eq. 10.1

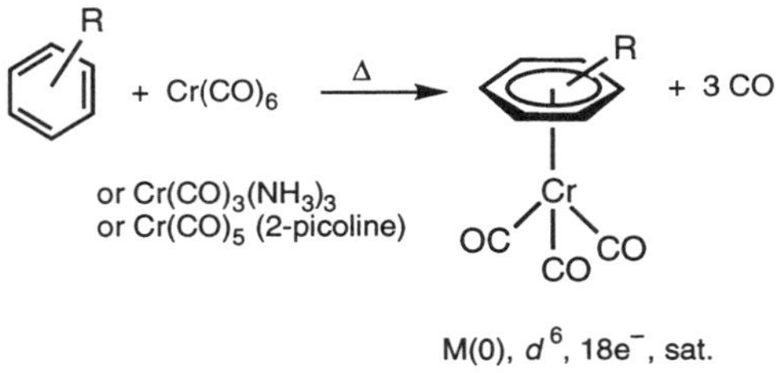

$$M(0), d^6, 18e^-, sat.$$

Eq. 10.2

$$(Naphthalene)Cr(CO)_3 + Ar'H \rightleftharpoons Ar'H\ Cr(CO)_3 + Naphthalene$$

The complexation of arenes to the chromium tricarbonyl fragment is facilitated by electron-donating groups on the arene ring, and, in general, electron-rich arenes readily form arenechromium tricarbonyl complexes, while electron-poor arenes react much more slowly or not at all. (The chromium tricarbonyl complex of nitrobenzene is unknown.) In some cases, discrimination between two similar arene rings in the same compound is possible, provided there is sufficient electronic difference between them.

Cationic cyclopentadienyliron arene complexes are most readily prepared by treatment of ferrocene with the arene in the presence of aluminum trichloride.[5] Although this places some restriction on the arenes to be complexed a modest range of complexes is available by this procedure (Eq. 10.3). Related ruthenium arene

Eq. 10.3

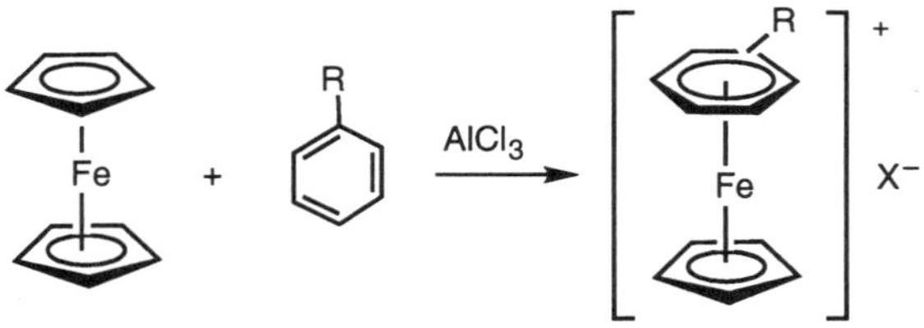

complexes are best made by the reaction of arenes with labile cyclopentadienyl ruthenium complexes (Eq. 10.4).[6] Cationic manganese arene complexes have also been prepared by ligand exchange processes (Eq. 10.5).[7]

Eq. 10.4

Eq. 10.5

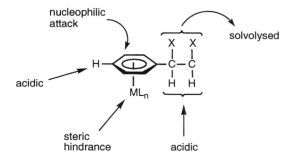

Complexation of arenes has a profound effect on their reaction chemistry (Figure 10.2).[8] Relative to the arene ring, the ML_n fragment is net electron-withdrawing, as evidenced by the high dipole moment (5.08D for benzene chromium

Figure 10.2   Effects of Complexation of Arenes to Metals

tricarbonyl), the increase in acidity of benzoic acid upon complexation ($pK_a = 4.77$ for the $Cr(CO)_3$ – benzoic acid complex vs. 5.75 for free benzoic acid) and the decrease in basicity for $Cr(CO)_3$ – complexed aniline ($pK_b = 13.31$ vs 11.70 for aniline itself). As a consequence, the arene ring becomes activated towards nucleophilic attack, rather than the normal electrophilic attack. In addition, the electron-deficient arene ring is better able to stabilize negative charge, thus both ring and benzylic deprotonation become more favorable. Finally, the metal-ligand fragment completely blocks one face of the arene, and directs incoming reagents to the face opposite the metal. All of these effects have been used to an advantage in organic synthesis.

b. Reactions of η^6-Arenemetal Complexes[9]

1. Nucleophilic Aromatic Substitution of Aryl Halides[1]

Although simple aryl halides are relatively inert to nucleophilic aromatic substitution, when complexed to appropriate transition metal fragments, this reaction is greatly facilitated. For example, the rate of substitution of chloride by methoxide in η^6-chlorobenzenechromium tricarbonyl approximates that of uncomplexed nitrobenzene,[10] an indication of the electron-withdrawing power of the chromium tricarbonyl fragment. Sodium phenoxide and aniline also readily effect this substitution.[11] η^6-Fluorobenzenechromium tricarbonyl is even more reactive, undergoing substitution by a range of nucleophiles including alkoxides, amines, cyanide[12] and stabilized carbanions (Eq. 10.3).[13]

Eq. 10.3

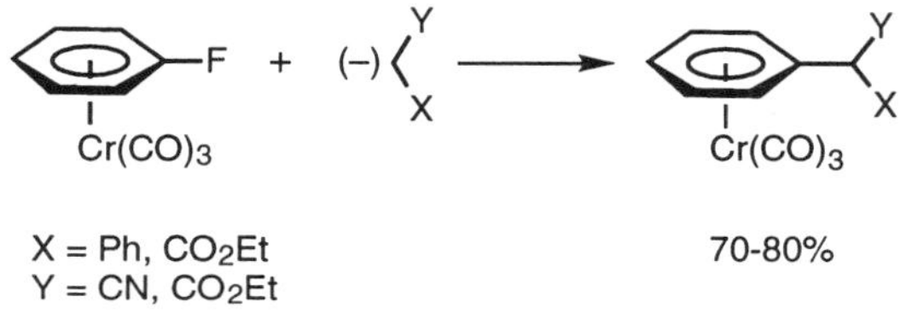

X = Ph, CO₂Et
Y = CN, CO₂Et

70-80%

With chromium arene complexes, these substitution reactions proceed by a two-step process, involving nucleophilic attack of the arene ring from the face opposite the metal, forming an anionic η^5-cyclohexadienyl complex, followed by rate-limiting loss of the halide from the endo side of the ring (Eq. 10.4). The displacement of halide by carbanions is limited to stabilized carbanions capable of

Eq. 10.4

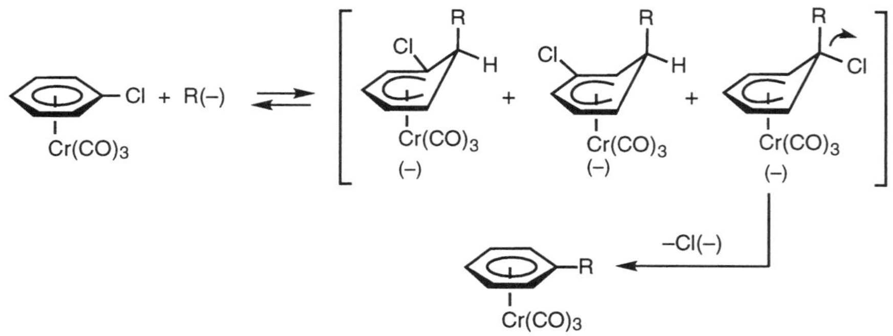

adding reversibly to the complexed arene. More reactive carbanions, such as 2-lithio-1,3-dithiane, attack the complexed arene *ortho*, and *meta* to the halide, producing η^5-cyclohexadienyl complexes which cannot directly lose chloride, and furthermore, cannot rearrange (equilibrate) to the η^5-cyclohexadienyl complex, which can (Eq. 10.4).[14] Stabilized carbanions also initially attack complexed chlorobenzene *ortho* and *meta* to the chloride, but the addition is reversible, and, upon equilibration, eventually attack at the halide-bearing position occurs, followed by loss of halide, resulting in substitution.

Although these nucleophilic aromatic substitution reactions are most extensively studied for the arenechromium tricarbonyl complexes, it is cationic areneruthenium complexes that have found the most use in complex organic synthesis, particularly in forming the aromatic ether linkage common to vancomycin analogs (Eq. 10.5).[15] Cationic manganese arene complexes have been used for similar purposes (Eq. 10.6).[16] Aryl polyethers were made using multiple aryl chloride substitutions with both ruthenium (Eq. 10.7)[17] and iron arene complexes (Eq. 10.8).[18]

Eq. 10.5

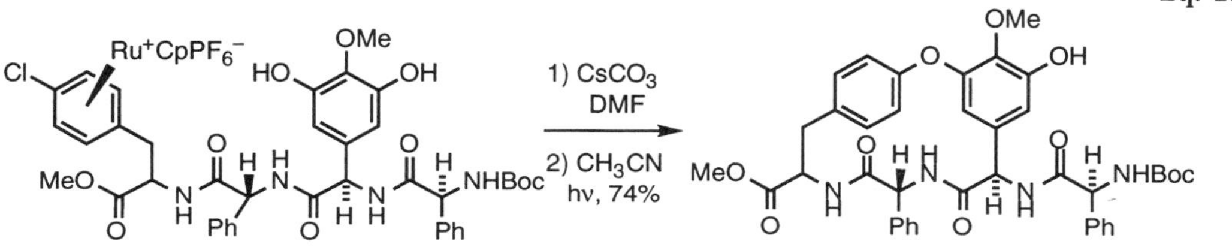

Eq. 10.6

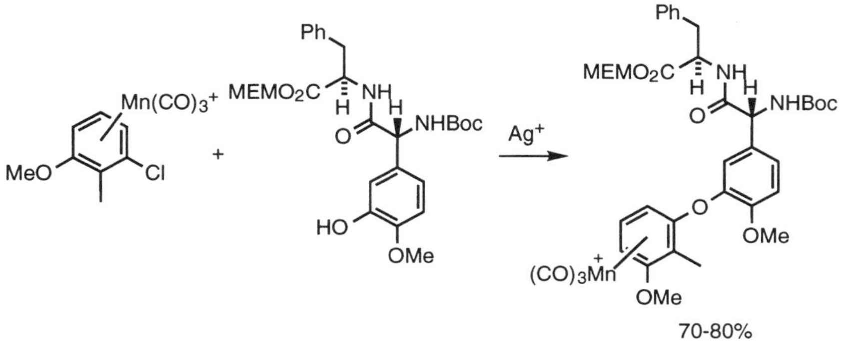

Eq. 10.7

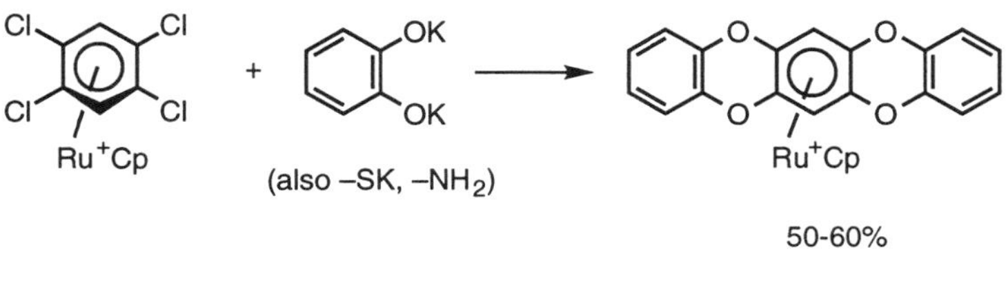

Eq. 10.8

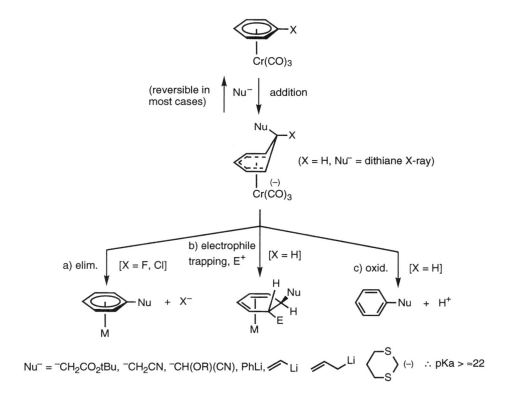

2. Addition of Carbon Nucleophiles to Arenechromium Complexes[19]

A range of carbanions attack chromium-complexed arenes from the face opposite the metal producing anionic η^5-cyclohexadienyl complexes. For all but the most nucleophilic carbanions, this process is reversible and the initial site of attack (kinetic) may not correspond to the alkylation site in the final product (thermodynamic). This η^5-cyclohexadienyl intermediate (which has been characterized by x-ray crystallography in the case of lithiodithiane as the carbanion[20]) can react further, along three different pathways (Figure 10.3).

Figure 10.3   Reactions of Arenechromium Complexes with Carbanion

Path (a) has already been discussed above. Trapping of the η^5-cyclohexadienyl complexes with electrophiles (path b) is a potentially useful but quite restricted reaction manifold. With carbanions that add reversibly, the only electrophile that undergoes efficient reaction is the proton (CF$_3$COOH) leading to cyclohexadienyl complexes. All other electrophiles react preferentially with the free carbanion in equilibrium with the η^5-cyclohexadienyl complex, resulting in regeneration of the starting η^6-arene complex. It is only with carbanions that react irreversibly that general trapping by electrophiles is efficient, and this process has proven useful in several ways (see below). Oxidation of the η^5-cyclohexadienyl intermediate (path c) is the most extensively-developed process, is efficient with all carbanions that add, and results in overall nucleophilic aromatic substitution. This is by far the most extensively utilized of the three processes and will be considered first.

Because of the reversibility of alkylation with most carbanions, the regioselectivity in the product is complex. For monosubstituted arenes bearing a resonance-donor substituent, (MeO, Me$_2$N, F) meta attack is strongly favored, while acceptor substituents, such as TMS or CF$_3$, direct attack to the para position. (π-Accepting substituents such as acyl or nitrile groups undergo competitive alkylation.) The situation with methyl and chloro-substituted arenes is more complex, and substantial amounts of ortho-alkylation can be observed depending on the carbanion.

Regioselectivity with o-disubstituted arene complexes can be high, and often is the result of attack of the sterically less hindered position meta to the strongest donor (Figure 10.4).[1a] The complexity of regioselectivity in the context of indole alkylation is seen in Figure 10.5,[21] in which steric, thermodynamic or kinetic factors may dominate, depending on how the reaction is run.

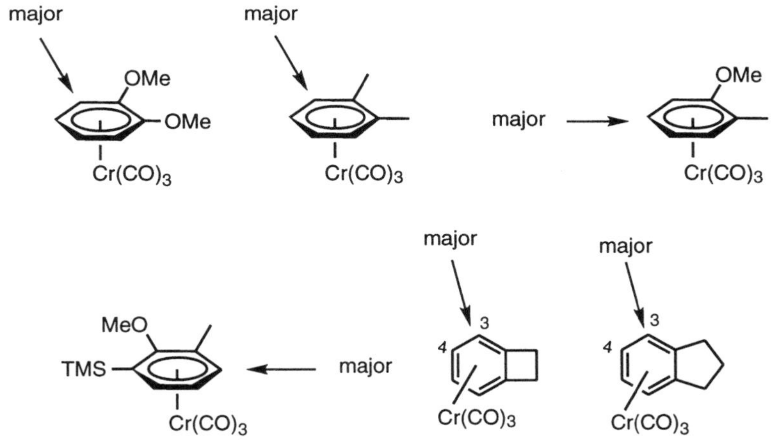

Figure 10.4 Regioselectivity of Arene Alkylation

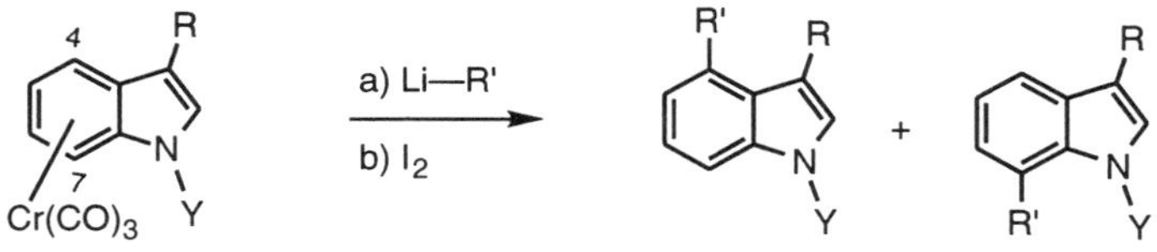

a. R' = C(Me)₂CN; R = H; Y = Me (thermodynamic)   99   :   1 (92%)
b. R' = (1,3-dithianyl); R = H; Y = Me (kinetic)   14   :   86 (68%)
c. R' = C(Me)₂CN; R = CH₂SiMe₃ (steric)   17   :   83 (82%)
d. R' = C(Me)₂CN; R = CH₂SiMe₃; Y = Si(tBu)(Me)₂ (steric) 95   :   5 (78%)

Figure 10.5   Regioselectivity of Indole Alkylation

Despite these complexities, alkylation reactions of arenechromium tricarbonyl compounds have been used in a number of complex systems (Eq. 10.9),[22] (Eq. 10.10),[23] and (Eq. 10.11).[24]

Eq. 10.9

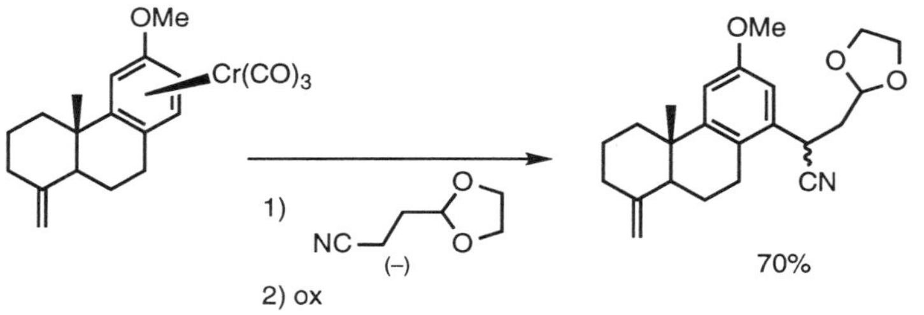

Eq. 10.10

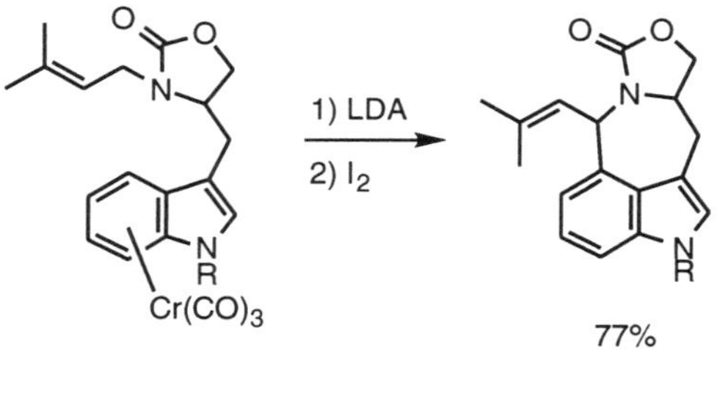

Eq. 10.11

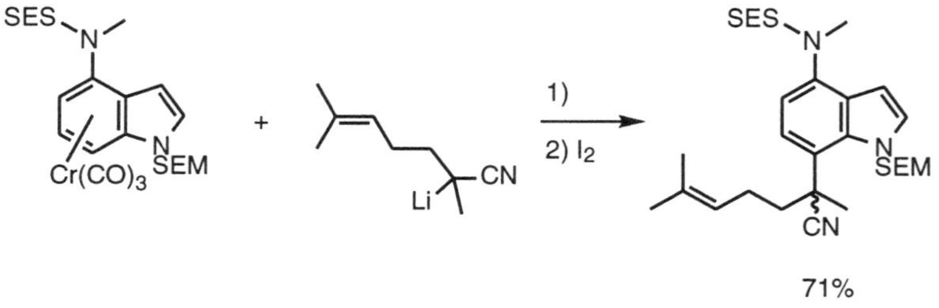

Although release of the alkylated arene from the η^5-cyclohexadienyl complex is usually accomplished by oxidation, it is also possible to effect a hydride abstraction utilizing trityl cation. This results in regeneration of the (now alkylated) arene complex, permitting further use of the $Cr(CO)_3$ fragment (Eq. 10.12).[25] In this case, ortho-alkylation is almost surely directed by the nitrogen in the benzylic position.

Eq. 10.12

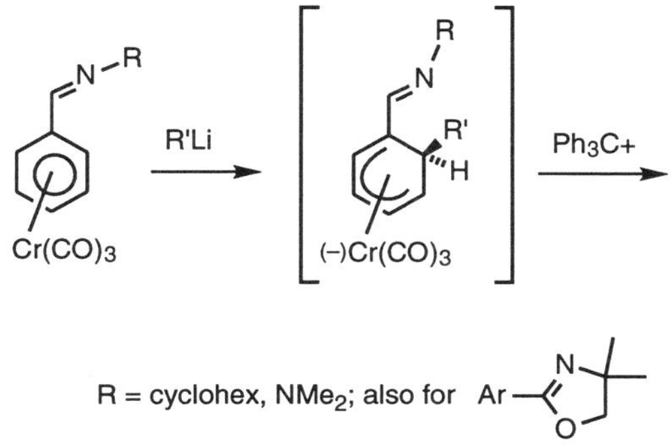

R = cyclohex, NMe_2; also for Ar

Unsymmetrically 1,2- and 1,3-disubstituted η^6-arenechromium tricarbonyl complexes are chiral, and are enantiomeric on the basis of which face the metal fragment occupies (Figure 10.6). They can often be resolved, and, using the chemistry in Eq. 10.12, they can be synthesized directly with high enantiomeric excess. This has been achieved both by using optically active hydrazone (SAMP) directing groups (Eq. 10.13),[26] or by the use of chiral ligands *for the organolithium reagent*, to direct it to one of the two prochiral ortho positions (Eq. 10.14).[27]

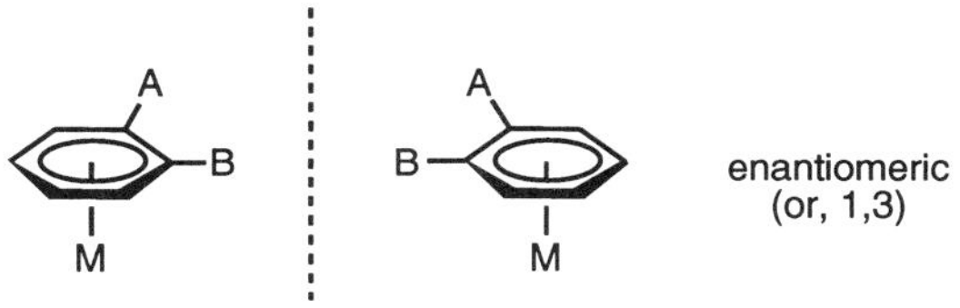

enantiomeric
(or, 1,3)

Figure 10.6   Chirality in Disubstituted Arene Complexes

Reaction of η^5-cyclohexadienyl complexes with electrophiles is also an efficient process with many synthetic applications. With carbanions which add reversibly, the electrophile must be a proton. With anisole as substrate this is a useful procedure for the synthesis of substituted cyclohexenones (Eq. 10.15).[28] This ability of chromium to activate arenes towards nucleophilic attack, to direct the attack to specific positions on the arene ring, and to permit protolytic cleavage of the η^5-cyclohexadienyl complex to produce cyclohexenones has been used in a noteworthy total synthesis of acorenone B (Eq. 10.16).[29] Again, protons are the only electrophiles that cleave the η^5-cyclohexadienyl complex. Since most alkylations (see exceptions below) are reversible, other electrophiles react preferentially with the free

Eq. 10.13

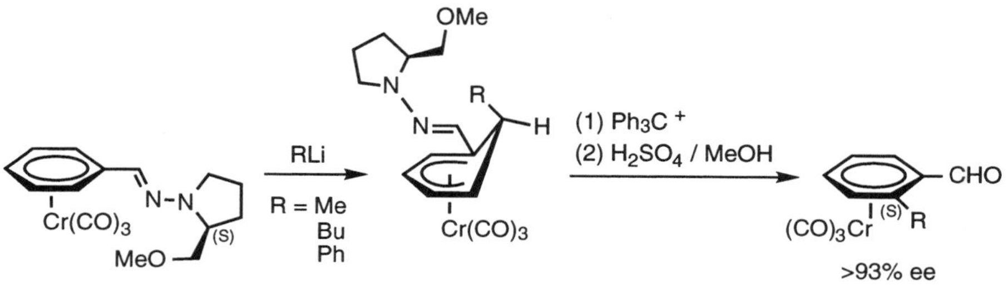

Eq. 10.14

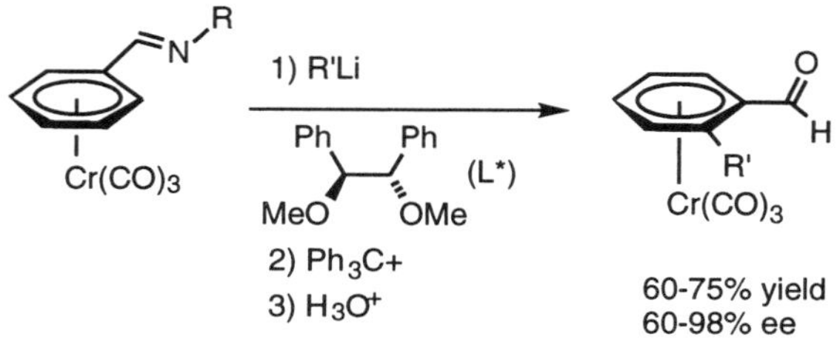

60-75% yield
60-98% ee

Eq. 10.15

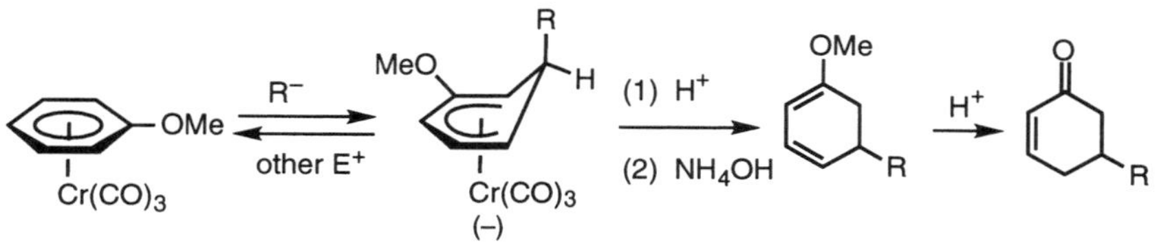

Eq. 10.16

(±)-Acorenone B

296

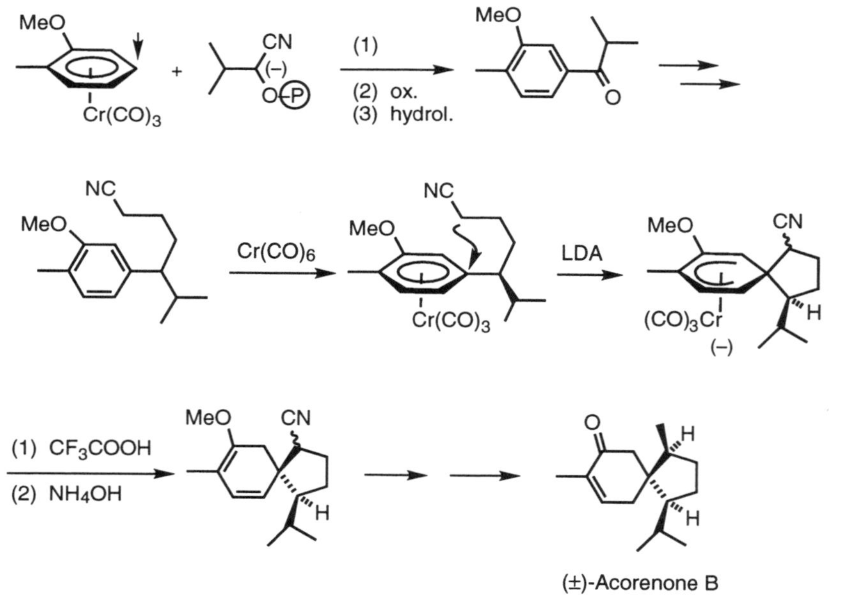

carbanion, regenerating the η^6-arenechromium tricarbonyl complex, and alkylating the electrophile. By incorporating a chiral alcohol by nucleophilic displacement of a fluoro group, asymmetry has been induced in the alkylation/protonation (Eq. 10.17).[30]

Eq. 10.17

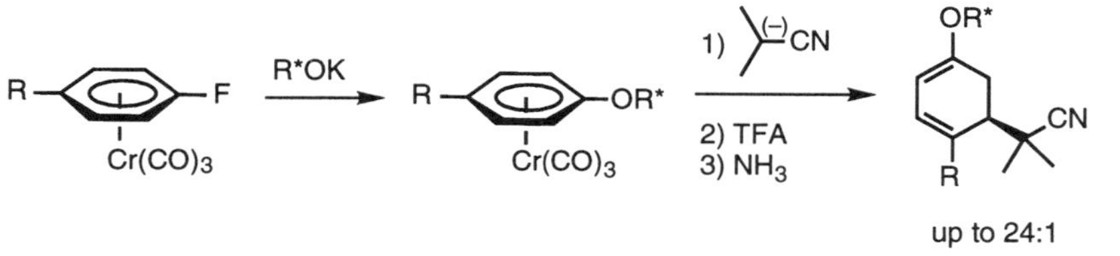

up to 24:1

Since dithiane anion adds irreversibly to the η^6-arenechromium tricarbonyl complex, it should, in principle, be possible to trap the resulting anionic η^5-cyclohexadienyl complex with electrophiles other than a proton. Indeed, treatment of this complex with reactive alkyl halides does result in a clean reaction but, surprisingly, an *acyl* group rather than the corresponding alkyl group is introduced, and from the same face as the metal, resulting in clean *trans* difunctionalization of the initial arene (Eq. 10.18).[31] To account for this, initial nucleophilic attack by the dithiane had to occur from the face opposite the metal, as expected. This is followed

Eq. 10.18

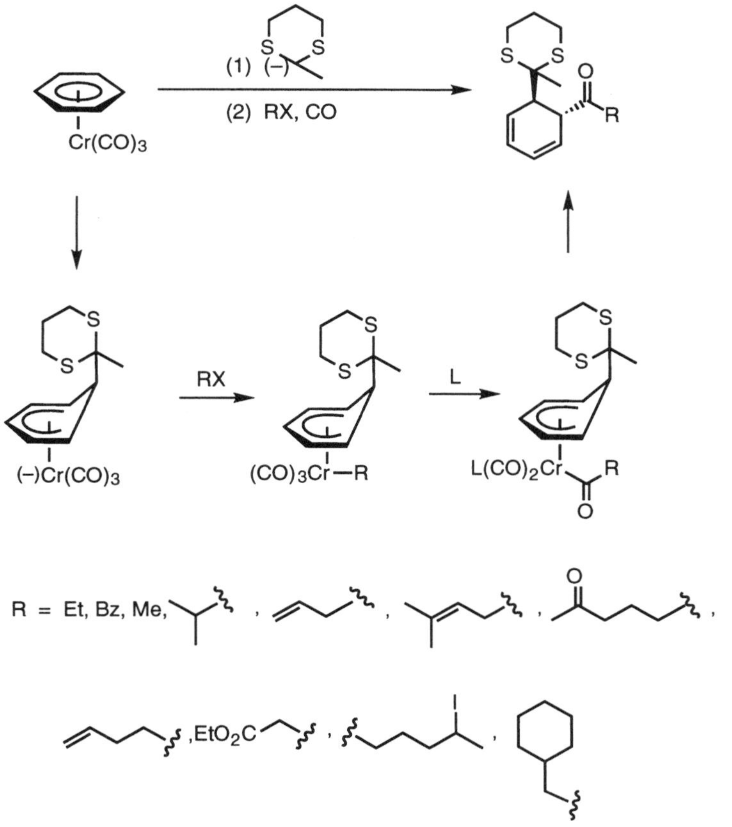

by alkylation at the *metal*, then migration to an adjacent carbonyl group to produce an acyl intermediate. Reductive elimination from this complex would deliver the acyl group from the *same* face as the metal, resulting in the clean *trans* disubstitution observed. This chemistry works equally well with substituted naphthalenes (Eq. 10.19).[32] This one-pot addition of two carbon substituents across an arene double bond serves as the key step in a synthesis of the aklavinone AB ring (Eq. 10.20).[32]

Eq. 10.19

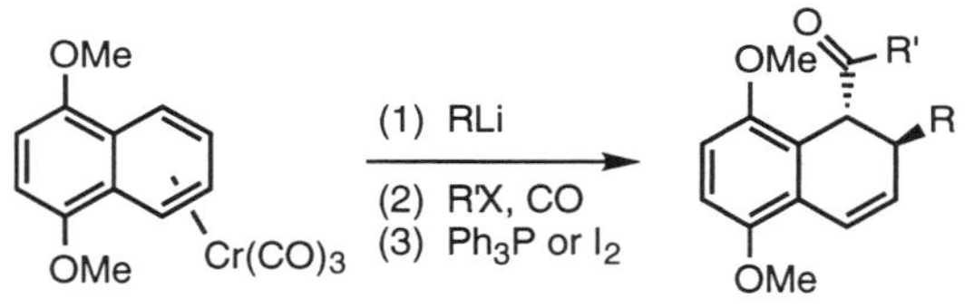

Eq. 10.20

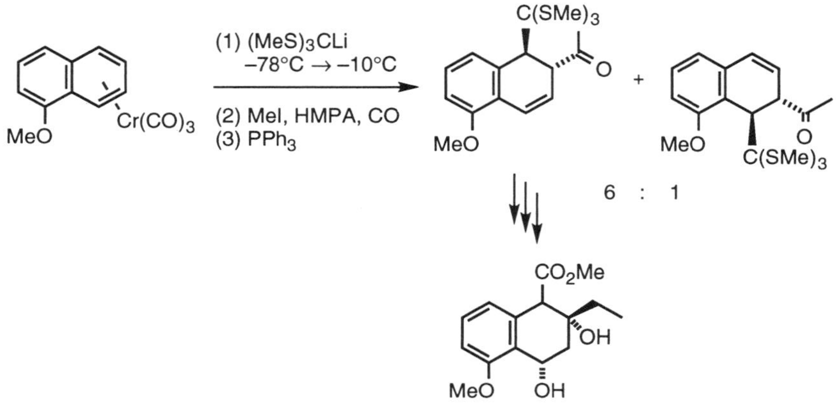

Asymmetry has been induced into this dialkylation process in two ways. Alkylation of complexed arenes having a chiral oxazoline directing group results in the conversion of two aromatic carbons to two new stereogenic centers with excellent de (Eq. 10.21).[33] In this case, asymmetric induction results from alkylation of the arene from the face *opposite* the large isopropyl group. Note that allyl, benzyl, and propargyl groups are transferred from the metal without CO insertion, a consequence of their lower aptitude toward migration. The second approach parallels that used in Eq. 10.14, and utilized a chiral ligand *for the organolithiation reagent* to direct alkylation to one of the two prochiral ortho positions (Eq. 10.22).[34]

3. Lithiation of Arenechromium Tricarbonyl Complexes[35]

Complexation of an arene to chromium activates the ring towards lithiation, and ring-lithiated arenechromium tricarbonyl complexes are readily prepared, and generally reactive towards electrophiles (Eq. 10.23).[36] With arenes having substituents with lone pairs of electrons, such as MeO, F, or Cl, lithiation always

Eq. 10.21

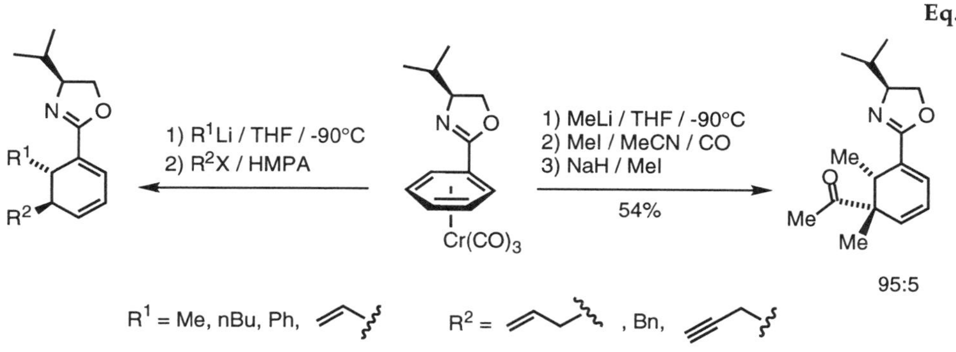

R^1 = Me, nBu, Ph, ⟋⟍ R^2 = ⟋⟍ , Bn, ⟋⟍

via

Eq. 10.22

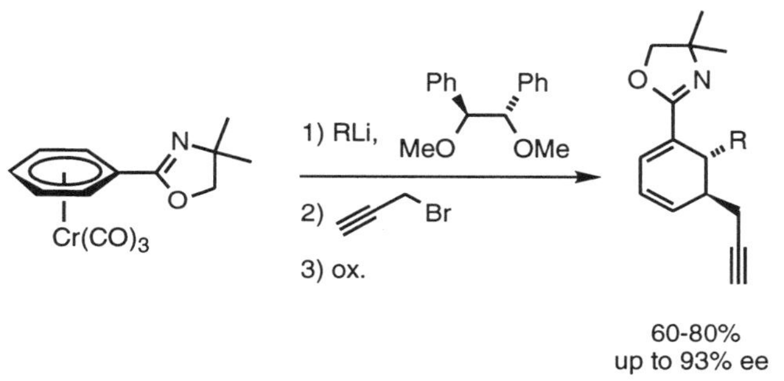

60-80%
up to 93% ee

Eq. 10.23

Y = H, OMe, F, Cl
E^+ = CO_2, MeI, PhCHO, TMSCl, ⟍

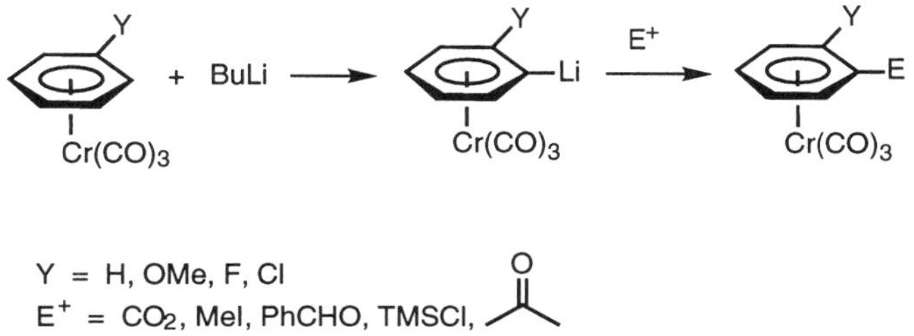

occurs *ortho* to the substituted position. This process is efficient even with quite complex systems, and high selectivity can be achieved. For example, treatment of dihydrocryptopine with chromium hexacarbonyl resulted in exclusive complexation of the slightly more electron-rich dimethoxybenzene ring, perhaps with some assistance from the benzylic hydroxy group. Lithiation occurred exclusively *ortho* to the methoxy group and peri to the benzylic hydroxy group, resulting in regiospecific alkylation (Eq. 10.24).[37]

Eq. 10.24

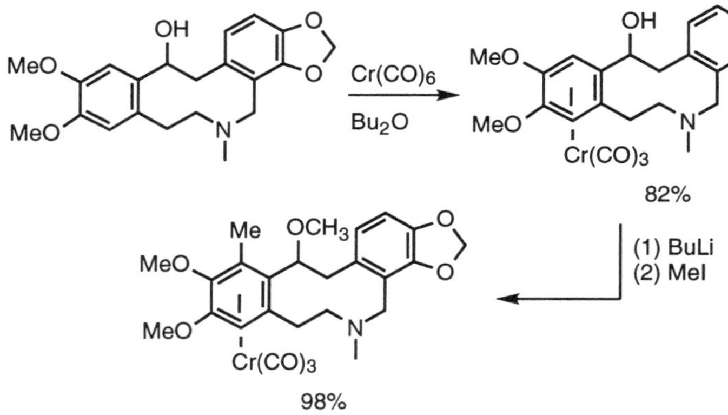

Complexed indoles are lithiated in the normally unreactive 4-position provided a large protecting group is present on nitrogen to suppress lithiation at the 2-position (Eq. 10.25).[38]

Eq. 10.25

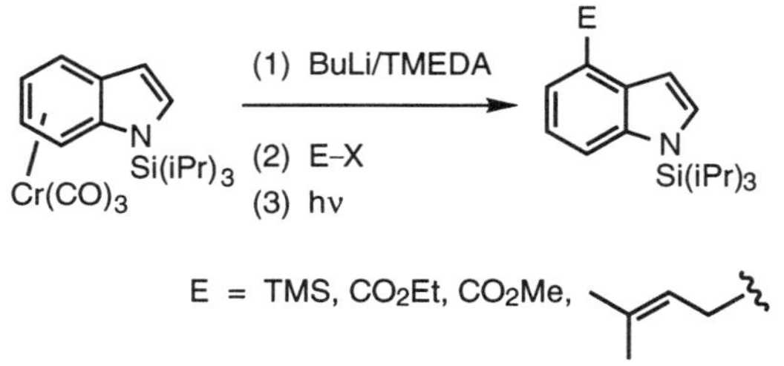

$E = TMS, CO_2Et, CO_2Me,$

Recall that unsymmetrically disubstituted arene complexes are chiral. Ortholithiation/alkylation generates just such complexes, and, as above, asymmetry can be induced[14] several different ways. Arene complexes having a chiral benzylic carbon bearing a donor substituent (MeO, OCH_2OCH_3, NMe_2) undergo clean ortholithiation/alkylation in excellent yield and with high diastereoselectivity. Asymmetric induction is thought to result from ligand-directed o-metallation to the ortho position which allows the bulky benzyl substituent to be on the face of the complex *opposite* the metal (Eq. 10.26).[39] By using chiral ketals as directing groups, optically active o-substituted benzaldehyde complexes were available (Eq. 10.27).[40] An interesting use in synthesis is seen in Eq. 10.28.[41]

A more versatile approach to asymmetric induction in the o-lithiation step involves the use of chiral organolithium reagents. In this instance, lithiated C-2 symmetric dibenzylamine efficiently discriminated between the prochiral ortho positions, resulting in good yields and high enantiomeric excesses (Eq. 10.29).[42]

Eq. 10.26

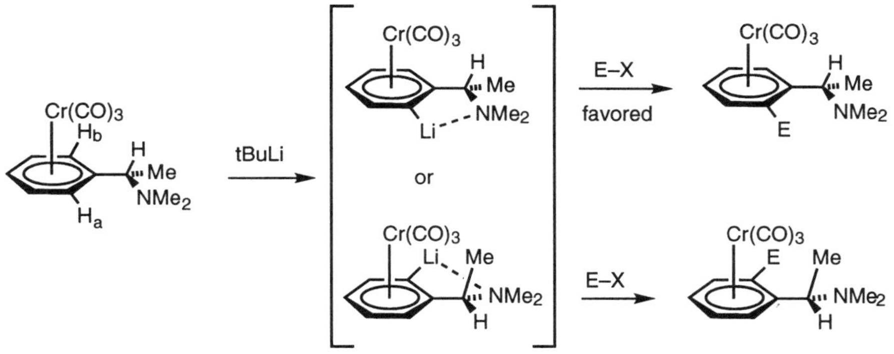

E–X = MeOSO$_2$F, Me$_3$SiCl, Ph$_2$PCl: also RCHO, R$_2$CO

Eq. 10.27

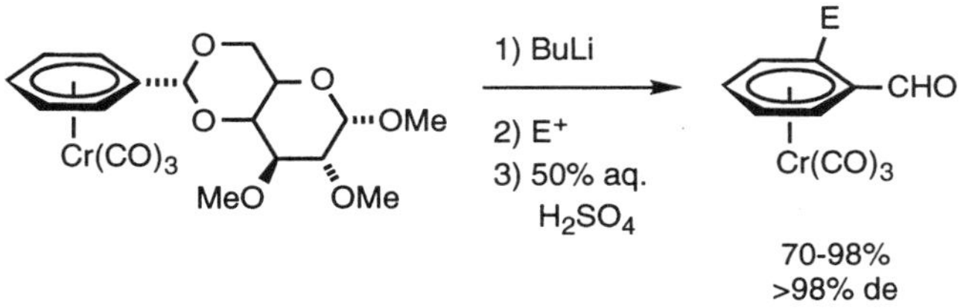

70-98%
>98% de

Eq. 10.28

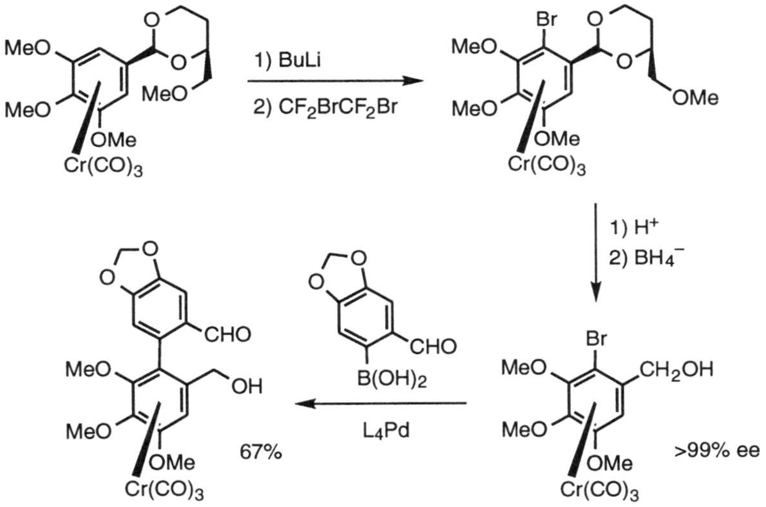

(CHO away from Cr(CO)$_3$)

Eq. 10.29

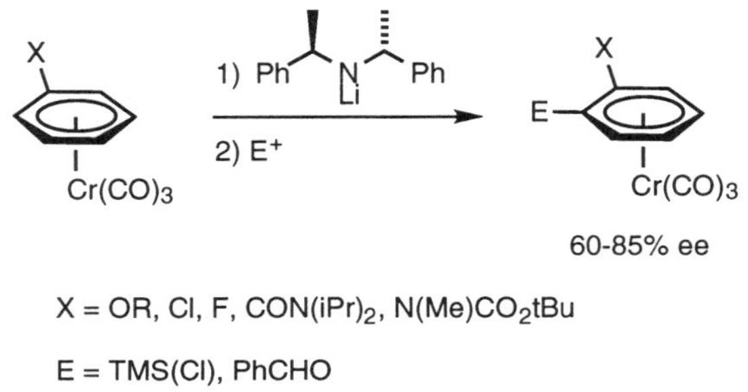
X
1) Ph N Ph
Li
2) E+
Cr(CO)3
E
60-85% ee
X = OR, Cl, F, CON(iPr)2, N(Me)CO2tBu
E = TMS(Cl), PhCHO

4. Side Chain Activation and Control of Stereochemistry[43]

Another manifestation of the "electron withdrawing" properties of the metal complexed to arenes is its ability to stabilize negative charge at the benzylic position of alkyl side chains. This allows facile benzylic alkylation of complexed arenes with a wide range of electrophiles (Eq. 10.30). As expected, benzylic functionalization occurs from the face *opposite* the metal with a high degree of stereo control. This is

Eq. 10.30

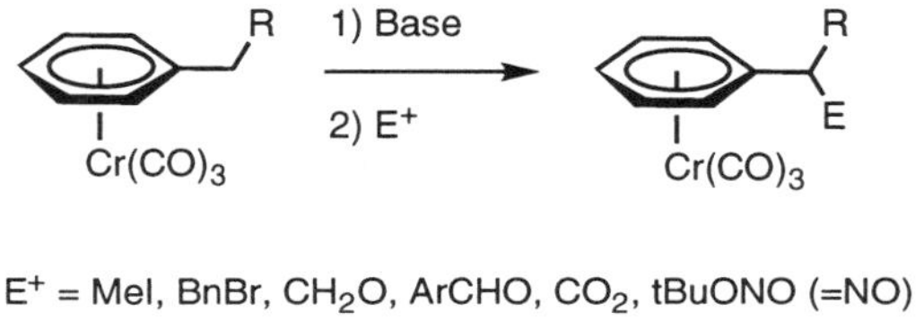
R
1) Base
2) E+
Cr(CO)3
E
E+ = MeI, BnBr, CH2O, ArCHO, CO2, tBuONO (=NO)

most apparent with cyclic systems, since the benzylic protons are inequivalent because of restricted rotation (Eq. 10.31).[44] Even acyclic systems undergo benzylic alkylation with a high degree of stereo control provided there is an ortho-substituent

Eq. 10.31

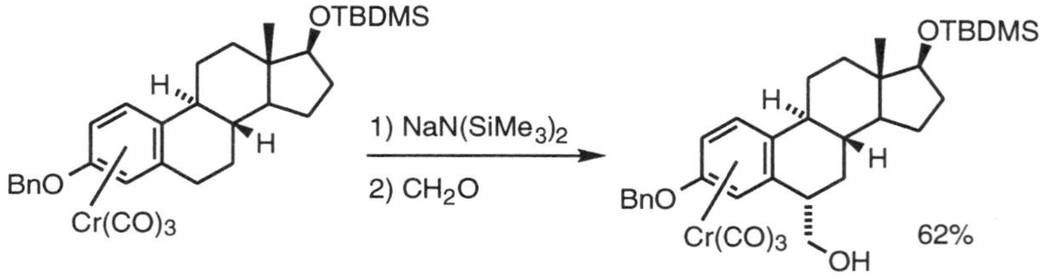
OTBDMS
H
BnO
Cr(CO)3
1) NaN(SiMe3)2
2) CH2O
OH
62%

to force the benzylic substituent to adopt a single rotomeric isomer (Eq. 10.32).[45] In some cases, ring lithiation competes with benzylic lithiation. When this occurs, the ring site can be blocked by silylation, which can then be removed after successful benzylic alkylation (Eq. 10.33).[46] This entire process has been used in the synthesis of natural products (Eq. 10.34).[47]

Eq. 10.32

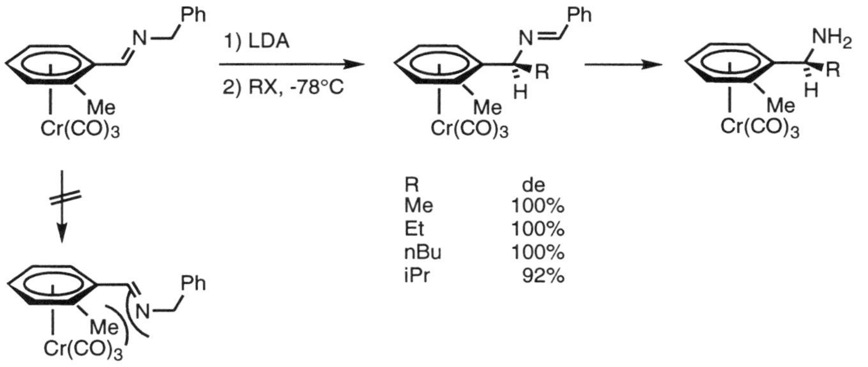

R	de
Me	100%
Et	100%
nBu	100%
iPr	92%

Eq. 10.33

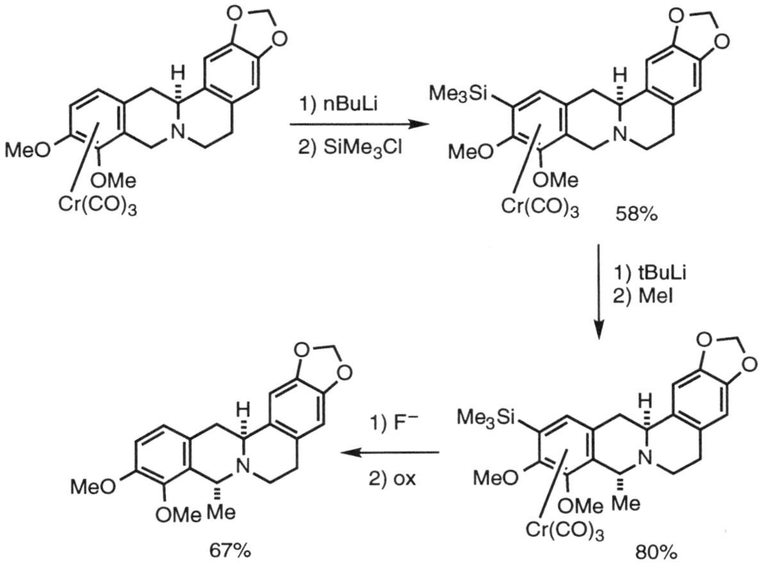

58%

80%

67%

Eq. 10.34

60%

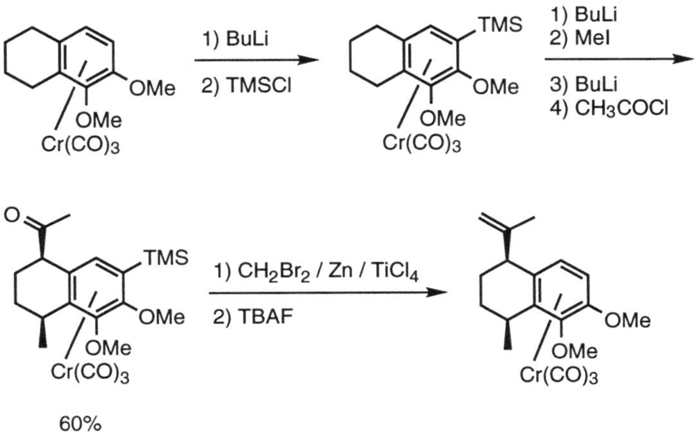

Benzylic lithiation/alkylation is not restricted to arenechromium tricarbonyl complexes. Cationic iron arene complexes undergo multiple alkylations to give highly branched compounds (Eq. 10.35).[48]

Eq. 10.35

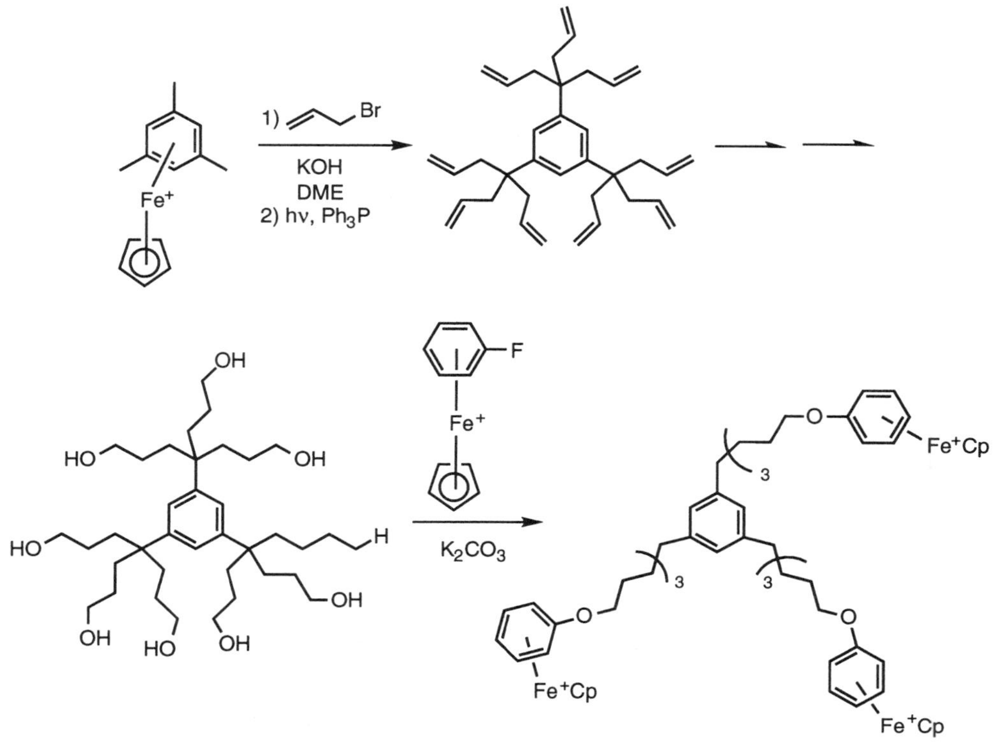

Most remarkably, benzyl *cations* are also stabilized by chromium tricarbonyl fragments. Complexed benzylic alcohols produce complexed cations resulting from acid-assisted loss of the OH group, from the face opposite the metal, to give a cation that is slow to racemize and which can react with a variety of nucleophiles, again from the face opposite the metal, resulting in overall retention (Eq. 10.36).[49] If, as can

Eq. 10.36

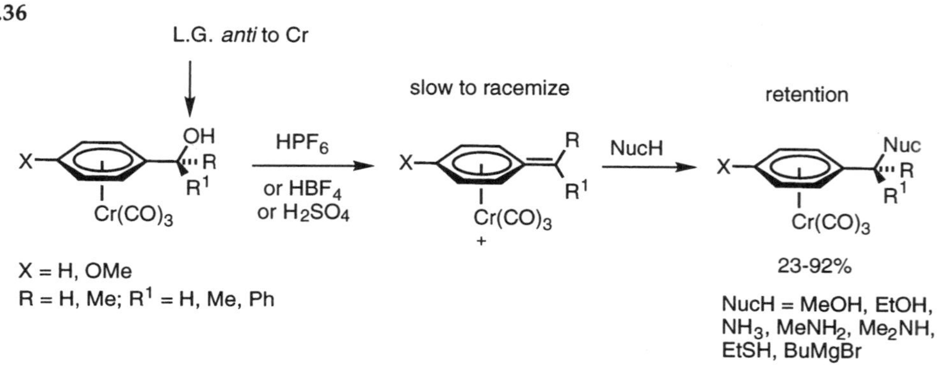

be the case in cyclic systems, the leaving group cannot get anti to the metal, the replacement reaction still proceeds, albeit more slowly, and nucleophilic attack again occurs from the face opposite the metal, resulting in overall inversion. This reaction chemistry is quite efficient with complex systems (Eq. 10.37)[50] and (Eq. 10.38).[51]

Eq. 10.37

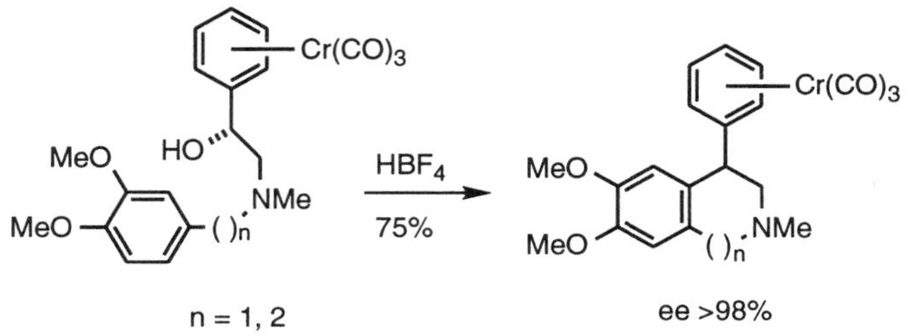

n = 1, 2   75%   ee >98%

Eq. 10.38

61%

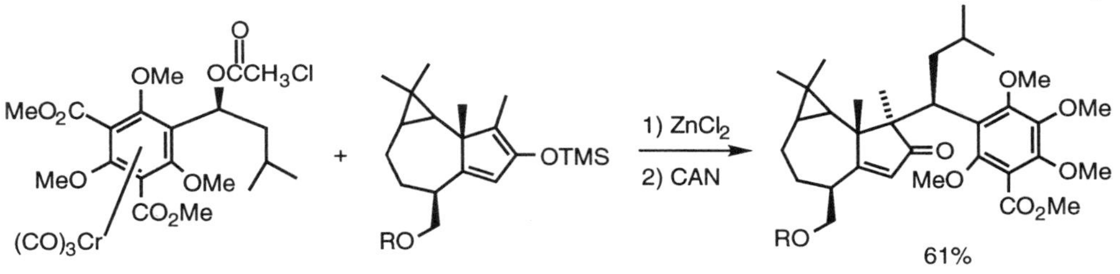

In all of the above η^6-arene complexes, the metal occupies a single face of the arene, and directs reaction chemistry to the opposite face of the molecule. This results in a very high degree of stereocontrol, and has found many uses in organic synthesis. A very early example of this is seen in Eq. 10.39,[52] in which α-alkylation of an enone by complexed α-methyl indanone occurred exclusively from the face opposite the metal. Removal of an activated benzylic proton permitted intramolecular alkylation of the keto group, again from the face opposite the metal, resulting in clean *cis* stereochemistry.

Eq. 10.39

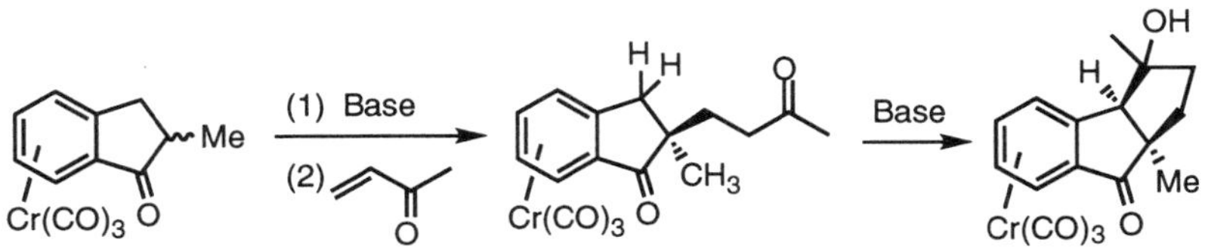

Racemic 2-indanol undergoes facially specific complexation, from the same face as the OH group, presumably directed there by complexation to the oxygen. These enantiomeric complexes can be resolved, and oxidation of one of the enantiomers produces an indanone complex which is optically active solely by virtue of the face of the arene occupied by the $Cr(CO)_3$ fragment (see above). Since reactions of this optically active complex occur exclusively from the face opposite the metal, alkylation by Grignard reagents produces a single enantiomer of the indanol, α-alkylation a single enantiomer of the indanone, and α-alkylation followed by reduction a single enantiomer of the α-alkyl indanol (Eq. 10.40).[53] Similar transformations are feasible with tetralone.[54]

Eq. 10.40

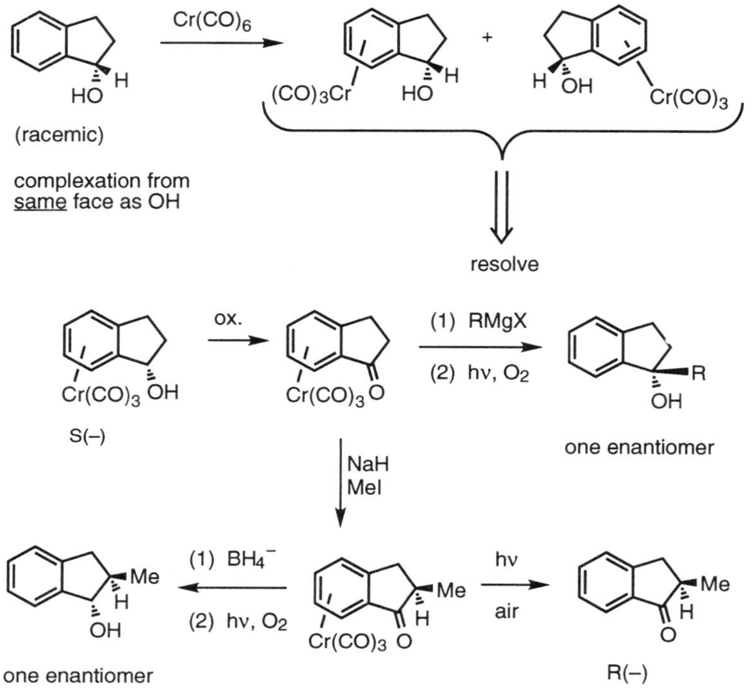

As shown above, unsymmetrically 1,2- and 1,3-disubstituted arenechromium tricarbonyl complexes are chiral and are enantiomeric on the basis of which face of the arene the chromium tricarbonyl fragment occupies (Figure 10.5). These can be resolved,[55] or in some cases prepared by asymmetric synthesis.[56] Since reactions occur from the face opposite the metal, highly enantioselective reactions can be achieved.

For example, nucleophilic addition to chromium-complexed, ortho-substituted aryl aldehydes proceeds with a high degree of stereoinduction resulting from addition to the aldehyde carbonyl group from the face opposite the metal. A single prochiral face of the aldehyde is presented to the attacking nucleophile because the ortho substituent strongly favors the less hindered aldehyde rotamer (Eq. 10.41).[56] Similar factors result in high diastereoselectivity in other reactions of complexed aldehydes including imine alkylation (Eq. 10.42),[57] (Eq. 10.43),[58] and aza Diels-Alder reactions (Eq. 10.44).[59]

Eq. 10.41

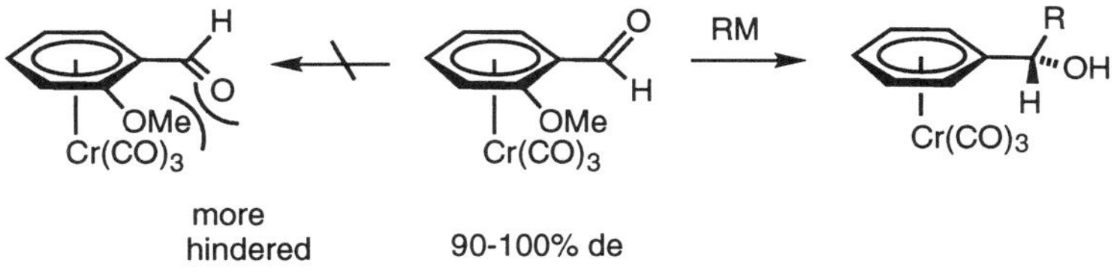

H
O
OMe
Cr(CO)₃
more
hindered

O
H
OMe
Cr(CO)₃
90-100% de

RM

R
H OH
Cr(CO)₃

Eq. 10.42

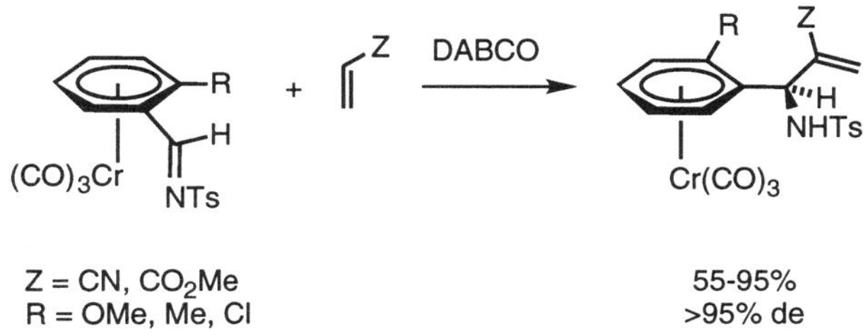

R
H
(CO)₃Cr
NTs

+ Z

DABCO

R Z
H
NHTs
Cr(CO)₃

Z = CN, CO₂Me
R = OMe, Me, Cl

55-95%
>95% de

Eq. 10.43

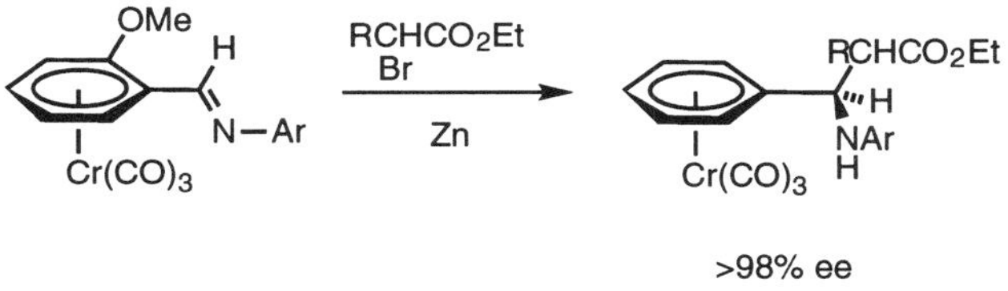

OMe
H
N—Ar
Cr(CO)₃

RCHCO₂Et
Br

Zn

RCHCO₂Et
H
NAr
Cr(CO)₃ H

>98% ee

Eq. 10.44

R
H
NR'
Cr(CO)₃

+

OTMS
OMe

SnCl₄

O
N
R
Cr(CO)₃

>98% de

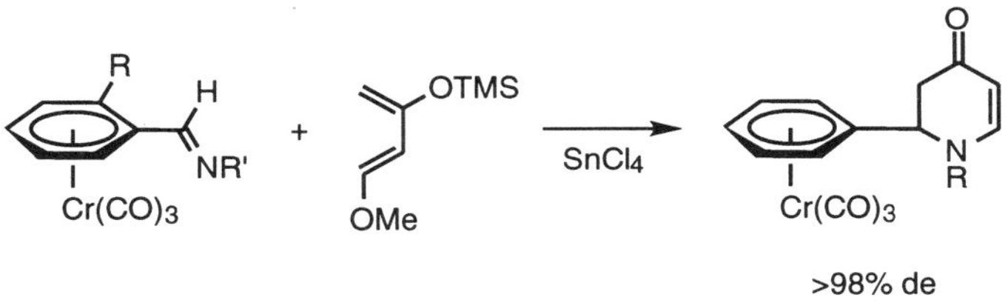

A very nice synthetic example utilizing a majority of the features found in arenechromium chemistry is seen in Eq. 10.45.[60]

10.3 η^2-Arenemetal Complexes[61]

The transition metals which form η^6-arene complexes are all strong π-acceptor metal-ligand systems, binding to the electron-rich π-donor-cloud of the arene and acting as an electron-withdrawing group. In contrast, the metal-ligand systems which form η^2-arene complexes, particularly $[Os(NH_3)_5]^{2+}$, have powerful σ-donor ligands which are incapable of appreciable π-interaction with arenes. These complexes show a strong preference for π-*acceptor* ligands such as alkenes, nitriles, aldehydes and alkynes, and will bind to these ligands in preference to donor ligands such as amines, esters, ethers, alcohols and amides. The remarkable thing about

Eq. 10.45

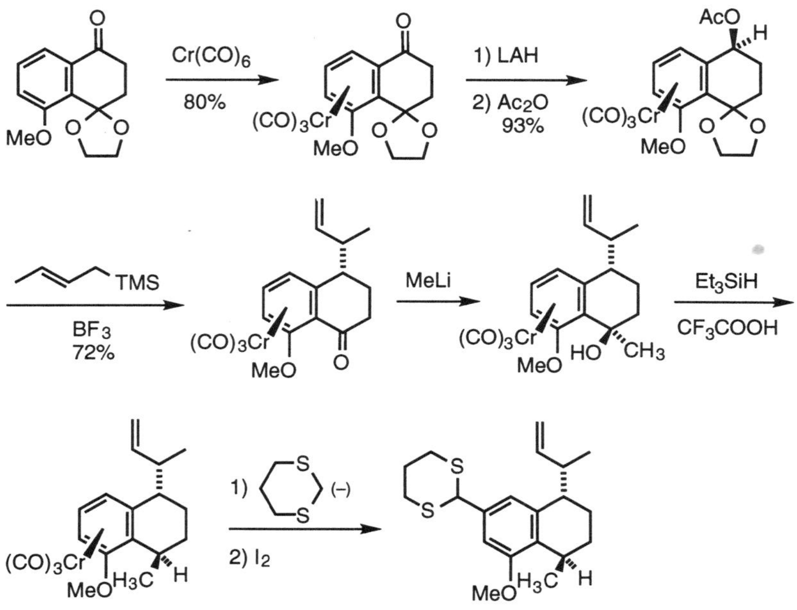

$Os(NH_3)_5]^{2+}$ is that it treats arenes as a source of an "olefinic" ligand, complexing a 6π system in an η^2-fashion, essentially deconjugating the arene. Complexation to arenes occurs at the site which results in the minimum disruption of the π-system of the arene, and an external olefin (as in styrene) will complex in preference to an internal one.

These complexese are prepared by reducing $Os(NH_3)_5^{3+}$ in the presence of the arene. (This procedure requires "glove box" conditions.) Decomplexation can be achieved by oxidation with DDQ or CAN under relatively severe conditions (Eq. 10.46).

Eq. 10.46

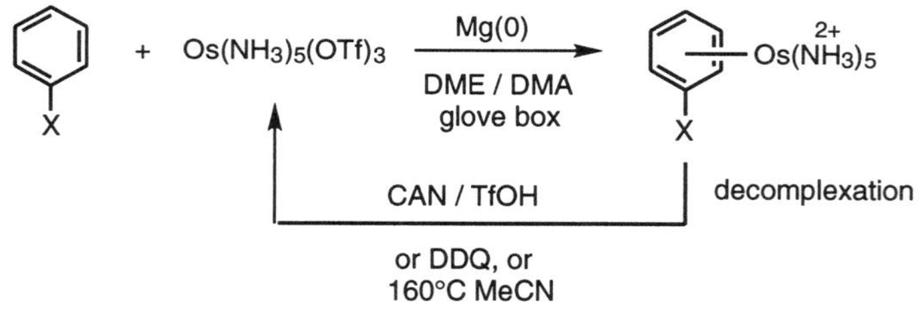

η^2-Complexation of arenes can result in some remarkable reaction chemistry. η^2-Complexation of phenols to $[Os(NH_3)_5]^{2+}$ allows them to react like the γ-enol of enones, and to perform Michael additions to conjugated enones (Eq. 10.47) and (Eq. 10.48).[62] Anilines behave in a similar manner (Eq. 10.49).[63]

Eq. 10.47

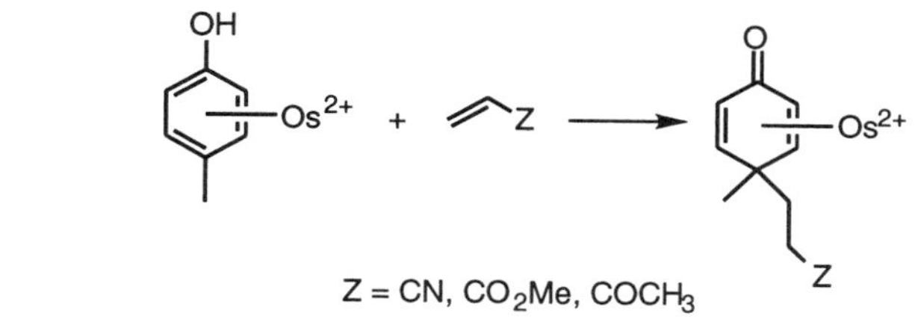

Z = CN, CO$_2$Me, COCH$_3$

Eq. 10.48

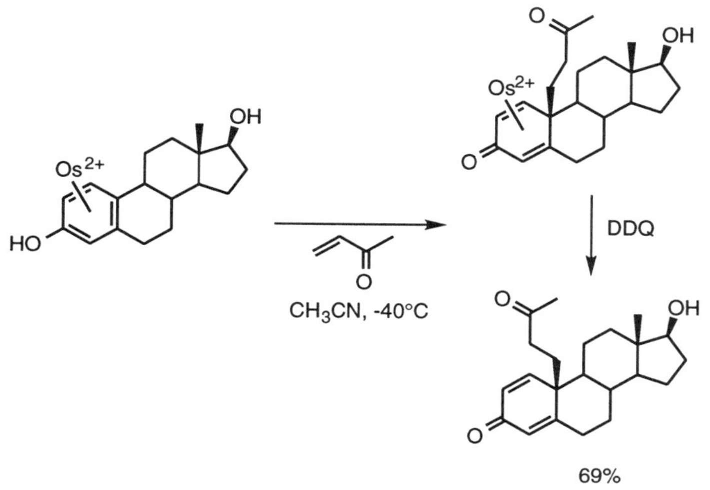

69%

Eq. 10.49

60-90%

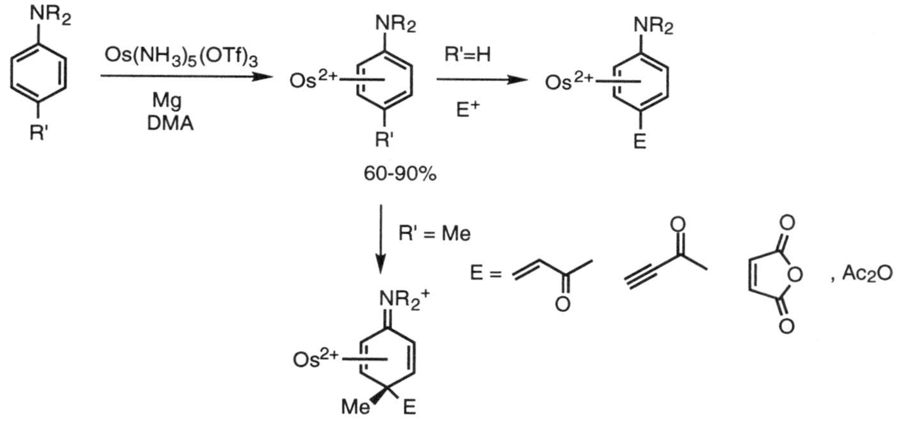

The 4,5-η^2-complex of pyrrole is the most stable, and as expected, it undergoes reactions typical of enamines, addition of electrophiles at the 3-position, with reclosure to the 2-position with acetylene dicarboxylates. The initially formed 3H-pyrroline complexes can also undergo nucleophilic attack by external agents as well (Eq. 10.50).[64]

Although osmium preferentially complexes the 4,5-position of pyrroles, the 3,4 position is also accessible. Complexation here generates an azomethine ylide, which, in the presence of 1,3-dipolarophiles, undergoes facile cycloaddition (Eq. 10.51).[65]

Eq. 10.50

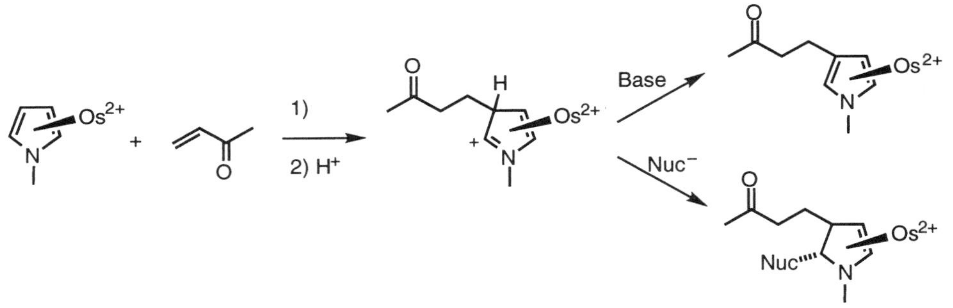

Eq. 10.51

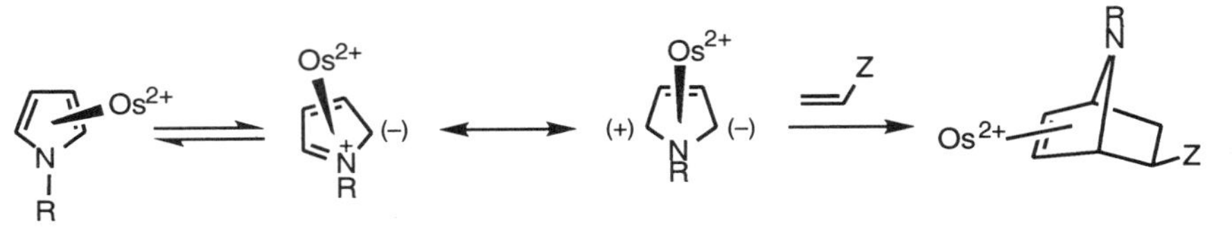

Reactions of Os^{2+} complexes of furans are similar to those of pyrroles.[66] Electrophilic attack at the 4-position is favored, but because of the instability of the thus formed 4,5-η^2-3H furanium species, the outcome is more complex (Eq. 10.52),[66] and depends on the location of substituents on the furan.

Eq. 10.52

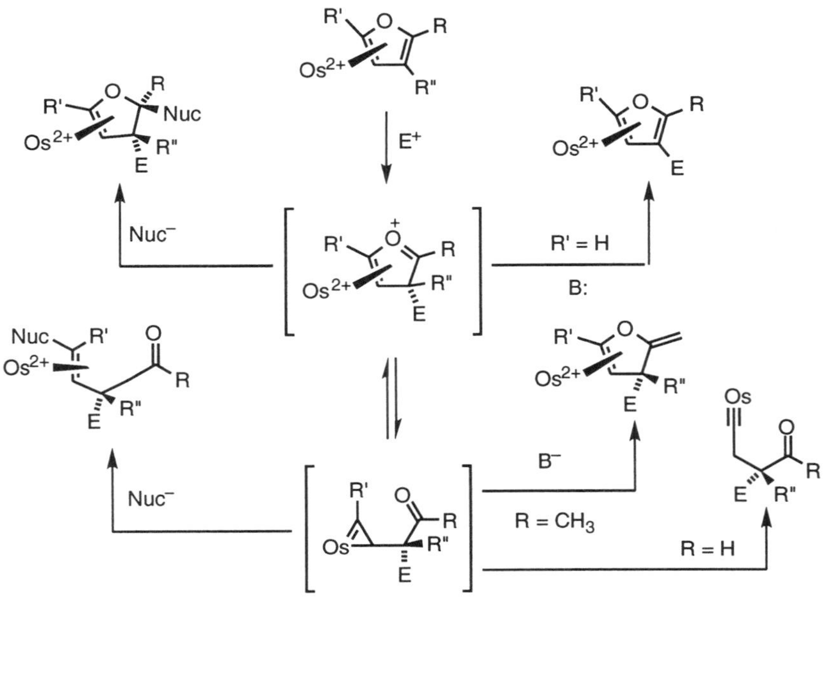

At present, η^2-arene complexes have found little use in complex organic syntheses, since the reactions are stoichiometric in osmium, and some of the laboratory procedures are cumbersome. However, as the methodology develops, its unique transformation chemistry should join the arsenal of the synthetic organic chemist, along with other organometallic processes originally considered too arcane for use in organic synthesis, as organopalladium chemistry was a decade ago.

References

1. For a review see: Semmelhack, M.F., "Transition Metal Arene Complexes: Nucleophilic Addition" in "Comprehensive Organometallic Chemistry II, Wilkinson, G.; Stone, F.G.A.; Abel, E.W. eds., Pergamon, New York, 1995, vol. 12, pp. 929-1015.
2. (a) Nicholls, B.; Whiting, M.C. *J. Chem. Soc.* **1959**, 551. (b) Mahaffy, C.A.L.; Pauson, P. *Inorg. Synth.* **1979**, *XIX*, 154.
3. (a) Mosher, G.A.; Rausch, M.D. *Synth. React. Inorg. Metal. Org. Chem.* **1979**, 4, 38. (b) Rausch, M.D. *J. Org. Chem.* **1974**, *39*, 1787.
4. (a) Kundig, E.P.; Perret, C.; Spichiger, S.; Bernardinelli, G. *J. Organomet. Chem.* **1985**, *286*, 183. (b) For preparation of the starting complex see: Desobry, V.; Kündig, E.P. *Helv. Chim. Acta* **1981**, *64*, 1288.
5. (a) Khand, I.U.; Pauson, P.L.; Watts, W.E. *J. Chem. Soc. C* **1968**, 2257. (b) Astruc, D.; Dabard, R. *J. Organomet. Chem.* **1976**, *111*, 339.
6. Kudinov, A.M.; Rybinskaya, M.I.; Struchkov, Y.T.; Yanovskii, A.I.; Petrovskii, P.V. *J. Organomet. Chem.* **1987**, *336*, 187.
7. Bhasin, K.K.; Balkeen, W.G.; Pauson, P.L. *J. Organomet. Chem.* **1981**, *204*, C25.
8. This figure was adopted from an excellent (dated) review: Semmelhack, M.F. *J. Organomet. Chem. Libr.* **1976**, *1*, 361.
9. For a review see: (a) Kalinin, V.N. *Russ. Chem. Rev.* **1987**, *56*, 682. (b) Uemura, M. *Adv. Met. Org. Chem.* **1991**, *2*, 195.
10. Brown, D.A.; Raju, J.R. *J. Chem. Soc. A* **1966**, *40*, 1617.
11. Rosca, S.J.; Rosca, S. *Rev. Chim.* **1974**, *25*, 461.
12. Mahaffy, C.A.L.; Pauson, P.L. *J. Chem. Res.* **1979**, 128.
13. Baldoli, C.; DelButtero, P.; Licandro, E.; Maiorana, S. *Gazz. Chim. Ital.* **1988**, *118*, 409.
14. Semmelhack, M.F.; Hall, H.T., Jr. *J. Am. Chem. Soc.* **1974**, *96*, 7091, 7092.
15. (a) Pearson, A.J.; Chelliah, M.V. *J. Org. Chem.* **1998**, *63*, 3087 and references therein. (b) Janetka, J.W.; Rich, D.H. *J. Am. Chem. Soc.* **1997**, *119*, 6488 and references therein.
16. Pearson, A.J.; Shin, H. *J. Org. Chem.* **1994**, *59*, 2314.
17. Dembek, A.A.; Fagan, P.J. *Organometallics* **1996**, *15*, 1319.
18. (a) Abd-El-Azia, A.S.; de Denus, C.R.; Zaworotko, M.J.; Sharma, C.V.K. *J. Chem. Soc., Chem. Comm.* **1998**, 265. See also (b) Pearson, A.J.; Gelormini, A.M. *J. Org. Chem.* **1994**, *59*, 4561. (c) Pearson, A.J.; Gelormini, A.M. *Tetrahedron Lett.* **1997**, *38*, 5123.
19. For a review see: Semmelhack, M.F. in "Comprehensive Organic Synthesis", Trost, B.M., Ed., Pergamon Press, 1991, Vol. 4, pp. 517-29. See also ref. 1 above.
20. Semmelhack, M.F.; Hall, H.T., Jr.; Farina, R.; Yoshifuji, M.; Clark, G.E.; Bargar, T.; Hirotsu, K.; Clardy, J. *J. Am. Chem. Soc.* **1979**, *101*, 3535.
21. Semmelhack, M.F.; Wulff, W.; Garcia, J.L. *J. Organomet. Chem.* **1982**, *240*, C5.
22. Cambie, R.C.; Rutledge, P.S.; Stevenson, R.J.; Woodgate, P.D. *J. Organomet. .Chem.* **1994**, *471*, 133; 149 (1994).
23. Semmelhack, M.F.; Knochel, P.; Singleton, T. *Tetrahedron Lett.* **1993**, *34*, 5051.
24. Semmelhack, M.F.; Rhee, H. *Tetrahedron Lett.* **1993**, *34*, 1399.
25. Fretzen, A.; Ripa, A.; Liu, R.; Bernardinelli, G.; Kündig, E.P. *Chem. Eur. J.* **1998**, *4*, 102.
26. Kündig, E.P.; Liu, R.; Ripa, A. *Helv. Chim. Acta* **1992**, *79*, 2657.
27. Fretzen, A.; Kündig, E.P. *Helv. Chim. Acta* **1997**, *80*, 2023.
28. (a) Semmelhack, M.F.; Harrison, J.J.; Thebtaranonth, Y. *J. Org. Chem.* **1979**, *44*, 3275. (b) Boutonnet, J.C.; Levisalles, J.; Normant, J.M.; Rose, E. *J. Organomet. Chem.* **1983**, *255*, C21.
29. Semmelhack, M.F.; Yamashita, A. *J. Am. Chem. Soc.* **1980**, *102*, 5924.
30. Pearson, A.J.; Gontcharov, A.V. *J. Org. Chem.* **1998**, *63*, 152.

31. Kündig, E.P.; Cunningham, A.F., Jr.; Paglia, P.; Simmons, D.P.; Bernardinelli, G. *Helv. Chim. Acta* **1990**, *73*, 386.

32. Kündig, E.P.; Inage, M.; Bernardinelli, G. *Organometallics* **1991**, *10*, 2921.

33. (a) Kündig, E.P.; Bernardinelli, G.; Liu, R.; Ripa, A. *J. Am. Chem. Soc.* **1991**, *113*, 9676. (b) Kündig, E.P.; Ripa, A.; Bernardinelli, G. *Angew. Chem. Int. Ed.*, **1992**, *31*, 1071. (c) Kündig, E.P.; Qualtropani, A.; Inage, M.; Ripa, A.; Dupre, C.; Cunningham, A.F., Jr.; Bourdin, B. *Pure Appl. Chem.* **1996**, *68*, 97.

34. (a) Amurrio, D.; Khan, K.; Kündig, E.P. *J. Org. Chem.* **1996**, *61*, 2258. (b) Beruben, D.; Kündig, E.P. *Helv. Chim. Acta.* **1996**, *79*, 1533. (c) Kündig, E.P.; Amurrio, D.; Anderson, G.; Beruken, D.; Khan, K.; Ripa, A.; Longgang, L. *Pure Appl. Chem.* **1997**, *69*, 543.

35. For a review see: Semmelhack, M.F., "Transition metal Arene Complexes: Ring Lithiation," in "Comprehensive Organometallic Chemistry II", Wilkinson, G.; Stone, F.G.A.; Abel, E.W. eds., Pergamon, New York, 1995, vol. 12, pp. 1017-1037.

36. (a) Semmelhack, M.F.; Bisaha, J.; Czarny, M. *J. Am. Chem. Soc.* **1979**, *101*, 768. (b) Card, R.J.; Trahanovsky *J. Org. Chem.* **1980**, *45*, 2560. (c) Gilday, J.P.; Negri, J.T.; Widdowson, D.A. *Tetrahedron* **1989**, *45*, 4605. (d) Dickens, P.J.; Gilday, J.P.; Negri, J.T.; Widdowson, D.A. *Pure Appl. Chem.* **1990**, *62*, 575. (e) Kündig, E.P.; Perret, C.; Rudolph, B. *Helv. Chim. Acta* **1990**, *73*, 1970.

37. Davies, S.G.; Goodfellow, C.L.; Peach, J.M.; Waller, A. *J. Chem. Soc., Perkin I* **1991**, 1019.

38. (a) Nechvatal, G.; Widdowson, D.A. *J. Chem. Soc., Chem. Comm.* **1982**, 467. (b) Beswick, P.J.; Greenwood, C.S.; Mowlem, T.J.; Nechvatal, G.; Widdowson, D.A. *Tetrahedron* **1988**, *44*, 7325. (c) Masters, N.F.; Mathews, N.; Nechvatal, G. Widdowson, D.A. *Tetrahedron* **1989**, *45*, 5955. (d) Dickens, M.J.; Mowlen, T.J.; Widdowson, D.A.; Slawin, A.M.Z.; Williams, D.J. *J. Chem. Soc., Perkin Trans I* **1992**, 323.

39. (a) Uemura, M.; Miyake, R.; Nakayama, K.; Shiro, M.; Hayashi, Y. *J. Org. Chem.* **1993**, *58*, 1238. (b) Christian, P.W.; Gil, R.; Muñoz-Fernandez, K.; Thomas, S.E.; Wierzchleyski, A.T. *J. Chem. Soc., Chem. Comm.* **1994**, 1569.

40. (a) Han, J.; Son, S.V.; Chung, Y.K. *J. Org. Chem.* **1997**, *62*, 8264. For related examples see: (b) Alexakis, A.; Kanger, T.; Mangenez, P.; Rose-Munch, F.; Perrotey, A.; Rose, E. *Tetrahedron Asymmetry* **1995**, *6*, 47; 2135.

41. (a) Uemura, M.; Daimon, A.; Hayashi, Y. *J. Chem. Soc., Chem. Comm.* **1995**, 1943. (b) For related processes to make single atrop isomers of aryl naphthalene systems see: Watanabe, T.; Uemura, M. *J. Chem. Soc., Chem. Comm.* **1998**, 871.

42. (a) Ewin, R.A.; MacLeod, A.M.; Price, D.A.; Simpkins, N.S.; Watt, A.P. *J. Chem. Soc. Trans Perkin I* **1997**, 401. (b) Price, D.A.; Simpkins, N.S.; MacLeod, A.M.; Watt, A.P. *J. Org. Chem.* **1994**, *59*, 1961. (c) Kündig, E.P.; Quattrepani, A. *Tetrahedron Lett.* **1994**, *35*, 3497.

43. For a review see: Davies, S.G.; McCarthy, T.D., "Transition Metal Arene Complexes. Side Chain Activation and Control of Stereochemistry," in "Comprehensive Organometallic Chemistry II", Wilkinson, G.; Stone, F.G.A.; Abel, E.W. eds., Pergamon, New York, 1995, vol. 12, pp. 1039-1070.

44. Arzeno, H.B.; Barton, D.H.R.; Davies, S.G.; Lusinchi, X.; Meunier, B.; Pascard, C. *Nouv. J. Chim.* **1980**, *4*, 369.

45. Solladie-Cavallo, A.; Farkhani, D. *Tetrahedron Lett.* **1986**, *27*, 1331.

46. Baird, P.D.; Blagg, J.; Davies, S.G.; Sutton, K.H. *Tetrahedron* **1988**, *44*, 171.

47. Schmalz, H.C.; Arnold, M.; Hollander, J.; Bats, J.W. *Angew. Chem. Int. Ed. Engl.* **1994**, *33*, 108.

48. Mowlines, F.; Djakovitch, L.; Boese, R.; Gloaguen, B.; Theil, W.; Fillant, J-L.; Delville, M-H.; Astruc, D. *Angew. Chem. Int. Ed. Engl.* **1993**, *32*, 1075.

49. For a review see: Davies, S.G.; Donohoe, T.J. *Synlett* **1993**, 323.

50. (a) Coote, S.J.; Davies, S.G.; Middlemiss, D.; Naylor, A. *Tetrahedron Asymmetry* **1990**, *1*, 33. (b) Coote, S.J.; Davies, S.G.; Middlemiss, D.; Naylor, A. *J. Chem. Soc., Perkin Trans 1* **1985**, 2223. (c) Coote, S.J.; Davies, S.G.; Middlemiss, D.; Naylor, A. *Tetrahedron Lett.* **1989**, *30*, 3581.

51. Tamaka, T.; Mikamiyama, H.; Maeda, K.; Ishida, T.; In, Y.; Iwata, C. *J. Chem. Soc., Chem. Comm.* **1997**, 2401.

52. Jaouen, G.; Meyer, A. *Tetrahedron Lett.* **1976**, 3547.

53. Meyer, A.; Jaouen, G. *J. Chem. Soc., Chem. Comm.* **1974**, 787.

54. Schmalz, H.G.; Millies, B.; Bats, J.W.; Dürner, G. *Angew. Chem. Int. Ed. Engl.* **1992**, *31*, 631.

55. (a) Solladie-Cavallo, A.; Solladie, G.; Tsamo, E. *J. Org. Chem.* **1979**, *44*, 4189. (b) Davies, S.G.; Goodfellow, C.L. *J. Chem. Soc. Perkin Trans I* **1989**, 192.

56. (a) Bromley, L.A.; Davies, S.G.; Goodfellow, C.L. *Tetrahedron Asymmetry* **1991**, *2*, 139. (b) Uemura, M.; Miyake, R.; Nakayama, K.; Shiro, M.; Hayashi, Y. *J. Org. Chem.* **1993**, *58*, 1238.

57. Kündig, E.P.; Xu, L.H.; Schnell, B. *Synlett* **1994**, 413.

58. Baldoli, C.; Del Buttero, P.; Licandro, E.; Papagni, A.; Pilate, T. *Tetrahedron* **1996**, *52*, 4849.

59. Kündig, E.P.; Xu, L.H.; Romanens, P.; Benardinelli, G. *Synlett* **1996**, 270.

312

60. Uemura, M.; Nishmura, H.; Minami, T.; Hayashi, Y. *J. Am. Chem. Soc.* **1991**, *113*, 5402.
61. For a review see: Harman, W.D. *Chem. Rev.* **1997**, *97*, 1953.
62. (a) Kopach, M.E.; Harman, W.D. *J. Am. Chem. Soc.* **1994**, *116*, 6581. (b) Kopach, M.E.; Kolis, S.P.; Liu, R.; Robertson, J.W.; Chordia, M.D.; harman, W.D. *J. Am. Chem. Soc.* **1998**, *120*, 6199. (c) Kolis, S.P.; Kopach, M.E.; Liu, R.; Harman, W.D. *J. Am. Chem. Soc.* **1998**, *120*, 6205.
63. Kolis, S.P.; Gonzalez, J.; Bright, L.M.; Harman, W.D. *Organometallics* **1996**, *15*, 245.
64. Hodges, L.M.; Gonzalez, J.; Koonts, J.I.; Myers, W.H.; Harman, D. *J. Org. Chem.* **1995**, *60*, 2125.
65. Gonzalez, J.; Koontz, J.I.; Myers, W.H.; Hodges, L.M.; Sabat, M.; Nilsson, K.R.; Neelz, L.K.; Harman, W.D. *J. Am. Chem. Soc.* **1995**, *112*, 3405.
66. Chen, H.; Liw, R.; Myers, W.H.; Harman, W.D. *J. Am. Chem. Soc.* **1998**, *120*, 509.

Index

nucleophilic attack of zerovalent metal, 130–131
carbonylative coupling:
σ-acyl complexes, 189
via transmetallation, 78–79
carbonyls:
α, β-unsaturated, reduction, Group 6 catalyst precursor, 51
asymmetric hydrogenation:
heteroatoms, 47
ruthenium BINAP catalyst, 47
from carbon monoxide insertion, 93–94
insertion, to σ-acyl complexes, 78–79
ligand:
bonding mode, 6
exchange, first-order, 16
in migratory insertion, 24
order, 15
removal by amine oxides, 17
α, β-unsaturated carbonyls:
1,4-addition, diorganocuprates, 59
reduction, Group 6 catalyst precursor, 51
see also carbon monoxide
carboxylic acid:
conjugated, reaction failure with lithium diorganocuprate, 60
derivative:
from acyl/electrophile complexes, 26
from ketene complexes, 166
from oxidation of carbenes, 151
protection by allylic carbonates and esters, 265
cerium, oxidation of carbenes, 151
chirality, in metal complexes, induced asymmetry, 169
CHIRAPHOS, catalyst, 42–43
chloride, ligand order, 15
chlorides, nucleophilic attack by, 29
chloroacetoxylation, palladium-catalyzed, dienes, 262
chloromethylation, of (η^4-cyclobutadiene)iron tricarbonyl, 208
chloropalladation, α-methylene lactones synthesis, 226
cholestanone, reaction, with η^3-allylnickel halides, 281
chromium:
η^6-arene complexes, preparation, 288
arene metal complexes:
preparation, 287
stable, 287
arenechromium tricarbonyl:
activation of benzylic position, 302
addition of carbon nucleophiles, 292–298
complexation mechanism, 288
from arene exchange with (naphthalene)Cr(CO)$_3$, 288
lithiation, 298–302

negative charge stabilization, benzylic group, 302
nucleophilic attack, 31
preparation, 288
reactions, 290–307
η^6-arenechromium tricarbonyl complex:
chirality, 295
dithiane anion addition, 297
arenechromium tricarbonyl complexes, chirality, 306
η^6-chlorobenzenechromium tricarbonyl, substitution rates, 290
chromium(0) complexed trienes, cycloaddition, [6 + 2], 218
chromium alkoxycarbenes, photolysis, 163
chromium carbene complexes:
photolysis, to esters, 166
photolytic oxidation, 163
chromium heteroatom carbene complexes:
bond configurations, 163
molecular orbitals, 163
photochemistry, 163–164
chromium hexacarbonyl:
ligand exchange, 16
recovery in Staudinger reaction, 164–165
chromium pentacarbonyl:
for carbene complexes, 145
reactivity, 150
Fischer carbene complexes, 143–144
η^6-fluorobenzenechromium tricarbonyl, substitution rates, 290
hydroquinone chromium carbonyl, in Dötz reaction, 157
K$_2$Cr(CO)$_5$, reaction with amides, 164
chromium carbene complexes:
reactions, 163–167
[2 + 2] cycloaddition of ketenes, 165
"masochistic stereoinduction", 165
preparation of β-lactams, 164–165
preparation of dipeptides, 167
synthesis of amino acids, 166–167
chromium heteroatom carbene complexes, reactions, 163–167
cis insertion, σ-alkylpalladium(II) complexes, 71
citronellol, synthesis via asymmetric hydrogenation, 46
cobalt:
alkyne dicobalt, in Pauson–Khand reaction, 230–231
alkynecobalt complexes, chiral boron enolates, 228
η^3-allylcobalt complexes, 278
insertion reactions, 278
cobalt phthaloyl:
mechanism of formation, 240–241
preparation, 240–241
reactions, 240–241

cyclopropanation: (*continued*)
  metathesis "cascade" cyclization reactions,
    171
  olefin:
    with diazo compounds, 170
    metal carbene intermediates, 170
  suppression by CO pressure, 152
cyclopropanes, synthesis, from
    η^3-allylpalladium, 30

D
d electron configuration, transition elements,
    3–4
d orbital, symmetry, 9
decarbonylation:
  metal carbonyl, 124
  reactions, 134–136
dendrimers, iodoarene/ethynylbenzene
    coupling, palladium catalysts, 92
diallyl *bis* sulfone:
  cyclization, Pd(dba)$_2$-catalyzed, 266
  synthesis, from allylic alkylation of allylic
    halides, palladium-catalyzed, 266
diazo compound:
  catalytic cyclopropanation with, 170
  metal-catalyzed decomposition, via
    unstabilized electrophilic carbenes,
    170–176
diazoalkenes, unsaturated carbene complexes,
    cycloadditions, 150
α-diazocarbonyl compound, catalytic
    cyclopropanation, 170
α, α'-dibromoketones, α, α'-disubstituted,
    reaction with iron tetracarbonyl, 125
dichlorocyclobutene, reaction with Fe$_2$(CO)$_9$,
    207
dicyclohexylamide, preparation of thermally
    stable cuprates, 61
Diels–Alder reactions, diiron bridged acyl
    complex, 139–140
dienals, reaction with iron nonacarbonyl, 204
diene:
  catalytic cyclopropanation, with diazo
    compounds, 170
  chloroacetoxylation, Pd(OAc)$_2$-catalyzed, for
    1-acetoxy-4-chlorobut-2-ene, 267
  complexation with iron tricarbonyl, 205–206
  cyclic, σ-acyl iron catalyst, 208–209
  cycloaddition, to bridge bicyclic system,
    125–126
  from allyls, via palladium(0) complexes,
    264–265
  from cyclodimerization of alkenes and
    alkynes, 114
  heterocyclic, neutral η^3-allylmolybdenum,
    reaction, 210–211
  insertion, into η^3-allyl palladium, for diene
    acetates, 267–268

ligand, bonding mode, 6
monocyclopropanation, 151–152
transition metal complexes, 204–220
η^4-diene:
  hydride abstraction, 31
  reactivity to nucleophiles, 30
η^5-diene, cationic, nucleophilic addition, 31
diene, alkoxylation, palladium-catalyzed, 262
diene, chloroacetoxylation, palladium-catalyzed,
    262
1,3-diene:
  alkylation-acylation, η^3-allylcobalt complexes,
    278
  hydrozirconation, 73
  reaction with Fe(CO)$_4$, 204
  1,4-reduction to Z-alkene, Group 6 catalyst
    precursor, 51
  telomerization, transition-metal catalyzed,
    246–250
1,5-diene:
  alkene insertion, via η^3-allyl palladium, 251
  Cope rearrangement, 200
α, Ω-diene:
  and butylmagnesium chloride, with
    dicyclopentadienylzirconium(IV)
    dichloride catalyst, 111
  and n-butylmagnesium chloride, with
    dicyclopentadienylzirconium(IV)
    dichloride catalyst, 110–111
diene acetates, synthesis, insertion, η^3-allyl
    palladium, 267–268
1,3-dienes, cyclooligomerization, nickel(0)
    catalyst, 248
dienes, intramolecular, cyclization, 108
dienyls, transition metal complexes, 211–214
dihydrocryptopine, reaction, with chromium
    hexacarbonyl, complexation sites, 299–300
dimethoxyethene:
  complex with CpFe(CO)$_2$(isobutene)+:
    reaction with 2,3-butandiol, for dioxene
      complexes, 202
    reactivity, 202
dimethyldioxirane, oxidation of carbenes, 151
α, α'-dimethylketone, stereospecific reduction,
    205–206
1,3-diols, η^3-allyl molybdenum catalyst, aldol
    condensation, 276
DIOP, catalyst, 42–43
dioxyfulvenes, cycloaddition, metal-catalyzed,
    [3 + 3], 126
DIPAMP, catalyst, 42–43
dipeptides:
  "double diastereoselection", 167
  preparation via carbene complexes, 167
diphenylphosphide, preparation of thermally
    stable cuprates, 61
dissociative process, ligand substitution,
    14

olefin: (*continued*)
  hydrogenation:
    competetion with isomerization, 41
    homogeneous catalyst, 40
  ligand:
    in migratory insertion, 24–25
    order, 15
    π-acceptor, 5
  metathesis, in carbene complexes, 154
  order of reactivity, 170
  oxidative addition/insertion, 95–96
  prochiral, asymmetric hydrogenation, 42
  tetrasubstituted:
    noncyclic, and Wilkinson's complexes, 42
    reduction catalysts, 50
    trisubstituted, noncyclic, and Wilkinson's complexes, 42
  Wilkinson's complex, reactivity, 42
  see also alkene
oligofurans, via transmetallation, 81
oligophenyls, synthesis via Suzuki coupling, 85
oligo(thiophene ethynylenes), synthesis via palladium catalysts, 92
oligothiophenes, via transmetallation, 81
organic halides:
  η^3-allylnickel halides, reaction mechanism, 281
  functionalized:
    reactivity, 65
    synthesis of organocopper reagents, 65
  oxidative addition, to low valent transition metals, 80
organic iodides, functionalized, with diethylzinc, 65
organoboron compound, transmetallation, to palladium, 83
organolithium reagent:
  and neutral metal carbonyl, 25
  to generate organocopper reagent, 64
  transmetallation, 81
organometallic:
  ligands, classes, 4–8
  reaction mechanism, 13–37
    β-hydride elimination, 25
    ligand exchange processes, 13–18
    migratory insertion, 24–25
    oxidative addition, 18–23
    reductive elimination, 22
  synthesis, migratory insertion, unsaturated ligand, 24
organometallic complex:
  geometries:
    five-coordinate, 10
    four-coordinate, 10
    octahedral, 10
    six-coordinate, 10
    square-planar, 10
    tetrahedral, 10

trigonal bipyramidal, 10
"ortho lithiation", 105
  arene metallation, 85
orthopalladation:
  alkene insertion, 105
  carbonyl insertion, 105
  ligand directed, 104–105
osmium:
  arene metal complexes, stable, 287
  osmium furan complexes, 310
  $[Os(NH_3)_5]_2+$, η^2-arene complex formation, 307
  4,5-η^2-pyrrole complex, cycloaddition, 309–310
osmium tetroxide, olefin *cis* hydroxylation, iron(III) tricarbonyl, 204
oxazolines, chiral, *trans* difunctionalization, diastereoselectivity, 298
oxidation state, definition, 2–3
oxidative addition, 18–23
  alkynes and alkenes, via hydroboration, 82–83
  mechanism, 18–23
  palladium(0) catalyst, 79
  radical chain, aryl halides to biaryls, 22
  reactivity, electron-rich substrates, 78
  to transition metal hydride, 39
  transition metals, 78
oxo reaction, 127
oxyCope rearrangement, palladium(II) catalyst, 200
oxygen heterocycles:
  five-membered, aldehyde cycloaddition, palladium(0)-catalyzed, 273
  from σ-alkylpalladium(II) and alcohol, 192

P

π^* orbital, symmetry, 9
π-back bonding:
  ligands:
    longitudinal acceptors, 9
    perpendicular acceptors, 9
π-bonding, details, 9
π-unsaturated hydrocarbon, nucleophilic attack, 28
palladium:
  σ-alkenylpalladium(II) complexes, protolytic cleavage, 226
  σ-alkyl-π-allyl palladium(II) complexes, reductive elimination rate, 256
  σ-alkylpalladium, intramolecular formation, 266
  σ-alkylpalladium(II) complexes:
    for alcohol addition to alkenes, 194
    amine reactions, 196
    β-elimination, 71, 198
    carbon monoxide trapping, 198
    from intramolecular alcohol addition, for oxygen heterocycles, 192